River Basin Disaster Resilience and Sustainability by All

9th International Conference on Flood Management (ICFM9)

Proceedings of the International Association of Hydrological Sciences (PIAHS) Volume 386, 2024

Tsukuba, Japan
18-22 February 2023

ISBN: 979-8-3313-0728-8

Printed from e-media with permission by:

Curran Associates, Inc.
57 Morehouse Lane
Red Hook, NY 12571

Some format issues inherent in the e-media version may also appear in this print version.

International Association of Hydrological Sciences
Centre for Ecology and Hydrology
Wallingford, Oxfordshire OX10 8BB
United Kingdom

Phone: 441 491 692 442
Fax: 441 491 692 448

http://iahs.info/

Additional copies of this publication are available from:

Curran Associates, Inc.
57 Morehouse Lane
Red Hook, NY 12571 USA
Phone: 845-758-0400
Fax: 845-758-2633
Email: curran@proceedings.com
Web: www.proceedings.com

TABLE OF CONTENTS

Controlling Sedimentation Through Regulating the River by Thalpitigala Reservoir Project 1
Thanura Lasantha Babarande Guruge, Bandara Palugaswewa, Shinji Egashira

Building Disaster Prevention Capacity in a Continental Scale Country: The Trajectory and Experience of the Geological Survey of Brazil 7
Leandro G. Kuhlmann, Mamoru Miyamoto, Julio C. Lana, Tiago Antonelli

Estimating the Rate of Change in Heavy Rainfall Intensity for Flood Control Planning in Japan.................... 13
Yuta Maeda, Tetsuya Takeshita, Masaki Kawasaki

Development of a Virtual Flood Experience System and Its Suitability as a Flood Risk Communication Tool 21
Masatoshi Denda, Masakazu Fujikane

Necessary Storage for Managing Early Flash Flood to Save Crops in the North-Eastern Region of Bangladesh 27
Muhammad Masood, Abdul Wahid Mohamed Rasmy

Projection of Climate Change Effects on Flood Inundation in the Gin River Basin, Sri Lanka........................ 33
Jayapadma Mudalige Miyuru Uthpala Jayapadma, Kazuyoshi Souma, Hiroshi Ishidaira, Jun Magome, Thushara Navodani Wickramaarachchi

Performance Analysis of Physically-Based (HEC-RAS, CADDIES) and AI-Based (LSTM) Flood Models for Two Case Studies 41
Marina Batalini De Macedo, Nikunj K. Mangukiya, Maria Clara Fava, Ashutosh Sharma, Roberto Fray Da Silva, Ankit Agarwal, Maria Tereza Razzolini, Eduardo Mario Mendiondo, Narendra K. Goel, Mathew Kurian, Adelaide Cássia Nardocci

Temporal and Spatial Distributions of Flood Water Storage Volume of the Main River and Tributaries in the Ishikari River Basin............................ 47
Shoji Fukuoka, Yutaro Ishii

Investigation of Uncertainties in Multi-Variable Bias Adjustment in Multi-Model Ensemble.......................... 55
Saurabh Kelkar, Koji Dairaku

Investigation of Transition Signals from Single Cell to Multicell Thunderstorms in the Kinki Region, Japan by Using Vertical Vorticity and Multi-Parameter Radar Analysis.......................... 61
Fauziana Ahmad, Kosei Yamaguchi, Eiichi Nakakita, Tomoki Ushiyama

Calculating Flood Probability in Obihiro Using a Probabilistic Method: Incorporating the Probability of Dike Failure with Uncertainty 69
Fumihiko Uemura, Guus Rongen, Shigekazu Masuya, Takatoshi Yoshida, Tomohito J. Yamada

Uncertainty in Simulated Streamflow Using Runoff Driven by the Outputs of a High-Resolution Regional Climate Model............................ 75
Aulia Febianda Anwar Tinumbang, Kazuaki Yorozu, Yasuto Tachikawa, Yutaka Ichikawa, Hidetaka Sasaki, Tosiyuki Nakaegawa

Artificial Neural Networks Applied for Flood Forecasting in Ungauged Basin – the Paranaíba River Study Case 81
Abderraman R. A. Brandão, Frederico C. M. De Menezes Filho, Paulo T. S. Oliveira, Maria C. Fava

Investing in Resilience: Estimating Financial Needs and Benefits of Flood Protection in Developing Asian Countries ... 87
Mikio Ishiwatari, Daisuke Sasaki

Hydrodynamic Investigation of Laguna Lake, Philippines for Water Security and Flood Risk Management of Metro Manila ... 95
Eugene C. Herrera, Kensuke Naito

A Method for Municipalities to Enhance Disaster Resilience for Potential Future Floods Using a Climate Projection Database ... 101
Akihiro Suzuki, Fumihiko Uemura, Tsuyoshi Hoshino, Taro Yamamoto, Tomohito J. Yamada

Sediment Transport Processes in the Sangu River Basin Using a Rainfall-Sediment Runoff Model for Sustainable River Management ... 109
Md. Majadur Rahman, Daisuke Harada, Shinji Egashira

Flood Simulation Studies with Rainfall-Runoff-Inundation (RRI) Model, Over Yangon City, Myanmar ... 115
Pyae Pyae Phyoe, Tatsuhiko Uchida

Hydraulic Analysis of the Marikina River Floodplain During Typhoon Vamco Using Numerical Modelling ... 121
Jonathan S. Serrano, Eugene C. Herrera, Kensuke Naito

Assessment of Flood Damage to Agricultural Crops Under Climate Change Scenarios Using MRI-AGCM Outputs in the Solo River Basin of Indonesia ... 127
Badri Bhakta Shrestha, Mohamed Rasmy, Tomoki Ushiyama, Ralph Allen Acierto, Takatoshi Kawamoto, Masakazu Fujikane, Hiroyuki Ito, Takafumi Shinya

Evaluation of Climate Change Impacts on Urban Flooding Using High-Resolution Rainfall Data ... 133
Hideo Amaguchi, Jonas Olsson, Akira Kawamura, Yoshiyuki Imamura

Dual-Stage Attention-Based LSTM Network for Multiple Time Steps Flood Forecasting ... 141
Fan Wang, Weiqi Wang, Wuxia Bi, Wenqing Lin, Dawei Zhang

Development of Model Predictive Controls for Flood Control and Power Generation of Reservoir Cascade in Mahaweli Basin, Sri Lanka to Mitigate Energy and Economic Crisis Provoked by the COVID-19 Pandemic ... 147
Ranapura Dewage Thilini Kaushalya, Sajana Pramudith Hemakumara, Selvarajah Hemakanth, Seenipellage Chaminda Sugeeshwara

Study on a Flood Damage Reduction Countermeasure by Installing a Spillway on a Riverine Levee for Promoting the Initiative of River Basin Disaster Resilience and Sustainability by All ... 153
Osamu Itagaki, Miho Ohara, Toshio Koike

Methods to Create Hazard Maps for Flood Disasters with Sediment and Driftwood ... 159
Daisuke Harada, Shinji Egashira

Investment Decision Support for Business Sustainability Improvement Based on Flood Inundation Simulations with a Variety of Extreme Rainfall Scenarios ... 165
Saki Nonaka, Kazuki Suzuki, Nobuyuki Iwamae, Daisuke Nohara, Yoriyuki Yamada

Flood Resilient Landscapes: Area-Based Solutions Combine Added Value for Society with Flood Risk Management ... 173
Annemargreet De Leeuw, Ellen Tromp, Femke De Boer

Flood Damage Model Bias Caused by Aggregation ... 181
Seth Bryant, Heidi Kreibich, Bruno Merz

Designing and Developing Dynamic Decision Support Information for Disaster Response ... 189
Makoto Hanashima, Hiroaki Sano, Yuichiro Usuda

An Approach to Assess Sediment Disaster Risk Using Soil Moisture Deficit: A Case Study About Sediment Disasters Occurred During 2010–2011 in Niigata Prefecture, Japan ... 197
Minjiao Lu, Takahiro Ogura, Thandar Tun Zin, Shigeaki Oka, Shun-Ichi Onitsuka, Naoaki Kawamura, Takahiro Yamamoto, Koichi Komiyama

Satellite and UAV Derived Seasonal Vegetative Roughness Estimation for Flood Analysis ... 203
Andre Araujo Fortes, Masakazu Hashimoto, Keiko Udo, Ken Ichikawa

Attributing Weather Patterns to Davao River Extreme Rainfall from Reanalysis and GCM ... 209
Ralph Allen Acierto, Tomoki Ushiyama, Toshio Koike

Long-Range Streamflow Prediction Using a Distributed Hydrological Model in a Snowfed Watershed ... 217
Abdul Moiz, Akiyuki Kawasaki

Evaluation of Satellite Precipitation Products for Real-Time Extreme River Flow Modeling in Data Scarce Regions ... 223
Mihretab G. Tedla, Mohamed Rasmy, Toshio Koike, Li Zhou

Application of Entrainment Velocity Concept for Evaluation of River Morphology: A Numerical Model Study ... 229
Robin K. Biswas, Shinji Egashira, Daisuke Harada

A Tale of Two Floods: Hawkesbury-Nepean Valley Floods of February 2020 and March 2021 ... 237
Wendy Sharples, Katayoon Bahramian, Kesav Unnithan, Christoph Rüdiger, Jiawei Hou, Christopher Pickett-Heaps, Elisabetta Carrara

Countermeasures Against Flood Damage of Buildings and Their Applicability ... 251
Nozomu Kiuchi, Taku Nakano

A Study on Improving Disaster Mitigation Awareness by Simulated Flood Experience Using VR Videos ... 259
Tsuyoshi Koyabu, Masatoshi Denda

Developing a Flood Monitoring System by Utilizing Real-Time Satellite Rainfall Estimates and Water Energy Budget-Based Rainfall-Runoff Inundation Model in West Africa ... 265
Abdul Wahid Mohamed Rasmy, Maksym Gusyev, Katsunori Tamakawa, Miho Ohara, Toshio Koike

Developing an Integrated Flood Management Plan Under Climate Change for Grand River North West River Basin, Mauritius ... 271
Akshay Kowlesser, Mohamed Rasmy, Toshio Koike

Flood Risk Management in Planning and Construction of City: The Guangzhou Experience ... 277
Minyi Kuang, Yanyan Zheng, Xingdong Deng, Yukui Yang, Jingchen Wang, Xintian Sui, Yuan Peng

Working Together on Civil-Engineering Structures Within the Dutch Flood Protection Programme: Challenges and Opportunities ... 285
Ellen Tromp, Joost Breedeveld, Johan Apperloo, Jaap Kottier

Significance of Having Integrated Water Resource Plan in a Complex Watershed System for Better Water Management During Covid Pandemic: The Case of Mahaweli River Basin 291
Roshan Indika Jayasinghe, Mohamad Rasmy, Toshio Koike

Assessment of Flood Risk in a Coastal City Considering Multiple Socio-Economic Vulnerability Scenarios 299
Mousumi Ghosh, Subimal Ghosh, Subhankar Karmakar

Global Experiences in Flood Management – Perspectives Through ICFM Webinar Series.. 307
Cheng Zhang, Wuxia Bi

How to Avoid Heavy Life Losses in Liulin-Like Flash Flood Events.. 313
Changzhi Li, Miao Zhang, Qing Li, Bingshun He

The Role of Construction Consultants in Disaster Recovery 321
Minoru Kamoto, Akira Nakamura, Masashi Ogura

4 "R" Approach of Catastrophic Disaster of Melamchi Flood: A Sustainable Urban Planning 327
Tapendra Bahadur Khadka

Regional Characteristics of Shape and Stagnation of Rainfall in Japan Using Radar/Raingauge Analyzed Precipitation Product.. 333
Yuta Ohya, Tomohito J. Yamada

An Integrated Design Discharge Calculation System for Small to Mid-Sized Ungauged Catchments in Serbia.. 339
Nikola Zlatanovic, Mamoru Miyamoto

Author Index

Proc. IAHS, 386, 1–6, 2024
https://doi.org/10.5194/piahs-386-1-2024

Open Access PIAHS

Controlling Sedimentation through Regulating the River by Thalpitigala Reservoir Project

Thanura Lasantha Babarande Guruge[1], Bandara Palugaswewa[1], and Shinji Egashira[2]
[1]Ministry of Irrigation, Colombo 07, Sri Lanka
[2]ICHARM, Tsukuba, Japan

Correspondence: Thanura Lasantha Babarande Guruge (thanuralasantha@gmail.com)

Received: 20 July 2023 – Revised: 3 October 2023 – Accepted: 14 November 2023 – Published: 19 April 2024

Abstract. Mahaweli is the main river in Sri Lanka and it has many main multipurpose reservoirs, especially to cater power and irrigation requirement of the country. Reservoir capacity loss which is the major problems causing to the effectiveness of the major reservoirs in Sri Lanka, is mainly caused by sedimentation due to upstream erosion.

Hence proper monitoring, calculating and adopting measures are very important in planning a reservoir. But due to the complex nature of factors effect on sedimentation and lack of long-term data make calculation difficult.

This study discusses how different data obtained from numerical modelling and bathymetric surveys of different locations of the catchment and those are used to select and verify the empirical equation, calculation of capacity loss and make proposals to minimise sedimentation for proposed Thalpitigala Multipurpose Project in Mahaweli upper catchment.

The outcomes indicate high erosion ($690\,m^3\,km^{-2}$) and sediment transportation making rapid loosing of the capacity (5.13 MCM in 15 years) of the proposed reservoir. Thus, upper catchment protection with reforestation to reduce erosion, introduction of check dams in the upstream flow to trap the transported sediment and introduction of proper removal measures of deposited sediment are discussed for the proposed reservoir project.

1 Introduction

Sediment load is one of the most important parameters required for the selection of the dam type, the estimation of dead storage and for the design of proper sediment flushing facilities.

Among the catchments in Sri Lanka, the highest sediment yield has been reported from the Uma Oya catchment by a study done in 2019 (Dias, 2019). It is shown that erosion is 3 to 130 times faster than the soil loss tolerance ($5\,t\,ha^{-1}\,annum^{-1}$) in Sri Lanka.

Uma Oya catchment (between 6°45′23″ N, 81°03′08″ E and 6°59′29″ N, 81°00′58″ E), a sub basin of Mahaweli river basin is located in the South-Eastern slopes of the central hills of Sri Lanka with an area of $720\,km^2$ (Fig. 1)

The last reservoir about to construct in this catchment is Thalpitigala which is considered as a multipurpose reservoir project.

Figure 2 shows the Uma Oya catchment ($720\,km^2$) and locations of proposed reservoir- Thapitigala, and other considered points.

The upstream reservoirs of the catchment, Puhulpola reservoir (Catchment area $180\,km^2$) and Dyaraba reservoir (Catchment area $158\,km^2$), may not cut down in full the suspended solids passing to the Thalpitigala reservoir, since those reservoirs have bottom outlets to flush the sediments, hence it is a commendable to consider nearly 80 % of the catchment sediments are to be come into the proposed reservoir.

In contrast, Thalpitigala reservoir can't eject trapped sediments at once to the river downstream as there are set of reservoirs in Mahaweli system. Consequently, it is obvious that the sediment management in the proposed reservoir is a key design factor to maintain the wellbeing and the effectiveness of the reservoir in the future.

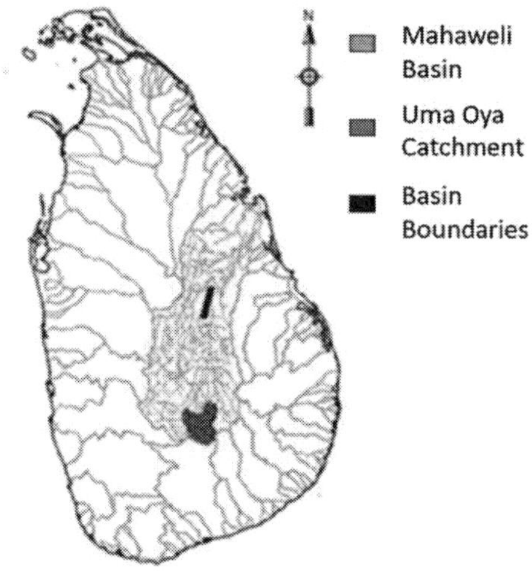

Figure 1. Mahaweli River Basin and Uma Oya Sub Basin.

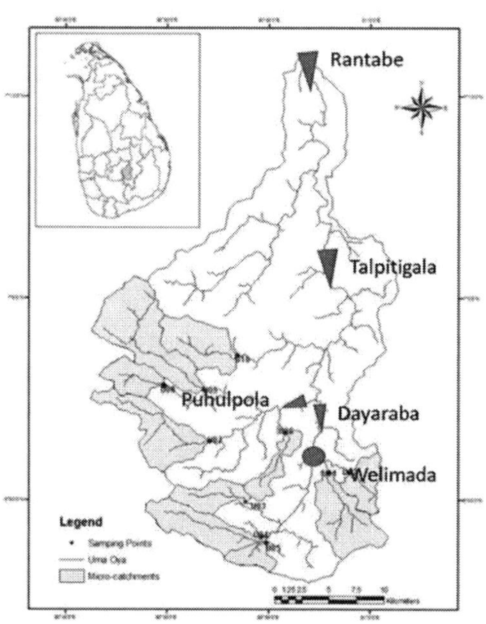

Figure 2. Area Map of Uma Oya Catchment.

In Figs. 3 and 4, the arial photos in rainy and dry seasons in the proposed reservoir stretch.

The magnitude of sediment transportation in the water can be evidently seen by comparing the dry and flood flows shown in following Figs. 3 and 4. The spill water characteristic of upstream Dayraba reservoir in rainy seasons is shown in Fig. 5.

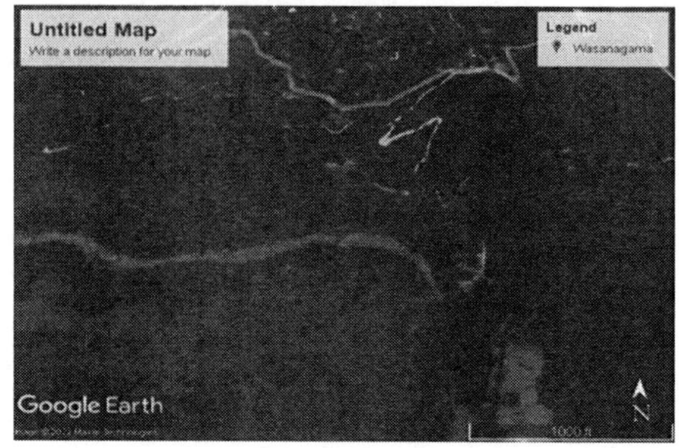

Figure 3. Arial photo of Uma Oya in rainy season (©Google Earth).

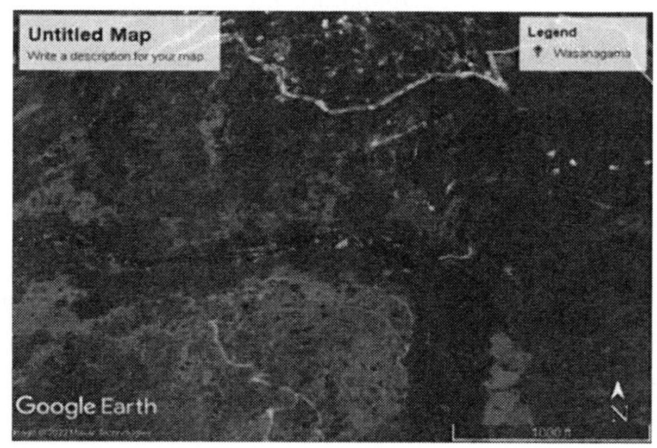

Figure 4. Arial photo of Uma Oya in dry season (©Google Earth).

2 Characteristics of sediment yield in study basin

The sediment quantity transported with the upstream flows deposited and travelled in a reduced rate in the reservoir system governing the distance by the decreased water velocity, making complex analyses in sediment studies.

In the study of the erosion rates in the catchment, the land cover properties, land use properties, geography, soil parameters etc. have to be considered.

When we have to study the sediment inflow, settling, and degradation courses relation with known factors such as water velocity, particle size, particle type, turbidity, cross flows etc. The difficulty of analysing the Sedimentation processes in a reservoir can be reduced by truncating minor factors and considering most influencing factors,

Because of this complexity, many mathematical models and empirical relationships are being developed to simulate the physical processes of sediment transport and deposition in reservoirs.

The results of modelling and simulations can be verified by bathymetric measurements of sediment loads in known

Figure 5. Photo of Run off of upper catchment of Uma Oya-Dayaraba Dam.

points and volumetric measurements of partials in runoff water. Those can be spatially and temporally analysed to understand the variation of sediment transportation throughout the river flow.

Herewith we are going to discuss some empirical measurements and simulations used to determine the reservoir sedimentation, life reduction and measures intended to implement to minimize the vulnerability.

The following calculations show three different methods used in three different occasions, for three different locations of the same catchment.

2.1 Calculation of Sediment Delivery Ratio and Sediment Yield

The sediment delivery ratio (SDR) of a river basin reflects the conveyance status of sediment which occurs due to erosion of soil, through runoff of the stream flow. The sediment delivery ratio calculation models have certain related conditions such as environmental factors, rainfall, land topography, soil type and texture etc. The drainage basin area method (Vanoni, 1975) is used to determine SDR considering the similarities of above factors to model catchment.

$$\mathrm{SDR} = 0.42A^{-0.125}$$

Total catchment area of Uma Oya at the confluence of Mahaweli (Upstream of Ranteme reservoir) is 720 km^2; Catchment Area = 281.25 sq. miles; Therefore SDR = 0.21

$$\mathrm{SDR} = \frac{\text{Sediment yield}}{\text{Gross Erosion}}$$

Universal soil loss equation

$$A = R \times K \times \mathrm{LS} \times C \times P$$

Table 1. Specific sediment yield at Welimada Station.

t km^{-2}		Max	Avg	Min
Months	Jan	644	160	10
	Feb	302	92	0
	Mar	67	15	1
	Apr	65	25	0
	May	248	44	2
	June	87	19	1
	July	137	25	1
	Aug	253	46	1
	Sept	91	27	7
	Oct	322	112	2
	Nov	1145	264	11
	Dec	367	146	12
Annual		1983	974	454

Erosive rainfall forces	$(R) = 120$
Inherent erodibility of the soil	$(K) = 0.67$
Hillside length – slope factor	$(\mathrm{LS}) = 1.91$
Land cover factors	$(C) = 0.5$
Land Use factors	$(P) = 0.75$
Long term soil loss rate $(A) = 57.59\,\mathrm{t\,Acre^{-1}\,annum^{-1}}$	
Soil gross erosion	$= 144.00\,\mathrm{t\,ha^{-1}\,annum^{-1}}$
Annual sediment yield	$= 30.0\,\mathrm{t\,ha^{-1}\,annum^{-1}}$
Annual Sediment Yield	$= 11.28\,\mathrm{m^3\,ha\,annum^{-1}}$
Annual Sediment Yield	$= 1128\,\mathrm{m^3\,km^{-2}\,annum^{-1}}$

2.2 Annual Sedimentation for Proposed Reservoir Location.

As per the sediment analyses done by Ministry of Irrigation, the following results have been obtained in feasibility report. The analyses have been done through bathymetric mapping done for downstream of Uma Oya for a catchment area of 556 km^2 at Bathmedilla weir section (SDR = 0.22) in 2013.

Annual sediment yield = $18.30\,\mathrm{t\,ha^{-1}\,annum^{-1}}$

Annual sediment yield = $690\,\mathrm{m^3\,km^2\,annum^{-1}}$

2.3 Annual sedimentations at Puhulpola and Dayaraba Reservoirs

The study done by Ministry of Irrigation and Farab in 2010 in above reservoirs in the upstream of Uma Oya using EPM method and sediment sampling data for 8 years, shows the results as follows.

This is a very comprehensive study which has considered most of the complex parameters in simulation of mathematical model. Table 1 shows the summary of compiled data of 8 years

Table 2 shows the results obtained for upper catchment reservoirs from simulation of sediment bulk density.

Table 2. Results of sediment yield in Uma Oya catchment.

Location	SDR	Annual Sediment Yield ($m^3\,km^{-2}\,annum^{-1}$)
Puhulpola	0.30	661
Dyaraba	0.32	724

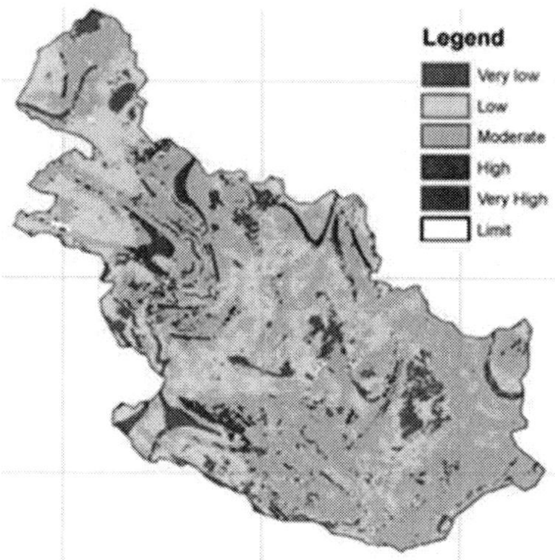

Figure 6. Erosion rate in Uma Oya basin.

Table 3. Comparison of results with calculated results using universal soil loss equation.

Location	Catchment Area (km^2)	SDR	Annual Sediment		
			USLE	BM	SBD
Puhulpola	180	0.30	1630		661
Dayaraba	157	0.32	1739		720
Thalpitigala	556	0.22	1195	690	
Rantebe	720	0.21	1128		

USLE – Universal soil loss equation. BM – Bathymetric Mapping. SBD – Simulation of Sediment Bulk Density.

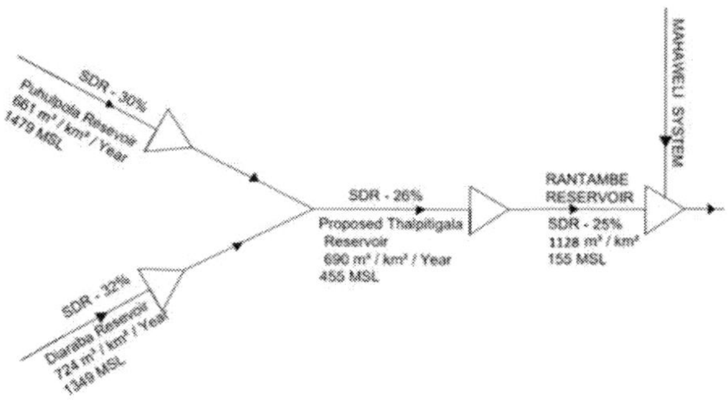

Figure 7. Schematic diagram of Reservoir system in Uma Oya.

Figure 6 shows the variation of erosion rate in Uma Oya basin according to the simulation done using EPM model.

Herewith, as indicated in Table 3 the practical results obtained from BM and SBD are considered for further analyses as USLE shows fairly high values than experienced.

Accordingly, the specific sediment yield in each location can be showed in following Fig. 7.

Considering the annual sediment yield to proposed Thalpitigala Reservoir, the deposition was projected.

Initial Reservoir capacity = 15.56 MCM

Project Life time = 50 years

Trap Efficiency

Assume percentage filled with$t$ years; 0.43 %

First C/T = 0.066

First Trap Efficiency = 82 %

Final C/T = 0.037

Final Trap Efficiency = 73 %

Average TE = 77.5 %

Time period = 14.4 years

Sediment yield = 5.13 MCM

Accordingly, within one third of life time, the reservoir losses one third of its capacity which implies the total dead storage vanishes. Hence sediment management is very essential

3 Sediment Control and Management Methods

In controlling and managing the sedimentation, it is needed to outline an allowable sediment discharge from the point we consider, to the downstream. Then the sediment volume should be controlled at the considered point.

In order to reduce reservoir sedimentation, we may have two possible methods: One is to reduce the inflow rate of sediment to the reservoir, and the other is to reduce sedimentation within the reservoir.

The prior methods implicate various erosion controls by means of structural and non-structural measures by using check dams, soil conservation performances, etc. The latter is to reduce the sediment deposition within the reservoir by means of dam operations such as sediment flushing, sediment sluicing, sediment bypassing, etc. Figure 8 illustrates a sample of the check dam to reduce sediment erosion.

4 Conclusion and Recommendations

Among the said measures, for this project the following measures are considered.

- Reduce sediment yield from upstream – Introduce check dams to capture sediments

 https://doi.org/10.5194/piahs-386-1-2024

Figure 8. Closed type check dam.

- Introduce upper catchmant management to reduce erosion. – Reforestration

- Route Sediments/Remove sediments – Introduce two bottom outlets to the dam flush through separate channel to adjacent basin- Wela Oya which do not have reservoirs in down stream

4.1 Check dams

In this regard our main intention is to trap sediments. Accordingly, 50 % of sediment specially the bead load is intended to be trapped and removed by dredging. Therefore, the closed type check dams (about 3 in selected locations) are recommended. The collected or dredged sediments can be used for other purposes such as construction sand from course material and fertilizer by fine organic layer after drying. Those options should be checked and verified with annual collected sediments volume, and verify the feasibility.

4.2 Bottom outlets

Two bottom outlets are suggested for the dam, from each side of the spillway. The sediment flushing is suggested while the reservoir water level is at low levels in order to increase the efficiency if sediments removing. As mention early when the sediment flushing is in operation the sediments could be mixed and disturbed via mechanical methods in order to increase the sediment outflow. Then the trapped or deposited sediments also could be deposited and removed. However frequent and timely flushing can reduce the requirement for mechanical mixing. Those should be carefully mentioned in operation manuals.

In a flushing process, mainly the finer particles are extracted. However, the flushed sediment can't be allowed to transport in the same basin further downstream river flow since there are some important reservoirs in Mahaweli system. Therefore, the flushed bed load could be trapped just downstream of the dam with a structure such as sedimentation tank, small check dam type walls or cascade, etc. The collected sediments could be removed and used as suggested in check dams.

The suspended sediment can be transported with the flush water to the adjacent catchment Wela Oya which doesn't have any reservoir in its downstream.

When these structures are designed, design Sediment discharges, flood volumes and levels, required sediment removing frequency, the access and machineries to be used, etc. should be thoroughly examined and analysed in crucial locations.

The monitoring of sediment is very important even after the construction of structures. Hence periodical bathymetric mapping in the reservoir and check dams are essential for timely removals of sediment. The volumetric measurement of sediment yield in identified control points in daily basis is required for integrated sediment management of total reservoir system.

4.3 Upper Catchment Protection Measures

The most sustainable sedimentation control measure among all is the upper catchment development to minimize the erosion. Accordingly, reforestation in steep slopes of the catchment is introduced with soil protection methods. 200 ha new forest plantations have been introduced as an initial step of this.

Code availability. The software EPM is not used in this research. Uma Oya multipurpose project used it for analysing their data. Their results were taken only to verify this research results. Reference is given for that.

Data availability. This study is a part of designing of Thalpitigala Reservoir Project. The designing process is not yet finished. So, some of the data is still not published.

Author contributions. TLBG: Conceptualization, Formal analysis, Investigation, Methodology, Project administration, Resources, Validation, Visualization, Writing – original draft preparation, BP: Data curation, Funding acquisition, SE: Supervision, Writing – review & editing.

Competing interests. At least one of the (co-)authors is a guest member of the editorial board of *Proceedings of IAHS* for the special issue "ICFM9 – River Basin Disaster Resilience and Sustainability by All". The peer-review process was guided by an independent editor, and the authors also have no other competing interests to declare.

https://doi.org/10.5194/piahs-386-1-2024

Disclaimer. Publisher's note: Copernicus Publications remains neutral with regard to jurisdictional claims made in the text, published maps, institutional affiliations, or any other geographical representation in this paper. While Copernicus Publications makes every effort to include appropriate place names, the final responsibility lies with the authors.

Special issue statement. This article is part of the special issue "ICFM9 – River Basin Disaster Resilience and Sustainability by All". It is a result of The 9th International Conference on Flood Management, Tsukuba, Japan, 18–22 February 2023.

Acknowledgements. We would like to express our special thanks to ICHARM for giving us this opportunity through ICFM-9, guiding the report and funding. We extend our thanks to Ministry of Irrigation, Department of Irrigation, Central Engineering Consultancy Bureau of Sri Lanka and Uma Oya Multipurpose project for providing data, support and guidance.

Review statement. This paper was edited by Kensuke Naito and reviewed by two anonymous referees.

References

Dias, B. A. R. H. (E.U.): Assessment of Soil Erosion in Uma Oya Catchment, J. Environ. Profession. Sri Lanka, 8, 39–51, 2019.

Ministry of Irrigation and Water Resources Management of Sri Lanka, Farab, Poiry: Report on Metrology, Hydrology and Sedimetology, 70–82, 2010.

Ministry of Irrigation and Water Resources Management of Sri Lanka: Feasibility report Lower Uma Oya Reservoir Project (Thalpitigala Reservoir), 41 pp., 2015.

Ouyang, D. and Bartholic, J.: Institute of Water Research, Michigan State University, East Lansing, MI Predicting Sediment Delivery Ratio in Saginaw Bay Watershed, 659–671, 1997.

U.S Department of Interior B.O: Reservoir Sedimentation – Technical Guideline for Bureau of Reclamation, 2–30, 1982.

Vanoni, V. A.: Sedimentation Engineering Practice. American Society of Civil Engineers, Manuals and Reports on Engineering Practice, 1975.

Proc. IAHS, 386, 7–11, 2024
https://doi.org/10.5194/piahs-386-7-2024

Open Access PIAHS

Building disaster prevention capacity in a continental scale country: the trajectory and experience of the Geological Survey of Brazil

Leandro G. Kuhlmann[1], **Mamoru Miyamoto**[2], **Julio C. Lana**[1], **and Tiago Antonelli**[1]

[1]Geological Survey of Brazil, Rio de Janeiro, 22290-240, Brazil
[2]International Centre for Water Hazard and Risk Management, Tsukuba, 305-8516, Japan

Correspondence: Leandro G. Kuhlmann (leandro.kuhlmann@sgb.gov.br)

Received: 21 July 2023 – Revised: 16 October 2023 – Accepted: 15 November 2023 – Published: 19 April 2024

Abstract. The goal of this paper is to share the trajectory and current results of the training program on disaster risk reduction from the Geological Survey of Brazil (GSB) as well as the challenges and perspectives for future development. The risks faced by a society are determined by the hazards presented by nature and by the cultural norms of how society decides to deal with them. Historically Brazil has had spontaneous urban growth, which led to the occupation of naturally hazardous areas. The National Plan of Protection and Civil Defense Act of 2012 led to an increase in responsibility of the central government to support local governments in disaster prevention. In 2007 the SGB established its training program for disaster prevention capacity building, and a decade and a half later around 3500 people have received entry level education in risk management. Due to the restrictions imposed by the COVID-19 pandemic, the SGB implemented distance learning courses, with significant impact.

1 Introduction

In the international literature, the term "Geological Risks" is usually restricted to earthquake and volcanic risks, however the Geological Survey of Brazil (SGB) uses the term to a broader extent including flood and sediment disasters (Sampaio et al., 2013), which was chosen in this paper to keep coherence with the institutional decision.

In 2006 two geologists and one engineer from the SGB promoted the first Geological Risks Perception Course (GRPC) through a pioneer partnership with the Municipality of Nova Friburgo/RJ, which included a training program for local officials, technical contributions for the Municipal Plan for Risk Reduction and an educational booklet for children and teenagers. In 2012 the Law 12.608 brought substantial modifications to the disaster management in Brazil (Sampaio et al., 2013). Those modifications led to an increase in the responsibilities of the National Government and subsequentially to the SGB. The general curriculum of the original GRPC was adapted into a permanent project with a team. Eventually around 15 people engaged in the GRPC, while still doing other activities such as risk and hazard mapping. The organization of classes for the GRPC often depended upon support from various local and regional Civil Defense Offices and their internal capabilities.

Reinforcing capacity at the national and local levels is considered central a major factor in disaster preparedness (Matsuoka, 2016). Sharing the expertise and the experience in capacity development is expected from the international community as a contribution to implement the Sendai Framework worldwide (Koike et al., 2018). The goal of this paper is to share the expertise and experience of the SGB in the capacity building in Brazil.

2 Background and context

2.1 Legal framework of the courses

The National Plan for Protection and Civil Defense (Law 12.608/2012) defines the responsibilities of each sphere of government. Brazil has a Central Government, 27 states (including the Federal District) and a total of 5570 municipalities, with a median population for of 11 732 people (Brazilian Institute for Geography and Statistics, 2021).

Published by Copernicus Publications on behalf of the International Association of Hydrological Sciences.

The Municipality level holds the greater authority in terms of DRM. The National Government is responsible for supplying financial and technical support to State and local governments through courses, risk mapping and others.

The main legal instruments responsible for turning general governmental directives into concrete measures and services are the Multi-annual Plan and the Annual Budget Law. Since the year of 2012, those instruments comprise the Governmental Action 20LA, that allocates the budget and responsibility to the SGB to provide educational material and courses in DRR; the Budget Law also defines that the courses should be directed to public managers, public servants and for the population in general.

2.2 The characteristics of the Municipality Civil Defense Offices

The Project Elos (National Secretariat for Protection and Civil Defense, 2021) supplied a comprehensive range of statistics that describe the profile of the Municipality Civil Defense institutions (MCDs). The report highlights the challenges faced by the Municipalities such as being undermanned, undertrained, underequipped, and commonly hiring through nominations, a choice often criticized because of the vulnerability of such nominees to political instability (Lopez and Praça, 2018).

Even in developed countries, the smaller municipalities are often found lacking in capacity to manage all facets of disaster management (Ubaura, 2017). In Brazil, the large number of municipalities, with small populations and scarce resources, make it unlikely that most of them would have financial or managerial resources to hire highly skilled professionals, those able to deal with the complexity of the studies necessary and with the political acumen to navigate the intricate system of governance to implement technical solutions.

Brazil is the 5th largest country in the world, the distribution of thousands of municipalities in the continental scale country poses a challenge to the mission of the SGB, to provide training courses for those authorities. Not only some locations are remote and have difficult access, but they are also unique in terms of their needs due to their highly variable social and physical characteristics. While certain municipalities are used to slow paced floods of mega rivers, others might be in mountainous areas or in a part of the coastline with strong interference of the tidal effects.

3 The Geological Risks Perception Course – GRPC

3.1 The Standard GRPC

The Geological Risks Perception Course – GRPC is structured to provide basic knowledge on the process of flood disaster and sediment disaster and is divided into four modules. In the first module, the GRPC explains the basic terminology, where concepts such as risk, hazard, and susceptibility. Simple examples are provided for the students to understand how to differentiate such classifications.

In the second module the course presents the concept of flooding, which is explained as part of a natural geological process. The GRPC presents the hydrological cycle using the examples of geomorphological patterns that usually occur across a fluvial system: from a river's steep headwaters and their valleys sculpted by erosion down to their flat river mouths overflowing with sediments. The concept of return periods and flood frequency is presented in a simplified manner, to provide understanding of the logical aspect behind occurrence of extreme events.

In the third module the GRPC presents the contents on risks associated with erosion. The GRPC presents examples of the geomorphological features of ravines, gullies, rivers, and coastal erosions. The goal is to allow the student to be able to identify erosional processes and understand their dynamic nature.

The fourth module concerns landslides and debris flow. The GRPC presents concepts and examples of the processes of landslide and debris flow. The GRPC focuses not only on the typical geomorphological settings of landslide prone areas but also on the inadequate construction methods that are popular in Brazil, such as cut and fill on steep slopes, without the supervision of a professional.

The fifth module focuses on hard and soft measures in disaster prevention. The GRPC presents a broad, simplified view of the types of construction works that can be considered as alternatives for risk reduction and disaster mitigation. The GRPC also presents soft measures such as alert systems, evacuation drills and community engagement are also presented.

The sixth module is an explanation on how SGB uses those concepts to map high and very high-risk areas. The methodology of the risk sectoring is quite simple and intuitive and is based mostly on visual identification of signs of terrain instability for landslides and interaction with local populations for floods. While certain projects of the SGB incorporate numerical modelling, they are not presented by the GRPC due to their complexity. The objective is to encourage local public officials to prepare their own assessments, instead of continuously requesting support from the central government as their main strategy.

Lastly, the seventh module involves how to read and understand the risk assessments provided by the SGB and using the information to pursue approval of local policies that strengthen the community against natural hazards.

3.2 The Fully Asynchronous Course – FAC

The pandemic of COVID19 presented profound difficulties to the continuity of the GRPC program. Travel restrictions and subsequent inability to gather the participants together completely disrupted the normal workflow of the program. The first response was an instant shift to remote

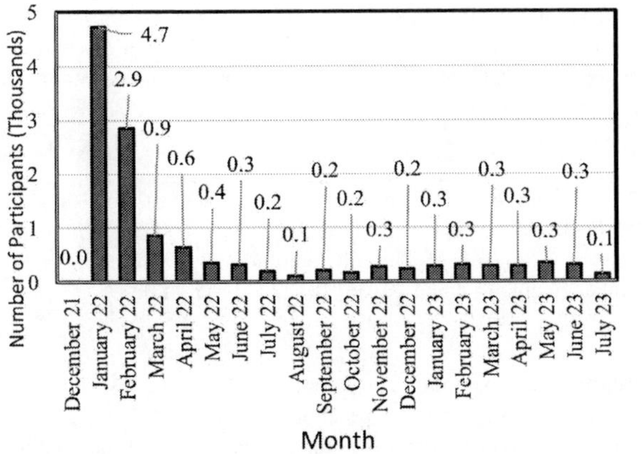

Figure 1. Evolution of the number of Participants of the FAC course from December 2021 to October 2022.

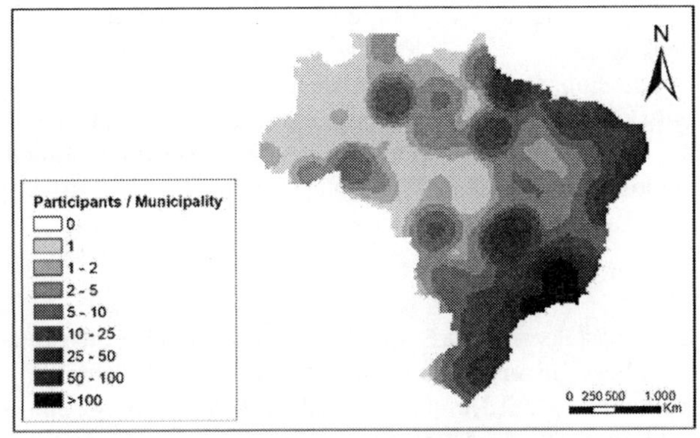

Figure 2. Spatial distribution of the participants (National School of Public Administration, 2023).

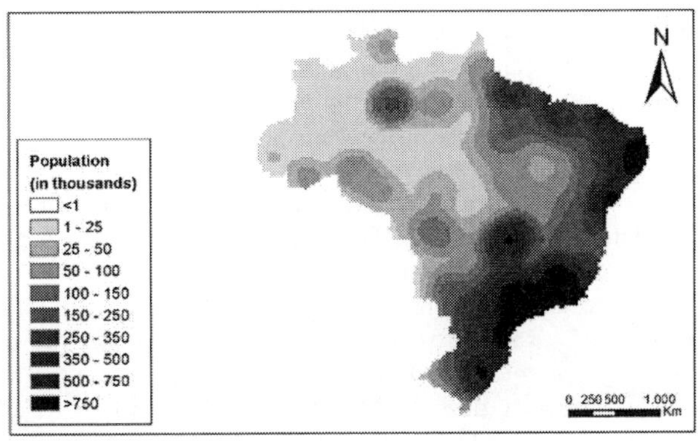

Figure 3. Approximate population distribution in Brazil (Brazilian Institute for Geography and Statistics, 2021).

classes through videoconferences. The reduced cost of remote classes and ability to organize classes made it possible to surpass the official goals in terms of both course frequency and participant numbers. The remote classes did, however, highlight the difficulty in engaging in active discussion with the participants. Subsequently and significantly, there was a feeling that the courses were less efficient in meeting their goals – primarily in terms of supplying quality education.

As the concept of distance learning gained acceptance, a partnership was successfully established by the SGB with the National School of Public Administration and allowed to host a Fully Asynchronous Course (FAC). The contents of the course stayed the same as the standard GRPC, but its presentation was simplified and made more dynamic with videos and illustrated content. The course was released by the SGB for free and is available for public servants, and for the population in general, with no prerequisites for participation.

The FAC has had an unexpectedly high number of participants that peaked soon after its release (Fig. 1) and has reached a total of 13 000 participants from December 2021 to July 2023. Of those participants, 39 % were able to successfully answer the questionnaires and acquire certification. Nonetheless, all participants are considered by the SGB as beneficiaries of the course – because of the expected results to promote a disaster resilient culture.

Public servants constituted 50 % of the participants, with the other 50 % being non-governmental participants.

The spatial distribution of the participants in Brazil (Fig. 2) is aligned with the actual population distribution in Brazil (Fig. 3), with some participants even from remote areas, with difficult access. Among municipalities with a population of at least 50 000 people 87 % had at least one participant, with an average of 15 participants.

3.3 Perspectives of future development of the program

From the duties defined by Law 2012 and subsequently by Action 20LA of the Federal Government, the most important gap that remained unaddressed by the SGB was the education of government managers. There was a certain apprehension that government managers (e.g., mayors, lawmakers, secretariats, advisors) might have minimal interest in engaging in actual technical discussion. Mobilizing and gathering municipal authorities, which tend to be busy, also was always considered a greater challenge in comparison to mobilizing their technical staff. For those reasons, the SGB decided to establish another partnership, with the Getulio Vargas Foundation (FGV), a national reference in terms of public management education. The FGV was responsible for preparing the educational material, hosting the course, and issuing certificates; while the SGB supplied the financing, the general guidelines, and the final revision of the course.

The course entitled Public Policies for Disaster Risk and Response Management at Municipality Level was released in August 2023 and within the first two months had over 800 participants.

The course developed for public managers supplies not only knowledge of the legal aspects of disaster management, but also to serve as an appeal for greater engagement of local authorities in disaster prevention policies.

4 Conclusion

In the past decade, the SGB has envisioned significant efforts to address the mission presented by the National Plan of Protection and Civil Defense. Nonetheless, the employment of public servants in Municipality Civil Defense offices has a dynamic that does not tend to align and fixate them for long term careers in disaster management, therefore, the maintenance of the GRPC program, as it has been operating for the past 15 years, is not expected to be sufficient for the country's required disaster prevention education.

New strategies have been applied in the past two years and the implementation of remote learning platforms has been accepted by users. This has enhanced the ability of the small SGB team to reach more citizens and more public servants. The total number of participants from both the standard GRPC and the FAC programs correspond to 17 000 people in the past 15 years.

The initiative of supplying a dedicated course to public managers is expected to raise awareness of the importance of building strong local institutions.

The trajectory described in this paper shows how small groups or even individual initiatives may be scaled-up into significant projects that impact thousands of people.

Data availability. The data used in this manuscript is publicly available from the addresses: https://www.ibge.gov.br/estatisticas/sociais/populacao/9103-estimativas-de-populacao.html?=&t=downloads (Brazilian Institute for Geography and Statistics, 2021) and https://emnumeros.escolavirtual.gov.br/dados-abertos/ (National School of Public Administration, 2023).

Author contributions. LGK: Conceptualization; Writing – original draft preparation; Visualization. MM: Supervision. TA: Project administration. JCL: Resources.

Competing interests. At least one of the (co-)authors is a guest member of the editorial board of *Proceedings of IAHS* for the special issue "ICFM9 – River Basin Disaster Resilience and Sustainability by All". The peer-review process was guided by an independent editor, and the authors also have no other competing interests to declare.

Disclaimer. Publisher's note: Copernicus Publications remains neutral with regard to jurisdictional claims made in the text, published maps, institutional affiliations, or any other geographical representation in this paper. While Copernicus Publications makes every effort to include appropriate place names, the final responsibility lies with the authors.

Special issue statement. This article is part of the special issue "ICFM9 – River Basin Disaster Resilience and Sustainability by All". It is a result of The 9th International Conference on Flood Management, Tsukuba, Japan, 18–22 February 2023.

Acknowledgements. The authors would like to acknowledge the Japanese International Cooperation Agency – JICA, for promoting and financing the cooperation and partnerships that long led to the preparation of this manuscript.

The International Centre for Water Hazard and Risk Management – ICHARM, for the incentive of continuous partnerships.

The International Conference on Flood Management Ad-hoc Committee and Organizing Committee for the constant support and coverage of submission and publication charges.

The Geological Survey of Brazil, for continuously engaging in the partnerships that have made the partnerships that led to this publication.

Review statement. This paper was edited by Svenja Fischer and reviewed by two anonymous referees.

References

Brazilian Institute for Geography and Statistics: Estimativas da Populacao Residente para os Municipios Brasileiros com data de referencia em 1 de julho de 2021, Rio de Janeiro RJ, Brazil, Brazilian Institute for Geography and Statistics [data set], https://www.ibge.gov.br/estatisticas/sociais/populacao/9103-estimativas-de-populacao.html?=&t=downloads (last access: 1 February 2023), 2021.

Koike, T., Hayashi, H., Satake, K., Tsukahara, K., Kawasaki, A., Amano, Y., Takara, K., Saya, S., Nishiguchi, N., Nishikawa, S., Tamura, K., Hiroki, K., Shaw, R., and Ikeda, T.: Role Played by Science and Technology in Disaster Risk Reduction: From Framework Planning to Implementation, J. Disaster Res., 13, 1222–1232, https://doi.org/10.20965/jdr.2018.p1222, 2018.

Lopez, F. G. and Praça, S.: Cargos de confianca e politicas publicas no Executivo federal, in: Burocracia e politicas publicas no Brasil: interseções analíticas, edited by: Pires, R., Lotta, G., and Elias de Oliveira, V., Instituto de Pesquisa Economica Aplicada, Brasilia, Brazil, 141–160, ISBN 978-85-7811-331-5, 2018.

Matsuoka, K.: Promptness and sustainability in town reconstruction following the 2011 Great East Japan Earthquake, in: Asian Law in Disasters: Toward a Human-Centered Recovery, 1st edn., edited by: Kaneko, Y., Matsuoka, K., and Toyoda, T., Routledge, 263–275, https://doi.org/10.4324/9781315680323, 2016.

National School of Public Administration: Escola Virtual em Numeros: Dados Abertos, National School of Public Administration [data set], https://emnumeros.escolavirtual.gov.br/dados-abertos/ (last access: 1 February 2023), 2023.

National Secretariat for Protection and Civil Defence: Diagnostico de capacidades e necessidades municipais em protecao e defesa civil, Technical Report, Brasilia – DF, Brazil, 83 pp., ISBN 78-65-84510-08-1, 2021.

Sampaio, T. Q., Pimentel, J., Silva, C. R., and Moreira, H. F.: A Atuação do Serviço Geologico do Brasil (CPRM) na Gestão de Riscos e Resposta a Desastres Naturais, 6th National Congress of Public Management, 16–18, National Council of State Administration Secretariats, April 2013, Brasilia, Brazil, 26 pp., https://www.consad.org.br/documentos (last access: 10 December 2023.), 2013.

Ubaura, M.: Challenges in Residents-Centered Reconstruction Process from a Major Disaster: Lessons Learned from the Great East Japan Earthquake, in: Community – Based Reconstruction of Society University Involvement and Lessons from East Japan Compared with Those from Kobe, Aceh, and Sichuan, edited by: Hokugo, A. and Kaneko, Y., Springer, Singapore, 3–13, https://doi.org/10.1007/978-981-10-5463-1, 2017.

Proc. IAHS, 386, 13–19, 2024
https://doi.org/10.5194/piahs-386-13-2024

Estimating the rate of change in heavy rainfall intensity for flood control planning in Japan

Yuta Maeda, Tetsuya Takeshita, and Masaki Kawasaki

National Institute for Land and Infrastructure Management, 1 Asahi, Tsukuba, Ibaraki 305-0804, Japan

Correspondence: Yuta Maeda (maeda-y88pn@mlit.go.jp)

Received: 19 July 2023 – Accepted: 15 November 2023 – Published: 19 April 2024

Abstract. Heavy rainfall leading to disasters is projected to intensify and become more frequent as the average global temperature continues to rise. To manage flood risk, flood control plans that account for future climate change are urgently needed in Japan. We have estimated the ratio of the intensity of heavy rainfall in the future to that in the past (hereinafter referred to as "rainfall change rate"), using large-ensemble climate model predictions. The results show that the average rainfall change rate across Japan is 1.1, with an increase in temperature of 2 °C. This phenomenon is higher in Hokkaido, the northernmost island in Japan. This seems to be due to the large increase in sea surface temperature around Hokkaido as well as the high rate of increase in vapour pressure in lower-temperature areas, based on the Tetens equation. The Japanese government has set the target rainfall change rate for the flood control planning to 1.15 in Hokkaido and 1.1 for Japan excluding Hokkaido.

1 Introduction

Heavy rainfalls leading to disasters are projected to intensify and become more frequent as the average global temperature continues to rise. To manage flood risk, flood control plans that account for future climate change are urgently needed in Japan. In recent years, an ensemble climate-prediction database has been developed, in which perturbations are applied to the boundary conditions of climate models in order to perform a large number of calculations. Utilizing this database, it is possible to evaluate probabilistically extreme events such as heavy rains in the future.

For this reason, the National Institute for Land and Infrastructure Management (NILIM) is conducting research on future changes in rainfall by analysing large-ensemble climate model predictions from the perspective of developing flood control planning methods that take climate change into account.

As part of that research, NILIM calculated the rainfall change rate, which plays an important role in flood control planning based on climate change, clarified the distribution trend of values by regional divisions, and provided the technical data to the advisory committee of the Ministry of Land, Infrastructure, Transport and Tourism (MLIT) (hereinafter referred to as "the committee"). The results were analysed in terms of the relationship between temperature and saturated water vapour pressure, sea surface temperature, and other factors, and the technical data were provided to the committee. Then, based on these and other data, the committee set the target rainfall change rates to be used in flood control planning based on climate change (MLIT, 2021).

This paper describes the method for calculating the technical data prepared by NILIM that were provided to the committee, from the perspective of the provider.

2 Methods

2.1 Characteristics of rainfall change rate

Climate change is projected assuming different levels of impacts depending on socio-economic development and radiative forcing scenarios. Therefore, in developing flood control plans based on climate change, it is not possible to rely solely on historical data; instead, climate-change predictions based on models that take various scenarios into account must also be considered.

However, the predictions of such models are known to contain some biases. In utilizing climate-change prediction models for flood control planning, there are several possible ways to address these biases.

One method is to directly evaluate the forecast outputs from climate-change prediction models with bias correction. Another method is to extract climate-change impacts as indices from the forecast results and apply them as changes to existing planning methods.

The rainfall change rate addressed in this paper belongs to the latter method. One drawback of this approach is that it cannot fully utilize changes such as spatiotemporal rainfall patterns, but it is advantageous in that it provides a stable indicator by capturing the global change trend due to climate change, and the factors of change can be easily explained.

2.2 Calculation of rainfall change rate

2.2.1 Large-ensemble climate model predictions

Large-ensemble climate model predictions, which are known as the database for policy decision-making for future climate changes (d4PDF, 2015), are computed at a horizontal resolution of 20 km under the representative concentration pathways (RCP) 8.5 scenario, which is a high-emissions scenario that assumes no policy mitigation measures.

In this paper, we used two different outputs derived from d4PDF, which were further dynamically downscaled to a horizontal resolution of about 5 km, using the Nonhydrostatic Regional Climate Model of the Meteorological Research Institute.

One, the dataset (Yamada, 2021) we refer to as d4PDF (5 km, Yamada), is the downscaled output for Hokkaido as the target area in the Social Implementation Program on Climate Change Adaptation Technology (SI-CAT) of the Ministry of Education, Culture, Sports, Science and Technology (MEXT), using the Earth Simulator of the Japan Agency for Marine-Earth Science and Technology (JAMSTEC) (Yamada et al., 2018; Hoshino et al., 2020). The other, the dataset (Sasai, 2019) we refer to as d4PDF (5 km, SI-CAT), is the downscaled output for Japan excluding Hokkaido in SI-CAT, also using the Earth Simulator (Sasai et al., 2019).

2.2.2 Calculation process

The following is the process for calculating the rainfall change rate.

1. Japan was divided into 15 areas, considering the possibility that the rainfall change rate may differ from region to region.

2. Among the ensemble climate-prediction data, the relationships among accumulated rainfall, rainfall area, and rainfall duration were organized for each year for the data of past experiments (corresponding to climate change from 1981 to 2010) and those of future experiments (corresponding to a point in time when the global average temperature rises 2 and 4 °C above the pre-industrialization level in the mid-18th century). The number of years of data is 360 years of past experiments (30 years × 12 perturbations) and 360 years of future experiments (30 years × 6 sea surface temperature [SST] patterns ×2 perturbations).

   Here, 6 SST patterns refers to the six SST patterns (denoted CC, GF, HA, MI, MP, and MR) selected to avoid similarity among them, based on a cluster analysis of 28 models among the coupled atmosphere–ocean models submitted to the Fifth Coupled Model Intercomparison Project (CMIP5) of the Global Climate Research Program.

3. We conducted a statistical analysis of the maximum annual accumulated rainfall corresponding to the rainfall area and duration obtained in (2) above and calculated the ratio between the future and past experiments for rainfall with a 1/100 annual exceedance probability. The free software R was used to compute the statistical parameters. In organizing the data above, the average values were calculated for a total of nine patterns, including three rainfall area patterns (400, 1600, 3600 km^2) and three rainfall duration patterns (12, 24, 48 h), assuming a large river basin. In addition, we studied the case of a small rainfall area (less than 400 km^2) and the case of a short rainfall duration (less than 12 h), assuming a small river basin.

Figure 1. Regional division.

3 Results and discussion

3.1 Rainfall change rate for large river basin

The results averaged for 9 different rates – for rainfall durations of 12, 24, and 48 h and rainfall areas of 400, 1600, and 3600 km^2 – are shown in Figs. 2 and 3. At a 4 °C increase in the global average temperature, the arithmetic mean for all regional divisions is about 1.25×, but the results for north

Table 1. Sea surface temperature patterns (from d4PDF website).

CMIP5 model	Abbreviation	Organization(Nation)
CCSM4	CC	National Center for Atmospheric Research (USA)
GFDL-CM3	GF	NOAA Geophysical Fluid Dynamics Laboratory (USA)
HadGEM2-AO	HA	Met Office Hadley Centre (UK)
MIROC5	MI	AORI, NIES, JAMSTEC (Japan)
MPI-ESM-MR	MP	Max Planck Institute for Meteorology (Germany)
MRI-CGCM3	MR	Meteorological Research Institute (Japan)

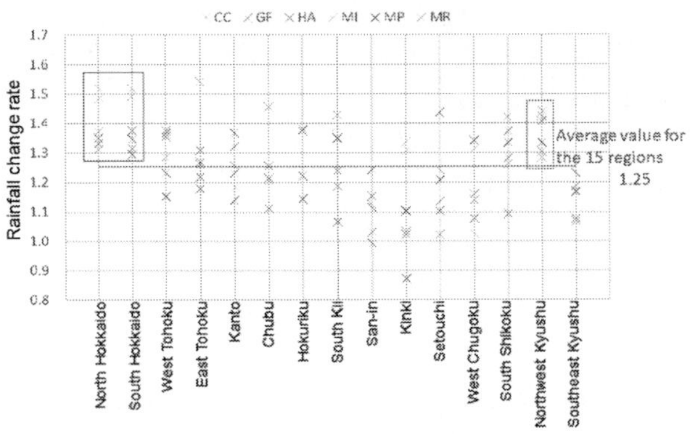

Figure 2. Rainfall change rate for each regional division and SST pattern (4 °C increase).

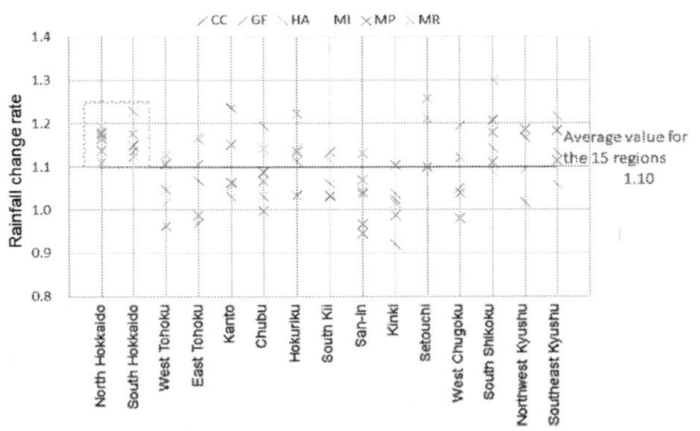

Figure 3. Rainfall change rate for each regional division and SST pattern (2 °C increase).

and south Hokkaido and Northwest Kyushu exceed the national average in each of the 6 SST cases.

At a 2 °C increase in the average global temperature, the arithmetic mean for all regional divisions is about 1.10×, but the results for north and south Hokkaido exceed the national average in each of the 6 SST cases.

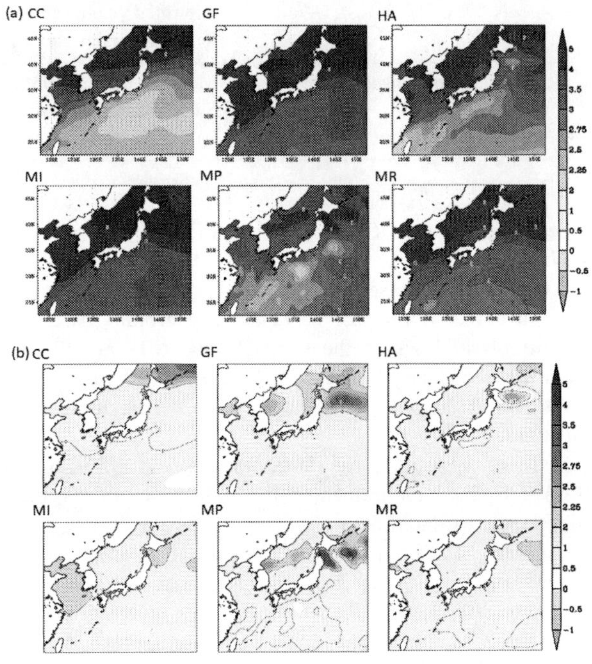

Figure 4. Temperature increases for each SST pattern ((**a**) 4 °C increase (**b**) 2 °C increase) (Modified by the author based on the data provided by Meteorological Research Institute).

3.1.1 Discussion of sea surface temperature

The high value of the rainfall change rate for Hokkaido at both a 4 and 2 °C increase in the average global temperature is discussed from the perspective of SST.

The SSTs in the d4PDF future projections are calculated by adding the future SST change pattern based on the 6 main SST models of the CMIP5 to the observed SSTs minus the trend component. Figure 4 show the temperature increase for each SST model in d4PDF (20 km) for summer (June–August) when the average global temperature increases by 4 and 2 °C, respectively.

At a 4 °C increase in the average global temperature (Fig. 4a), the amount of SST increase is larger around Hokkaido in all models. In contrast, in areas along the Pacific Ocean on Japan's other three main islands, the differ-

Table 2. Saturated water vapour content at rising temperature calculated by Tetens equation (MLIT, 2021).

		Present	RCP2.6	RCP8.5
Sapporo < in North Hokkaido>	Mean air temperature (Jun–Oct)	18.2 °C	19.6 °C	23.2 °C
	Saturated water vapor pressure	20.9 hPa	22.8 hPa (+9 %)	28.4 hPa (+36 %)
	Saturated water vapor content	15.6 $g\,m^{-3}$	16.9 $g\,m^{-3}$ (+8 %)	20.8 $g\,m^{-3}$ (+33 %)
Tokyo <in Kanto>	Mean air temperature (Jun–Oct)	24.2 °C	25.5 °C	28.6 °C
	Saturated water vapor pressure	30.2 hPa	32.6 hPa (+8 %)	39.2 hPa (+30 %)
	Saturated water vapor content	22.0 $g\,m^{-3}$	23.7 $g\,m^{-3}$ (+8 %)	28.2 $g\,m^{-3}$ (+28 %)
Fukuoka <in Northwest Kyushu>	Mean air temperature (Jun–Oct)	25.3 °C	26.6 °C	29.5 °C
	Saturated water vapor pressure	32.3 hPa	34.8 hPa (+8 %)	41.2 hPa (+28 %)
	Saturated water vapor content	23.5 $g\,m^{-3}$	25.2 $g\,m^{-3}$ (+7%)	29.6 $g\,m^{-3}$ (+26%)
Okinawa <in Nansei Shoto>	Mean air temperature (Jun-Oct)	27.4 °C	28.5 °C	30.9 °C
	Saturated water vapor pressure	36.5 hPa	38.9 hPa (+7 %)	44.7 hPa (+22 %)
	Saturated water vapor content	26.4 $g\,m^{-3}$	28.0 $g\,m^{-3}$ (+6 %)	31.9 $g\,m^{-3}$ (+21 %)

ence in the amount of SST increase is larger, depending on the model.

At a 2 °C increase in the average global temperature (Fig. 4b), the amount of SST increase around Hokkaido is particularly large in the CC, GF, and MP models, and there are also areas of large SST increases in southeastern Hokkaido in the MI model and in eastern Hokkaido in the MR model.

It is possible that the large increase in SST around Hokkaido may have influenced the high value of the rainfall change rate for Hokkaido.

Similarly, the high value of the rainfall change rate for northwest Kyushu only in the 4 °C-increase condition may be due to difference in the amount of SST increase between the 4 and 2 °C conditions. However, a more precise analysis of the details, including changes in precipitation mechanisms associated with changes in SST, is required.

3.1.2 Discussion of the relationship between temperature and saturated water vapour pressure

As shown in the results in Sect. 3.1, the rainfall change rate for north and south Hokkaido exceed the national average.

Utsumi et al. (2011) reported that the relationship between extreme 1-h precipitation intensity and surface temperature shows a positive relationship up to a certain temperature in the Japan region. They also reported that in winter, the rate of increase in precipitation intensity with increasing temperature in the area comprising Honshu and Hokkaido was distributed around a value close to the saturated water vapour rate of change (7 % $°C^{-1}$) expected from the Clausius–Clapeyron relationship.

Therefore, the high value of the rainfall change rate in Hokkaido is discussed in terms of the rate of change of saturated water vapour pressure/volume with temperature increase expected from the Clausius–Clapeyron relationship, and in terms of the amount of temperature increase.

Equation (1) is Tetens Equation (Tetens, 1930), which shows the relationship between air temperature and saturated water vapour pressure, which is used as a practical experimental formula for the Clausius–Clapeyron relationship.

$$E = 6.11 \times 10^{\frac{7.5T}{T+237.3}} \quad (1)$$

(E: saturated water vapour pressure (hPa), T: air temperature (°C))

In Tetens equation, the present air temperature and saturated water vapour pressure are respectively T1 (°C) and E1 (hPa), and the future air temperature and saturated water vapour pressure are $T2 = T1 + K$ (°C) and E2 (hPa). The rate of change in saturated water vapour pressure (E2/E1) is transformed as follows.

$$\begin{aligned} E_2/E_1 &= \left(6.11 \times 10^{\frac{7.5T_2}{T_2+237.3}}\right) / \left(6.11 \times 10^{\frac{7.5T_1}{T_1+237.3}}\right) \\ &= 10^{\frac{7.5\times K\times 237.3}{(T_1+K+237.3)(T_1+237.3)}} \end{aligned} \quad (2)$$

Therefore, if the temperature rise K (°C) is constant, the lower the current temperature $T1$ (°C), the larger the rate of change in saturated water vapour pressure.

Next, Table 2 shows that higher-latitude regions in Japan are generally projected to experience larger increases in annual mean temperature, with future changes in annual mean temperature at RCP 8.5 being 5.0 °C in Sapporo, 4.4 °C in Tokyo, 4.2 °C in Fukuoka, and 3.5 °C in Okinawa. In addition to lower present temperatures in high-latitude regions, higher temperature increases due to climate change may result in higher rates of increase in saturated water vapour in Hokkaido compared with other regions (Sapporo: +33 %, Tokyo to Okinawa: +28 % to +21 % at RCP 8.5) as well as a high rainfall change rate.

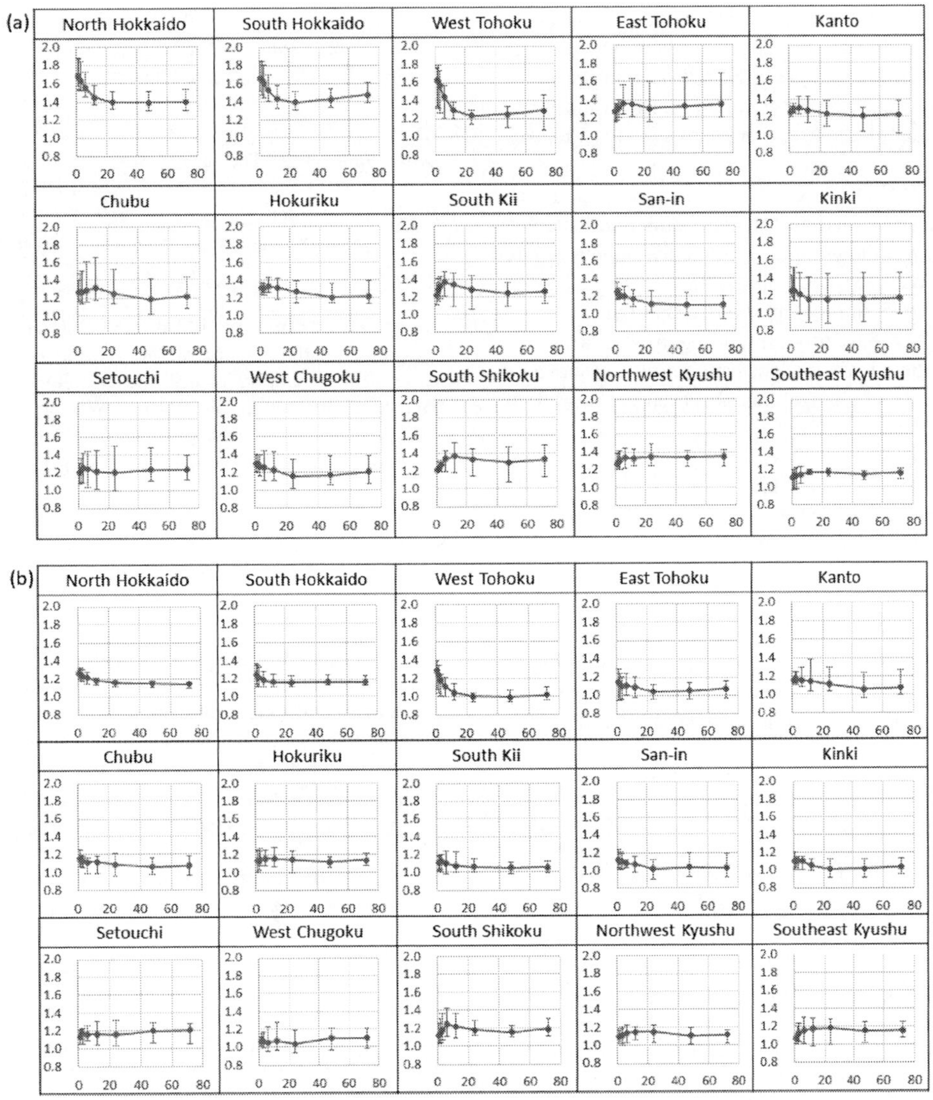

Figure 5. Relationship between rainfall duration (hour) (horizontal axis) and the rainfall change rate (vertical axis) **(a)** 4 °C increase **(b)** 2 °C increase.

3.2 Rainfall change rate for small river basin

3.2.1 Rainfall change over a short period and small area due to climate change

In the committee materials (MLIT, 2019), offered by Yamada in 2019, the effect of climate change on rainfall was suggested to be stronger when the time of rainfall events is shorter and the area covered is smaller. It was also suggested that the trend of spatiotemporal concentration of rainfall would be more pronounced at a 4 °C increase in the average global temperature compared with a 2 °C increase.

Therefore, in Sect. 3.2.2, we evaluate the trend of change in rainfall change rate for a range of short rainfall durations (12 h or less) and small rainfall areas (400 km^2 or less), assuming small river basins.

3.2.2 Rainfall change rate in short time and small area

In this section, we focus on rainfall duration and calculate the rainfall change rate for each rainfall duration for a rainfall area of 400 km^2 and an annual exceedance probability of 1/100 for a 4 and 2 °C increase in the average global temperature (Fig. 5).

At a 4 °C increase, for rainfall durations of less than 12 h, the increasing trend of the rainfall change rate becomes more pronounced, especially in north and south Hokkaido and in western Tohoku. Among the data for short rainfall durations, the 3- and 6 h rainfall data in particular show an increasing trend compared with the 12–48 h rainfall change rate for all regional divisions. The rainfall change rate for rainfall duration of 3–6 h to that of 12–48 h averaged over all 15 regional divisions was 1.05×.

However, at a 4 °C increase, for rainfall durations of 1 and 2 h, the results were divided into an increasing trend and a decreasing trend, depending on the regional division, compared with the rainfall change rate for 12–48 h.

At a 2 °C increase, the results showed that the magnitude of change for a short duration rainfall (3–6 h) was not as pronounced as that for a 4 °C increase. No significant increase or decrease in the rate of change was observed for rainfall for 1–2 h.

We also confirmed the rainfall change rate calculated by d4PDF for rainfall areas of 400 km^2 or less, under conditions of 3, 6, and 12 h of rainfall duration for a 4 °C increase and a 2 °C increase. In both models, there was no significant increase or decrease in the rainfall change rate due to differences in rainfall area. Therefore, the rainfall change rate calculated for rainfall areas of 400 km^2 or more are considered to be similar for smaller areas. However, it should be noted that this is not a sufficient evaluation, especially for rainfall areas of less than 100 km^2, due to the d4PDF (5 km) resolution (25 km^2 per mesh).

4 Conclusions

We have described the concepts and calculation methods for rainfall change rate submitted by NILIM to the Japanese government's advisory committee. The committee set the target rainfall change rate for flood control planning to 1.15 in Hokkaido and 1.1 for Japan excluding Hokkaido, assuming a 2 °C increase in the global average temperature.

Since the committee's recommendation (revised in April 2021) was issued, five fundamental river management policies overseen by the government have been revised to use the rainfall change rate described in this paper.

The rainfall change rate is considered a valid approach for addressing the issue of flood control planning under climate change by taking into account conventional flood control planning methods and recent developments in climate change forecasting technology.

However, the rainfall change rate is an indicator that focuses on rainfall and does not fully utilize the output of climate-change prediction models, including the spatiotemporal distribution of rainfall. In flood control planning and river management under climate change, it is therefore necessary to continue examining the issue from various perspectives in conjunction with the future development of climate change forecasting technology.

Code availability. The software R is freely available (https://cran.r-project.org/). The code for computing the statistical parameters is available from the corresponding author on request.

Data availability. The dataset we refer to as d4PDF (5 km, SI-CAT) is available at https://doi.org/10.20783/DIAS.562 (Sasai, 2019). The dataset we refer to as d4PDF (5 km, Yamada) is available at https://doi.org/10.20783/DIAS.647 (Yamada, 2021).

Author contributions. YM did the Data curation, Formal analysis, Investigation, Visualization, Writing – original draft preparation. The co-authors contributed as follows: TT participated in Project administration, Supervision, Writing – review and editing; MK participated in Conceptualization, Methodology, Project administration, Supervision, writing – review and editing.

Competing interests. The contact author has declared that none of the authors has any competing interests.

Special issue statement. This article is part of the special issue "ICFM9 – River Basin Disaster Resilience and Sustainability by All". It is a result of The 9th International Conference on Flood Management, Tsukuba, Japan, 18–22 February 2023.

Acknowledgements. This paper utilized dynamical downscaling data, which are produced from d4PDF using the Earth Simulator, from the Social Implementation Program on Climate Change Adaptation Technology (SI-CAT), which is sponsored by the Ministry of Education, Culture, Sports, Science and Technology (MEXT).

Review statement. This paper was edited by Mamoru Miyamoto and reviewed by two anonymous referees.

References

CRAN: The Comprehensive R Archive Network, https://cran.r-project.org/ (last access: 6 March 2024), 2024.

d4PDF: Database for Policy Decision-Making for Future Climate Change, https://www.miroc-gcm.jp/d4PDF/index_en.html (last access: 14 October 2022), 2015.

Hoshino, T., Yamada, T. J., and Kawase, H.: Evaluation for Characteristics of Tropical Cyclone Induced Heavy Rainfall over the Sub-basins in The Central Hokkaido, Northern Japan by 5-

km Large Ensemble Experiments, Atmosphere-Basel., 11, 1–11, https://doi.org/10.3390/atmos11050435, 2020.

MLIT (Ministry of Land, Infrastructure, Transport and Tourism): The expert meeting on flood control planning to cope with climate change: Flood control planning considering climate change (Appendix), 78, https://www.mlit.go.jp/river/shinngikai_blog/chisui_kentoukai/pdf/06_sankousiryou.pdf (last access: 14 October 2022), 2019.

MLIT (Ministry of Land, Infrastructure, Transport and Tourism), The expert meeting on flood control planning to cope with climate change: Flood control planning considering climate change (Recommendations), https://www.mlit.go.jp/river/shinngikai_blog/chisui_kentoukai/index.html (last access: 14 October 2022), 2021.

Sasai, T.: Dynamical downscaling data for near future atmospheric projection (from Tohoku to Kyushu) by SI-CAT, Data Integration and Analysis System (DIAS) [data set], https://doi.org/10.20783/DIAS.562, 2019.

Sasai, T., Kawase, H., Kanno, Y., Yamaguchi, J., Sugimoto, S., Yamazaki, T., Sasaki, H., Fujita, M., and Iwasaki, T.: Future projection of extreme heavy snowfall events with a 5-km large ensemble regional climate simulation, J. Geophys. Res-Atmos., 124, 13975–13990, 2019.

Tetens, O.: Uber einige meteorologische Begriffe, Z. Geophys., 6, 297–309, 1930.

Utsumi, N., Seto, S., Kanae, S., and Oki, T.: Relations of hourly precipitation extremes and temperature over Japan based on ground observational records, J. Jap. Soc. Civil Eng. B1, 67, I_307-I_312, 2011.

Yamada, T. J.: 5-km resolution ensemble climate data around Hokkaido (heavy rainfall event), Data Integration and Analysis System (DIAS) [data set], https://doi.org/10.20783/DIAS.647, 2021.

Yamada, T. J., Hoshino, T., Masuya, S., Uemura, F., Yoshida, T., Omura, N., Yamamoto, T., Chiba, M., Tomura, S., Tokioka, S., Sasaki, H., Hamada, Y., and Nakatsugawa, M.: The Influence of Climate Change on Flood Risk in Hokkaido, Adv. River Eng., 24, 391–396, 2018.

Proc. IAHS, 386, 21–26, 2024
https://doi.org/10.5194/piahs-386-21-2024

Development of a virtual flood experience system and its suitability as a flood risk communication tool

Masatoshi Denda and Masakazu Fujikane

Water-related Hazard Group, Public Works Research Institute, Tsukuba city, 305-8516, Japan

Correspondence: Masatoshi Denda (denda-m933jp@pwri.go.jp)

Received: 29 July 2023 – Revised: 27 November 2023 – Accepted: 30 November 2023 – Published: 19 April 2024

Abstract. Preventing late evacuation has been a challenge in Japan's flood management. Our hypothesis is that people often have little idea about how dangerous a flood can be and thus do not understand the importance of early evacuation. To help them realize the possible risk of flooding, we developed the Virtual Flood Experience System (VFES) using virtual reality (VR) technology. This tool is also expected to improve flood risk communication and flood evacuation training. We evaluated the realness of a virtual flood reconstructed by VFES and conducted an evacuation behaviour experiment using VFES. This paper reports an overview of VFES, including the results of the evaluation and experiment conducted in Aga Town, Niigata Prefecture, Japan. The results confirmed that VFES can successfully reconstruct the flood conditions caused by Typhoon No.19 in 2019 and quantitatively record the difference in evacuation behaviour and the time required to select appropriate evacuation behaviour between individuals with knowledge of flood risk and those without, indicating its high potential as an effective risk commutation tool capable of providing virtual flood experiences and assisting behavioural psychology experiments.

1 Introduction

Preventing late evacuation has been a challenge in Japan's flood management. Evacuation training can be more practical if participants can imagine what flooding is really like, but it is often hard for them to do so. However, it becomes possible with virtual reality (VR) technology, which can provide a vivid image of flooding (Calil et al., 2021).

Fujimi and Fujimura (2020) studied how people would feel and act during flooding by creating a virtual flood situation in a virtual town. They found that a scene indicating the immediate danger of flooding is an important clue people use to start evacuation and that other people's behaviour, such as evacuation, also affects one's decisions about what to do during flooding. Though addressing some important aspects of human behaviour, the study did not quantitatively create a flood situation using numerical flood simulation.

To further investigate why people are sometimes too late to evacuate, flooding must be quantitatively reproduced using numerical simulation. If a town and flooding are realistically reproduced, people can virtually experience a flood situation that is likely to happen in their residential areas, and thus can more directly feel the level of danger they could be exposed to. To this end, we developed the Virtual Flood Experience System (VFES) using 3D spatial information, hydrologic and hydraulic numerical simulation, and VR technology. In this paper, we outline the system, including its flood reproduction ability and experiments of virtual flood evacuation.

2 Overview of VFES

2.1 System components

VFES is composed of three subsystems (Fig. 1): (1) the spatial information subsystem, (2) the hydrologic and hydraulic subsystem, and (3) the virtual reality integration subsystem.

2.2 Spatial information subsystem

The spatial information subsystem creates the basic spatial information for VFES using the Fundamental Geospatial Data (FGD) provided by the Geospatial Information Authority of Japan (GSI). FGD includes terrain data (5 m-grid digital elevation model), the benchmarks of field mea-

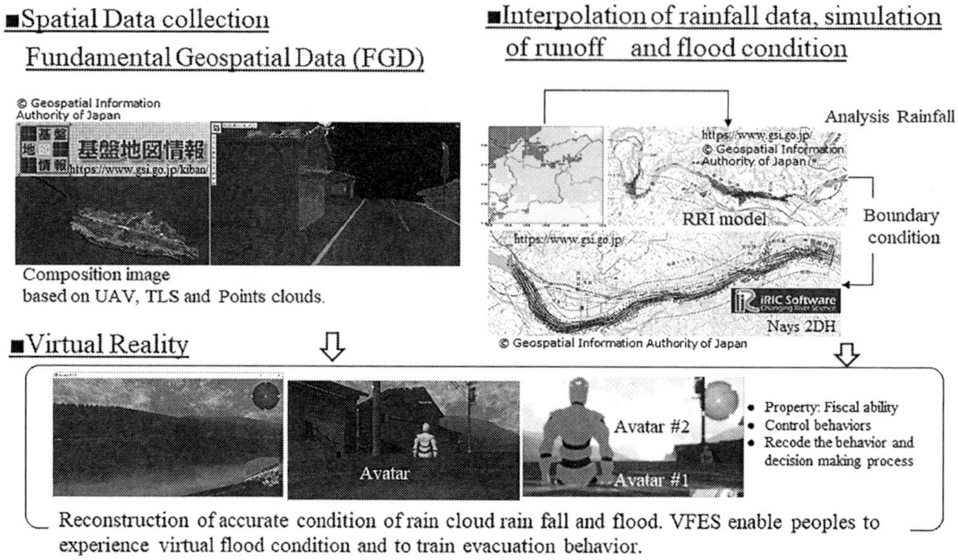

Figure 1. Overviews of VFES.

surement, shorelines, administrative boundaries and present points, road edges, and building perimeters, and can be used to create the base map of the study site.

Although FGD provides enough information to develop the base map of the study site, it does not guarantee sufficient accuracy needed to produce a precise map of the study site. For example, FGD indicates the locations of some buildings slightly off from their actual positions. It does not provide accurate vegetation information about plants and trees, either.

FGD also does not provide enough data to accurately illustrate roads running through tunnels three-dimensionally. If VFES uses FGD's road data without any additional processing, roads in tunnels will be illustrated as if they were pasted on mountain slopes. Therefore, highly accurate field surveys are necessary to reconstruct such structural features with high accuracy. VFES can import point cloud data collected by unmanned aerial vehicles (UAVs) and terrain laser scanners (TLSs) to accurately express roads in tunnels and vegetation on a map.

Although point clouds are powerful field measurement data, they also do not provide enough information to accurately reconstruct the planes, shapes, and textures of buildings. For their accurate reconstruction, 3D CAD is used to improve the buildings' shapes, and UAV-based aerial photographs are used to modify their textures. In this study, we also employ 3D CAD and photogrammetry for these purposes.

2.3 Hydrologic and hydraulic subsystem

The hydrologic and hydraulic subsystem are composed of the hydrologic and hydraulic units. The hydrologic unit simulates runoff and sets the input condition of the upper boundary discharge for 2D hydraulic simulation. Rainfall data observed at rain-gauge stations are interpolated and used to calculate the distribution of rainfall data, which are input to the Rainfall-Runoff-Inundation (RRI) model. The RRI model simulates basin-scale runoff and are calibrated with flow discharge. iRIC (Solver: Nays2DH), commonly used in Japan as a numerical simulation tool for river flow, flood prediction, and other similar purposes, is employed to perform 2D hydraulic simulations. Spatial information, such as topography and buildings, is imported into iRIC for 2D hydraulic simulation.

2.4 Virtual reality integration subsystem

VFES uses VR software to integrate spatial data reconstructed by the spatial information subsystem and flood conditions simulated by the hydrologic and hydraulic subsystem. Unity or Unreal Engine (UE), famous game engines, are mainly utilized for this purpose. These types of VR software have the powerful ability to reconstruct flood conditions with a high level of reality. Game engines enable users to apply virtual reality effects (VRE) to sky, light, rain, flood, and other features. People feel fear and sense the danger of flooding not only from how the surrounding landscape changes but also from lightning and thunder, for example. VRE can help create cloudy and rainy skies and turbulences on a smooth water surface using numerical simulation data.

VR software is also excellent at providing virtual experiences. The software allows users to create an avatar, which is their virtual self in virtual space. It lets people create avatars with different features and capabilities and control them as they like. For example, a healthy young adult with no physical problems can create an avatar who needs support in evacuation during a flood event and experience possible problems they may face in an emergency.

VFES can be an effective tool to overcome late evacuation. The system can quantitatively reconstruct dangerous flooding conditions and provide "lesson" VR video clips created based on real flood experiences, from which people without flood experiences can learn about flooding and evacuation. If people learn through such clips about how dangerous flooding can be, what actions they should take, and how critical early evacuation is to avoid possible dangers, even those without flood experiences should be able to act appropriately and decide to evacuate early.

To confirm this possibility, we conducted a feasibility study on the potential of VFES as a risk communication tool for flood risk reduction. We first applied VFES to an area in eastern Japan that suffered severe flood damage when Typhoon No.19 hit the region in 2019. Second, we evaluated the simulated flood conditions for the degree of realness by comparing them with the actual conditions. Third, we interviewed local people living in the flood-affected area about lessons they learned from the flood and actions they now consider appropriate and produced lesson VR video clips. Finally, we conducted an experiment with two groups of subjects by reference to methods of psychological experiments to investigate what decision they would make in a virtual flood situation.

3 Case study

3.1 Study site

We selected the Agano River basin and Aga Town as the study area. The Agano River, located in central Japan, is 210 km long with a basin area of 7710 m^2. The Agano River is the second biggest river in Japan in terms of discharge, and the basin is considered one of the leading hydropower generation areas in the country.

The rich discharge supports abundant harvests in the area. For the same reason, large-scale flooding has repeatedly occurred in the Agano River basin in the past 30 years: the largest in 1988 and one of a similar magnitude in 2019 both caused severe damage to local communities along the river. We specifically selected Sanekawajima, a district severely affected by both events, as the study area in consultation with Aga Town officials.

3.2 Reproduction of the past flood condition using VFES and river engineering

We conducted field surveys in 2019 and 2020 to collect the spatial data of Aga Town. The surveys using UAV and TLS were conducted in 2019. Photogrammetry was conducted in 2020. The RRI model simulated the flood conditions in 2019 using rainfall data in October 2019. Two-dimensional flood simulations were conducted using RRI-model runoff results as the boundary condition. The spatial data and the flood simulation data were integrated using Unreal Engine (UE) and the field survey results.

3.3 Hearing of public comments

We had a meeting with Sanekawajima residents on 12 December 2021, to evaluate the realness of images created by VFES and hear their life-threatening stories related to the study floods.

3.4 Creation of Lesson VR

We modified VR images based on the stories collected from the district's residents to improve the level of realness that the images convey. Also, based on their stories, we created VR video clips that contain the lessons the residents learned from their experiences. Using UE's movie camera function, we produced a number of flood images and edited them according to the scenarios prepared for the lesson VR video clips.

3.5 Effect of lesson VR video clips on evacuation behaviour

To evaluate whether lesson VR video clips improve evacuation behaviour, we conducted a virtual flood evacuation experiment using VFES with two groups of subjects by reference to psychological experiments. One group watched lesson VR video clips #1 to #3 ($n = 12$), while the other watched none ($n = 12$). The two groups answered questions about appropriate evacuation behaviour. The answers were analysed to find out the difference the correct answers number between the groups and the time required to answer each question.

4 Results

4.1 Evaluation of reproduction accuracy of the past flood condition using VFES

Figure 2 shows a bird's-eye view of a past flood scene reproduced using VFES. The blue line in the figure indicates the water edge line observed by Niigata Prefecture. The water edge line estimated by VFES matches the observed water edge line within a few meters. The comparison shows that VFES accurately reproduces the flood scenes. Utilizing 3D spatial data and hydrologic and hydraulic simulation technology, VFES successfully expresses the landscape and flood conditions in Sanekawajima District, including water levels and waves. Although the houses are not reproduced in detail to protect the residents' privacy, the distant landscape is expressed accurately, including house shapes, ridgelines, and

Figure 2. Comparison in inundation status between a photograph and a VR image.

the textures of structures and mountains. The results confirmed that VFES is fully capable of reproducing flood situations accurately enough to help people realize the danger of flooding.

4.2 Creating of lesson VR

Figure 3 shows the storyboards of the lesson VR video clips. We selected three topics from the stories we heard from Sanakawajima residents.

Lesson VR video clip #1 is designed to convey the message, "Experience sometimes fails you," based on a resident's story that flooding occurred after a dam discharge. The main idea of lesson VR video clip #2 is "Evacuation by car can be life-threatening", derived from another story about a car caught in flooding and soon submerged when evacuating at night. This clip primarily focuses on the possible danger of evacuation in a vehicle but also warns about night-time evacuation when people do not have good visibility. Lesson VR video clip #3's message is "Do not return home for things you left", which is from an episode in which a resident went back home to get something he forgot to take with him when he evacuated, only to be exposed to violent flooding. In addition to the message, it also points out that floodwaters can pose a life-threatening risk when they reach a certain depth.

4.3 Influence of lesson VR knowledges on evacuation behaviour

Table 1 and Fig. 4 compare the responses and time required to answer the questions between two groups of subjects, one of whom watched the lesson video clips while the other did not. Significantly more subjects provided correctly in the video-clip group than in the no-video-clip group in case 1. The difference between the two groups in terms of cases 2 and 3 is not very clear. This tendency is also the case with the comparison between the two groups in the time they took to answer the questions.

Table 1. Difference in answers between the "Experience" group and the "Non-experience" group on lecture VR.

Questions	Correct answer	Correct answer numbers (%)	
		Lesson VR experiencers	Lesson VR non-experiencers
1	Non	67	33
2	Non	92	83
3	Non	100	92

Question 1: Would you like to go see the river?
Question 2: Evacuation by car can be life-threatening?
Question 3: Do not return home for things you left?

5 Discussion

The work indicates the high potential of VFES as a risk communication tool. In order to achieve effective risk communication, it is crucial to share and communicate experiences and knowledge among people in a community.

In previous research and activities on disaster risk communication, flood experiences and knowledge are shared mostly orally with some photos from survivors. This approach has been found limited in communicating experiences and knowledge to other people or passing them own from generation to generation. However, VFES can revolutionarily change this situation since it quantitatively reconstructs both near and distant landscapes and creates flood conditions using VR images for locations where actual flood conditions were not observed during the event. In addition, VFES can dynamically create and show scenes in a flood situation from various viewpoints.

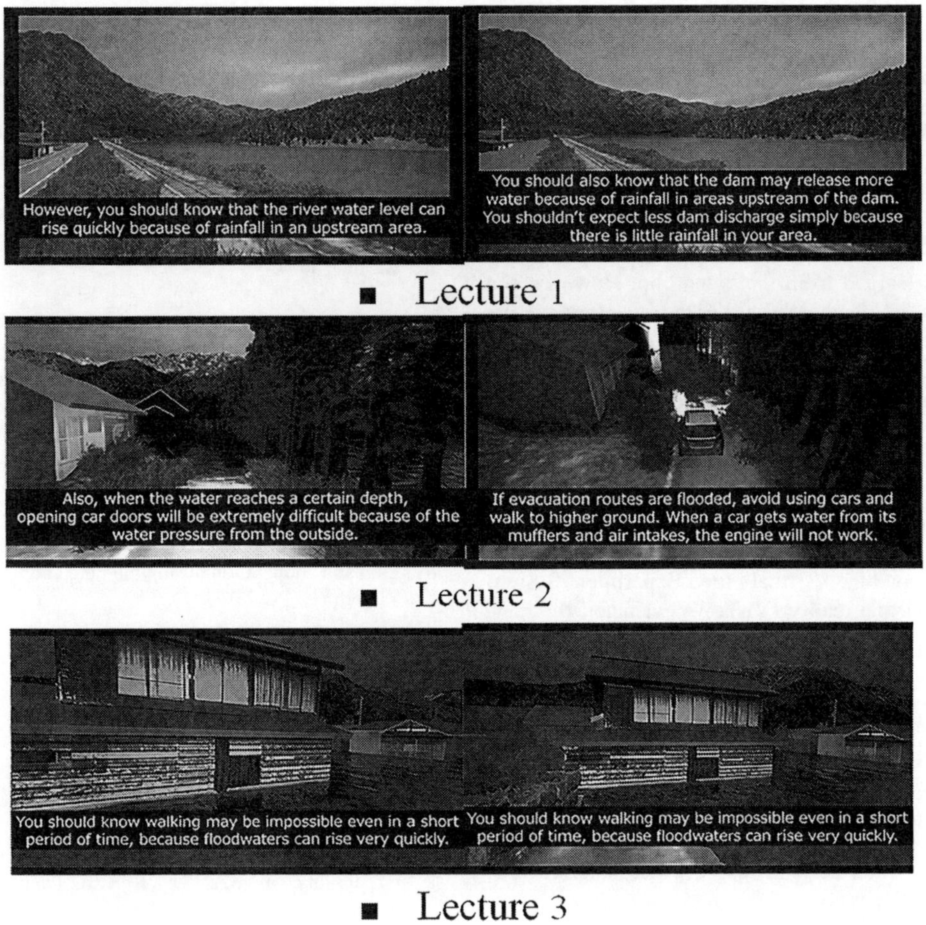

Figure 3. Examples of lesson VR video clips.

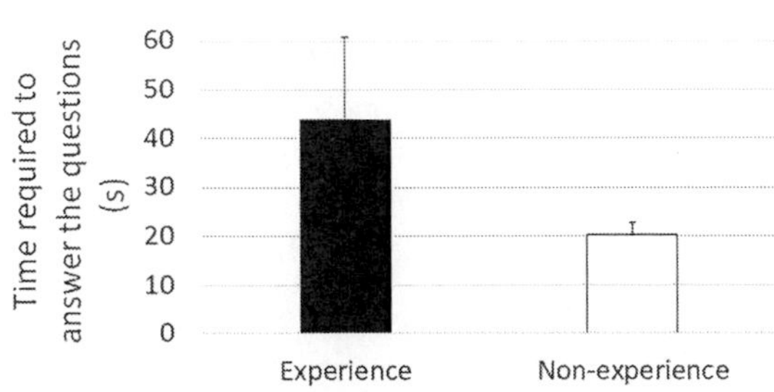

Figure 4. Comparison of response time to answer the questions.

The VFES-driven video clips can provide innovative, original information. Oral storytelling is not exactly suitable for describing a situation in detail, thus failing to elaborate on the actual conditions of flooding and conveying the real danger of the situation. This weakness often prevents people with no flood experience from having a clear idea about how dangerous flooding can be. VFES can help overcome this challenge.

Our research has also demonstrated the promising potential of VFES as a research tool. Fujimi and Fujimura (2020) indicated the potential roles that VR can play in providing virtual flood experiences and assisting experiments. As the research progresses, we have found more research topics in recognition of the potential of VR.

The following summarizes our research hypotheses at this moment, though it is still at a very early stage.

First, we hypothesize that visual information is highly likely to communicate the danger of flooding to people. The more visual information people have about possible dangers, the more likely they are to sense them and take appropriate actions. In cases 2 and 3 of our experiment, there was not much difference between the two groups of subjects in the percentage of answering correct answers and the time required to answer the questions. This is probably because both cases provide high-quality visual information by successfully reproducing the scenes that warn people about the danger of flooding at night.

Our second hypothesis is that knowledge matters when people have no access to visual information, such as being unable to see possible hazards. In case 1 of our experiment, in which the subjects could not see how critical the flood situation was, they were asked if they would go to the river to check the situation and decide whether to evacuate. The subjects who watched the lesson video clip were far more likely to choose the answer "No, I won't." than those who did not. This is probably because they chose the answer based on the knowledge they learned from the video clip. However, the video-clip group took longer to answer the question than the no-video-clip group. We assume that they still wondered if they should go and check the river while knowing from the video clip that it is something they should not do.

Our research has just begun, and we will continue conducting more experiments to test these hypotheses.

6 Conclusions

This study developed the "Virtual Flood Experiment System (VFES)" using virtual reality (VR). We evaluated the realness of a virtual flood reconstructed by VFES and conducted evacuation behaviour experiments using VFES. The results confirmed that VFES can reconstruct high-quality flood conditions and quantitatively record the difference in evacuation behaviour, indicating its high potential as an effective risk commutation tool.

Code availability. The code is not publicly accessible. VFES is currently under development.

Data availability. The data is not publicly accessible. These data are about the psychological processes of individual subjects and must be treated with extra care in terms of research ethics.

Author contributions. Conceptualization, Data curation, Formal analysis, Funding acquisition, Investigation, Methodology, Resources, Validation, Visualization, Writing – original draft preparation, Writing – review & editing: MD, Supervision: MF.

Competing interests. The contact author has declared that neither of the authors has any competing interests.

Disclaimer. Publisher's note: Copernicus Publications remains neutral with regard to jurisdictional claims made in the text, published maps, institutional affiliations, or any other geographical representation in this paper. While Copernicus Publications makes every effort to include appropriate place names, the final responsibility lies with the authors.

Special issue statement. This article is part of the special issue "ICFM9 – River Basin Disaster Resilience and Sustainability by All". It is a result of The 9th International Conference on Flood Management, Tsukuba, Japan, 18–22 February 2023.

Acknowledgements. We would like to thank the local government staff and residents in the Aga town for their tremendous contribution and cooperation during the field and interview surveys.

Review statement. This paper was edited by Svenja Fischer and reviewed by two anonymous referees.

References

Calil, J., Fauville, G., Queiroz, A., Leo, K., Mann, A., Wise-West, T., Salvatore, P., and Bailenson, J.: Using Virtual Reality in Sea Level Rise Planning and Community Engagement – An Overview, Water-Basel, 13, 1142, https://doi.org/10.3390/w13091142, 2021.

Fujimi, T. and Fujimura, K.: Testing public interventions for flash flood evacuation through environmental and social cues: The merit of virtual reality experiments, Int. J. Disaster Risk Reduct., 50, 101690, https://doi.org/10.1016/j.ijdrr.2020.101690, 2020.

Proc. IAHS, 386, 27–32, 2024
https://doi.org/10.5194/piahs-386-27-2024

Open Access PIAHS

Necessary storage for managing early flash flood to save crops in the north-eastern region of Bangladesh

Muhammad Masood[1] **and Abdul Wahid Mohamed Rasmy**[2]

[1]Design Circle-9, Bangladesh Water Development Board (BWDB), 72, Green Road, Dhaka, 1215, Bangladesh
[2]International Centre for Water Hazard and Risk Management (ICHARM), Tsukuba, Japan

Correspondence: Muhammad Masood (masood35bd@yahoo.com)

Received: 28 July 2023 – Revised: 20 November 2023 – Accepted: 4 December 2023 – Published: 19 April 2024

Abstract. Flash floods result from the abundant precipitation occurring in the Meghna basin, leading to significant losses of agriculture (specially Boro rice, the only crop of the farmers) in the north-eastern region of Bangladesh. Present practice to save the crop is compartmentalization of depressed area by submersible embankment. After harvesting, water is allowed to enter for fish cultivation with having objective to maintain biodiversity of that area. However, the crop is frequently damaged due to early flash flood before harvesting. Allocating necessary storage to provide spaces for excess water can save the crop. The present study calculates the necessary storage with a new method: Flood duration curve (FDC) to ameliorate floods and estimates water storage capacity of 6 compartments (locally termed as "Haor") by using satellite-based citizen science haor volume estimation technique. According to the capacity it is observed that early flash flood can be managed by sacrificing crops of one or two haors.

1 Introduction

Abundant precipitation in the Meghna basin specially in Meghalaya state of India has remarkable consequences for flash floods in depressed areas (locally termed as "Haor") of the north-eastern region of Bangladesh. Moreover, as a result of climate change, the more frequent occurrence of intense precipitation may further amplify both the frequency and intensity of flash floods (Masood and Takeuchi, 2016; Masood et al., 2015). This unique climatic characteristics in that area often result in agricultural losses.

In April–June 2022 a devastating flood occurred in this area which caused about 7.2 million people severely affected, 254 251 ha^2 of croplands damaged (HCTT, 2022) and estimated total damage at USD 547.6 million (PDNA, 2023). Remarkable characteristics of this flood was its prolonged period due to slower releases of water through the Surma river. After the devastating event, a 9 members' technical committee has been formed by the ministry of water resources to investigate the probable causes of delayed flood release and to recommend feasible solutions. Major two recommendations were (a) Keeping few haors out of crop cultivation and converting them into permanent reservoirs; (b) Identifying some suitable haors where early harvesting rice variety or any other crop can be cultivated and in emergency case these haros can be sacrificed by allowing flood, so that other areas will be saved from risk. In brief, by converting some haors as reservoirs early flash flood can be managed in that area (Masood et al., 2022).

Present practice to save the crop from flash floods is compartmentalization of depressed area by submersible embankment. The height of submersible embankment is designed in such a way that the embankment can protect flood up to harvesting of rice crops (usually up to 15 May). After harvesting, water is allowed to enter in each compartment/haor for fish cultivation with having objective to maintain biodiversity of that area. However, if pre-monsoon comes earlier (before harvesting) these embankments cannot save the crops as water level exceeds design level. In that case, storing the excess water in a reservoir could potentially save the crops as recommended by the technical committee.

This study estimates necessary storage at three stations using Flood duration curve (FDC) method (Takeuchi and Masood, 2017; Masood and Takeuchi, 2015). The study also estimates water storage capacity of 6 haors by using satellite-based citizen science haor volume estimation technique (Ahmad et al., 2020; Khan et al., 2023). Finally, the study investigates the viability of the strategy of deliberately allowing a haor to be inundated as a means of safeguarding other haors.

2 Study area

The downstream of the Meghna basin, in north-eastern Bangladesh is particularly recognized as a flash flood prone region. The Meghna river initiates in Manipur, India, where it is known as the Barak, and upon entering Bangladesh, it splits into two branches. The northern branch, known as Surma, flows northwards along the eastern side of Bangladesh, passing by Sylhet town. This basin encompasses two of the world's wettest places, Mawsynram and Cherrapunji, both situated in Meghalaya state, India, with average annual precipitation of 11 871 and 11 777 mm, respectively (Masood and Takeuchi, 2016). 2458 mm of rainfall was recorded in just three days from 15–17 June 2022 in Cherrapunji which is the highest precipitation in 122 years' history since the rainfall recording has started (PDNA, 2023).

The main crop in this region is Boro rice, a variety cultivated during the dry season in lower lands (Kamruzzaman and Shaw, 2018). The annual production of Boro rice from this area exceeds 3 million metric tons, accounting for approximately 17 % of the country's total production. Notably, Local Boro rice, a native variety, contributes around three-fourths of the total production (Masood and Takeuchi, 2016). Unfortunately, this crop often faces challenges due to flash floods, caused by the influx of river water from the mountainous regions in India.

The key geo-physical characteristic of the area is its lower topography compared to adjacent area; locally which is called "Haor". In brief, a haor is a wetland ecosystem in the north eastern part of Bangladesh which physically is a bowl or saucer shaped shallow depression. There are 373 haors comprising an area of about 8590 km^2 (CEGIS, 2012). However, up to June 2022 about 99 haors are protected by about 2728 km length submersible embankment which are constructed by Bangladesh Water Development Board (BWDB) with the assistance of government of Bangladesh and foreign aided fund (BWDB, 2022).

3 Data to be analysed

Observed river water level (daily) and discharge data (weekly) from 1980 to 2022 at three stations of Surma river, Kanairgat_SW266, Sylhet_SW267 and Sunamganj_SW269 were collected from the Hydrology Division, BWDB. For better result, here, daily water level data was used to draw FDC and the available discharge and corresponding water level was used to establish the stage-discharge relationship (H-Q rating equation) in each station.

4 Methodology

First, the FDC was potted from observed water level to estimate necessary storage.

4.1 Plotting of FDC

FDC presents the extreme values of moving averages of discharge over m-period (days) estimated from the observed time series. In this study, m was varied from 1 to 365 d. Generalized extreme value (GEV) distribution has been applied on this series to estimate extreme values for 5, 10, 20 and 50-years return periods. Detail procedure of plotting FDC with mathematical expression is presented by Masood and Takeuchi (2015).

In general, duration curves are drawn with considering annual maxima (maximum in monsoon period) from daily discharge data. However, in this study duration curves are drawn (a) with water level data instead of discharge data because of the availability of daily water level data and (b) with considering pre-monsoonal maxima (maximum in pre-monsoon period; March–May) series instead of annual maxima series as the main focus of this study is to alleviate pre-monsoonal early flash flood.

4.2 Estimation of necessary storage

The necessary storage indicates the volume of water necessary to store somewhere for flood control at the beginning of flood season. This storage can be calculated from FDC as illustrated by Takeuchi and Masood (2017). In short, the necessary storage is obtained from the largest inner rectangular area that just fits to the area surrounded by a duration curve, the target discharge level and the vertical axis at the origin. Rectangular area under FDC is expressed as volume of water [L^3] which is multiplication of x-axis; m [T] and y-axis; discharge [$L^3\,T^{-1}$]. In our study, to draw FDC we choose water level instead of discharge. Therefore, to get storage volume we had to estimate discharge from corresponding water level, which is calculated from H-Q rating equation of each gauge station. And, in our case, the submersible embankment crest level is chosen as the target level. Because, when river water level exceeds that level water overtops the embankment and causes flood in haor area.

4.3 Estimation of reservoir volume of haors

Satellite-based citizen science haor volume estimation technique has been used to estimate reservoir volume of haors. The technique is developed by the University of Washington SASWE Research Group under NASA-supported project led by University of North Carolina. With the technique, volumetric storage for 6 haors is calculated by employing extensive remote sensing data, which included information on water surface extent and elevation. Additionally, citizen-contributed gauge data was incorporated into the analysis.

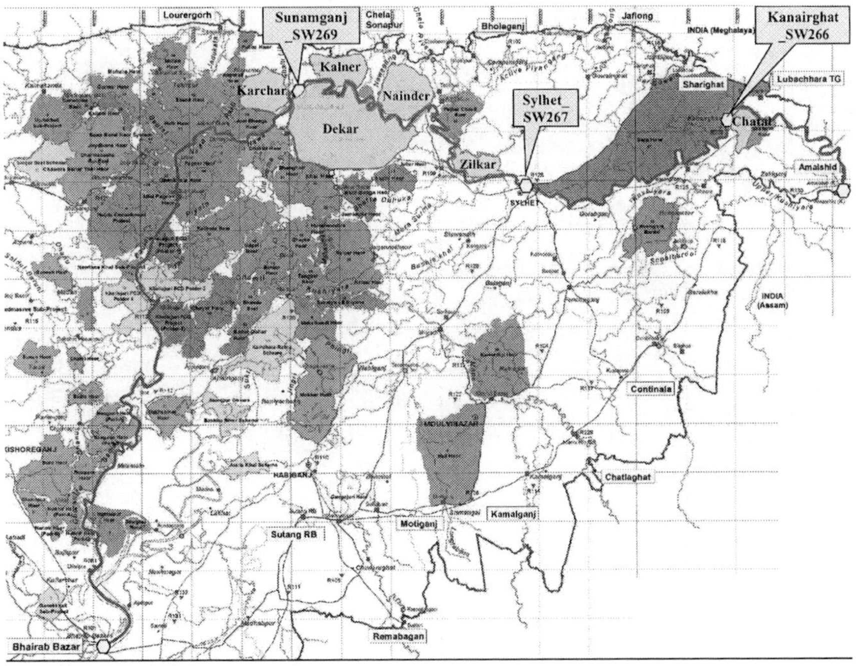

Figure 1. Haor map of north-eastern region of Bangladesh. The study haors are yellow marked.

A trapezoidal bathymetry was assumed, and an area-volume relationship was established for specific haors. This relationship was then applied to extrapolate volumetric estimates across all haors in the region. The detail methodology is described by Ahmad et al. (2020).

5 Results and discussions

5.1 Hydrologic characteristics

Using the 42-year (1980–2022) daily water level data for three stations Fig. 2 plots (a) box-and-whisker plot depicting the range of variability for each month, (b) pre-monsoonal daily water levels of historical floods in 1998, 2007, 2017 and 2022. The major observations from the figure are as follows:

- In the boxplots for all stations it is observed that variation of water level is high in April–May compared to that of other months. This higher variability of water level is due to higher variability in rainfall pattern in those months. In pre-monsoon season often high precipitation occur in Meghalaya region of India which cause early flash flood in haor area.
- It is also observed in the boxplots that number of outlier in February–April is higher than that of in other months. Outliers is often regarded as the presence of extreme values which may underestimate or overestimate a study because it lies at an abnormal distance from other values in a random sample from a population. These outliers in higher ranges represents very high extreme water level due to flash floods resulting from high precipitation in upstream.
- Pre-monsoonal daily water levels show how water rises suddenly in late-March to mid-April. In particular, during the devastating flood of 2017, water overtopped almost all embankments, resulting in the loss of all crops for that season. It is also observed that water levels are significantly higher than pre-monsoonal monthly mean of all stations.

5.2 Necessary storage

FDCs of three stations for 5, 10, 20, and 50-year return period have been plotted in Fig. 3. Largest rectangle within the FDC and above the thick line shown as dotted line in the figure. Area of the rectangle represents volume of storage needed to manage flood. The estimated necessary storages at three stations for different return period are listed in the Table 1. It is observed that necessary storage increases with the increase of return period. This is attributed to the elevated magnitude of floods associated with longer return periods. Consequently, a greater storage is essential for effectively managing floods of higher magnitudes. It is also observed that necessary storage at Sunamganj station is higher than that of other two stations. For example, necessary storage at Sunamganj is $1100\,\mathrm{M\,m^3}$

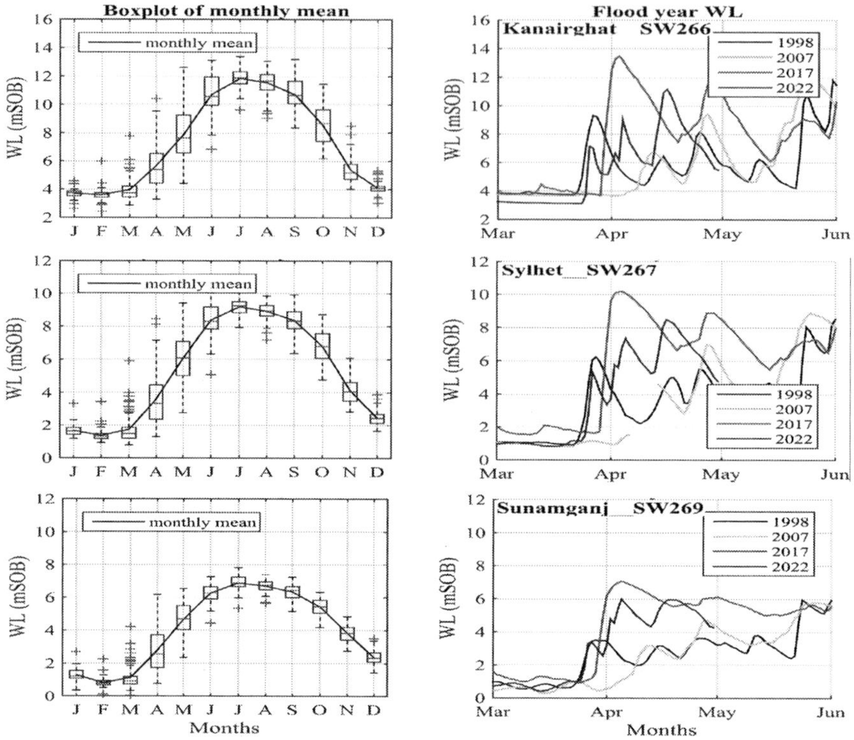

Figure 2. Box-and-whisker plot of monthly mean and pre-monsoonal daily water levels of historical floods.

Table 1. Necessary Storage of three stations for different return period.

Return period	Necessary storage ($M\,m^3$)		
	Kanairghat	Sylhet	Sunamganj
5	2	62	475
10	23	168	780
20	65	300	1100
50	155	498	1449
100	244	646	2126

while that at Kanairghat and Sylhet is 65 and 300 $M\,m^3$, respectively for 20-year return period. It is obvious that higher storage is necessary at downstream than that at upstream to manage flood as accumulated discharge increases towards downstream.

5.3 Water storage capacity of haors

The selection of 6 haors took into account the following factors: (a) adjacent to the Surma River to facilitate a quicker impact on flood with rapid water release, and (b) size of haor ranging from smaller to larger.

Reservoir volume of these haors have been calculated by the technique described in Sect. 4.3 and listed in Table 2. Figure 4 shows time series (from January 2017 to October 2022) of volume of water stored in Kalner haor which is located just upstream of Sunamganj station. It shows that water stored in the haor is nearly zero during the dry and pre-monsoon season. Because, at that time the haor is bounded by embankment to protect crops. After harvesting (usually after 15 May) water is overtopped the embankment and haor is submerged quickly. In general, haor reaches its highest capacity in monsoon (June–July). Maximum volume of water stored in Kalner haor is about 172 $M\,m^3$. However, at water level near the crest level of submersible embankment (end of May) volume of water stored in Kalner haor is near about 100 $M\,m^3$. Time series plots of water storage capacity of these haors are presented in the Supplement.

5.4 Early flash flood management by sacrificing any haor

The necessary storage denotes the volume of water that needs to be stored at upstream at the onset of the flood season for the purpose of flood control at a specific location. Therefore, it also indicates the relative difficulty or ease of flood management. In our study area, there is insufficient space to store excess water during floods unless a haor is deliberately al-

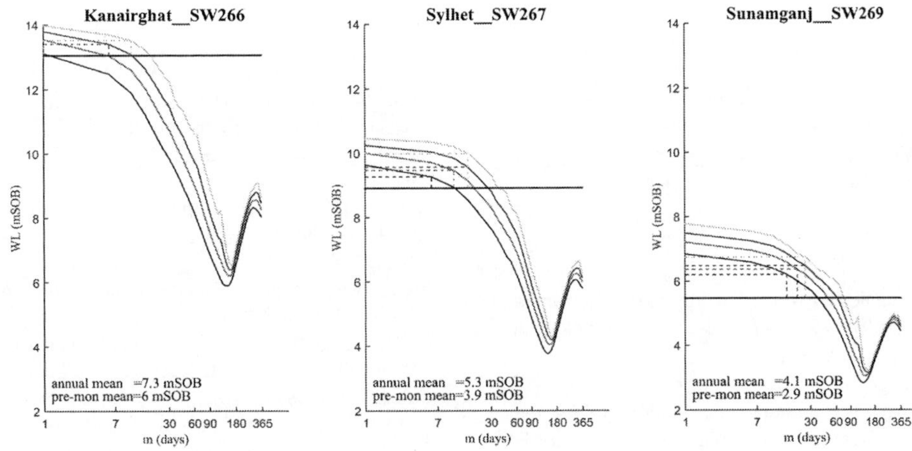

Figure 3. FDCs for three stations for 5 (black), 10 (red), 20 (blue), 50 (green)-year return period (thick line indicates embankment crest level). [SOB = Survey of Bangladesh].

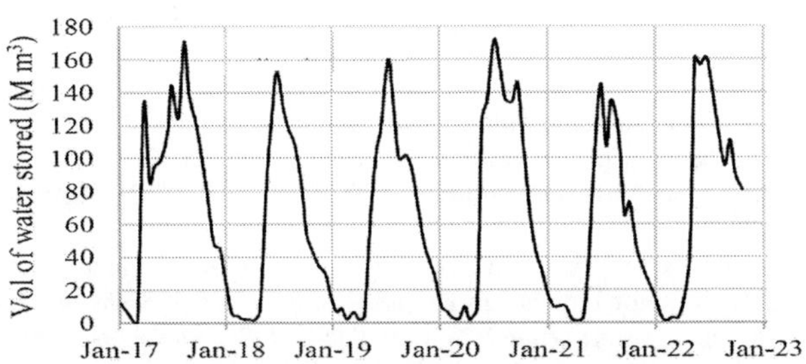

Figure 4. Time series of water volume stored in Kalner haor.

Table 2. Volume of water storage capacity of different haors.

Name of haor	Area (km^2)	Volume of water stored ($M m^3$)	
		Monsoon	Pre-monsoon
Chatal	24.7	144	48
Zilkar	46.0	111	25
Dekar	251.3	1239	512
Karchar	64.1	238	178
Kalner	84.9	172	100
Nainder	93.4	159	76

lowed to be inundated. Thus, opting to sacrifice the crops in one haor may be a viable solution to protect the crops in other haors.

From Table 1, it is observed that management of early flash flood at Kanairghat and Sylhet is relatively easier than that at Sunamganj as necessary storages are lower. The flood of those area can be relieved by sacrificing crops (i.e by allowing to be submerged) of one or two medium sized haors. However, for flash flood management at Sunamganj (necessary storage is $1100\,M\,m^3$ considering flood of 20-year return period), two large haors (like Dekar haor whose water storage capacity is $512\,M\,m^3$) need to be sacrificed.

This is an alternative solution for early flash flood management. However, if such a scenario is necessary, the government must provide compensation to farmers for their crop losses; otherwise, it could lead to a social crisis and conflicts among farmers.

6 Conclusions

This study presents an innovative methodology to estimate necessary storage for early flash flood management by FDC using long-term time series of water level data at three stations. It also calculates water storage capacity of 6 haors using satellite-based citizen science haor volume estimation technique. According to the capacity it is observed that early flash flood can be managed by sacrificing crops of one or two haors. Nevertheless, in order to prevent social conflicts and adequately compensate farmers for their losses, the government should implement necessary measures.

The study still has some limitations which can be addressed in future studies. For example, the uncertainty in estimating the storage, linked to the frequency distribution method, H-Q rating equation and availability of satellite data, requires attention. In this study, the investigation into an alternative approach for early flash flood management focuses solely on six haors. Including a larger number of haors in the analyses can enhance the comprehensiveness of the study, providing a more thorough understanding and presenting improved alternatives for selecting a viable solution.

Code availability. The python code to estimate reservoir volume of haor from remote sensing data is available from Khan et al. (2023).

Data availability. The data that support the findings of this study are available from the corresponding author ([Muhammad Masood) upon reasonable request.

Supplement. The supplement related to this article is available online at: https://doi.org/10.5194/piahs-386-27-2024-supplement.

Author contributions. The authors' contributions to the paper are as follows: Conceptualization, Data curation, Methodology, Formal analysis – MM; Supervision and Validation – AWMR; Writing – original draft preparation – MM. Both authors critically reviewed the results and have approved the final version of the manuscript.

Competing interests. At least one of the (co-)authors is a guest member of the editorial board of *Proceedings of IAHS* for the special issue "ICFM9 – River Basin Disaster Resilience and Sustainability by All". The peer-review process was guided by an independent editor, and the authors also have no other competing interests to declare.

Special issue statement. This article is part of the special issue "ICFM9 – River Basin Disaster Resilience and Sustainability by All". It is a result of The 9th International Conference on Flood Management, Tsukuba, Japan, 18–22 February 2023.

Acknowledgements. The authors express their gratitude to Faisal Hossain and Shahzaib Khan from the University of Washington, USA, for providing the code used to estimate the reservoir volume. The first author acknowledges the support extended by The International Centre for Water Hazard and Risk Management under the auspices of UNESCO (ICHARM) for attending ICFM9. Special thanks are given to ICFM9 for covering the publication charges.

Review statement. This paper was edited by Svenja Fischer and reviewed by Yukiko Hirabayashi and one anonymous referee.

References

Ahmad, S. K., Hossain, F., Pavelsky, T., Parkins, G. M., Yelton, S., Rodgers, M., Little, S., Haldar, D., Ghafoor, S., Khan, R. H., Shawn, N. A., Haque, A., and Biswas, R. K.: Understanding volumetric water storage in monsoonal wetlands of northeastern Bangladesh, Water Resour. Res., 56, https://doi.org/10.1029/2020WR027989, 2020.

BWDB: Annual Report 2021–22, Bangladesh Water Development Board, Dhaka, Bangladesh, 120 pp., 2022.

CEGIS: Master Plan of Haor Area, Summary Report, 1, Bangladesh Haor and Wetland Development Board, 2012.

HCTT: Flash Flood 2022 Humanitarian Response Plan, UN Bangladesh Coordinated Appeal, Humanitarian Coordination Task Team, 28 pp., 2022.

Kamruzzaman, M. and Shaw, R.: Flood and sustainable agriculture in the Haor Basin of Bangladesh: A review paper, Universal J. Agr. Res., 6, 40–49, https://doi.org/10.13189/ujar.2018.060106, 2018.

Khan, S., Hossain, F., Pavelsky, T., Parkins, G. M., Lane, M. R., Gomez, A. M., Minocha, S., Das, P., Ghafoor, S., Bhuyan, M. A., Haque, M. N., Sarker, P. K., Borua, P. P., Cretaux, J., Picot, N., Balakrishnan, V., Ahmad, S., Thapa, N., Bhattarai, R., Hasan, F., Fatima, B., Ashraf, M., Ahmad, S. K., and Compin, A.: Understanding Volume Estimation Uncertainty of Lakes and Wetlands Using Satellites and Citizen Science, IEEE J. STARS, 16, 2386–2401, https://doi.org/10.1109/JSTARS.2023.3250354, 2023.

Masood, M. and Takeuchi, K.: Climate Change Impact on the Manageability of Floods and Droughts of the Ganges-Brahmaputra-Meghna Basins Using FDCs and DDCs, J. Disaster Res., 5, 991–1000, https://doi.org/10.20965/jdr.2015.p0991, 2015.

Masood, M. and Takeuchi, K.: Climate change impacts and its implications on future water resource management in the Meghna Basin, Futures, 78, 1–18, https://doi.org/10.1016/j.futures.2016.03.001, 2016.

Masood, M., Yeh, P. J.-F., Hanasaki, N., and Takeuchi, K.: Model study of the impacts of future climate change on the hydrology of Ganges–Brahmaputra–Meghna basin, Hydrol. Earth Syst. Sci., 19, 747–770, https://doi.org/10.5194/hess-19-747-2015, 2015.

Masood, M., Ullah, M. E., and Das, S. C.: Causes of Early Flash-Flood in Haors of Bangladesh and Recommendations for Flood Management to Save Crops and to Maintain Biodiversity, IPWE-2023, Dhaka, 4–6 January 2023, ISBN 978-984-35-3438-5, 2022.

PDNA: Post Disaster Needs Assessment Bangladesh: Floods 2022, Ministry of Disaster Management and Relief, GoB, 50 pp., 2023.

Takeuchi, K. and Masood, M.: Necessary storage as a signature of discharge variability: towards global maps, Hydrol. Earth Syst. Sci., 21, 4495–4516, https://doi.org/10.5194/hess-21-4495-2017, 2017.

Proc. IAHS, 386, 33–40, 2024
https://doi.org/10.5194/piahs-386-33-2024

Open Access PIAHS

Projection of climate change effects on flood inundation in the Gin River basin, Sri Lanka

Jayapadma Mudalige Miyuru Uthpala Jayapadma[1], Kazuyoshi Souma[2], Hiroshi Ishidaira[2], Jun Magome[2], and Thushara Navodani Wickramaarachchi[3]

[1]Special Educational Program on River Basin Environmental Science, Integrated Graduate School of Medicine, Engineering, and Agricultural Sciences, University of Yamanashi, 4-3-11 Takeda, Kofu, Yamanashi, 400-8511, Japan

[2]Interdisciplinary Centre for River Basin Environment, University of Yamanashi, 4-3-11 Takeda, Kofu, Yamanashi 400-8511, Japan

[3]Department of Civil and Environmental Engineering, Faculty of Engineering, University of Ruhuna, Hapugala, Galle, Sri Lanka

Correspondence: Jayapadma Mudalige Miyuru Uthpala Jayapadma (miyurujphp@gmail.com)

Received: 6 July 2023 – Revised: 1 December 2023 – Accepted: 14 December 2023 – Published: 19 April 2024

Abstract. Human-induced climate change has altered the frequency and severity of heavy precipitation, droughts, and floods worldwide. Sri Lanka, a developing country in the Indian Ocean, has suffered frequent floods, affecting over 64 % of the population in the past decade. The Gin River, the primary drinking water source of the densely populated Galle district with over 1 million residents, faces recurrent flood disasters. Therefore, this study employed a multi-model ensemble to simultaneously simulate the impact of climate change on both river flow and inundation extent in the frequently flooded Gin River basin located in humid tropical monsoon region. Bias-corrected climate projections of an ensemble of five General Circulation Models (GCMs) from the Coupled Model Inter-comparison Project Phase 5 (CMIP5) under Representative Concentration Pathway (RCP) 4.5 emission scenario were used to project future river flow and flood inundation in the Gin River basin, Sri Lanka. The validated Rainfall–Runoff–Inundation (RRI) model projected a 5 % increase in high flows (flows exceeding a 5 % probability of exceedance ($<Q_5$)) at the downstream (Baddegama) gauging station and a 16 % increase at the upstream (Thawalma) gauging station in the future (2040–2060) in comparison to the past (1980–2000), which can be attributed to the elevated rainfall projected during the two monsoon seasons (southwest monsoon (May–September) and northeast monsoon (December–February)) by the GCM ensemble. In addition to the projected changes in high flows revealed by the RRI simulations, the model also yielded critical information about the expansion of annual maximum inundated extent in the basin by 3.5 % that will aid flood risk projection and sustainable management in the basin.

1 Introduction

Human-induced climate change is predicted to intensify average and heavy precipitation over Asia with high to medium confidence, which will pose severe threats to human livelihoods in the region in terms of disaster preparedness under all emission scenarios until at least the middle of the 21st century (IPCC, 2023). Damage due to floods is expected to increase with further economic growth, urbanization, and climate change, emphasizing the importance of understanding the driving factors of flood risk required for effective adaptation (Winsemius et al., 2016; Kreibich et al., 2017).

General Circulation Models (GCMs) are crucial tools for projecting future climate variables at regional to continental scales in climate-impact assessments. However, predicting future streamflow at a basin-scale under a changing climate comprises multiple stages, including the selection of emission scenarios, GCM selection, spatial disaggregation, bias correction, and hydrological modeling. The different combinations of scenarios, models, and methods at each stage yield varying projections, contributing to total uncertainty. Extensive research has consistently identified GCM selection as the major source of this uncertainty, recommending the use of a multi-model ensemble to minimize the uncertainty stemming from GCMs (Nijssen et al., 2001; Zhang et al., 2011; Bastola et al., 2011; Lee et al., 2021).

Sri Lanka is a developing country located in the Bay of Bengal, where more than 64 % of the island's population has suffered from flooding of major river basins in the past decade, including the Gin River basin. The Gin River is the main water source of drinking water in the Galle district and the district is home to more than 1 million people. Recurrence of flood disasters will significantly affect their livelihoods as well as the economic development of the area. Therefore, elucidation of the driving factors of floods in this river basin is urgently needed to develop effective flood mitigation plans. To address this need, this study assesses the impact of climate change on both future river flow and inundation using a multi-GCM ensemble, which is a key aspect of this study, as the initial step of an integrated flood risk assessment for the study area.

This study evaluated the impact of climate change on future flooding in the Gin River basin, Sri Lanka, using an ensemble of general circulation models (GCMs) obtained from the Coupled Model Inter-comparison Project Phase 5 (CMIP5). The precipitation projected under the Representative Concentration Pathway 4.5 (RCP4.5) emission scenario using GCMs (2040–2060) was input to the flood model developed using the Rainfall–Runoff–Inundation (RRI) model to project river flow and inundated area. The changes were assessed by comparing the river flow and inundated area for the future to those simulated using historical precipitation (1980–2000) generated by the GCMs.

Study area

The Gin River basin is one of the most frequently flooded river basins in Sri Lanka. It has a length of 113 km and a catchment area of 932 km^2, with annual average rainfall of over 2500 mm. The Galle district includes around 83 % of the catchment area of the Gin River. The catchment is located around 80°08″ to 80°40″ E and 6°04″ to 6°30″ N. The Gin River basin is the major water resource supporting the Galle district's water supply system. Figure 1 shows the location and the topography of the basin based on a digital elevation model (DEM). The upstream, midstream, and downstream sub-catchments were delineated based on the location of river gauging stations.

2 Materials and Methods

2.1 Climate change projection data

CMIP5 provides a framework for the execution of climate change experiments in accordance with emission scenarios driven by the RCPs defined in the Fifth Assessment Report of the Intergovernmental Panel for Climate Change (Taylor et al., 2012). Five GCMs based on the CMIP5 datasets were obtained from the Data Integration and Analysis System (DIAS) developed by the Japan Agency for Marine-Earth Science and Technology and partner universities (Kawasaki et al., 2017) to generate the inputs required by the hydrological model for simulation of future hydrological processes in the basin corresponding to the RCP4.5 emission scenario. RCP 4.5 emission scenario was selected in this study as it is considered a "stabilization" scenario in which greenhouse gas emissions peak and then decrease in the second half of the 21st century (IPCC, 2014; Taylor et al., 2012; Rogelj et al., 2016).

These GCMs were selected based on a scoring system that evaluates model performance at simulating major regional-scale climate variables through comparison with a reference dataset for the period of 1981–2000 in the study area. The precipitation and outgoing longwave radiation (OLR) at the ground surface, sea level pressure, sea surface temperature (SST), air temperature, specific humidity, and zonal and meridional winds at 850 hPa were selected as key variables. The reference datasets for precipitation, SST, and OLR were from the Global Precipitation Climatology Project (GPCP) (Adler et al., 2003), Hadley Centre (Rayner et al., 2003) and National Oceanic and Atmospheric Administration (NOAA) (Liebmann and Smith, 1996), respectively, while Japanese 25-year ReAnalysis (JRA-25) data (Onogi et al., 2007) were used as reference datasets for other climate variables. The monthly averaged spatial correlation and root mean square error (RMSE) of key variables were calculated using the tools available in DIAS. The model selection procedure is detailed in Nyunt et al. (2016).

Compared with observed rainfall from the Asian Precipitation–Highly Resolved Observational Data Integration Towards Evaluation of Water Resources (APHRODITE) project (Yatagai et al., 2012), the GCM rainfall projections exhibit underestimation of extreme values, poor seasonal simulation, and a high frequency of wet day errors. APHRODITE precipitation data was chosen due to the unavailability of continuous ground-gauged data for the period from 1980 to 2000. The biases inherent in GCM projections were corrected using the three-step bias correction method adopted by Nyunt et al. (2016).

2.2 Model and data used

The two-dimensional RRI model, which can simultaneously simulate rainfall-runoff and inundation (Sayama et al., 2012),

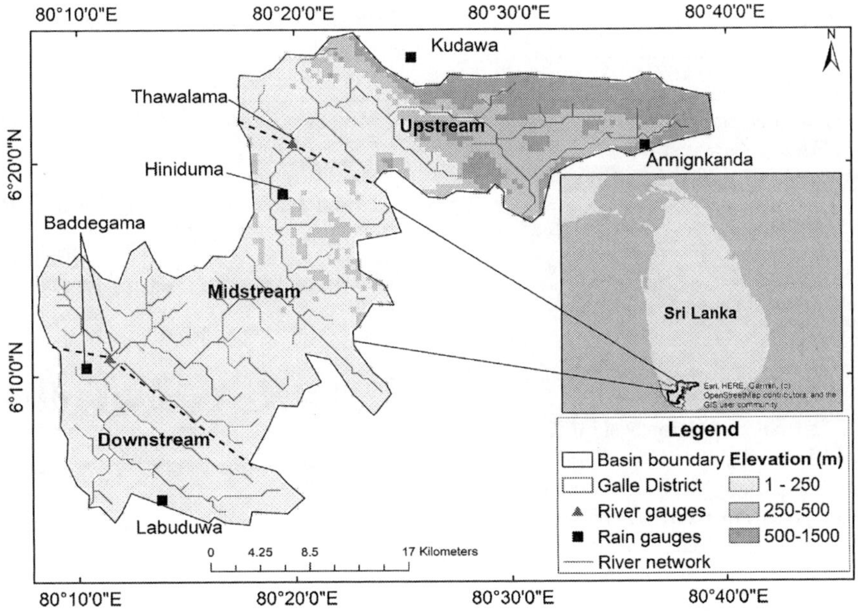

Figure 1. Location of the Gin River basin and topography of the basin based on the DEM. Dotted lines denote the sub-catchments delineated based on the location of river gauging stations.

was applied to the Gin River basin in this study. The governing equations of the RRI model are derived from the mass balance equation, Eq. (1), and momentum equation for gradually varied unsteady flow, Eqs. (2), and (3) with diffusive wave approximation where inertia terms of Eqs. (2) and (3) will be neglected, which would yield Eqs. (4) and (5). Further details about the model structure were reported by Sayama et al. (2012).

$$\frac{\partial h}{\partial t}+\frac{\partial q_x}{\partial x}+\frac{\partial q_y}{\partial y}=r-f \quad (1)$$

$$\frac{\partial q_x}{\partial t}+\frac{\partial u q_x}{\partial x}+\frac{\partial v q_x}{\partial y}=-gh\frac{\partial H}{\partial x}-\frac{\tau_x}{\rho_\mathrm{w}} \quad (2)$$

$$\frac{\partial q_y}{\partial t}+\frac{\partial u q_y}{\partial x}+\frac{\partial v q_y}{\partial y}=-gh\frac{\partial H}{\partial y}-\frac{\tau_y}{\rho_\mathrm{w}} \quad (3)$$

$$q_x=-\frac{1}{n}h^{\frac{5}{3}}\sqrt{\left|\frac{\partial H}{\partial x}\right|}\operatorname{sgn}\left(\frac{\partial H}{\partial x}\right) \quad (4)$$

$$q_y=-\frac{1}{n}h^{\frac{5}{3}}\sqrt{\left|\frac{\partial H}{\partial y}\right|}\operatorname{sgn}\left(\frac{\partial H}{\partial y}\right) \quad (5)$$

The variable h indicates the water surface height from the ground surface; q_x and q_y indicate the unit width discharges in the directions of the x and y axes, respectively; r denotes rainfall intensity; f indicates the infiltration rate; H indicates the water surface height from the datum; ρ_w represents the density of water; sgn indicates the signum function; and g is gravitational acceleration. The terms τ_x and τ_y indicate the shear stresses in the directions of the x and y axes, respectively.

The Shuttle Elevation Derivatives at multiple Scales (HydroSHEDS) data of 15 arcsec resolution (around 500 m) provided by the U.S. Geological Survey (Lehner et al., 2008), was used for DEM, flow accumulation, and flow direction data. The Global Land Data Assimilation System (GLDAS) (Rodell et al., 2004) at 0.25° resolution was used to estimate daily evapotranspiration. The CMIP5 monthly evaporation projections from the Copernicus Climate Data Store were used for the future flood simulations (Copernicus Climate Change Service (CDS), 2018). Land-use maps were obtained from the land cover project of the European Space Agency's Climate Change Initiative (Copernicus Climate Change Service (CDS), 2019). The spatial resolution of the model matched that of the DEM data (500 m). Simulations were conducted with a 600 s initial time step for slope-river interaction and a 60 s initial time step for the river, given that river calculations typically necessitate smaller time steps. The fifth-order Runge-Kutta method with adaptive time-step control, employed in RRI, shortens the time steps if necessary. The discharge simulated by RRI using daily rainfall was then compared against the observed daily discharge. The Nash–Sutcliffe Efficiency (NSE) was used to evaluate the accuracy of simulated discharge. NSE varies from $-\infty$ to 1, where a value of 1 indicates that the simulated hydrographs agree perfectly with the observed hydrographs.

Table 1. Calibrated river parameters.

River Parameter	Value
Manning's roughness coefficient [$s\,m^{-1/3}$]	0.04
Width parameter (C_w)	6.50
Width parameter (S_w)	0.30
Depth parameter (C_d)	0.70
Depth parameter (S_d)	0.25

Table 2. Calibrated slope parameters.

Slope Parameter	Value
Manning's roughness coefficient [$s\,m^{-1/3}$]	0.6
Soil depth [m]	1.0
Porosity (gamma)	0.471
Lateral saturated hydraulic conductivity (k_a) [$m\,s^{-1}$]	0.1

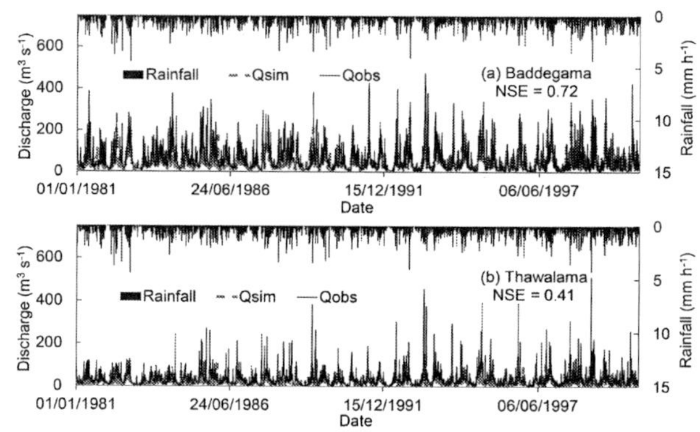

Figure 2. Comparison of simulated (Qsim) and observed (Qobs) hydrographs corresponding to the historical climate. **(a)** Downstream gauging station. **(b)** Upstream gauging station.

3 Results and Discussion

3.1 Setup of the RRI model

The RRI model was first calibrated for a significant flood event in May 2003 and was validated for flood events in May 2016 and May 2017, using observed precipitation data from rain gauges, through comparison with observed river discharge and inundation extent data (Jayapadma et al., 2022). The same model validated by Jayapadma et al. (2022), was verified for continuous long-term simulations incorporating observed evaporation data for the period of 1981–2000 and precipitation data from APHRODITE. Through comparison of the simulated discharge with observed discharge, the model parameters were optimized via trial and error. Observed river discharge records at upstream and downstream gauging stations (Thawalama and Baddegama, respectively) were provided by the Department of Irrigation of Sri Lanka. The optimized parameters of river and slope obtained from previous RRI model applications (Sayama et al., 2012; Jayapadma et al., 2022) are summarized in Tables 1 and 2.

A comparison of simulated and observed hydrographs based on daily APHRODITE rainfall, daily evaporation from GLDAS and the 1992 land-use map at the two gauging stations is shown in Figure 2 for the historical period of 1980–2000. All other model settings matched those used for the event-based validation described by Jayapadma et al. (2022). The index values for simulated river discharge at the downstream station (Baddegama; NSE = 0.72, RMSE = 46.32 $m^3\,s^{-1}$; Fig. 2a) indicated a reasonable agreement with the observed river discharge for the period of 1980–2000, especially based on the NSE values. However, the peak flows were underestimated. This discrepancy can be attributed to the average approximation of soil parameters and rainfall (intensity and amount) in the catchment, because the peak flow in the catchment is primarily generated by the dominant saturated subsurface and saturated excess overland flows, as elaborated by Jayapadma et al. (2022).

Contrasting model performance was found at the Thawalama gauging station, where both peak flows and low flows were underestimated, perhaps due to the coarser resolution of the input rainfall and evaporation data (NSE = 0.41, RMSE = 30.58 $m^3\,s^{-1}$; Fig. 2b). In addition, the coarser-resolution of topographic data can lead to the loss of fine-scale topographic details and misinterpretation of the steep, meandering river network in the upstream area of the basin. This, in turn, may result in the model underestimating peak flows, as it might not accurately capture the complex flow paths and localized flood-prone areas within the steep mountainous upstream sub-catchment. Furthermore, the coarse resolution of topography data can impact the estimation of low flows by misrepresenting the contribution of groundwater and small streams. However, the model performance was sufficient for its application in climate impact assessment within the basin, particularly in heavily flooded downstream areas of the basin (NSE = 0.72).

3.2 GCM selection and bias correction

The five GCMs selected for this study based on the scoring system were ACCESS1.0 (Bi et al., 2013), CESM.CAM5 (Neale et al., 2012), CMCC.CMS (Scoccimarro et al., 2011), CNRM-CM5 (Voldoire et al., 2013), and MPI.ESM.LR (Giorgetta et al., 2013). The three-step bias correction method employed in this study effectively removed the biases present in monthly rainfall projections relative to APHRODITE rainfall data.

3.3 Impact of climate change on monthly rainfall

The basin will receive increased rainfall with greater variability in the months from August to November in the future un-

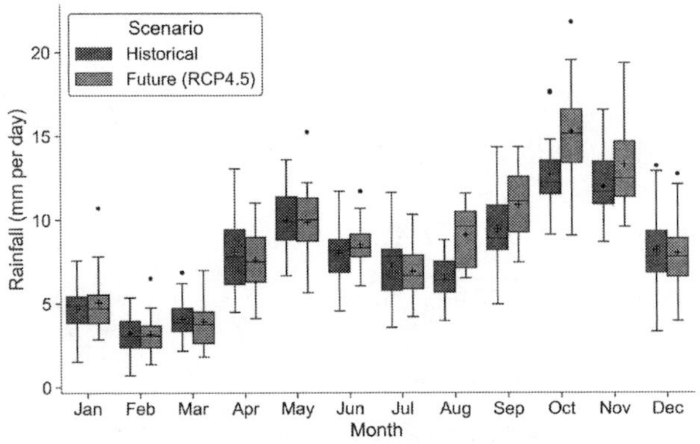

Figure 3. Comparison of basin-averaged monthly rainfall projected by the GCM ensemble with historical data. The bottom and top of the box show the first and third quartiles, the "+" indicates the mean, and the flat line inside the box is the median. The whiskers represent the minimum and maximum values. Black dots denote outliers.

der the RCP4.5 emission scenario, as projected by the GCM ensemble (Fig. 3). In addition, rainfall during the southwest monsoon season (May–September), when the basin historically received the heaviest rainfall, will intensify by 9.7 %. In contrast, the northeast monsoon (December–February) season showed no significant change (0.85 % increase) in rainfall under the RCP4.5 emission scenario. The projected intensification of rainfall during monsoons may affect water availability and the occurrence of extreme events in the basin.

3.4 Impacts of climate change on future floods

3.4.1 Impact of climate change on river flow

The intensification of rainfall projected by the GCM ensemble (average of five GCM outputs) will increase high flows (5 % probability of exceedance; Q_5) by 16 % at the upstream gauging station and by 5 % at the downstream gauging station, which may exacerbate flooding. Meanwhile, mean annual discharge of the river is projected to decrease by 34 % and 25 % at the downstream and upstream gauging stations, respectively (Fig. 4), which may influence water availability in the basin.

3.4.2 Impact of climate change on inundation

The 20-year average of the annual maximum inundated area projected by the GCM ensemble (average of five GCM outputs) for the future was 185.71 km^2, representing an increment of 3.5 % compared to the historical average of 180.16 km^2. All GCMs except the CESM and CNRM models predicted expansion of the inundated area throughout the basin, whereas the CESM and CNRM models predicted no future expansion of inundated area in the downstream region under the RCP4.5 emission scenario (Fig. 5). Similarly, the 20-year average of the annual maximum inundated area simulated using the rainfall projections from the GCM ensemble for the future was compared with the values simulated under historical climate in the upstream, midstream, and downstream sub-catchments of the basin. The inundated area in the middle and downstream regions is predicted to increase by 2 % and 1 %, respectively, compared to a predicted increase of 14 % in the upstream sub-catchment relative to the past. The difference in the upper sub-catchment may be attributed to the elevated river discharge projected at the upstream gauging station, as well as the steeper slope and narrower river channel (Jayapadma et al., 2022), which could facilitate higher flow velocities and result in greater inundated areas in the upstream sub-catchment.

4 Conclusion, limitations, and future directions

The performance of the calibrated RRI model in the Gin River basin was adequate for continuous long-term simulation and application to climate-change projection. This study revealed that the basin is to receive increased rainfall with greater variability in the months from August to November in the future under the RCP4.5 emission scenario. Additionally, rainfall during the southwest monsoon season (May–September), when the basin historically received the heaviest rainfall, is to intensify further by 9.7 %. The intensified monsoon rainfall projected under the RCP4.5 emission scenario will significantly increase future high flows (16 % upstream, 5 % downstream) as well as the peak inundation extent in the basin by 3.5 %, which requires the attention of stakeholder institutions for effective disaster preparedness. Furthermore, the modeling framework employed in this study can be applied to modeling the impact of climate change in similar river basins.

The study specifically focused on assessing the impact of climate change on floods in the study area. Uncertainties in climate-impact assessments can arise at various stages, including GCM uncertainty, bias correction, and hydrological modeling. A multi-model ensemble was employed to minimize GCM and bias correction uncertainties. Even though this study focused on assessing the impact of climate change on floods, it's crucial to acknowledge that uncertainties in low-flow estimation may have arisen during hydrological modeling, due to factors such as the coarser resolution of topography, rainfall data, and the average approximation of catchment parameters. In addition, evapotranspiration processes were not incorporated directly in our hydrological simulations (RRI) to assess low-flow variations in a warmer climate. Addressing these limitations could enhance the accuracy of low-flow estimations.

Therefore, we recommend that future research directions focus on investigating low-flow impacts, considering evapotranspiration processes directly. In addition, this study could be extended to assess the possible impacts of projected land-

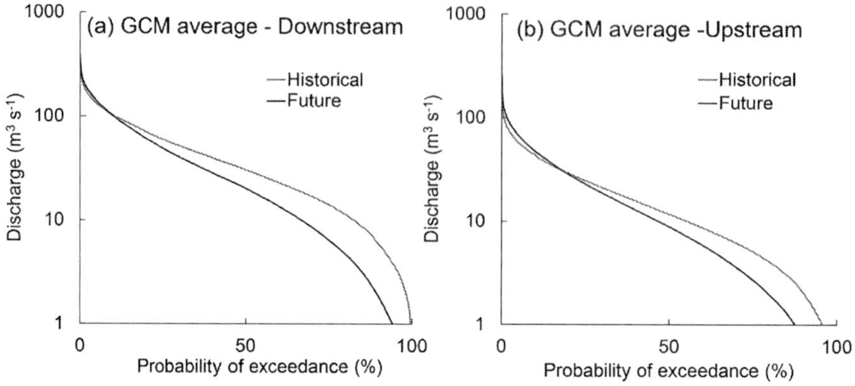

Figure 4. Comparison of projected flow duration curves with historical data at the two gauging stations.

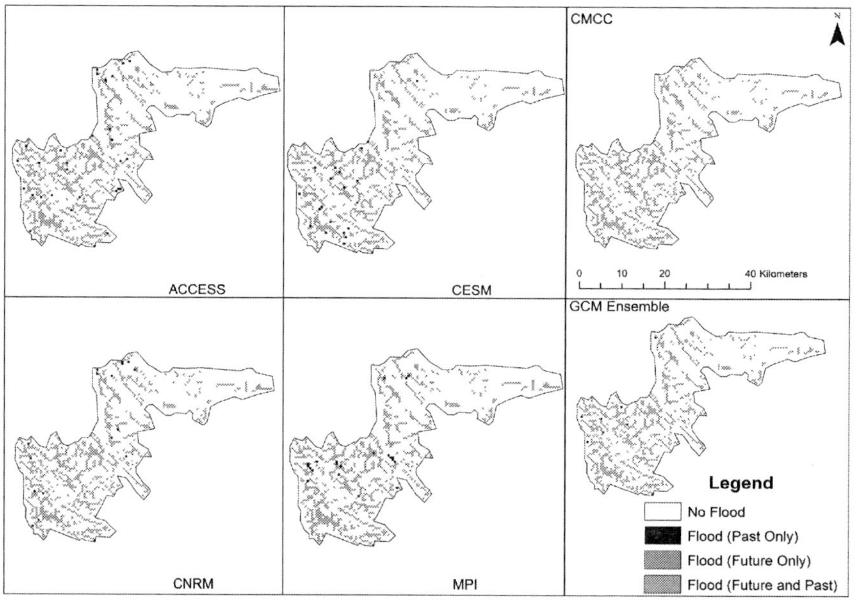

Figure 5. Comparison of 20-year average values of annual maximum inundated area. "Flood Past (only)" represents the inundated area simulated exclusively during the period of 1980–2000. The "Flood Future (only)" represents the areas that are projected to flood in the future (2040–2060) but were not in the past. The "Flood (Future and Past)" represents the intersection of flooded areas projected in both the past and the future.

use and climate changes under various emission scenarios to generate useful information that will support effective flood risk management.

Data availability. All raw data can be provided by the corresponding authors upon request.

Author contributions. JMMUJ, KS, HI, JM planned the study. JMMUJ performed simulations and analyzed the data under the supervision of all co-authors. JMMUJ wrote the manuscript draft with contributions from all authors. KS, HI, JM, and TNW reviewed and edited the manuscript.

Competing interests. The contact author has declared that none of the authors has any competing interests.

Disclaimer. Publisher's note: Copernicus Publications remains neutral with regard to jurisdictional claims made in the text, published maps, institutional affiliations, or any other geographical representation in this paper. While Copernicus Publications makes ev-

ery effort to include appropriate place names, the final responsibility lies with the authors.

Special issue statement. This article is part of the special issue "ICFM9 – River Basin Disaster Resilience and Sustainability by All". It is a result of The 9th International Conference on Flood Management, Tsukuba, Japan, 18–22 February 2023.

Financial support. This research has been supported by the Japan Society for the Promotion of Science (grant nos. JP19H02246 and JP19H01378), the Nagoya University (High Performance Computing Research Project for Joint Computational Science), and the Ministry of Education, Culture, Sports, Science and Technology (HPCI System Research Project (Project ID: hp210291)).

Review statement. This paper was edited by Tomoki Ushiyama and reviewed by two anonymous referees.

References

Adler, R. F., Huffman, G. J., Chang, A., Ferraro, R., Xie, P. P., Janowiak, J., Rudolf, B., Schneider, U., Curtis, S., Bolvin, D., Gruber, A., Susskind, J., Arkin, P., and Nelkin, E.: The version-2 global precipitation climatology project (GPCP) monthly precipitation analysis (1979–present), J. Hydrometeorol., 4, 1147–1167, https://doi.org/10.1175/1525-7541(2003)004<1147:TVGPCP>2.0.CO;2, 2003.

Bastola, S., Murphy, C., and Sweeney, J.: The role of hydrological modelling uncertainties in climate change impact assessments of Irish river catchments, Adv. Water Resour., 34, 562–576, https://doi.org/10.1016/j.advwatres.2011.01.008, 2011.

Bi, D., Dix, M., Marsland, S. J., O'Farrell, S., Rashid, H. A., Uotila, P., Hirst, A. C., Kowalczyk, E., Golebiewski, M., Sullivan, A., Yan, H., Hannah, N., Franklin, C., Sun, Z., Vohralik, P., Watterson, I., Zhou, X., Fiedler, R., Collier, M., Ma, Y., Noonan, J., Stevens, L., Uhe, P., Zhu, H., Griffies, S. M., Hill, R., Harris, C., and Puri, K.: The ACCESS coupled model: Description, control climate and evaluation, Aust. Meteorol. Oceanogr. J., 63, 41–64, https://doi.org/10.22499/2.6301.004, 2013.

Copernicus Climate Change Service (CDS): CMIP5 monthly data on single levels, Copernicus Climate Change Service (CDS) [data set], https://doi.org/10.24381/cds.9d44a987, 2018.

Copernicus Climate Change Service (CDS): Land cover classification gridded maps from 1992 to present derived from satellite observation. Copernicus Climate Change Service (C3S) Climate Data Store (CDS) [data set], https://doi.org/10.24381/cds.006f2c9a, 2019.

Giorgetta, M. A., Jungclaus, J., Reick, C. H., Legutke, S., Bader, J., Böttinger, M., Brovkin, V., Crueger, T., Esch, M., Fieg, K., Glushak, K., Gayler, V., Haak, H., Hollweg, H.-D., Ilyina, T., Kinne, S., Kornblueh, L., Matei, D., Mauritsen, T., Mikolajewicz, U., Mueller, W., Notz, D., Pithan, F., Raddatz, T., Rast, S., Redler, R., Roeckner, E., Schmidt, H., Schnur, R., Segschneider, J., Six, K. D., Stockhause, M., Timmreck, C., Wegner, J., Widmann, H., Wieners, K.-H., Claussen, M., Marotzke, J., and Stevens, B.: Climate and carbon cycle changes from 1850 to 2100 in MPI-ESM simulations for the Coupled Model Intercomparison Project phase 5, J. Adv. Model. Earth Syst., 5, 572–597, https://doi.org/10.1002/jame.20038, 2013.

IPCC: Climate Change 2013 – The Physical Science Basis, Cambridge University Press, Cambridge, 1535 pp., https://doi.org/10.1017/CBO9781107415324, 2014.

IPCC: Summary for Policymakers, in: Climate Change 2021 – The Physical Science Basis, Cambridge University Press, 3–32, https://doi.org/10.1017/9781009157896.001, 2023.

Jayapadma, J. M. M. U., Souma, K., Ishidaira, H., Magome, J., and Wickramaarachchi, T. N.: The Effect of Incorporation of Embankment Information for Flood Simulation of the Gin River, Sri Lanka, J. Disaster Res., 17, 475–486, https://doi.org/10.20965/jdr.2022.p0475, 2022.

Kawasaki, A., Yamamoto, A., Koudelova, P., Acierto, R., Nemoto, T., Kitsuregawa, M., and Koike, T.: Data Integration and Analysis System (DIAS) Contributing to Climate Change Analysis and Disaster Risk Reduction, Data Sci. J., 16, 41, https://doi.org/10.5334/dsj-2017-041, 2017.

Kreibich, H., Di Baldassarre, G., Vorogushyn, S., Aerts, J. C. J. H., Apel, H., Aronica, G. T., Arnbjerg-Nielsen, K., Bouwer, L. M., Bubeck, P., Caloiero, T., Chinh, D. T., Cortès, M., Gain, A. K., Giampá, V., Kuhlicke, C., Kundzewicz, Z. W., Llasat, M. C., Mård, J., Matczak, P., Mazzoleni, M., Molinari, D., Dung, N. V., Petrucci, O., Schröter, K., Slager, K., Thieken, A. H., Ward, P. J., and Merz, B.: Adaptation to flood risk: Results of international paired flood event studies, Earth's Future, 5, 953–965, https://doi.org/10.1002/2017EF000606, 2017.

Lee, S., Qi, J., McCarty, G. W., Yeo, I. Y., Zhang, X., Moglen, G. E., and Du, L.: Uncertainty assessment of multi-parameter, multi-GCM, and multi-RCP simulations for streamflow and non-floodplain wetland (NFW) water storage, J. Hydrol., 600, 126564, https://doi.org/10.1016/j.jhydrol.2021.126564, 2021.

Lehner, B., Verdin, K., and Jarvis, A.: New global hydrography derived from spaceborne elevation data, Eos, Trans. AGU, 89, 93–94, https://doi.org/10.1029/2008EO100001, 2008.

Liebmann, B. and Smith, C. A.: A description of a complete (interpoled) outgoing longwave radiation dataset., B. Am. Meteorol. Soc., 77, 1275–1277, 1996.

Neale, R. B., Chen, C.-C., Gettelman, A., Lauritzen, P. H., Park, S., Williamson, D. L., Conley, A. J., Garcia, R., Kinnison, D., Lamarque, J.-F., Marsh, D., Mills, M., Smith, A. K., Tilmes, S., Vitt, F., Morrison, H., Cameron-Smith, P., Collins, W. D., Iacono, M. J., Easter, R. C., Ghan, S. J., Liu, X., Rasch, P. J., and Taylor, M. A.: Community Atmosphere Model 5.0 (CAM5) – Scientific Description, https://ncar.github.io/CAM/doc/build/html/cam5_scientific_guide/index.html (last access: 9 June 2022), 2012.

Nijssen, B., O'donnell, G. M., Hamlet, A. F., and Lettenmaier, D. P.: Hydrologic sensitivity of global rivers to climate change, Climatic Change, 50, 143–175, https://doi.org/10.1023/A:1010616428763, 2001.

Nyunt, C. T., Koike, T., and Yamamoto, A.: Statistical bias correction for climate change impact on the basin scale precipitation in Sri Lanka, Philippines, Japan and Tunisia, Hydrol. Earth Syst. Sci. Discuss. [preprint], https://doi.org/10.5194/hess-2016-14, 2016.

Onogi, K., Tsutsui, J., Koide, H., Sakamoto, M., Kobayashi, S., Hatsushika, H., Matsumoto, T., Yamazaki, N., Kamahori, H., Takahashi, K., Kadokura, S., Wada, K., Kato, K., Oyama, R., Ose, T., Mannoji, N., and Taira, R.: The JRA-25 reanalysis, J. Meteorol. Soc. Japan, 85, 369–432, https://doi.org/10.2151/jmsj.85.369, 2007.

Rayner, N. A., Parker, D. E., Horton, E. B., Folland, C. K., Alexander, L. V., Rowell, D. P., Kent, E. C., and Kaplan, A.: Global analyses of sea surface temperature, sea ice, and night marine air temperature since the late nineteenth century, J. Geophys. Res. Atmos., 108, 4407, https://doi.org/10.1029/2002jd002670, 2003.

Rodell, M., Houser, P. R., Jambor, U., Gottschalck, J., Mitchell, K., Meng, C.-J., Arsenault, K., Cosgrove, B., Radakovich, J., Bosilovich, M., Entin, J. K., Walker, J. P., Lohmann, D., and Toll, D.: The Global Land Data Assimilation System, B. Am. Meteorol. Soc., 85, 381–394, https://doi.org/10.1175/BAMS-85-3-381, 2004.

Rogelj, J., den Elzen, M., Höhne, N., Fransen, T., Fekete, H., Winkler, H., Schaeffer, R., Sha, F., Riahi, K., and Meinshausen, M.: Paris Agreement climate proposals need a boost to keep warming well below 2 °C, Nature, 534, 631–639, https://doi.org/10.1038/nature18307, 2016.

Sayama, T., Ozawa, G., Kawakami, T., Nabesaka, S., and Fukami, K.: Analyse pluie-débit-inondation de la crue de 2010 au Pakistan dans le bassin de la rivière Kaboul, Hydrol. Sci. J., 57, 298–312, https://doi.org/10.1080/02626667.2011.644245, 2012.

Scoccimarro, E., Gualdi, S., Bellucci, A., Sanna, A., Fogli, P. G., Manzini, E., Vichi, M., Oddo, P., and Navarra, A.: Effects of tropical cyclones on ocean heat transport in a high-resolution coupled general circulation model, J. Climate, 24, 4368–4384, https://doi.org/10.1175/2011JCLI4104.1, 2011.

Taylor, K. E., Stouffer, R. J., and Meehl, G. A.: An overview of CMIP5 and the experiment design, B. Am. Meteorol. Soc., 93, 485–498, https://doi.org/10.1175/BAMS-D-11-00094.1, 2012.

Voldoire, A., Sanchez-Gomez, E., Salas y Mélia, D., Decharme, B., Cassou, C., Sénési, S., Valcke, S., Beau, I., Alias, A., Chevallier, M., Déqué, M., Deshayes, J., Douville, H., Fernandez, E., Madec, G., Maisonnave, E., Moine, M. P., Planton, S., Saint-Martin, D., Szopa, S., Tyteca, S., Alkama, R., Belamari, S., Braun, A., Coquart, L., and Chauvin, F.: The CNRM-CM5.1 global climate model: Description and basic evaluation, Clim. Dynam., 40, 2091–2121, https://doi.org/10.1007/s00382-011-1259-y, 2013.

Winsemius, H. C., Aerts, J. C. J. H., van Beek, L. P. H., Bierkens, M. F. P., Bouwman, A., Jongman, B., Kwadijk, J. C. J., Ligtvoet, W., Lucas, P. L., van Vuuren, D. P., and Ward, P. J.: Global drivers of future river flood risk, Nat. Clim. Change, 6, 381–385, https://doi.org/10.1038/nclimate2893, 2016.

Yatagai, A., Kamiguchi, K., Arakawa, O., Hamada, A., Yasutomi, N., and Kitoh, A.: APHRODITE: Constructing a Long-Term Daily Gridded Precipitation Dataset for Asia Based on a Dense Network of Rain Gauges, B. Am. Meteorol. Soc., 93, 1401–1415, https://doi.org/10.1175/BAMS-D-11-00122.1, 2012.

Zhang, H., Huang, G. H., Wang, D., and Zhang, X.: Uncertainty assessment of climate change impacts on the hydrology of small prairie wetlands, J. Hydrol., 396, 94–103, https://doi.org/10.1016/j.jhydrol.2010.10.037, 2011.

Proc. IAHS, 386, 41–46, 2024
https://doi.org/10.5194/piahs-386-41-2024

Performance analysis of physically-based (HEC-RAS, CADDIES) and AI-based (LSTM) flood models for two case studies

Marina Batalini de Macedo[1], Nikunj K. Mangukiya[2], Maria Clara Fava[3], Ashutosh Sharma[2], Roberto Fray da Silva[4], Ankit Agarwal[2], Maria Tereza Razzolini[5], Eduardo Mario Mendiondo[6], Narendra K. Goel[2], Mathew Kurian[7], and Adelaide Cássia Nardocci[5]

[1]Natural Resources Institute, Federal University of Itajuba, Itajubá, Brazil
[2]Department of Hydrology, Indian Institute of Technology Roorkee, Roorke, India
[3]Institute of Exact and Technological Sciences, Federal University of Viçosa, Rio Paranaíba, Brazil
[4]Institute of Advanced Studies, University of Sao Paulo, São Carlos, Brazil
[5]School of Public Health, University of Sao Paulo, São Paulo, Brazil
[6]Sao Carlos School of Engineering, Department of Hydraulics and Sanitation, University of Sao Paulo, São Carlos, Brazil
[7]Water-Energy-Food Nexus, Pennsylvania State University, State College, USA

Correspondence: Marina Batalini de Macedo (marinamacedo@unifei.edu.br) and Ashutosh Sharma (ashutosh.sharma@hy.iitr.ac.in)

Received: 11 July 2023 – Revised: 7 November 2023 – Accepted: 13 December 2023 – Published: 19 April 2024

Abstract. Megacities in developing countries are commonly affected by flooding events. The use of flood models can contribute to an evidence-based decision-making process. For a good representation, these models require physical data for catchment parameterization, and observed data for calibration and validation, which is often scarce. In this study, we analysed the performance results of physically-based (HEC-RAS, CADDIES) and AI-based (LSTM) flood models for two case studies: the Narmada basin in India and the Aricanduva catchment in Brazil. The models were evaluated for accuracy, interpretability, running time, and complexity.

1 Introduction

Flood models are good tools to guide public policies for mitigation and adaptation to extreme rainfall events since they allow simulation of different strategies scenarios. It is possible to separate the models into two main categories, those based on physical processes and characteristics of the area (here called physically-based) and those based on artificial intelligence (AI), which are mainly based on time series analysis of the data (here called AI-based).

Batalini de Macedo et al. (2022) performed a bibliometric review of flood models and observed a significant increase in their use since 2016, both for physically-based and AI-based. Furthermore, the main keywords identified in the literature were "HEC-RAS" for studies with physical models, and "machine learning" for studies with AI models, between them, "LSTM".

The different purposes when using flood modelling requires different results generated by the model (such as spatial-temporal resolution) and different processing time (e.g. early warning systems requires results in enough time to take response actions). Additionally, the monitoring conditions of the basin (e.g. ungauged or poorly gauged) may affect the applicability of a model and the quality of the results.

Thus, studies resembling these models for different scales and purposes of use can provide important information to help decision-makers choose which model to adopt.

Therefore, in this study, we analysed the performance of two physically-based flood models, with different levels of complexity and of a AI-based model. The comparison was made for two case studies in partially-gauged basins with different scales in India and Brazil.

2 Methodology

The models chosen for evaluation and the metrics used to analyse their performance (in terms of accuracy, interpretability, running time, and complexity) are briefly presented. The analysis was made for two case studies are presented, with different scales of application, located in India and Brazil.

2.1 Physically-based (PB) flood models

The physical models chosen for evaluation were the HEC-RAS and the CADDIES, both open-access. The HEC-RAS is a physical model developed by the US Army Corps of Engineers (HEC, 2018) that allows simulations of one-dimensional steady and unsteady flow and two-dimensional unsteady flow calculations (here called as HEC-RAS 2D). HEC-RAS uses the four-point implicit finite difference scheme. Its main input parameters for the development of flood maps are bathymetry data or Digital Elevation Models (DEM), roughness coefficients for each type of land use, and the series of stage and flow hydrographs as boundary conditions. In the absence of observed data downstream, the model also allows defining normal depth as a boundary condition.

CADDIES is a cellular automata model that allows discretely simulating shallow water equations (SWEs) in grids. Instead of using transition rules, this model is based on the previous state of the cell and neighbouring cells, with rules that control the evolution of the state of each cell (Guidolin et al., 2016). The model uses rainfall series, the DEM, and an average roughness coefficient for the catchment as input data, therefore, with a high simplicity for the user. Its professional version (CADDIES PRO) allows different roughness coefficients for each land use; however, it is not open access.

2.2 Artificial intelligence (AI) flood model

The AI model chosen was the LSTM. This model is a variation of the recurrent neural network (RNN), incorporating the ability to learn long-term dependencies, especially in sequence prediction problems (Shen, 2018). Compared to simple RNN, instead of having one state variable, the information in LSTM units is controlled by cell states and input, forget, and output gates. There are many variants of LSTM based on the interconnection of these gates, which control what information should be added, forgotten, and obtained as output from the LSTM cell. In this study, we evaluate the simple LSTM network for predicting flow in the river.

The LSTM model was trained for both study areas using daily meteorological forcing (rainfall, temperature, solar radiation, relative humidity, wind speed) and static watershed attributes (mean elevation and slope, land use land cover characteristics and soil characteristics). The RMSE between the observed and simulated streamflow was chosen as the loss parameter to be minimized across training sequences. As a pre-processing of the dataset, the daily streamflow and rainfall distributions were transformed to get the distribution close to normal since rainfall and streamflow typically have the Gamma distribution (Mangukiya et al., 2023). The hyperparameter of the LSTM model was manually optimized, and finally, 365 d of training instances, a hidden-state size of 256, and a dropout rate of 0.5 were used. As final result, the ensemble results from five simulations with different random seeds were obtained to reduce the uncertainty of the LSTM model output due to its stochastic nature. More details of the LSTM model strategy adopted can be read from Mangukiya et al. (2023).

2.3 Metrics for evaluation

To evaluate the applicability of the flood models for different study areas, their performance was compared in terms of accuracy, running time, complexity, and interpretability. For accuracy, the NSE (Eq. 1) and NRMSE (Eq. 2) of the water level were calculated.

$$\mathrm{NSE} = 1 - \frac{\sum_{t=1}^{T}\left(Q_\mathrm{o}^t - Q_\mathrm{m}^t\right)^2}{\sum_{t=1}^{T}\left(Q_\mathrm{o}^t - \overline{Q_\mathrm{o}}\right)^2} \tag{1}$$

$$\mathrm{NRMSE} = \frac{1}{\overline{Q_\mathrm{o}}}\sqrt{\frac{\sum_{t=1}^{T}\left(Q_\mathrm{m}^t - Q_\mathrm{o}^t\right)^2}{T}} \tag{2}$$

Where: Q_o is the observed discharges, Q_m is the modelled discharge, at times t. For HEC-RAS, Q_o and Q_m were obtained for a intermediate gauge station in the catchment, as for LSTM they were obtained for the exutory gauge station.

The accuracy was assessed just for HEC-RAS and LSTM simulations, as the CADDIES model generates only maximum flood depth instead of a time series; in the second case, a qualitative analysis was performed, comparing with pictures taken from citizens during the event modelled, for point known.

For the running time, it was presented the time taken for each model to provide the final flood map and a description of the CPU features used for each case. The complexity was evaluated from each model's input parameter and configurations. Finally, interpretability is a more subjective metric, including discussing the results provided by each model and the ease of understanding its results and applying them in the decision-making process.

2.4 Study areas

Two study areas were selected for this study, representing different contexts for floods.

The first case study is the Aricanduva catchment, located in São Paulo – SP, Brazil (Fig. 1), covering an area of

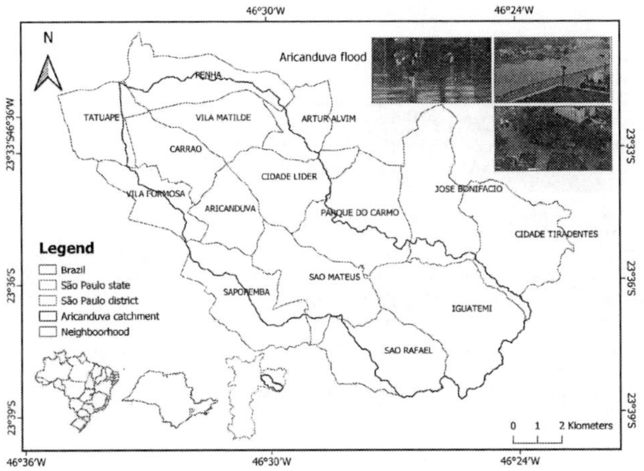

Figure 1. Location of Aricanduva catchment.

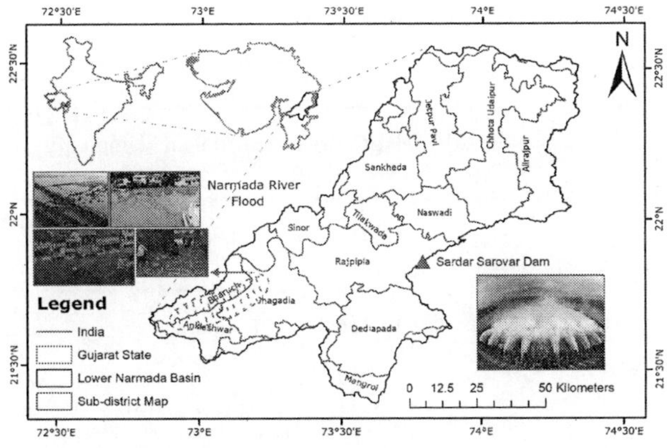

Figure 2. Location of Narmada basin.

102.5 km^2. This catchment is completely urbanized and has recurrent flood events, including flash floods. The population density is high and therefore it has high risks of people displacement, injuries, and economic losses (Simas, 2017). Additionally, this catchment is well monitored, with rainfall, stream flow, and stage data monitored each 10 min for 10 years by SAISP system (Barros et al., 2016; SAISP, 2023). However, data from flood extent and water depth outside the channel are not monitored, therefore, being considered a partially gauged catchment.

The simulations were performed for the extreme rainfall event that occurred the 16 February 2019, with observed rainfall data of 10 min being used as input for CADDIES and HEC-RAS and. One point of stream flow data was used as input boundary condition in HEC-RAS and input data in LSTM, and the other was used to verify of model accuracy. As for the DEM, the SRTM with 30 m resolution was used.

The second case study is the Narmada basin in India (Fig. 2). The Narmada River is India's sixth-longest and fifth-largest river, with a basin area of 98 796 km^2. The lower part of the basin is bounded by Sardar Sarovar Dam and the coastal region. The primary LULC class of the Lower Narmada basin is agriculture cropland (61 %), dense vegetation and forest (21 %), and urban land (8 %).

The river stretch in the lower Narmada basin is monitored by a gauge station at Garudeshwar weir (which observes water level and discharge) and Bharuch golden bridge (which observes water level) (IMD, 2023; India-WRIS, 2023).

The simulations were performed for the extreme rainfall event that occurred the 9 July 1994, with observed daily rainfall data being used for CADDIES and observed hourly streamflow as input for HEC-RAS and LSTM. One point of streamflow data was used as input boundary condition in HEC-RAS, and the other was used to verify of model accuracy. As for the DEM, the SRTM with 30 m resolution was used (USGS, 2023).

For both case studies, the models were not calibrated, since there was no spatial monitored data to calibrate CADDIES and HEC-RAS 2D. Additionally, the inflow data was used as input to LSTM and initial boundary condition to HEC-RAS. Even though, we have used a calibrated hyperparameter on the Manning's roughness coefficient, based on previous studies and land use characteristics.

3 Results and Discussion

Considering the different aspects of the two case studies, the flood model results were evaluated (Table 1). As seen from Fig. 3 and Table 1, the CADDIES model performed well in the Aricanduva basin, when comparing to HEC-RAS results. While for the Narmada basin, the flood extent and depth in the downstream part of the basin was severely underestimated (Fig. 4), when comparing to the results obtained by HEC-RAS.

The flood event in the Narmada basin is mainly a fluvial flood, that happens due to the incapacity of the river channel to conduct all the inflow that arrives from the Sardar Sarovar Dam, located upstream, during extreme rainfall events, overflowing. The HEC-RAS 2D can account for this overflow, since it takes the flow monitored coming from the dam as an upstream boundary condition. However, in the open-access version of the CADDIES model, the only input variable is the rainfall, therefore it cannot account with the inflow coming from the dam. This limitation explain the discrepancy between the Narmada case study results from HEC-RAS and CADDIES.

In the Aricanduva basin the CADDIES model performed well considering the simplicity of inputs. In this case, as the main problems related to flooding in the basin are due to rainfall, the model obtained good results. When analyzing the flood maps for the Aricanduva catchment using CADDIES and HEC-RAS, we can see that both generated very similar results. However, CADDIES takes much longer to run.

Table 1. Model performance metrics.

Model	Type of result	Case study	NSE/NRMSE	Running time
HEC-RAS	Spatial and temporal	Aricanduva	0.72/0.19	25 s[a]
		Narmada	0.85/0.15	31 min 15 s[a]
CADDIES	Spatial	Aricanduva	–	1 min 43 s[a]
		Narmada	–	10 h 1 min 33 s[a]
LSTM	Temporal	Aricanduva	0.79/0.88	57 min 21 s[b]
		Narmada	0.80/0.94	55 min 45 s[b]

[a] i7-8565U CPU@ 1.80 GHz 1.99 GHz, RAM 8 GB. [b] Intel Xeon CPU E5-1650 v3@3.50 GHz 3.50 GHz, RAM 8 GB, training time and test time considered.

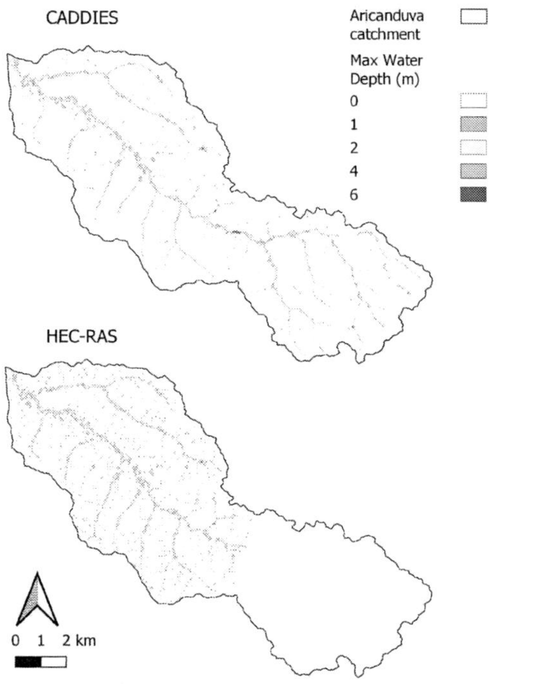

Figure 3. Flood maps results for Aricanduva catchment when using CADDIES and HEC-RAS models.

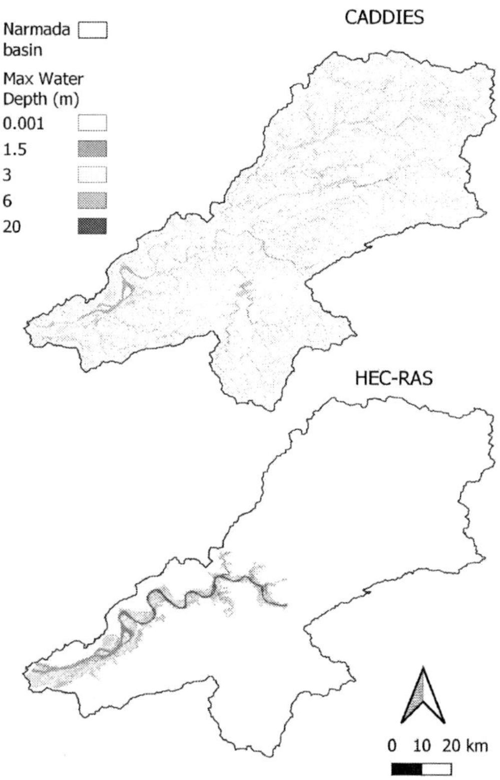

Figure 4. Flood maps results for Narmada basin when using CADDIES and HEC-RAS models.

Another aspect to consider about the HEC-RAS is the higher level of knowledge of hydrological variables required from the user and their influence on the results. Thus, small changes in the variables used can be very significant in the final values, so uncertainty needs to be considered. On the other hand, there is also greater flexibility in adjusting and calibrating the model until obtaining satisfactory results.

Figure 5 shows SAISP reported flood locations in the Aricanduva catchment and the detailing of a flooding area in Aricanduva avenue near the bridge Eng. Alberto Badra during the flood event simulated by the models. The extent and values of the flooded spot are very similar for both models, including the numerical values for the pixel, 5.3 and 5.1 m using the CADDIES and HEC-RAS model, respectively. Thus, although we do not have enough data to validate the generated flood maps, we can consider the results satisfactory since most of the points reported as flood sites had the highest depths reported in the maps generated by the models, in addition to the consistency between both results.

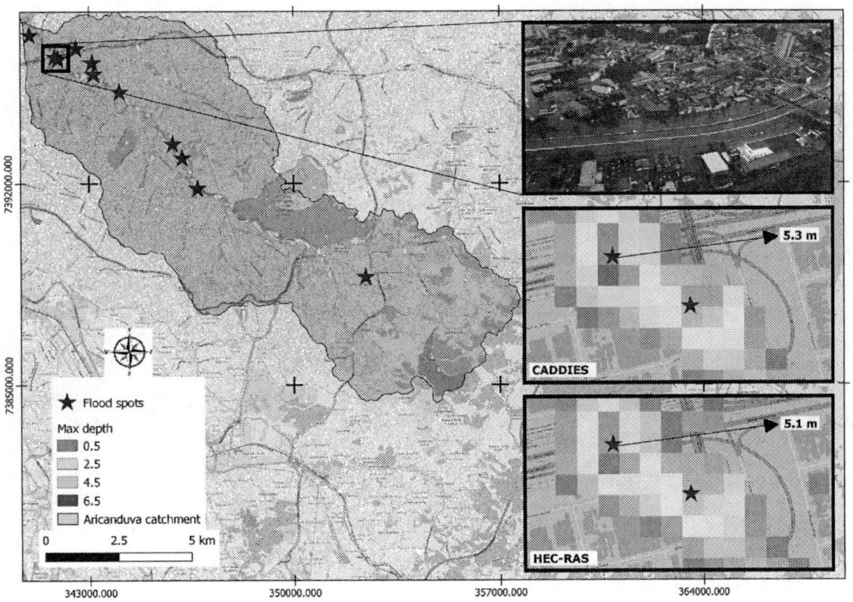

Figure 5. Flood spots reported by SAISP and the max depth results for the area using CADDIES and HEC-RAS. Picture source: Paulino (2020).

LSTM and HEC-RAS models can generate discretized results in time. However, LSTM does not perform hydrodynamic simulations. The LSTM made reasonable estimates for both case studies (Table 1), proving to be quite flexible for different applications. It also had lower execution time than CADDIES, but still above HEC-RAS. The running time for LSTM have considered both training time and test time since we were comparing the entire applicability of the models. However, for practical application of the LSTM, after the training, only the test time must be considered, which was 19 s for both cases.

4 Conclusions

This study evaluated the performance of physically-based and AI-based models in predicting floods at (i) an urban catchment and relatively small area in Brazil and (ii) a rural and huge catchment in India.

From the results, the only unsuitable outcome from the models was the one for the Narmada basin using the CADDIES model. From this, we can conclude that the model is most adequate for urban areas, especially for pluvial floods, without a considerable amount of flow insertion in the system from some other river or upstream structure.

CADDIES and HEC-RAS had very similar results for the Aricanduva catchment. The advantage of CADDIES is that it is much simpler to configure, does not require the insertion of boundary conditions, and uses only rainfall as an input variable. However, the CADDIES model takes much longer to run. HEC-RAS has the advantage of allowing us to evaluate results over time. Further, it allows the simulation of several processes, simpler to complex settings depending on the data availability, and structured processes for model calibration and validation.

When considering execution time, accuracy, and temporal results, the LSTM model and the HEC-RAS appear ahead. Depending on the application of the results, the LSTM model may be more suitable, for example, for flow prediction for the operation of reservoirs and drainage control structures, since the flood extent visualization is not required for these cases and has the advantage of being more easily applied in different case studies.

We strongly suggest performing uncertainty analysis for future studies, as data limitation leads to significant uncertainty beyond the complexities in all the models used. Moreover, several other model configurations, metrics, and statistics should be analyzed in-depth when evaluating models' performance.

Code availability. he HEC-RAS software is freely available from US Army Corps of Engineers Hydrologic Engineering Section (HEC, 2018, https://www.hec.usace.army.mil/software/hec-ras/). he CADDIES software is freely available from University of Exeter (Guidolin et al., 2016, https://www.exeter.ac.uk/research/centres/cws/resources/caddies/). The python code for developing the LSTM model can be accessed from https://github.com/mhpi/hydroDL (Feng et al., 2023), this code was developed from a third part.

Data availability. The digital elevation map is available from the Shuttle Radar Topography Mission dataset from USGS Earthexplorer (https://earthexplorer.usgs.gov/, USGS, 2023). The rainfall and streamflow datasets for the Narmada basin are available from India Meteorological Department (IMD) (2023, https://dsp.imdpune.gov.in/) and India Water Resource Information System (India-WRIS) (2023, https://indiawris.gov.in/wris/), respectively. The rainfall and streamflow datasets for the Aricanduva catchment are available from the São Paulo Alert System of Inundation (SAISP) (SAISP, 2023, https://www.saisp.br/estaticos/sitenovo/home.html, https://portal.inmet.gov.br/).

Author contributions. MBdM: Writing – original draft preparation, Writing – review & editing, Visualization, Investigation, Formal analysis, Conceptualization. NKM: Writing – original draft preparation, Writing – review & editing, Software, Investigation, Formal analysis, Conceptualization. MCF: Writing – original draft preparation, Writing – review & editing, Visualization, Software, Formal analysis, Conceptualization. AS: Writing – review & editing, Investigation, Conceptualization. RFdS: Methodology, investigation. AA: Supervision, Conceptualization. MTR: Supervision, Conceptualization. EMM: Resources, Supervision, Conceptualization. NKG: Project administration, Conceptualization. MK: Project administration, Conceptualization. ACN: Project administration, Conceptualization.

Competing interests. The contact author has declared that none of the authors has any competing interests.

Disclaimer. Publisher's note: Copernicus Publications remains neutral with regard to jurisdictional claims made in the text, published maps, institutional affiliations, or any other geographical representation in this paper. While Copernicus Publications makes every effort to include appropriate place names, the final responsibility lies with the authors.

Special issue statement. This article is part of the special issue "ICFM9 – River Basin Disaster Resilience and Sustainability by All". It is a result of The 9th International Conference on Flood Management, Tsukuba, Japan, 18–22 February 2023.

Financial support. This research has been supported by the Belmont Forum Disaster Risk Reduction (DRR) (FAPESP grant no. 2020/15729-0 and FAPESP grant no. 2019/23393-4).

Review statement. This paper was edited by Mohamed Rasmy and reviewed by three anonymous referees.

References

Barros, M. T., Conde, F., Andrioli, C. P., and Zambon, R. C.: Flood Forecasting System in a Mega City: Challenges and Results for the São Paulo Metropolitan Region, in: World Environmental and Water Resources Congress 2016, 10–19, https://doi.org/10.1061/9780784479889.002, 2016.

Batalini de Macedo, M., Fray da Silva, R., Fava, M. C., Sharma, A., K. Mangukiya, N., Sarmento Buarque, A. C., Razzolini, M. T., Mendiondo, E. M., Goel, N. K., Kurian, M., and Nardocci, A. C.: Modelling urban floods in megacities: a comparative bibliometric review of traditional physically based and artificial intelligence models, IAHS-AISH Scientific Assembly 2022, Montpellier, France, 29 May–3 June 2022, IAHS2022-687, https://doi.org/10.5194/iahs2022-687, 2022.

Feng, D., Liu, J., Bindas, T., and Fang, K.: HydroDL, Github [code], https://github.com/mhpi/hydroDL/, last access: 7 January 2023.

Guidolin, M., Chen, A. S., Ghimire, B., Keedwell, E. C., Djordjević, S., and Savić, D. A.: A weighted cellular automata 2D inundation model for rapid flood analysis, Environ. Model. Softw., 84, 378–394, https://doi.org/10.1016/j.envsoft.2016.07.008, 2016.

HEC – Hydrologic Engineering Center: HEC-RAS 5.0, Hydraulic Reference Manual, http://www.hec.usace.army.mil/software/hec-ras/documentation.aspx (last access: 1 January 2024), 2018.

IMD – India Meteorological Department: Data Supply Porta, IMD [data set], https://dsp.imdpune.gov.in/, last access: 7 January 2023.

India-WRIS – India Water Resource Information System: Data availability, India-WRIS [data set], https://indiawris.gov.in/wris/#/, last access: 7 January 2023.

Mangukiya, N. K., Sharma, A., and Shen, C.: How to enhance hydrological predictions in hydrologically distinct watersheds of the Indian subcontinent?, Hydrol. Process., 37, e14936, https://doi.org/10.1002/hyp.14936, 2023.

Paulino, E.: Alagamento interdita trechos da Avenida Aricanduva, Mobilidade Sampa: Brazil, https://mobilidadesampa.com.br/2020/03/alagamento-interdita-trechos-da-avenida-aricanduva/ (last access: 1 January 2024), 2020.

SAISP – São Paulo Alert System of Inundation, Public Products, SAISP [data set], https://www.saisp.br/estaticos/sitenovo/home.html, last access: 7 January 2023.

Shen, C.: A transdisciplinary review of deep learning research and its relevance for water resources scientists, Water Resour. Res., 54, 8558–8593, https://doi.org/10.1029/2018WR022643, 2018.

Simas, I. T. H.: Análise retrospectiva de episódios de inundações na Bacia Hidrográfica do Rio Aricanduva – São Paulo [Retrospective analysis of flooding episodes in the Aricanduva River catchment – São Paulo], Thesis, Universidade de São Paulo, São Paulo, https://doi.org/10.11606/D.8.2017.tde-21072017-162915, 2017.

USGS – US Geological Survey: EarthExplorer, USGS [data set], https://earthexplorer.usgs.gov/, last access: 7 January 2023.

Proc. IAHS, 386, 47–53, 2024
https://doi.org/10.5194/piahs-386-47-2024

Temporal and spatial distributions of flood water storage volume of the main river and tributaries in the Ishikari River basin

Shoji Fukuoka[1] **and Yutaro Ishii**[2]

[1]Research and Development Initiative, Chuo University, Tokyo, 112-8551, Japan
[2]Civil, Human and Environmental Engineering, Chuo University, Tokyo, 112-8551, Japan

Correspondence: Shoji Fukuoka (sfuku@tamacc.chuo-u.ac.jp)

Received: 21 July 2023 – Revised: 18 October 2023 – Accepted: 13 December 2023 – Published: 19 April 2024

Abstract. In recent years, Japan experienced frequent and intense heavy rainfalls, resulting in serious flood disasters in many areas. It is also necessary to control floods in tributaries and basins of these tributaries in addition to flood control of the main river. This will be accomplished by understanding flood formation process in the basins of tributaries as well as the main river. Integrated flood flow analysis for the main river and tributaries was conducted by the use of observed temporal changes in the water surface profiles of the 2016 flood in the Ishikari River, Hokkaido, Japan. Temporal and spatial water storage volume distributions in the Chitose River and Ishikari River basins give a new graphical representative method to visualize that when, where, how much volume of rainfall are stored in each basin, and make it possible to verify effects of flood control measures.

1 Introduction

In recent years, Japan experienced frequent and intense heavy rainfalls, resulting in serious flood disasters. To have an integrated flood control plan of the entire basin including the main river and tributaries is important for reducing effectively flood disasters. For this purpose, it is essential to estimate the flood flow dynamics in the main river and tributaries. There are two analysis methods for such studies: hydrological rainfall-runoff analysis and hydraulic flood flow analysis.

The former is a method for mainly representing the runoff mechanism of rainfall and estimating the discharge. The storage function method (Kimura, 1961; Hoshi and Yamaoka, 1982), tank models (Sugawara, 1972; Gotoh et al., 2019), and distributed models (Tachikawa et al., 2004) considering rainfall distribution and topographical and geological characteristics are used to estimate runoff discharge.

On the latter method, Fukuoka et al. (2004) proposed 2D flood flow analysis method to obtain flow velocity distributions and flood flow discharge hydrographs at any point in rives. Fukuoka (2017) also estimated the temporal and spatial distributions of water storage volume in the upstream basin, dams, rivers, and inundation area in the September 2015 flooding of the Kinu River basin and visualized water storage volumes in the basin. Mikami et al. (2021) evaluated effects of the discharge hydrographs of the primary and secondary tributaries on the flood discharge of the main river in the upper Tone River basin by 2D flood flow analysis.

In this study, analysis of flood flow of the 2016 typhoon was conducted to obtain the water surface profiles over time for the main river and tributaries in the lower Ishikari River basin shown in Figs. 1 and 2. The discharges from most of secondary tributaries are estimated by using the storage function method by Hoshi and Yamaoka (1982). We first examined characteristics of the three flood flows of 2016 typhoon and the effects of the discharge hydrographs in the Chitose River basin on the Ishikari River. Then, the temporal and spatial water storage volume distributions in each tributary and entire basin are discussed.

2 Characteristics of river basin and rainfall

The Ishikari River has the second largest basin in Japan and consist of a large number of tributaries. The Ikusyunbetsu River, Yubari River, and Chitose River are primary

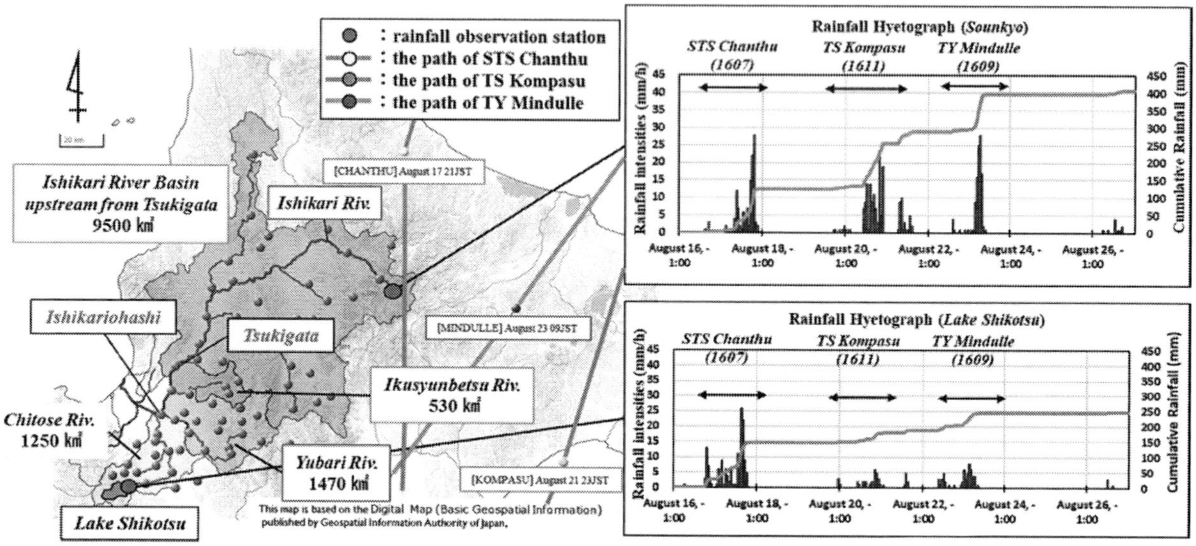

Figure 1. Basin map of the Ishikari River and the Chitose River and the rainfall hyetographs at the main stations in the upper Ishikari River basin and the Chitose River basin.

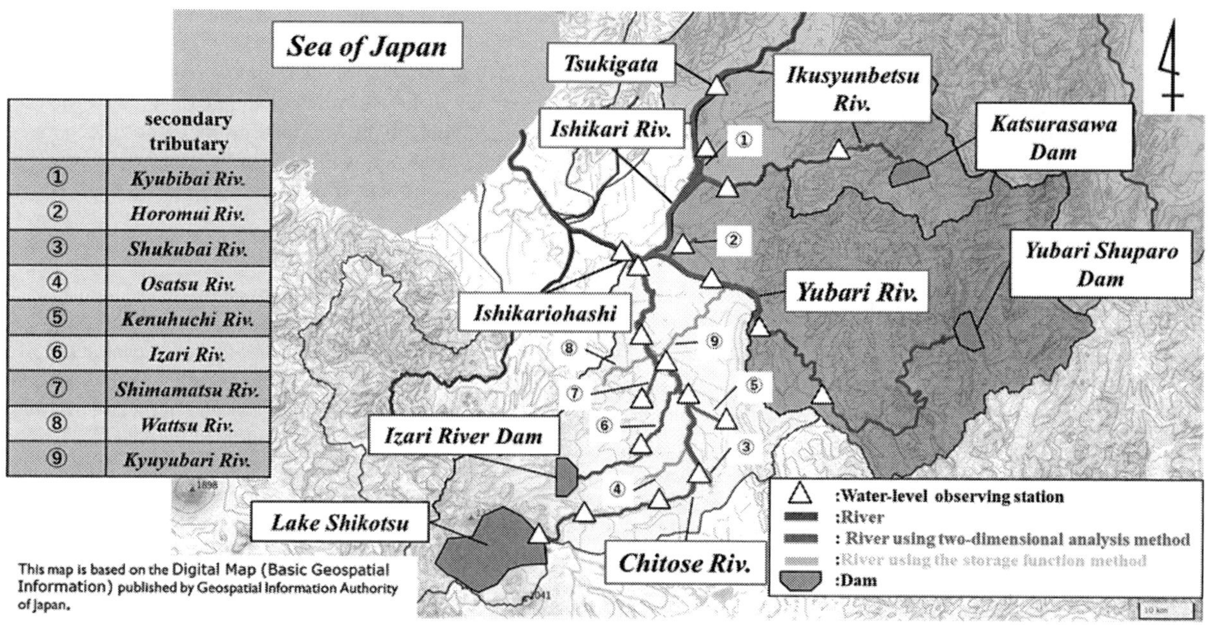

Figure 2. Basin map of the lower Ishikari River.

tributaries and confluent in the lower Ishikari River reach between Ishikariohashi (25 km from the river mouth, hereinafter shown as KP25) and Iwamizawaohashi (KP45). There are 83 rainfall observation stations in the lower Ishikari River basin as shown in Fig. 1. The hyetographs at the main stations in the upper Ishikari River basin and the Chitose River basin are also shown in Fig. 1. STS Chanthu, TS Kompasu, and TY Mindulle which are the names of typhoons hit Hokkaido during 17–23 August 2016, and brought successively extensive disasters in the Ishikari River basin. The three typhoons took different paths, resulting in different rainfall distributions. STS Chanthu brought the rain exceeding 20 mm h^{-1} in the Chitose River basin and TS Kompasu and TY Mindulle did the peak rain about 25 mm h^{-1} in the upper basin of the Ishikari River basin.

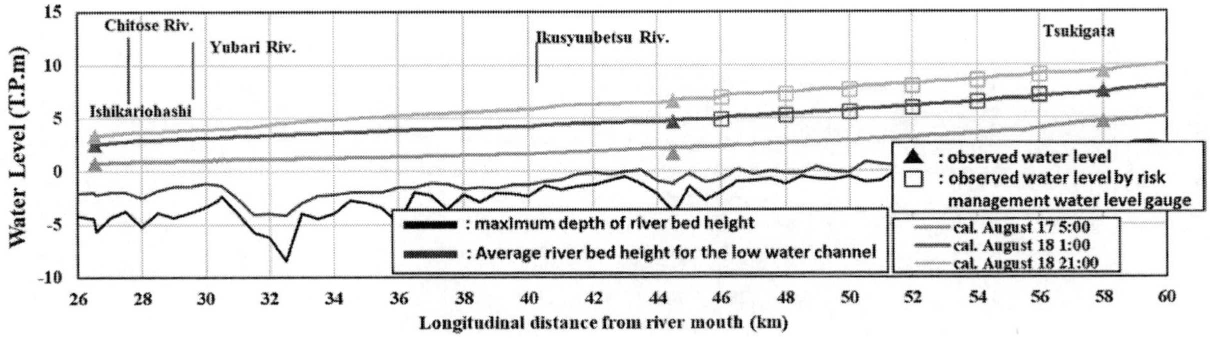

Figure 3. Flood surface profile of the Ishikari River in STS Chanthu.

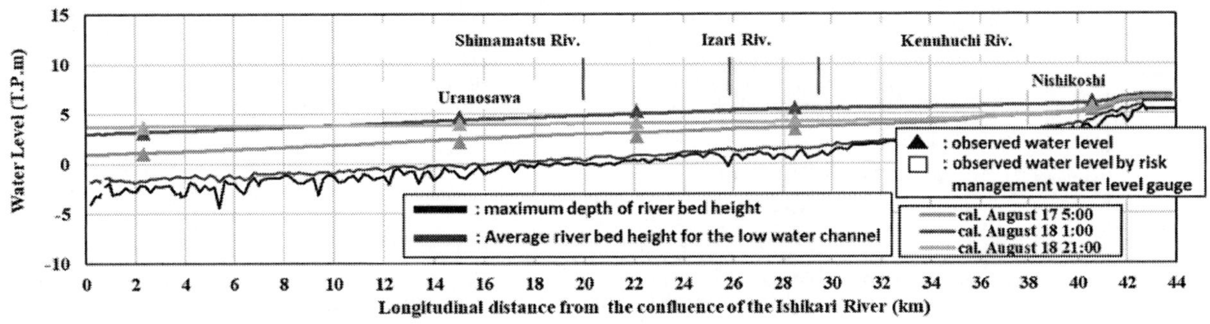

Figure 4. Flood surface profile of the Chitose River in STS Chanthu.

3 Analysis method

3.1 Flood flow analysis in the main river and primary tributaries

Parts of red line shown in Fig. 2 are river reaches analysed in the 2D flood flow analysis so as to match the observed water surface profiles. The upstream boundary condition for the analysis was given by observed water level. Velocity profiles were then calculated at each cross-section. The discharge hydrograph was obtained from calculated velocity distributions and channel area during flood.

3.2 Calculation method of runoff discharge in secondary tributaries

Flood flows of the secondary tributaries coloured by red in Fig. 2 were analysed in the same way as the flood calculation of the main and primary rivers. However, for the secondary tributaries coloured by orange in Fig. 2, water level data was not available because there were not water-level observation stations. Therefore, runoff volumes from such tributaries were estimated by the storage function method (Hoshi and Yamaoka, 1982) and added as lateral inflow to the 2D flood flow analysis. The residual areas were not included in the lateral inflow because their areas are very small.

4 Analysis results

Figure 3 shows the temporal variations of the water surface profiles of the Ishikari River floods due to STS Chanthu. Figure 4 also shows the temporal variations of the water surface profiles of the Chitose River. The peak discharge of Ishikari River occurred at 21:00 JST (Japan Standard Time), 18 August. On the other hand, the Chitose River reached its peak at 01:00 JST, 18 August. Backwaters from the Ishikari River to the Chitose River are comparatively small at the peak time.

TS Kompasu brought more rainfall in the upper reach of the Ishikari River than STS Chanthu. Figure 5 shows that the peak water level of the Ishikari River of TS Kompasu is higher than that of STS Chanthu. Figure 6 shows that the water level of the Chitose River is also rising due to the rise of water level of the Ishikari River (backwater). In particular, at 19:00 JST, 21 August, the backwater reaches up to 40 km from the confluence of the Ishikari River.

The discharge hydrographs of the Ishikari River are shown in Fig. 7. The calculated discharge hydrograph and the observed discharge at two stations of the Chitose River are compared in Fig. 8. STS Chanthu brought heavy rainfalls in the Chitose River basin, and the peak of flood flows of the Chitose River occurred about 1 h before of the Ishikari River. In particular, the peak at 01:00 JST, 18 August was 491 $m^3\ s^{-1}$ at Uranosawa.

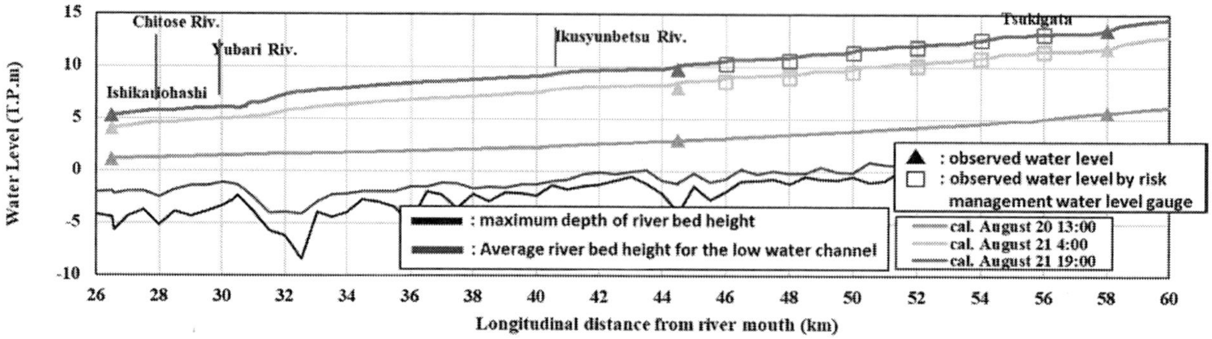

Figure 5. Flood surface profile of the Ishikari River in TS Kompasu.

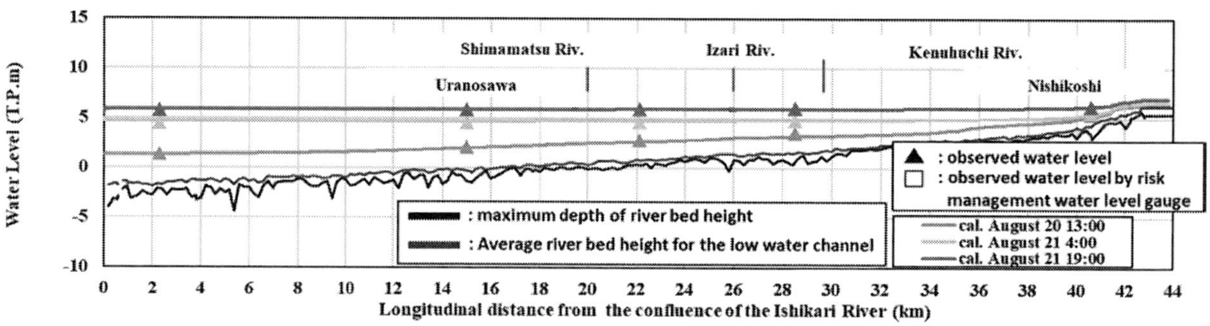

Figure 6. Flood surface profile of the Chitose River in TS Kompasu.

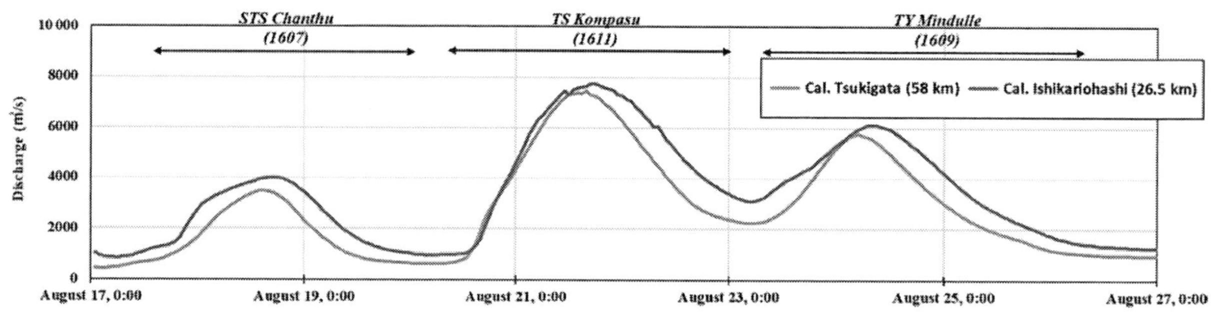

Figure 7. Discharge hydrograph of the Ishikari River.

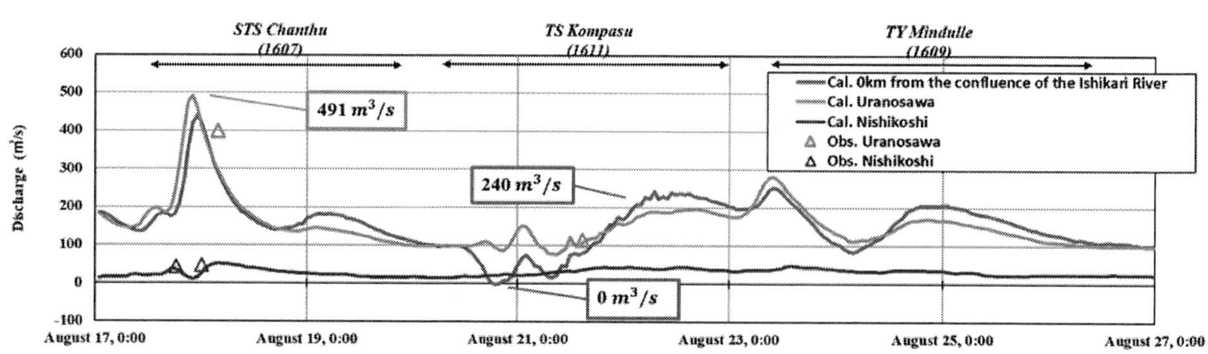

Figure 8. Discharge hydrograph of the Chitose River.

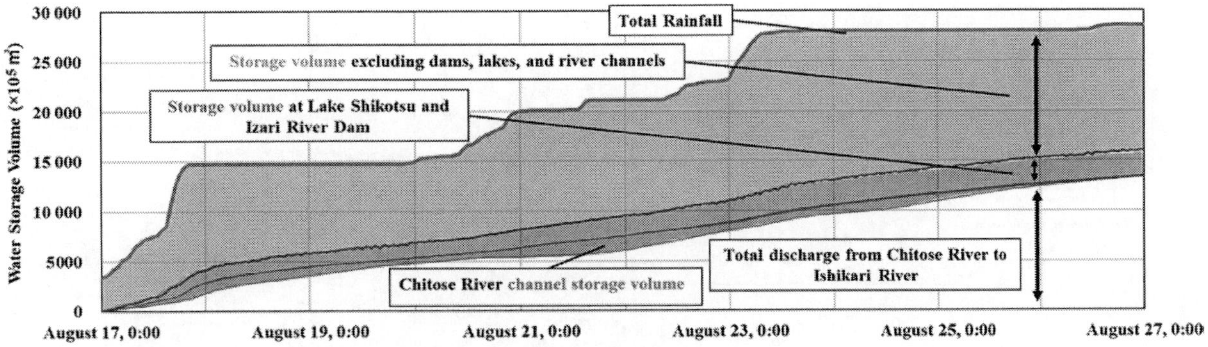

Figure 9. Graphical representation of temporal and spatial water storage volume distributions in the Chitose River basin.

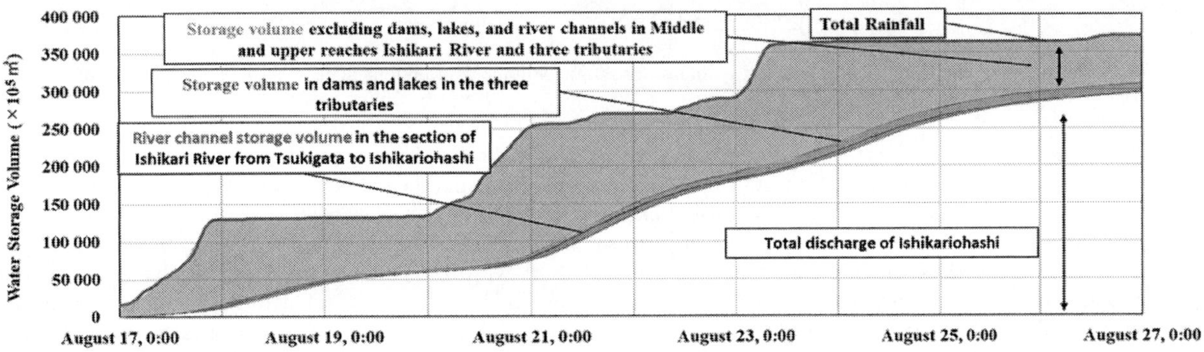

Figure 10. Graphical representation of temporal and spatial water storage volume distributions in the Ishikari River basin.

In the second, the Chitose River are strongly influenced by the backwater of the Ishikari River. At 19:00 JST, 21 August, the peak discharge time of the Ishikari River, the discharge at 0 km from the confluence of the Ishikari River is nearly zero. As the water level of the Ishikari River goes down, the discharge from the Chitose River to the Ishikari River increases, reaching $240\,\mathrm{m^3\,s^{-1}}$ at 07:00 JST, 22 August.

5 Graphical representations of the distribution of temporal and spatial water storage volume in the basin

Let us discuss the method how to obtain the temporal and spatial water storage distributions of the entire basin and each tributary basin. Water storage distributions and volumes are estimated by the 2D flood flow analysis determined to coincide with the observed water level of each river. The discharge hydrographs at each cross-section are determined from the velocity distributions and water level change obtained from the analysis. Then, the temporal and spatial river water storage volume, S_{river}, is calculated from Eq. (1).

$$\mathrm{d}S_{\mathrm{river}}/\mathrm{d}t = Q_{\mathrm{in}}(x) - Q_{\mathrm{out}}(x+\Delta x) \qquad (1)$$

$Q_{\mathrm{in}}(x)$ ($\mathrm{m^3\,s^{-1}}$) is the discharge at the upstream section x (m). $Q_{\mathrm{out}}(x+\Delta x)$ ($\mathrm{m^3\,s^{-1}}$) is the discharge at the downstream section $x+\Delta x$.

Then, the water storage volume of the river basin can be obtained by Eq. (2), storage volume equation, which is based on the rainfall volume in the basin, storage volumes by dams, retarding basins, etc., and river channel storage volume calculated by Eq. (1).

$$\mathrm{d}S_{\mathrm{basin}}/\mathrm{d}t + \mathrm{d}S_{\mathrm{dam}}/\mathrm{d}t + \mathrm{d}S_{\mathrm{river}}/\mathrm{d}t = \frac{1}{3.6}A \cdot R - Q_{\mathrm{out}} \qquad (2)$$

A ($\mathrm{km^2}$) is the basin area, R ($\mathrm{mm\,h^{-1}}$) is the rainfall in the basin, S_{dam} ($\mathrm{m^3}$) is the volume of water storage in dams, retarding basins and lakes, S_{basin} ($\mathrm{m^3}$) is the volume of water stored in the basin, and Q_{out} ($\mathrm{m^3}$) is the river discharge outgoing from the downstream end of the basin.

Calculated distributions of temporal and spatial water storage volume demonstrate when, where, how much rain waters are in the basin. In this study, the average rainfall for the basin was calculated by the Thiessen method using the observed rainfall data from the rainfall stations shown in Fig. 1. The storage capacity of the dam is obtained from the difference between the actual discharge and the inflow cal-

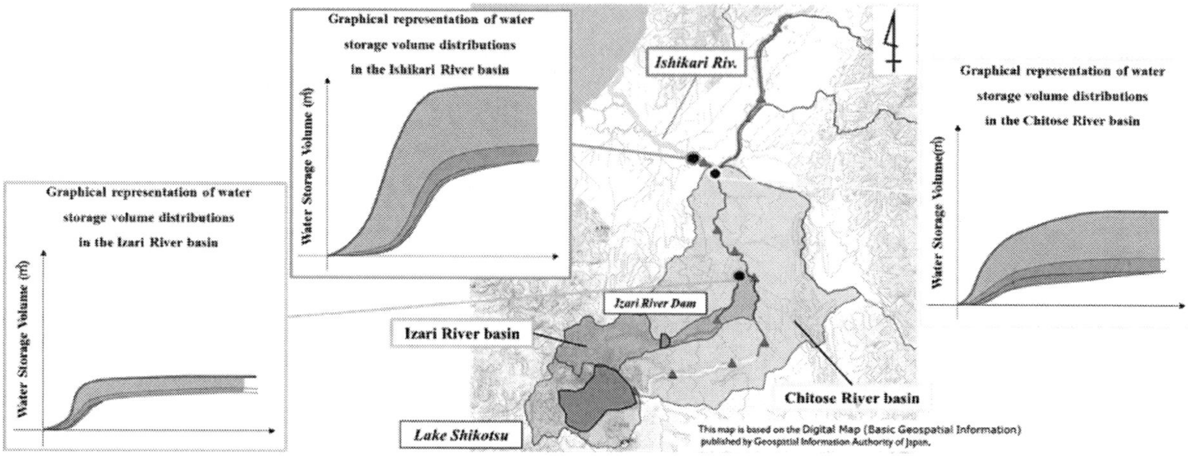

Figure 11. Graphical representations of temporal and spatial water storage volume distributions in each basin on the basin map.

culated from the water level change. The storage capacity of Lake Shikotsu is obtained from the product of the water level change and the surface area of Lake Shikotsu. Lake Shikotsu stored 2.65×10^7 m^3 at the peak. Izari River Dam stored 9.00×10^4 m^3 at the peak. Figure 9 shows the graphical representation of distribution of water storage volume in the Chitose River basin for three typhoons. In the lower basin of the Chitose River, there was a station where small-scale rainfall of 0.5 mm h^{-1} was measured for 1 h at 00:00 JST on 17 August, so that the figure shows the basin storage from the beginning of the graph. STS Chanthu brought a large amount of rainfall in the basin, and the flood flow in the Ishikari River was affected by inflow from the Chitose River flood. The storage volume in the Chitose River increased significantly due to the rising water level of the Ishikari River. This is evident by the water storage volume distributions described in the red area of Fig. 9. This shows that the flood of the Chitose River is markedly controlled by the water level of the Ishikari River. Figure 10 also shows the integrated water storage volume distributions in the Ishikari River basin for three typhoons.

In this way, the graphical representation of storage water volume distributions can clearly illustrate the characteristics of the basin. Therefore, as shown in Fig. 11, the characteristics of the basin can be represented according to each scale by overlaying them on the basin map. In addition, creating this map before and after the implementation of flood control measures in the basin makes it possible to verify effect of measures to be adopted. This is an essential task in order to know how flood control at each basin scale affects the entire basin.

In this study, the amount of water storage volume in the basin, S_{basin}, is not directly evacuated, but estimated from the difference between the total rainfall volume and the summation of water volume stored into the dams, lakes, and rivers and water volume outgoing from the basin. In the future, the analysis using the observation of water surface profile in the tributaries will be extended upstream to calculate properly the amount of runoff volume from upstream small basins. Then, we will investigate the runoff mechanism and storage conditions by using distributed tank model so as to fit the flow discharge hydrographs obtained in the tributary.

6 Conclusion

The flood flow analysis was conducted in the lower Ishikari River basin for three typhoons in 2016. Based on the analysis results, we investigated the flood formation process of the Chitose River and its effects on the flood flow of the Ishikari River. The temporal and spatial water storage volume distributions in the Ishikari River basin and the Chitose River basin were analysed. The graphical representation of storage water volume distributions presents a new way to visualize when, where, how much volume of rainfall exist in the basin.

Graphical representation of storage water volume distributions before and after the implementation of flood control measures provides important means for making a choice of the proper flood control in the river basin.

Code availability. The program was developed by Research and Development Initiative, Chuo University Fukuoka Unit. It is not open to the public.

Data availability. The data is provided by the Hokkaido Regional Development Bureau of the Ministry of Land, Infrastructure, Transport and Tourism. It is not open to the public.

Author contributions. SF: Conceptualization, Funding acquisition, Methodology, Project administration, Supervision, Validation, Writing-review & editing. YI: Formal analysis, Writing-original draft preparation, Investigation, Software, Data curation, Visualization.

Competing interests. The contact author has declared that none of the authors has any competing interests.

Disclaimer. Publisher's note: Copernicus Publications remains neutral with regard to jurisdictional claims made in the text, published maps, institutional affiliations, or any other geographical representation in this paper. While Copernicus Publications makes every effort to include appropriate place names, the final responsibility lies with the authors.

Special issue statement. This article is part of the special issue "ICFM9 – River Basin Disaster Resilience and Sustainability by All". It is a result of The 9th International Conference on Flood Management, Tsukuba, Japan, 18–22 February 2023.

Acknowledgements. We would like to thank the Hokkaido Regional Development Bureau of the Ministry of Land, Infrastructure, Transport and Tourism and the River and Erosion Control Division of the Construction Department of the Hokkaido Government providing the data of the Ishikari River Basin for us.

Review statement. This paper was edited by Mohamed Rasmy and reviewed by three anonymous referees.

References

Fukuoka, S.: Creating an idea of integrated river plan due to the use of widespread water surface profiles of floods in river basins, Adv. River Eng., 23, 251–256, 2017.

Fukuoka, S., Watanabe, A., Hara, T., and Akiyama, M.: Highly accurate estimation of hydrograph of flood discharge and water storage in rivers by using an unsteady two-dimensional flow analysis based on temporal change in observed water surface profiles, Journal of Japan Society of Civil Engineers, 761/II-67, 45–56, 2004.

Gotoh, T., Fukuoka, S., and Oosaku, K.: Identification of the tank model parameters by discharge hydrographs of a main river and tributaries obtained from time series of observed water surface profiles and flood forecasting using it, Journal of Japan Society of Civil Engineers, Ser.B1, 75, I_1291–I_1296, 2019.

Hoshi, K. and Yamaoka, I.: A relationship between kinematic wave and storage routing models, Annual Journal of Hydraulic Engineering, JSCE, 26, 273–278, 1982.

Kimura, T.: Flood runoff tracking by storage function method, Public Works Research Institute, Ministry of Construction, Japan, Public Works Research Institute of the Ministry of Construction, 1961.

Mikami, T., Mizukusa, K., Nishizawa, Y., Adachi, T., and Fukuoka, S.: The integrated flood flow analysis for the Tone River and its tributaries using observed water surface profiles, Journal of Japan Society of Civil Engineers, Ser.B1, 77, I_427–I_432, 2021.

Sugawara, M.: Runoff analysis method, Kyoritsu Shuppan CO., LTD., 1972.

Tachikawa, Y., Nagatani, G., and Takara, K.: Development of stage-discharge relationship equation incorporating saturated-unsaturated flow mechanism, Annual Journal of Hydraulic Engineering, JSCE, 48, 7–12, 2004.

Proc. IAHS, 386, 55–60, 2024
https://doi.org/10.5194/piahs-386-55-2024

Investigation of Uncertainties in Multi-variable Bias Adjustment in Multi-model Ensemble

Saurabh Kelkar and Koji Dairaku

Department of Engineering Mechanics and Energy, University of Tsukuba, Tsukuba, Japan

Correspondence: Saurabh Kelkar (kelkarsaurabh527@gmail.com)

Received: 22 July 2023 – Revised: 21 December 2023 – Accepted: 25 December 2023 – Published: 19 April 2024

Abstract. Post-processing methods such as univariate bias adjustment have been widely used to reduce the bias in the individual variable. These methods are applied to variables independently without considering the inter-variable dependence. However, in compound events, multiple atmospheric factors occur simultaneously or in succession, leading to more severe and complex impacts. Therefore, a multi-variable bias adjustment is necessary to retain the inter-variable dependence between the atmospheric drivers. The present study focuses on a multi-variable bias adjustment of surface air temperature and relative humidity in a multi-model ensemble. We investigated added values and biases before and after adjusting the variables. There are gains and losses throughout the process of adjustment. The bias adjustment effectively reduces bias in surface air temperature; however, it shows bias amplification for relative humidity at higher altitudes. Added values were improved at lower altitudes but showed reductions in surface air temperature at higher altitudes. Overall, the bias adjustment shows improvement in reducing bias over low-altitude urban areas, encouraging its application to assess compound events. These findings highlight a potential bias adjustment approach for the regions with a constraint on observational data.

1 Introduction

Quantifying the impact of climate change at a regional scale requires reliable, unbiased information. General circulation models (GCMs) have been our main source of knowledge on the past and future climate; however, regional-scale surface and atmospheric drivers of climate extremes are not well resolved in GCMs (Ayar et al., 2015). Therefore, regional climate models (RCMs) are widely used to dynamically downscale the GCMs. RCM products have been known to provide a clearer understanding of surface-induced processes compared to the parent model. However, they still have uncertainties, contributing to biased model output (Giorgi, 2019). Some of these uncertainties are related to the structural differences and internal variability in climate models, while some are related to the boundary conditions and bias propagation from GCM to RCM (Addor et al., 2016).

Bias Adjustment

Bias adjustment methods are the primary post-processing tools to reduce model bias. Many methods have been developed and applied to post-process climate simulations. Most bias adjustment methods reduce the bias by adjusting the statistical features, such as quantiles (Themeßl et al., 2012) and variance (Berg et al., 2012). Some methods focus on adjusting the boundary conditions of GCM to improve the RCM simulations (Bruyère et al., 2013; Kim et al., 2023). However, these methods are applied to single variables (univariate) at a time, i.e., independently (Hempel et al., 2013). Due to this limitation, the inter-variable dependence is often not considered, which might lead to misinterpretation of the results. Past research suggests that the single-variable bias adjustment methods suffice the specific regional impact studies (Casanueva et al., 2018). However, inter-variable dependence is significant for assessing compound events where multiple variables play a crucial role (Zscheischler and Seneviratne, 2017). Multi-Variable Bias Adjustment (MBA) methods were, thus, developed for consistent impact assessment.

However, similar to univariate bias adjustments, MBAs focus on correcting the statistical features and lack consideration of physical parameters such as underlying terrain. Topography is essential when the multiple surface-sensitive drivers regulate compound events or for understanding the hydrological processes over complex terrain. Some studies incorporated topography to downscaling reanalysis products and climate models (Gupta and Tarboton, 2016; Fiddes et al., 2022) and noted a relative advantage of acquiring reliable climate information over complex terrain. However, our study considers topography exclusively for bias adjustment purposes, focusing on the drivers of heat stress, the surface air temperature, and relative humidity (Davis et al., 2016), which are sensitive to the underlying terrain.

This study investigates the uncertainties associated with the bias adjustment method, based on topographical adjustment of surface air temperature and relative humidity in a multi-model ensemble of GCM-RCM. Therefore, this study refers "multi-variable" to model simulated surface air temperature and specific humidity (which has been used to estimate relative humidity).

2 Data and Method

We analyzed the 2 m surface air temperature ($T2$) and estimated relative humidity (RH) from an ensemble of Global- (GCM) and Regional-Climate Model (RCM) simulations performed by Nayak et al. (2018) over Japanese land. The ensemble includes three RCMs, NHRCM, -NRAMS, and WRF, forced by Japanese 25-year Reanalysis (JRA25) and three GCMs, CCSM4, MIROC5, and MRI-CGCM3. The RCM simulations are at 20 km horizontal grid spacing for the present climate (1981–2000) and future climate (2081–2100) under RCP4.5. The analysis in this study is based only on the present climate. For validation, the daily observation data on $T2$ (1981–2000), RH (2008–2020), and Elevation (DEM) are obtained from the NARO gridded (1 km grid spacing) dataset (Ohno et al., 2016). Hourly simulated climate data was aggregated to their daily means and were regridded on a reference Lat–Long grid with 20 km spacing with the inverse distance weighting method for a consistent comparison (Dodson and Marks, 1997).

The Multi-variable Bias Adjustment is based on the topographical adjustment of surface climate variables. We follow the Micromet framework detailed in Liston and Elder (2006). This method was initially developed to adjust individual variables from the weather station dataset; however, we adapted it to adjust multiple climate variables.

Ensemble simulations obtained from the study of Nayak et al. (2018) provide specific humidity instead of RH. Therefore, we obtain RH through the empirical method.

First, we calculate the mixing ratio and vapor pressure (hPa) from model simulated specific humidity and surface pressure (Wallace and Hobbs, 2006), which are used to calculate the dew point temperature as detailed in Liston and Elder (2006). The adjustment factor, h, is estimated using the lapse rate, Γ (°C km^{-1}), NARO elevation, h_{DEM} (in km), and model elevation data h_{model} (in km),

$$h = \Gamma(h_{\mathrm{DEM}} - h_{\mathrm{model}}) \qquad (1)$$

Once we obtain the adjustment factor, $T2$ and dew point temperature are adjusted and used for the RH estimations according to the Micromet framework. The Bias adjustment is applied to each of the GCMs, RCMs, and JRA25. Ensemble means of the GCMs, GCM-forced RCMs, and JRA25-forced RCMs are estimated before and after the adjustment. Here, we primarily focus on the ensemble means, but more information on biases in each model is shown in the Supplement.

We also assess the added value (AV) of the RCMs to examine the impact of bias adjustment. The added value (AV) issue has been central to the climate model community (Rummukainen, 2015). AV is found to depend on model grid resolutions (Lenz et al., 2017), physical processes such as con-

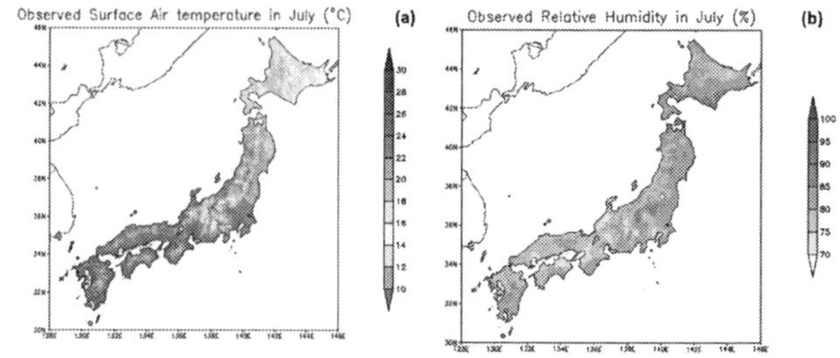

Figure 1. **(a)** Observed mean surface air temperature, $T2$ (°C), in July for the period 1981–2000; Observed mean relative humidity, RH (%), in July for the period 2008–2020.

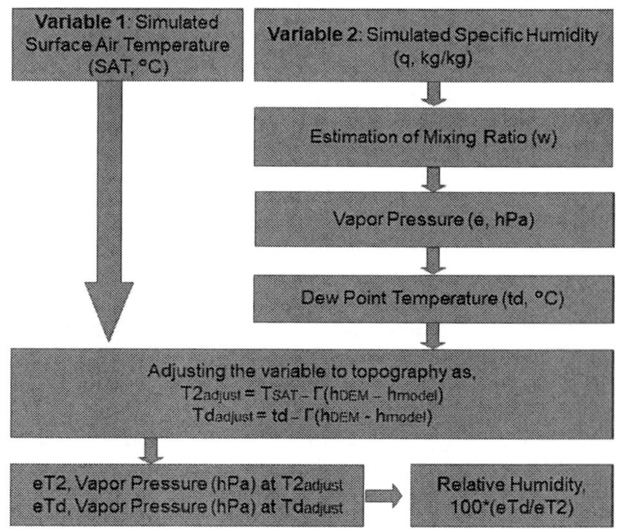

Figure 2. Process of RH estimations and adjustment to topography.

vective processes (Schaaf and Feser, 2018); as well as complex topography regions like mountainous terrains (Torma et al., 2015), which indicates the AV of the downscaled product is most likely when local and regional scale processes are essential for the climate and characteristics of the region. We use a metric of squared errors in climate variable (X) (Di Luca et al., 2013).

$$\mathrm{AV} = (X_{\mathrm{GCM}} - X_{\mathrm{Observation}})^2 - (X_{\mathrm{RCM}} - X_{\mathrm{Observation}})^2 \quad (2)$$

The added value defined here retains the units of the climate variables in the metric. It is positive whenever the biases in the RCMs are smaller than in the GCMs.

Before moving forward to the next section, we would like to note that this methodology adjusts the drivers to the underlying topography. Thus, it is irrelevant to the long-term trend and data training period.

3 Results and Discussion

Here, for brevity, we show the results for July, one of the hottest months in Japan. Figure 3 shows the mean bias in surface air temperature during July before (a, c) and after (b, d) applying the adjustment factor to the RCM ensemble – forced by three GCMs – MRI-CGCM3, CCSM4, and MIROC5 (Fig. 3a, b) and JRA25 reanalyses (Fig. 3c, d). Mean bias is a 20-year mean of difference between daily model and observational data.

Here, we examined the model ensemble mean only and observed negative bias over most of the Japanese land and weak positive bias over northern parts for the GCM forced RCM ensemble (Fig. 3a). On the other hand, the JRA25 forced RCM ensemble shows weak bias (Fig. 3c). Differences in bias before and after adjustment for each model are shown in Supplement Figs. S1–S6.

After applying the adjustment factor, the bias has prominently reduced, as seen in the rightmost panels of Fig. 3, which shows the difference in the absolute value of bias before and after adjustment.

The positive value indicates that the magnitude of bias before the adjustment was larger than after the adjustment, indicating bias reductions.

We indicate bias reductions from the perspective of absolute values as these biases are both positive and negative and would have obscured examining bias reductions otherwise.

As discussed earlier, the dependence structure between the climate variables is crucial for consistent assessment of compound events such as heat stress. The multi-variable bias adjustment includes the adjusted $T2$ while estimating the RH to consider the inter-variable relationship between $T2$ and RH. Here, the inter-variable dependence is considered through the empirical relationship between $T2$ and RH and not through the statistical features such as quantiles or probability distributions. In Fig. 4, we compare relative humidity during July before (b) and after (a), applying the adjustment factor to the RCM ensemble. We show the comparison for GCM forced ensemble because humidity data is available for GCM forced ensemble only.

Table 1. Bias at different altitude ranges (in meters) before (w/o) and after (w/) bias adjustment.

Altitude	0–500	500–1000	1000–1500	1500–2000	2000–2500
$T2$ Bias (°C) in GCM-forced RCM					
w/	−0.06	−0.89	−1.78	−3.21	−5.27
w/o	−0.45	0.23	0.56	1.78	2.51
$T2$ Bias (°C) in JRA25-forced RCM					
w/	0.25	−0.41	−1.14	−2.53	−4.50
w/o	−0.14	0.70	1.19	2.46	3.29
RH Bias in GCM-forced RCM					
w/	4.72	8.83	11.69	19.45	27.01
w/o	5.46	6.02	5.61	5.86	4.40

Intriguingly, after applying the adjustment factor (Fig. 4b), coastal regions and the Kanto plain of central Japan show a reduction in bias, yet mountainous areas of northern and central Japan display added bias. On the other hand, in the case of $T2$, we observe cold bias at high altitudes (Table 1). These results indicate that the impacting factors could originate from the empirical formulation of bias adjustment and calculation of RH.

It also points out that the bias reductions are also based on removing surface dependence of air temperature and relative humidity; however, it does not affect the bias originating with structural differences in each model. Since the structural difference arises from different physical schemes used in the model framework, the retained bias is also related to the precision of physical schemes in resolving atmospheric conditions.

Added value assessment contributes to our understanding of the model uncertainties. It highlights areas where models may excel and areas where they may fall short. In our study, we investigated altitudinal variation added value (Fig. 5). Positive added values are present before adjustment, indicating additional information provided by RCM. After the adjustment, RCMs still provide additional information but at a lesser magnitude. On the other hand, at lower altitudes, negative added values before adjustment are improved after adjustment (Fig. 5b, d), highlighting the combined skill of bias adjustment and RCM. It also indicates that the GCM biases have become comparable to RCMs after that adjustment, leading to significant improvement in values added at lower.

Similar to the estimation of added values for air temperature, added values for relative humidity were also calculated. The added bias in RH after adjustment at higher altitudes has impacted the overall representation of added values. The magnitude of added bias in the GCM ensemble was signifi-

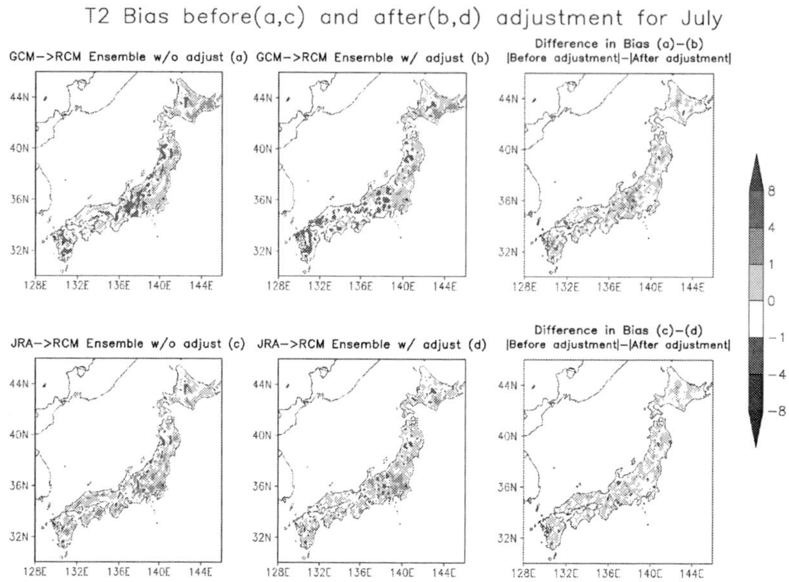

Figure 3. $T2$ bias (°C) before **(a, c)** and after applying adjustment factor **(b, d)** to GCM- **(a, b)** and JRA25-forced RCM ensemble simulations **(c, d)**.

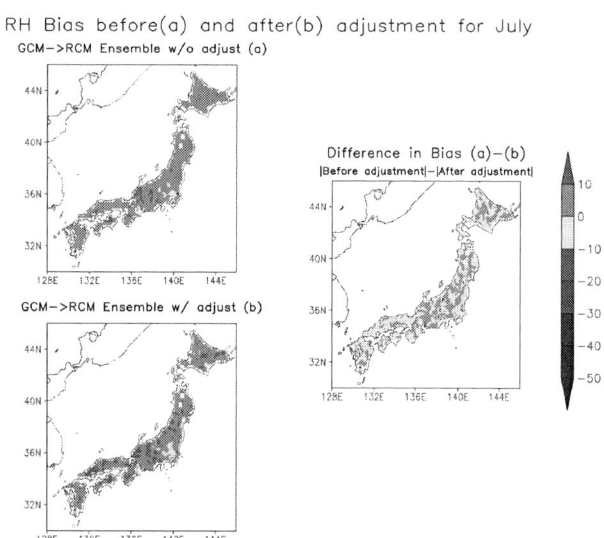

Figure 4. RH bias before **(a)** and after **(b)** applying adjustment factor to GCM forced RCM ensemble simulations.

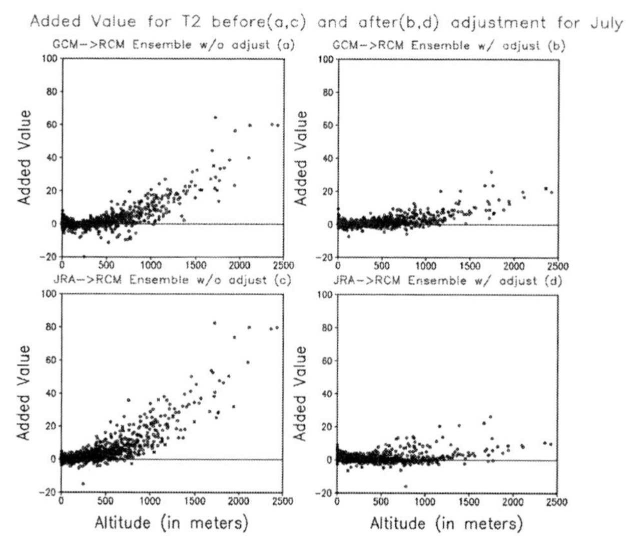

Figure 5. Comparison of added value in $T2$ in GCM forced **(a, b)** and JRA25 forced **(c, d)** ensemble simulations.

cantly larger than in RCMs, indicating strong positive added values (Fig. 6a, b).

In our study, the efficiency of Multi-Variable Bias Adjustment varied from variable to variable. In the case of air temperature, warm biases were reduced, but it retained a cold bias after the adjustment. As Multi-Variable Bias Adjustment is solely based on topography adjustments, the bias reductions are also based on removing the biases arising from orography representation in the climate model; however, it does not affect the bias originating from the structural differences in each model. Dissimilarity in physical schemes applied in each model to simulate surface-dependent convective processes would still contribute to total bias. In the added values, large added biases in GCMs after the adjustment resulted in strong positive added values, indicating that the interpretation of this type of added value might be related

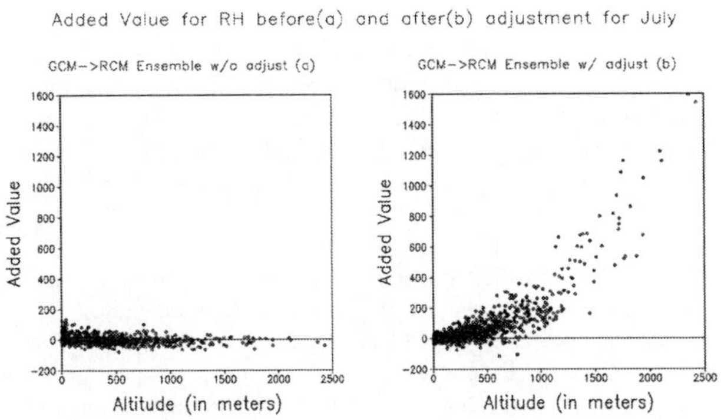

Figure 6. Comparison of added value in RH in GCM forced RCM ensemble simulations.

to the magnitude of the added bias and not reductions in biases after the adjustment.

4 Conclusions

The multi-variable bias adjustment shows a substantial reduction in $T2$ bias after the adjustment. Noticeably, the strong bias over the complex topography of central and northern Japan was reduced. Furthermore, the value added by the RCMs for $T2$ shows reduction at higher altitudes. In addition, the more substantial negatives observed over lower altitudes were improved to weaker negatives. On the other hand, the assessment of RH shows bias reductions over lower altitude areas ($<$ 500–700 m); however, the bias is retained over mountainous areas. Preserving inter-variable dependence is prominent for compound event assessment; however, in the case of a climate model, retaining that dependence would be challenging because of inherent biases. As highlighted earlier, the bias adjustment based on topographical adjustments reduces the effect of underlying topography but does not affect the bias originating from the structural differences inherent to the model. Therefore, another justification for amplified RH bias could be the combined effect of bias propagated from the parent model to RCMs, the inter-variable (while estimating RH from $T2$) bias transition, and the limitation of the bias adjustment method in considering the bias related to structural uncertainties. Overall, we observe bias reduction over low altitudes. The advantage of considering topography-based post-processing is it is independent of observational data, allowing its application when observational records are limited or lacking over a particular region (e.g., Tibetan plateau) or for long-term assessments (e.g., pre-industrial, future projections). Hydrological studies could also benefit from this approach, as processes such as rainfall run-off are highly sensitive to the terrain; however, they remain to be explored in future studies.

Code and data availability. Climate model data analyzed in this study is obtained from Nayak et al. (2018). The observation data is obtained from NARO (Ohno et al., 2016) and can be accessed upon registration from https://amu.rd.naro.go.jp/ (NARO, 2024). Data is analysed using CDO (https://code.mpimet.mpg.de/projects/cdo/embedded/index.html#x1-30001.1, Schulzweida, 2020) and visualized with GrADS (http://cola.gmu.edu/grads/downloads.php, Grid Analysis and Display System (GrADS), 2024).

Supplement. The supplement related to this article is available online at: https://doi.org/10.5194/piahs-386-55-2024-supplement.

Author contributions. SK conceptualized, analyzed, investigated, visualized the results, and wrote the manuscript. KD supervised the research.

Competing interests. The contact author has declared that neither of the authors has any competing interests.

Special issue statement. This article is part of the special issue "ICFM9 – River Basin Disaster Resilience and Sustainability by All". It is a result of The 9th International Conference on Flood Management, Tsukuba, Japan, 18–22 February 2023.

Acknowledgements. The authors thank reviewers and editors for their valuable feedback.

Financial support. This research has been supported by the Ministry of Education, Culture, Sports, Science, and Technology, Japan (Monbukagakusho Scholarship).

Review statement. This paper was edited by Tomoki Ushiyama and reviewed by two anonymous referees.

References

Addor, N., Rohrer, M., Furrer, R., and Seibert, J.: Propagation of biases in climate models from the synoptic to the regional scale: Implications for bias adjustment, J. Geophys. Res.-Atmos., 121, 2075–2089, https://doi.org/10.1002/2015JD024040, 2016.

Ayar, P. V., Vrac, M., Bastin, S., Carreau, J., Déqué, M., and Gallardo, C.: Intercomparison of statistical and dynamical downscaling models under the EURO- and MED-CORDEX initiative framework: present climate evaluations, Clim. Dynam., 46, 1301–1329, https://doi.org/10.1007/s00382-015-2647-5, 2015.

Berg, P., Feldmann, H., and Panitz, H.-J.: Bias correction of high resolution regional climate model data, J. Hydrol., 448–449, 80–92, https://doi.org/10.1016/j.jhydrol.2012.04.026, 2012.

Bruyère, C. L., Done, J. M., Holland, G. J., and Fredrick, S.: Bias corrections of global models for regional climate simulations of high-impact weather, Clim. Dynam., 43, 1847–1856, https://doi.org/10.1007/s00382-013-2011-6, 2013.

Casanueva, A., Bedia, J., Herrera, S., Fernández, J., and Gutiérrez, J. M.: Direct and component-wise bias correction of multi-variate climate indices: The percentile adjustment function diagnostic tool, Climatic Change, 147, 411–425, https://doi.org/10.1007/s10584-018-2167-5, 2018.

Davis, R. E., McGregor, G. R., and Enfield, K. B.: Humidity: A review and primer on atmospheric moisture and human health, Environ. Res., 144, 106–116, https://doi.org/10.1016/j.envres.2015.10.014, 2016.

Di Luca, A., de Elía, R. and Laprise, R.: Potential for small scale added value of RCM's downscaled climate change signal, Clim. Dynam., 40, 601–618, https://doi.org/10.1007/s00382-012-1415-z, 2013.

Dodson, R. and Marks, D.: Daily air temperature interpolated at high spatial resolution over a large mountainous region, Climate Res., 8, 1–20, https://doi.org/10.3354/cr008001, 1997.

Fiddes, J., Aalstad, K., and Lehning, M.: TopoCLIM: rapid topography-based downscaling of regional climate model output in complex terrain v1.1, Geosci. Model Dev., 15, 1753–1768, https://doi.org/10.5194/gmd-15-1753-2022, 2022.

Giorgi, F.: Thirty Years of Regional Climate Modeling: Where Are We and Where Are We Going next? J. Geophys. Res.-Atmos., 124, 5696–5723, https://doi.org/10.1029/2018JD030094, 2019.

Grid Analysis and Display System (GrADS): GrADS Software, http://cola.gmu.edu/grads/downloads.php, last access: 11 January 2024.

Gupta, A. S. and Tarboton, D. G.: A tool for downscaling weather data from large-grid reanalysis products to finer spatial scales for distributed hydrological applications, Environ. Modell. Softw., 84, 50–69, https://doi.org/10.1016/j.envsoft.2016.06.014, 2016.

Hempel, S., Frieler, K., Warszawski, L., Schewe, J., and Piontek, F.: A trend-preserving bias correction – the ISI-MIP approach, Earth Syst. Dynam., 4, 219–236, https://doi.org/10.5194/esd-4-219-2013, 2013.

Kim, Y.-I., Evans, J. P., and Sharma, A.: Multivariate bias correction of regional climate model boundary conditions, Clim. Dynam., 61, 3253–3269, https://doi.org/10.1007/s00382-023-06718-6, 2023.

Lenz, C. J., Früh, B., and Adalatpanah, F. D.: Is there potential added value in COSMO–CLM forced by ERA reanalysis data?, Clim. Dynam., 49, 4061–4074, https://doi.org/10.1007/s00382-017-3562-8, 2017.

Liston, G. E. and Elder, K.: A Meteorological Distribution System for High-Resolution Terrestrial Modeling (MicroMet), J. Hydrometeorol., 7, 217–234, https://doi.org/10.1175/JHM486.1, 2006.

NARO: The Agro-Meteorological Grid Square Data: https://amu.rd.naro.go.jp/, last access: 11 January 2024 (in Japanese).

Nayak, S., Dairaku, K., Takayabu, I., Suzuki-Parker, A., and Ishizaki, N. N.: Extreme precipitation linked to temperature over Japan: Current evaluation and projected changes with multi-model ensemble downscaling, Clim. Dynam., 51, 4385–4401, https://doi.org/10.1007/s00382-017-3866-8, 2018.

Ohno, H., Sasaki, K., Ohara, G., and Nakazono K.: Development of grid square air temperature and precipitation data compiled from observed, forecasted, and climatic normal data, Climate Biosphere, 16, 71–79, https://doi.org/10.2480/cib.J-16-028, 2016.

Rummukainen, M.: Added value in regional climate modeling, WIRES Clim. Change, 7, 145–159, https://doi.org/10.1002/wcc.378, 2015.

Schaaf, B. and Feser, F.: Is there added value of convection-permitting regional climate model simulations for storms over the German Bight and Northern Germany?, Meteorol. Hydrol. Water Manage., 6, 21–37, https://doi.org/10.26491/mhwm/85507, 2018.

Schulzweida, U.: Climate Data Operators (2.0.4), Zenodo [code], https://doi.org/10.5281/zenodo.1435454, 2020.

Themeßl, M. J., Gobiet, A., and Heinrich, G.: Empirical-statistical downscaling and error correction of regional climate models and its impact on the climate change signal,Climatic Change, 112, 449–468, https://doi.org/10.1007/s10584-011-0224-4, 2012.

Torma, C., Giorgi, F., and Coppola, E.: Added value of regional climate modeling over areas characterized by complex terrain—Precipitation over the Alps, J. Geophys. Res.-Atmos., 120, 3957–3972, https://doi.org/10.1002/2014jd022781, 2015.

Wallace, J. M. and Hobbs, P. V.: 3 – Atmospheric Thermodynamics, in: Atmospheric Science, 2nd Edn., edited by: Wallace, J. M. and Hobbs, P. V., Academic Press, 63–111, https://doi.org/10.1016/B978-0-12-732951-2.50008-9, 2006.

Zscheischler, J. and Seneviratne, S. I.: Dependence of drivers affects risks associated with compound events, Sci. Adv., 3, e1700263, https://doi.org/10.1126/sciadv.1700263, 2017.

Proc. IAHS, 386, 61–67, 2024
https://doi.org/10.5194/piahs-386-61-2024

Investigation of transition signals from single cell to multicell thunderstorms in the Kinki region, Japan by using vertical vorticity and multi-parameter radar analysis

Fauziana Ahmad[1], **Kosei Yamaguchi**[2], **Eiichi Nakakita**[2], **and Tomoki Ushiyama**[3]

[1]Malaysian Meteorological Department, Selangor, 46667, Malaysia
[2]Disaster Prevention Research Institute, Kyoto University, Kyoto, 611-0011, Japan
[3]ICHARM, 1–6 Minamihara, Tsukuba, Ibaraki, 305-8516, Japan

Correspondence: Fauziana Ahmad (fauziana@met.gov.my)

Received: 20 July 2023 – Revised: 30 November 2023 – Accepted: 20 December 2023 – Published: 19 April 2024

Abstract. Floods and landslides caused by extreme weather events, such as localized excessive rainfall in Japan, cause enormous devastation. Multicell clouds, on the other hand, have a common duration of over an hour and greater aerial coverage than single-cell clouds. For flood warning systems to function properly, the hazardous cloud must be detected quickly. Using pseudo and dual-Doppler vorticity approaches, the vertical vorticity estimate was extended from single-cell to multicell transitions case. According to the research, the single cells attained the peak of maximum vorticity of 0.08 and 0.01 s^{-1} for pseudo-vorticity and dual-Doppler vorticity, respectively. The maximum column of specific differential phase (K_{dp}) column above the melting layer indicated that the transition with 1 km depth as compared to differential reflectivity (Z_{dr}) column was not observed after the cell merging. However, it was identified 5 min after the cell merged. In contrast, the K_{dp} column was always identified after the cell merging, and the column showed an increase in intensity 5 min after the cells merged. Vertical vorticity and multi-parameter radar analysis provided an insight into the transition from single-cell to multicell formation.

1 Introduction

The localized heavy rainfall from cumulonimbus clouds, which commonly occurs in the summer season, is potentially hazardous. In 2008, a flash flood occurred in the Toga River located in Kobe, Japan due to the downpour of Guerilla-heavy rainfall from the development of isolated cumulonimbus clouds (Nakakita et al., 2013, 2017). The rapid rise of water level in the river basin caused the death of five people. The dissemination of early warnings could not be delivered promptly since the flood came rapidly. Furthermore, the single-cell clouds could merge into multicell clouds, which have a common lifetime of more than an hour and greater aerial coverage than a single-cell cloud. Small-to-medium-sized hail, small tornadoes, and heavy precipitation are all risks linked with multicell clouds. Heavy precipitation from multicell clouds thunderstorms might also intensify flash flood events, especially in urban areas. For example, the multicellular storm brought localized heavy rain near Zoshigaya, Tokyo on 5 August 2008, causing five sewage workers to be washed away by a flash flood (Kato and Maki, 2009; Hirano and Maki, 2010; Kim et al., 2012). The multicell formation studies in Japan are mostly investigated to understand the characteristics and its stage of formation. Previous research has focused on radar reflectivity (Nishiwaki et al., 2013; Shusse et al., 2005) based on the stage of multicell development. Kim et al. (2012) analyzed the precipitation core development and its structure by combining liquid water content (LWC) and K_{dp}. The study of the transition from single-cell to multicell is more concentrated on radar reflectivity in which the merging of echo cores involved the merger of updrafts (Wescott, 1994). They discovered that the enhancement of low-level convergence by storm outflows was essential in initiating the growth of new echo cores and increased the likelihood of a merger. According to a research by Vasiloff et al. (1986), the size and intensity of subsequent cells and updrafts increased due to a storm-environment feedback process involving updraft-downdraft interactions, as well as increased ambient vertical winds shear and buoyancy. In this study, we emphasized multicell development because the potential of cells merging related to the severity of thunderstorms is also important in the dissemination of early warnings.

Referring to the case of the Toga River basin, Nakakita et al. (2013, 2017) developed prediction methods on the first radar echo aloft that was described as a "baby-rain-cell" in a single cumulonimbus cloud by using Doppler weather radar observation. They confirmed the presence of a pair of positive and negative vertical vortex tubes as well as an updraft in the early stage of a rain-cell. According to the study, the vertical vorticity values should be greater than or equal to $0.03\,s^{-1}$ to identify the hazardous rain cell in the initiation stage before it developed to the cumulus stage. Since the successful lead time in the detection of "baby-rain-cell" in localized Guerilla-heavy rainfall, the vertical vorticity method is a novel approach to discover the potential of cell merging in the multicell thunderstorm analysis. Therefore, this study aimed to examine the availability of vertical vorticity to forecast the potential of cell merging related to heavy rain in the hazardous storm.

In addition, dual-polarization radar parameters are widely used for rainfall estimation, supercell signatures, and radar echo classification. Instead of utilizing horizontal polarization (Z_h), other radar parameters such as differential reflectivity (Z_{dr}) and specific differential phase (K_{dp}), are also mainly used in the detection of severe thunderstorms. For instance, Kim et al. (2012) analyzed the precipitation core development and its structure of multicell formation by combining liquid water content (LWC) and K_{dp}. The Z_{dr} column appears to be associated with rising motion within thunderstorms. Z_{dr} column analysis is strongly recommended in the operational and research application due to its identification of updraft location and strength (Kumjian and Ryzhkov, 2008). The Z_{dr} columns may also aid in tracking dominant updrafts in multicell storms or storm-merger situations. Furthermore, the positive values of K_{dp} observed above the melting level, which is labeled as the K_{dp} column, are usually associated with deep convection updraft cells. Many studies associate K_{dp} volumes with the updraft mass flux, lightning flash, and intense rainfall. Moreover, K_{dp} columns also reveal signatures of changing updraft properties in the evolution of storm events (van Lier-Walqui et al., 2016).

The vertical vorticity analysis is mostly investigated in supercell thunderstorms owing to the exhibition of updraft rotation. According to a study by Nakakita et al. (2017), a vorticity pair in the baby-rain-cells was detected, which was related to the existence of an updraft in the single-cell. Therefore, the characteristics and mechanisms for the development from single-cell to multicell thunderstorms, which contribute to localized heavy rainfall, motivated the authors to discover the significant criteria related to the severity of thunderstorms. The Z_{dr} column and K_{dp} column also indicated the strength and the location of the updraft, which were beneficial in the detection of hazardous clouds. Therefore, the study on polarimetric for the detection of potential heavy rainfall in multicell thunderstorms could provide insight into the understanding of the multicell characteristics from the initiation stage to the development stage of occurrence of heavy rainfall.

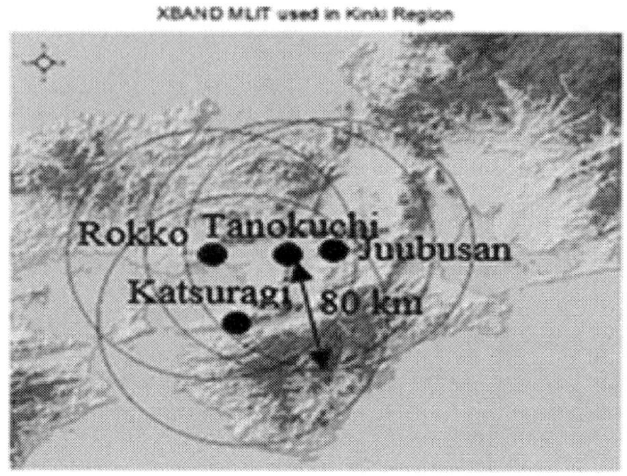

Figure 1. Four radar stations in the Kinki region, Japan (so-called Honshu), were used for the observation of the cell merging of the radar reflectivity. The black arrow shows the maximum range of 80 km between the radar stations.

2 Data and methodology

2.1 Data

This study used the XRAIN operated by the Ministry of Land, Infrastructure, Transport and Tourism (MLIT). In this event analysis, X-band polarimetric radar with parabolic antenna namely, Tanokuchi, Rokko, Juubusan, and Katsuragi radar stations, were selected, as illustrated in Fig. 1. The radars cover an 80 km radius and have a sampling resolution of 150 m.

In this study, the study area was located in Kinki region which includes an urban region located in western part of the main Japan island (so-called Honshu). The thunderstorms events were selected with the environmental conditions that support the multicell development as listed in Table 1. The atmospheric instability was indicated by using Convective Available Potential Energy (CAPE) and Bulk Richardson Number (BRN) from radiosonde data. The value of these parameters was taken at 09:00 Japan Standard Time (JST) and 21:00 JST observations at Shionomisaki station. The verti-

Table 1. List of selected events.

(Event)/Date	(1)/10 Sep 2014		(2)/13 Aug 2018	
CAPE ($J\,kg^{-1}$)	09:00 JST 0	21:00 JST 866	09:00 JST 2413	21:00 JST 1842
BRN	0	63	865	187
Vertical Wind Shear (s^{-1})	0.0048	0.0041	0.0012	0.0012

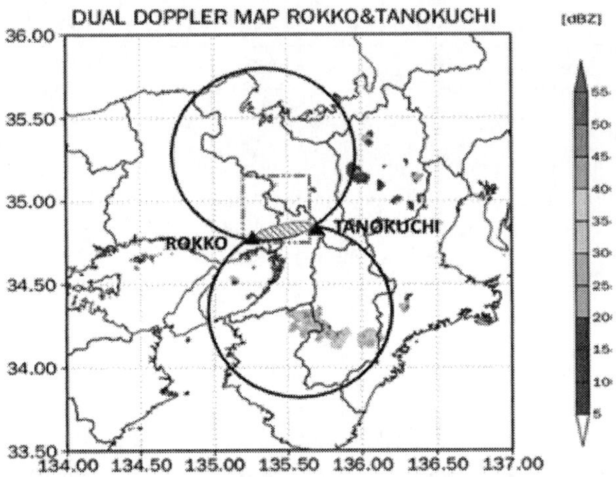

Figure 2. Radar observations for dual-Doppler radar with the red color box indicating the target area for analysis within the two solid circles, but excluding the hatched area.

cal wind profile from 1 to 3 km of height indicated a weak vertical shear that was appropriate for the multicell thunderstorms. The atmospheric instability and the environmental condition supported the development of multicell thunderstorms.

Dual-Doppler synthesis is the most important and widely used method for determining the air motions within precipitating cloud systems. The radar reflectivity (Z_h) and Doppler velocity obtained from two radars (Tanokuchi and Katsuragi for event 1, Tanokuchi and Rokko for event 2) were converted into 1 km × 1 km × 0.5 km of Constant Altitude Plan Position Indicator (CAPPI) by utilizing the variational method approach (Shimizu et al., 2008). Three wind components (u, v, and w) were retrieved on the Cartesian coordinate system within the solid circles, except for the hatched area in every 5 min. The radar observation used for the dual-Doppler analysis is shown in Fig. 2.

2.2 Methodology

The pseudo-vorticity observed at a single Doppler weather radar, and the radar selection is dependent on the distance from the radar to the target area. Dual-Doppler radar analysis is performed for retrieval of the three-dimensional (3D) wind fields to examine the kinematic mechanisms of cloud development by using two components of the Doppler velocity by two radars simultaneously. We compared both analyses to investigate the transition signals in cells merging since both methods produced the different temporal and spatial resolutions. The dual-polarization radar parameters, such as Z_{dr} and K_{dp} column were mainly used to investigate the strength and location of the updraft to examine the severity of thunderstorms (van Lier-Walqui et al., 2016). Therefore, the characteristic patterns of transition to multicell focused on the Z_{dr} column and K_{dp} column depth were investigated. We proposed the new method of analysis by emphasizing that the maximum of both columns to be calculated after the melting height because these columns reflected the considerable amounts of supercooled water were being lifted if the updraft occurs. The pseudo-vorticity analysis only implemented in event 1, meanwhile the DDA vorticity and multi-parameter radar analysis were conducted in event 2.

2.2.1 Dual-Doppler vorticity analysis

The dual-Doppler analysis (DDA) vorticity was estimated by applying Eq. (1) with a spatial resolution of 1 km × 1 km × 0.5 km of CAPPI 2 km at 5 min intervals. The core vorticity of the dual-Doppler analysis (DDA) was defined as the maximum intensity of the positive vorticity and the minimum intensity of the negative vorticity intensity, which was obtained for each CAPPI height.

In the Cartesian coordinate system, the vertical vorticity is expressed as in Eq. (1):

$$\zeta = \frac{\partial v}{\partial x} - \frac{\partial u}{\partial y} \quad (1)$$

where, u is zonal wind component, and v is meridional wind component in the x and y direction, respectively.

2.2.2 Pseudo-vorticity analysis

The pseudo-vorticity was estimated by applying the method proposed by Nakakita et al. (2017) with a spatial resolution of 50 m × 50 m in each elevation angle at 1 min intervals. Using the radial velocity speed, the vorticity can be calculated by using Eq. (2),

$$\zeta = 2 \times \frac{V_a - V_b}{2r} \quad (2)$$

where, V_a and V_b are radial velocities moving toward and away from the radar in the Cartesian coordinates, respectively, and $2r$ is total distance of radar pulse between the mesh center.

2.2.3 Z_{dr} column and K_{dp} column

The radar parameters obtained from each radar were converted to compose together into 0.25 km × 0.25 km × 0.25 km resolution of CAPPI by

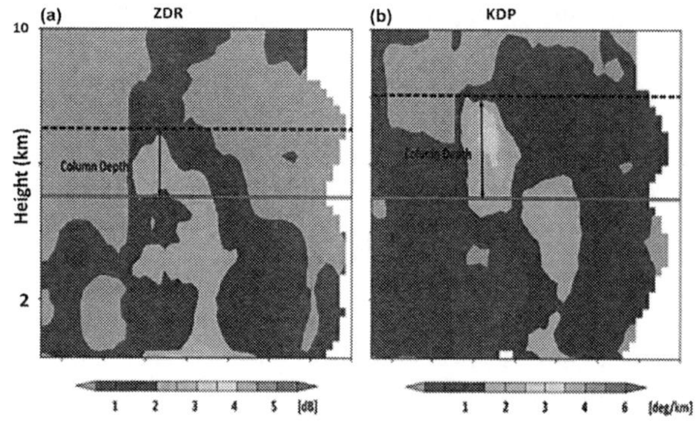

Figure 3. Illustration of: **(a)** Z_{dr} column depth, **(b)** K_{dp} column depth indicated by the solid arrow, and the melting level indicated by a red solid line.

Cressman interpolation (Masuda and Nakakita, 2014) by using four radars in every 5 min. Therefore, the characteristic patterns of transition focused on the Z_{dr} column and K_{dp} column depth were investigated as illustrated in Fig. 3a and b, respectively. The column depth was defined when the Z_{dr} (> 2 dB) and K_{dp} ($> 1.5° \text{km}^{-1}$) were above the melting height level. In this study, the melting level was observed at 5 km above ground level (a.g.l.) for both events, which was retrieved from the radiosonde data at Shionomisaki station.

3 Results and discussion

3.1 Vertical vorticity analysis

The vertical cross-section images of each single-cell that was investigated in this analysis were constructed as shown in Fig. 4. A single-cell was defined as cell developed within a range of 20 km from the center of the storm. As stated by Browning et al. (1976), the new cells that were developed from the parent cell should be 30 km from the core of Z_h. Hence, in this range, we determined the development of single-cell whether it was developed adjacent to the pre-existing cells by investigating the cross-section of radar reflectivity of dual-Doppler analysis. Therefore, the cross-section of each cell was constructed to identify its early stage and merging stage development. Since we were interested to distinguish the evolution of single-cell and multicell, the categorization of cells was divided with different labels of cells inspected. The classification of the single-cell label followed the cells' appearance (starting from A, B, and C), in contrast, the multicell label followed the cells merged starting from α, β, and γ in the order of formation. It was clearly observed that all single-cells merged with the existing multicell, and combined to become a more intense multicell. The time-series analysis of vertical vorticity was conducted to identify the temporal change of cells before and after single-cell merged with the multicell. The result showed that the intensity of the core vorticity reached its peak during the transition. In this study, the calculated average of vertical vorticity before merging was perceived at 0.002s^{-1}(DDA vorticity) and 0.02s^{-1} (pseudo-vorticity) as shown in Table 2.

Table 2. Peak of core vorticity intensity before merging into a multicell.

Single-Cell	Pseudo-Vorticity (s^{-1})		DDA Vorticity (s^{-1})	
A	0.033	−0.075	0.0023	−0.0020
B	0.022	−0.027	0.0008	−0.0003
C	0.017	−0.008	0.0019	−0.0014
D	0.021	−0.012	0.0014	−0.0007
E	0.020	−0.032	0.0029	−0.0020
F	0.016	−0.008	0.0014	−0.0017
Average	0.022	−0.027	0.0018	−0.0014

In this study, pseudo-vorticity presented the average vertical vorticity at 0.04s^{-1} for both single-cells and multicell. In contrast, the average of core vorticity for the transition from single-cell to multicell was attained in the range of 0.002 to 0.004s^{-1}. The mature vortex in the multicell was always discovered at the later stage that was defined as core vorticity maxima boundary at the threshold intensity of more than 0.005s^{-1} for DDA. These characteristic patterns could be observed by both methods by using pseudo-vorticity and DDA although the intensity of pseudo-vorticity was ten-times higher than DDA vorticity analysis.

The most important findings were made during the detection of maximum rainfall before it reaches the ground. It was quantified that the first detection of vertical vorticity in the single-cell could be referred to as an indicator of the peak of radar reflectivity after the cells merged into a multicell. In these new outcomes, the pseudo-vorticity quantified the lead time as an average of 17 min, and the DDA specified an average of 13 min before the peak of rainfall reaches the ground as shown in Fig. 5. These new findings could assist the forecasters in disseminating the early warnings to prevent more damage to infrastructure and casualties.

3.2 Multi-parameter radar analysis

Further investigation was conducted on the transition from single-cell to multicell in the multicell environment condition by selecting event 2, which was focused on the first single-cell development and transition into the multicell, as presented in Fig. 6. We focused on the multi-parameter radar analysis by examining the Z_{dr} and K_{dp} columns that are beneficial in evaluating the stage of multicell development. In addition, the peak of vertical vorticity analysis was investigated to verify the criterion index from the previous study. SC1 and SC2 showed the highest intensity for K_{dp} as shown in Fig. 6b and not presented at Z_{dr} as presented in Fig. 6a. The new cell developed at the south-western part from the

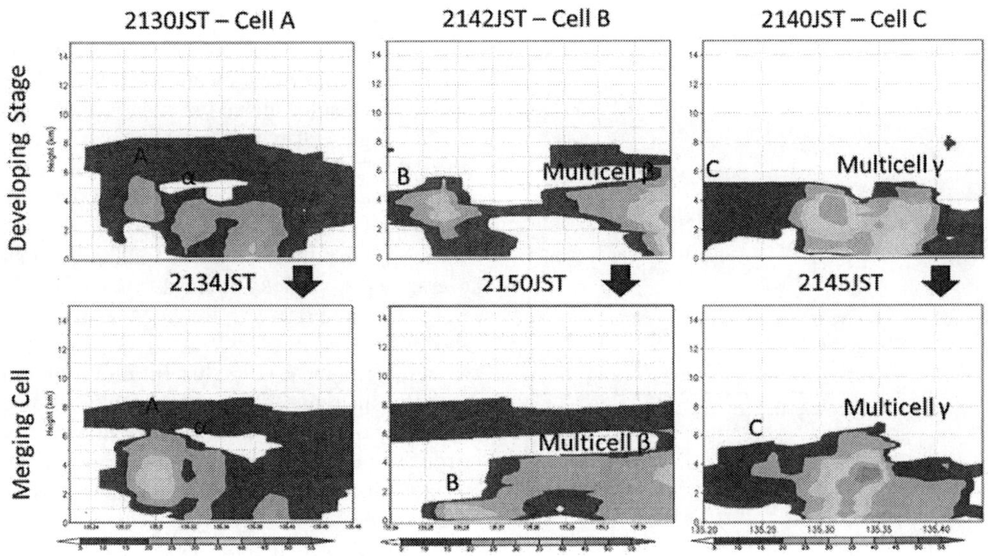

Figure 4. Vertical cross-section images of single-cell transformed into multicell with respect to time of developing stage and merged cells.

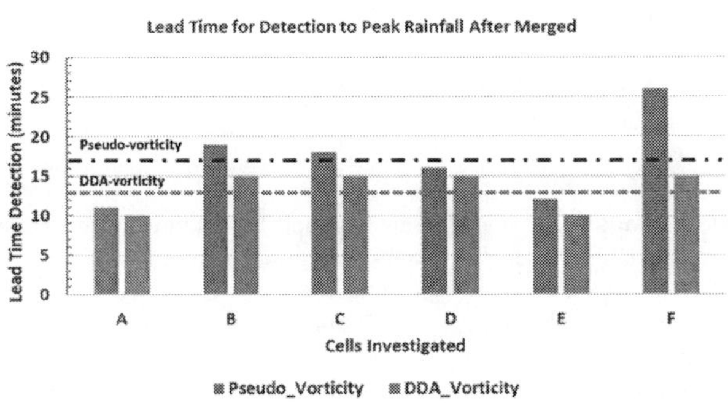

Figure 5. Lead time of the first detection of vertical vorticity before the peak of core Z_h after cells merged.

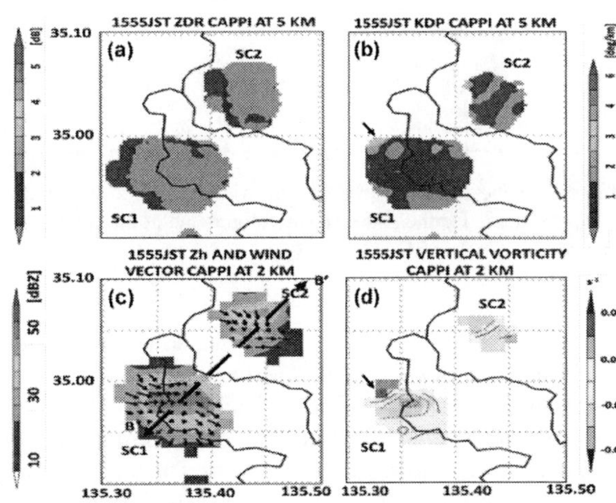

Figure 6. Horizontal distribution at 15:55 JST at for SC1 and SC2 for **(a)** Z_{dr}, **(b)** K_{dp} at height of 5 km, and for **(c)** Z_h, **(d)** vertical vorticity at height of 2 km.

core of Z_h in SC1. The horizontal wind distribution with radar reflectivity and vertical vorticity at 2 km of CAPPI height are presented in Fig. 6c and d.

In the meantime, the peak intensity of vertical vorticity of single-cell before transition into multicell was varied according to the influence of convergence and updraft that was supplied into the single-cell development as shown in Fig. 7. However, it was discovered that the peak core vorticity could be attained at $0.04\,s^{-1}$, which indicated the transition from single-cell to multicell in this study. As presented in Fig. 7a, the maximum Z_{dr} column (3–4 km) existed with similar time of updraft indicated that SC1 developed more intense as the peak updraft occurred 10 min after the development. The maximum K_{dp} column existed after 5 min of peak updraft, but a signature pattern that could be observed during the transition with maximum K_{dp} column of 1 km existed at 15:55 JST. As shown in Fig. 7b, the maximum column depth of Z_{dr} existed similar time with the peak updraft, with the highest column discovered at 16:15 JST, 15 min before the peak of updraft in the multicell. The K_{dp} column always existed in the single cell and multicell, however, the signature patterns during transition similar with SC1, which maximum K_{dp} column of 1 km discovered in this analysis.

The results indicated that the maximum K_{dp} column was always identified before and after cells merging, which 1 km of K_{dp} column always existed during transition from single-cell to multicell. Meanwhile, the maximum Z_{dr} column was discovered after 5 min of the transition. On average, the max-

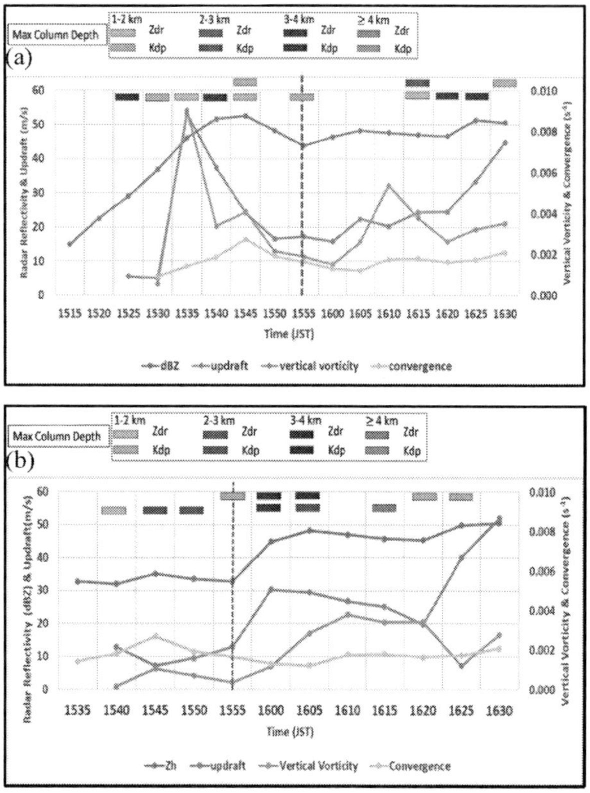

Figure 7. Time-series of averaged maximum reflectivity, updraft and convergence for the entire storm volume and at maximum vertical vorticity: **(a)** for SC1, **(b)** for SC2.

imum height of both columns was revealed at 3 km, at the most, from the melting level. The analysis proved that in the transition of single-cell to multicell, updraft frequently developed in the multicell formation with a strong updraft. The extended Z_{dr} and K_{dp} columns showed mainly the above melting level height with a similar position to the updraft formation. These results were similar to a previous study (van Lier-Walqui et al., 2016), which proved that both Z_{dr} and K_{dp} columns could be used to identify the updraft and correlated well with the intensity. We could examine the transition from single-cell to multicell results presented that the convergence mostly developed before the initiation of the updraft, and the peak of updraft intensity occurred consistent with the period of core vorticity for both cells. The convergence and updraft strength mainly affected the core vorticity intensity.

4 Conclusions

In conclusion, the study of transition signals from single-cell to multicell discovered that single-cell revealed a peak of core vorticity intensity after merging with pre-existing cells. The core vorticity was examined between each single-cell and multicell by applying the pseudo-vorticity and DDA methods, in which the core vorticity was defined as the maximum of positive vorticity and the minimum of negative vorticity. The results showed that the average of core vorticity in the single-cell before merging was $0.002\,s^{-1}$ for DDA, and $0.02\,s^{-1}$ for pseudo-vorticity analysis. This technique revealed that the peak of core vorticity intensity was mainly discovered after the cells merged. Both DDA and pseudo-vorticity found these patterns, and the study showed that the average core vorticity intensity of single-cell to multi-cell was 0.004 and $0.04\,s^{-1}$ for DDA and pseudo-vorticity, respectively. The findings from multi-parameter radar analysis revealed that K_{dp} is a useful proxy for deep convection updrafts since the life stages of multicell formation is controlled by the updraft and cells merging mainly related to the stronger updraft and downdraft of moist air in the presence of precipitation. In particular, Z_{dr} above melting height is a useful indicator of the early appearance of deep convection. The vertical vorticity and both Z_{dr} and K_{dp} columns analysis provided an insight into the transition from single-cell to multicell development, which is beneficial in the detection of severe weather resulting in flash floods.

Code availability. The code can be provided by the corresponding authors upon request.

Data availability. The data that support the findings of this study are available under the Data Integration and Analysis System (DIAS), which was developed and operated by a project supported by the Ministry of Education, Culture, Sports, Science and Technology. Restrictions apply to the availability of these data, which were used under license for this study. Data is available at https://auth.diasjp.net/cas/login (DIAS, 2022) with the permission of the DIAS Secretariat.

Author contributions. FA, KY, EN contributed to the design and implementation of the research; FA analyzed the data; FA, KY, EN, TU wrote the manuscript draft; all authors discussed the results and commented on the manuscript.

Competing interests. At least one of the (co-)authors is a guest member of the editorial board of *Proceedings of IAHS* for the special issue "ICFM9 – River Basin Disaster Resilience and Sustainability by All". The peer-review process was guided by an independent editor, and the authors also have no other competing interests to declare.

Disclaimer. Publisher's note: Copernicus Publications remains neutral with regard to jurisdictional claims made in the text, published maps, institutional affiliations, or any other geographical representation in this paper. While Copernicus Publications makes ev-

ery effort to include appropriate place names, the final responsibility lies with the authors.

Special issue statement. This article is part of the special issue "ICFM9 – River Basin Disaster Resilience and Sustainability by All". It is a result of The 9th International Conference on Flood Management, Tsukuba, Japan, 18–22 February 2023.

Acknowledgements. The authors express their sincere gratitude to Shingo Shimizu from the National Research Institute for Earth Science and Disaster Resilience (NIED), Tsukuba, Japan for the thoughtful discussions and for providing useful suggestions with the software developed regarding Dual Doppler radar analysis. Fauziana Ahmad acknowledges the award of the Malaysian Public Service Department Training (HLP) scholarship allowing the research to be undertaken.

Financial support. This research has been supported by the Japan Society for the Promotion of Science (grant nos. 15H05765 and 20H02258).

Review statement. This paper was edited by Tomoki Ushiyama and reviewed by two anonymous referees.

References

Browning, K., Frankhauser, J., Chalon, J., Eccles, P., Strauch, R., Merrem, F., Musil, D., May, E., and Sand, W.: Structure of an Evolving Hailstorm Part V: Synthesis and Implications for Hail Growth and Hail Suppression, Mon. Weather Rev., 104, 603–610, 1976.

DIAS: Data Integration and Analysis System (DIAS) [data set], https://auth.diasjp.net/cas/login (last access: 31 July 2022), 2022.

Hirano, K. and Maki, M.: Method of VIL calculation for X-band polarimetric radar and potential of VIL for nowcasting of localized severe rainfall – Case study of the Zoshigaya downpour, Scientific Online Letters on the Atmosphere, 6, 89–92, https://doi.org/10.2151/sola.2010-023, 2010.

Kato, A. and Maki, M.: Localized heavy rainfall near Zoshigaya, Tokyo, Japan on 5 August 2008 observed by X-band polarimetric radar: Preliminary analysis, Scientific Online Letters on the Atmosphere, 5, 89–92, https://doi.org/10.2151/sola.2009-023, 2009.

Kim, D., Maki, M., and Shimizu, S.: X-band Dual-Polarization Radar Observations of Precipitation Core Development and Structure in a Multicellular Storm over Zoshigaya, Japan, on August 5, 2008, J. Meteorol. Soc. Japan, 90, 701–719, 2012.

Kumjian, M. R. and Ryzhkov, A. V.: Polarimetric Signatures in Supercell Thunderstorms, J. Appl. Meteorol. Climatol., 47, 1940–1961, 2008.

Masuda, A. and Nakakita, E.: Development of a technique to identify the stage of storm life cycle using X-band Polarimetric Radar, J. Hydraul. Eng., 70, 493–498, 2014 (in Japanese).

Nakakita, E., Nishiwaki, R., Yamabe, H., and Yamaguchi, K.: Research on the prognostic risk of baby cell for guerilla-heavy rainfall considering by vorticity with doppler velocity, The Journal of Japan Society of Civil Engineers, 69, 325–330, 2013.

Nakakita, E., Sato, H., Nishiwaki, R., Yamabe, H., and Yamaguchi, K.: Early Identification of Baby-Rain-Cell Aloft in a Severe Storm and Risk Projection for Urban Flash Flood, Adv. Meteorol., 2017, 5962356, https://doi.org/10.1155/2017/5962356, 2017.

Nishiwaki, N., Misumi, R., Shimizu, S., Maesaka, T., Iwanami, K., Sakurai, N., Maki, M., Suzuki, S., Kato, A., and Yamaji, A.: Behavior and structure of convective clouds developing around a mountainous area observed by stereo photogrammetry and Ka-band and X-band radars: Case study of northern Kanto, Japan, J. Meteorol. Soc. Japan, 91, 609–626, 2013.

Shimizu, S., Uyeda, H., Moteki, Q., Maesaka, T., Takaya, Y., Akaeda, K., Kato, T., and Yoshizaki, M.: Structure and Formation Mechanism on the 24 May 2000 Supercell-Like Storm Developing in a Moist Environment over the Kanto Plain, Japan, Mon. Weather Rev., 136, 2389–2407, 2008.

Shusse, Y., Tsuboki, K., Geng, B., Minda, H., and Takeda, T.: Structure and Evolution of Deeply Developed Convective Cells in a Long-Lived Cumulonimbus Cloud under a Weak Vertical Wind-Shear Condition, J. Meteorol. Soc. Japan, 83, 351–371, 2005.

Van Lier-Walqui, M., Fridlind, A. M., Ackerman, A. S., Collis, S., Helmus, J., MacGorman, D. R., North, K., Kollias, P., and Posselt, D. J.: On Polarimetric Radar Signatures of Deep Convection for Model Evaluation: Columns of Specific Differential Phase Observed during MC3E, Mon. Weather Rev., 144, 737–758, 2016.

Vasiloff, S. V., Brandes, E. A., Davies-Jones, R. P., and Ray, P. S.: An Investigation of the Transition from Multicell to Supercell Storms, J. Appl. Meteorol. Climatol., 25, 1022–1036, 1986.

Westcott, N. E.: Merging of Convective Clouds: Cloud Initiation, Bridging, and Subsequent Growth, Mon. Weather Rev., 122, 780–790, 1994.

Proc. IAHS, 386, 69–74, 2024
https://doi.org/10.5194/piahs-386-69-2024

Open Access PIAHS

ICFM9 – River Basin Disaster Resilience and Sustainability by All

Calculating flood probability in Obihiro using a probabilistic method: incorporating the probability of dike failure with uncertainty

Fumihiko Uemura[1,2], **Guus Rongen**[3], **Shigekazu Masuya**[1], **Takatoshi Yoshida**[1], and **Tomohito J. Yamada**[2]

[1]River Environment Dept., Docon. Co. Ltd., Sapporo, 060-0042, Japan
[2]Faculity of Engineering, Hokkaido University, Sapporo, 060-8628, Japan
[3]Civil Engineering and Geosciences, Delft University of Technology, Delft, 2628 CN, the Netherlands

Correspondence: Fumihiko Uemura (fu1747@docon.jp)

Received: 28 July 2023 – Revised: 11 December 2023 – Accepted: 25 December 2023 – Published: 19 April 2024

Abstract. In Japan, estimating flood risk follows a deterministic approach. A probabilistic risk method, as adapted in the Netherlands, would be better suited for the quantitative evaluation of flood damage. This study applies such a method in Obihiro, Hokkaido, northern island of Japan. We modelled dike failure with the mechanism overtopping. The probability of dike failure is calculated with a Monte Carlo simulation, considering uncertainties in water levels, critical flow velocity, and dike heights. This results in more accurate failure probabilities compared to the deterministic approach. Additionally, we corrected the dike failure probability for upstream dike failures because these reduce downstream water levels. This conditional flood probability is about 1/10th of the independent situation, indicating a significant effect of considering dike failure in dependence on upstream failures.

1 Introduction

In assessing risk information, determining dike failure conditions is crucial, significantly influencing hazard evaluation. In Japan, the commonly used criterion for dike failure conditions is the Planned High Water Level (HWL) in flood control plans. However, actual dike failures can result from factors like overtopping, erosion, and seepage, differing from flood control plan criteria. Numerous cases have been reported where levees didn't breach despite water levels exceeding the HWL (MLIT, 2020).

Determining dike failure conditions is considered effective through probabilistic evaluation of multiple failure modes and uncertainties in dike failure. The United States and European countries have already incorporated a probabilistic assessment of failure modes into policies. In the Netherlands, a dike failure probability evaluation model accommodating multiple failure modes has been developed. Additionally, a risk-based approach that plans flood countermeasures based on floodplain risk has been introduced, indicating a growing integration of probabilistic considerations into disaster preparedness (Vergouwe, 2016).

This study aims to develop a Dike Failure Probability (DFP) evaluation model tailored to the target river's characteristics for the quantitative assessment of flood risk in Japan.

2 Method

The key points of our research approach are as follows:

1. We constructed a model that considers the temporal variation of hydraulic conditions and uncertainties inherent to the breach process. This is particularly relevant in Japan, characterized by steep-gradient rivers where water levels, their temporal changes, and associated uncertainties are significant.

2. We proposed a method that considers the presence or absence of upstream breaches in determining the downstream DFP. In Japan, the flooding pattern is characterized by flush flood, where damages vary at each breach

point, accounting for upstream overtopping or breaches and their impact on downstream discharge reduction is essential.

These two aspects form the foundation of our study, addressing the unique challenges posed by Japan's rivers, where damages vary at breach points, and the probability of downstream breaches changes due to upstream breaches.

2.1 Target area

This research focuses on the urban area of Obihiro City in Hokkaido, Japan. The DFP evaluation was conducted at 0.2 km intervals within the KP53.8 to KP62.8 section of the Tokachi River, which is adjacent to the study area. "KP" indicates the distance from the river mouth, and the Tokachi River is managed by the MLIT.

2.2 Model of dike failure due to overtopping

We developed the model to assess dike failure due to overtopping, which is the most prevalent failure mode in Japan. The process of failure due to overtopping is assumed to occur when the river water level exceeds the crest height, leading to the flow of floodwaters down the slope, resulting in erosion of the slope. An overview of the failure determination by the model is shown in Fig. 1, where (a) depicts the dike height and the time-series water level, and (b) illustrates the crest velocity and slope velocity. Flow velocity occurs during periods when the water level in (a) exceeds the dike height. (c) show the cumulative work on the slope and the threshold for dike failure determination. Erosion occurs in the range where slope velocity in (b) exceeds the critical erosion velocity. Dike failure is determined when the cumulative work (i.e., the total transferred energy) on the slope surpasses the threshold for failure determination.

Specifically, the dike failure model used Eqs. (1) and (2) to calculate overtopping discharge from the water level at the dike crest.

$$u_t = \sqrt{g h_t} \tag{1}$$

$$q_t = u_t h_t \tag{2}$$

The velocity at the toe of the slope $u_{\mathrm{slope,t}}$ is calculated based on the overtopping discharge q_t and the slope gradient α using Eq. (3), with a roughness coefficient f set at 0.08:

$$u_{\mathrm{slope,t}} = \left(\frac{8 g q_t \sin\alpha}{f}\right)^{\frac{1}{3}} \tag{3}$$

Subsequently, the occurrence of dike failure was determined using the cumulative value of the work done by the slope, which Dean et al. (2010) proposed as an indicator influencing dike erosion. The work done on the slope at a given time.

$$E_W(t) = K_W(W_t - W_c) = K_W \beta_W \left(u^3_{\mathrm{slope,t}} - u^3_{\mathrm{c,W}}\right) \tag{4}$$

here, K_W is the coefficient for the work done by erosion, β_W consolidates terms involving water mass density and shear stress, and $u_{\mathrm{c,W}}$ is the critical flow velocity at which work is done on the slope. Erosion occurs when $u_{\mathrm{slope,t}} > u_{\mathrm{c,W}}$. The threshold value for erosion occurrence under good turf growth conditions was determined as:

$$\sum\nolimits_{n=1}^{t} (E_W / K_W \beta_W) = 0.492 \times 10^6 \tag{5}$$

2.3 Probability evaluation using a Monte Carlo simulation

The DFP due to overtopping was calculated using a Monte Carlo (MC) simulation, following the approach of Tabata et al. (2015). In the MC simulation, uncertainties involved in the process are represented as variables with probability distributions and incorporated into the model. The DFP, P_{failure} was determined using Eq. (6):

$$P_{\mathrm{failure}} = n_{\mathrm{failure}} / N \tag{6}$$

here, N represents the total number of calculations in the MC simulation, and n_{failure} is the count of instances where dike failure is determined. In this study, the total number of calculations was set to 10 000. The boundary conditions for the model were defined by a hydrograph. Assuming a sine curve shape with a period of $T = 30$ hours the hydrograph was created using Eqs. (7), (8).

$$h_{\mathrm{peak}} = \sqrt{\frac{Q_{\mathrm{peak}}}{a}} - b \tag{7}$$

$$h_t = \left(\sin\frac{2\pi}{T}\cdot t\right) \times h_{\mathrm{peak}}, \left(0 - \frac{1}{4}T \le t \le T + \frac{1}{4}T\right) \tag{8}$$

Equation (7) represents the pre-established relationship for the river under consideration, denoted as the $H - Q$ relationship. The parameter Q_{peak} was varied in increments of $500\,\mathrm{m\,s^{-1}}$, and the DFP was assessed for each flow rate, resulting in a relationship between discharge and the DFP.

We assessed the uncertainties associated with elements within the developed model, focusing on water level, dike height, and erosion of the slope, as illustrated in Fig. 2.

2.3.1 Water level

In this study, we incorporated water level as a random variable in the MC simulation due to the non-uniform relationship between water level and discharge in actual flood events, influenced by changing channel morphology and water surface gradient. Figure 3 shows the variability in observed data around the $H - Q$ equation, indicating water level and discharge at each flood event's peak. To model water level as a random variable, we calculated the mean and variance of differences between water levels from the $H - Q$ equation and

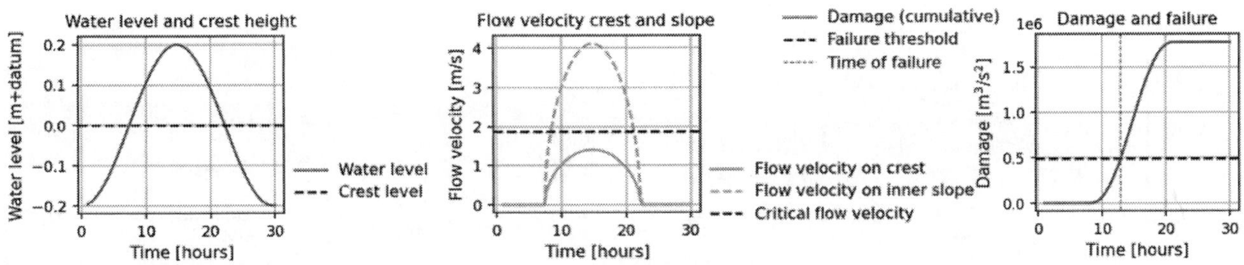

Figure 1. Overview of Dike Failure due to Overtopping in the Developed Model.

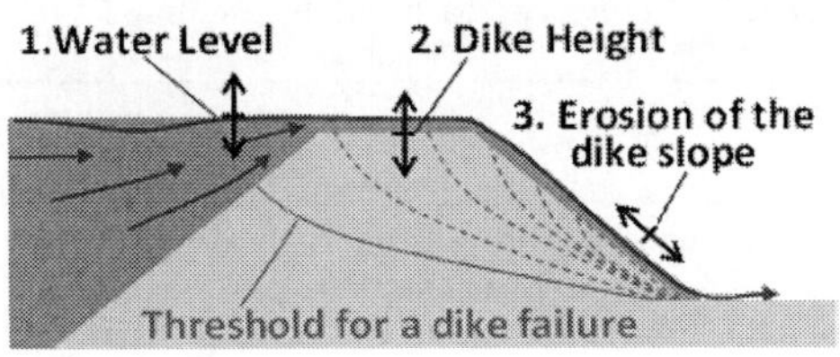

Figure 2. Uncertainties included in dike failure process.

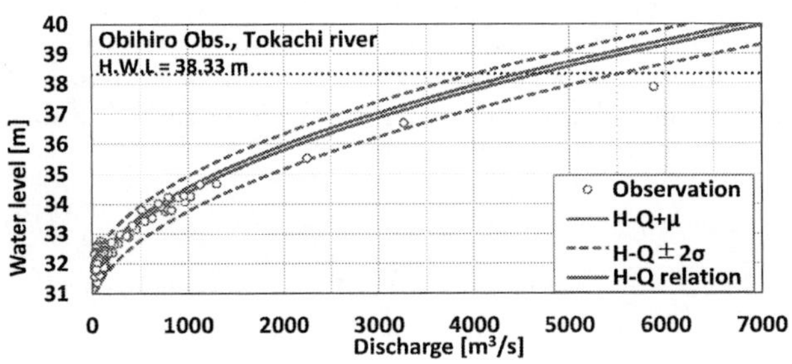

Figure 3. Comparison of $H - Q$ and Observations.

observed values using Eqs. (9), (10).

$$\mu_{\mathrm{WL}} = \frac{\sum\left(h_{\mathrm{Obs.}i} - h_{H-Q_i}\right)}{n},$$

$$\sigma_{\mathrm{WL}}^2 = \frac{\sum\left(h_{\mathrm{Obs.}i} - \mu_{\mathrm{WL}}\right)^2}{n} \quad (9)$$

$$h_{\mathrm{WL_{MCn}}} = h_{\mathrm{peak}} + h \sim N\left(\mu_{\mathrm{WL}}, \sigma_{\mathrm{WL}}^2\right) \quad (10)$$

here, $h_{\mathrm{Obs.}}$ represents the observed peak water level, h_{H-Q} denotes the water level based on the $H - Q$ equation, n is the number of observed samples, μ_{WL} represents the mean value of water level errors, σ_{WL}^2 is the variance, and $h_{\mathrm{WL_{MCn}}}$ indicates the water level for each MC simulation.

In the MC simulation, we assumed a normal distribution for the probability distribution, because conventionally the $H - Q$ equation has been derived by assuming a normal distribution for errors in observed water levels and discharges, determining coefficients "a" and "b" through a least squares minimization.

Additionally, the water level variability in this study excludes two elements: variation with discharge magnitude and temporal changes within a single event. Future verification should consider improvements in observation accuracy and data accumulation.

2.3.2 Dike height

In this study, we treated dike height as one of the random variables in the MC simulation. This is because dike height changes over time due to the heterogeneity of soil inside the dike and construction conditions. Figure 4 illustrates the planned dike height and the actual dike height. The actual dike height data were obtained at 10 m intervals in the longitudinal direction from the LP survey. To incorporate dike height as a random variable in the model, we calculated the mean and variance of the differences between the planned dike height and the actual dike height using Eqs. (11), (12).

$$\mu_{\mathrm{DH_{kp}}} = \frac{\sum\left(h_{\mathrm{actual}_i} - h_{\mathrm{plan}_i}\right)}{n},$$

$$\sigma_{\mathrm{DH_{kp}}}^2 = \frac{\sum\left(h_{\mathrm{actual}_i} - \mu_{\mathrm{DH_{kp}}}\right)^2}{n} \quad (11)$$

$$h_{\mathrm{DH_{MCn}}} = h_{\mathrm{plan_{kp}}} + h \sim N\left(\mu_{\mathrm{DH_{kp}}}, \sigma_{\mathrm{DH_{kp}}}^2\right) \quad (12)$$

Here, h_{actual} represents the actual levee height, h_{plan} is the planned levee height, n is the number of survey points in the 100 m section centered around each KP, $\mu_{\mathrm{DH_{kp}}}$ is the error in levee height at the evaluation point KP, $\sigma_{\mathrm{DH_{kp}}}^2$ is the variance, and $h_{\mathrm{DH_{MCn}}}$ denotes the levee height in each MC simulation run.

In the MC simulation, we assumed that the variance of dike height follows a normal distribution. Identifying the actual distribution is challenging, given the various factors influencing dike height variation, such as the heterogeneity of soil during construction, weather conditions, and construction methods. Verification through advancements in observational techniques for dike internal structures and data accumulation is anticipated in the future.

Additionally, it's essential to acknowledge that when defining dike height variability in this study, independence

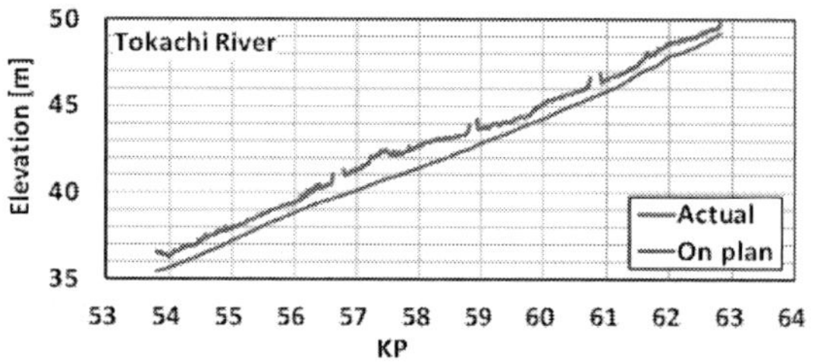

Figure 4. Comparison of Planned and Actual Embankment Heights.

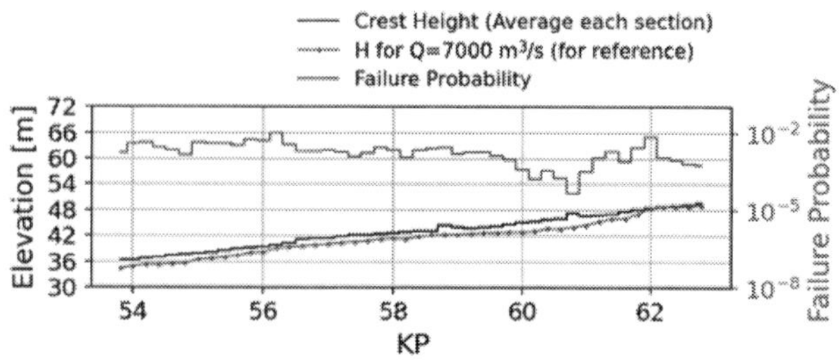

Figure 5. Calculation results for DFP.

Table 1. Critical velocity and variance on the dike slope.

Turf growth status condition	Threshold velocity u (m s^{-1})	Standard error, σ (m s^{-1})
Good	1.80	0.38
Average	1.30	0.12
Poor	0.76	0.04

Table 2. The integration approach and the resulting DFP.

Approach	Failure probability
Full dependence	0.00883
Applied method (partial dependence)	0.02564
Full independence	0.07476

among data intervals is unconfirmed if a singular event results in subsidence or damage surpassing a 10 m data interval. Future refinement of data intervals will require verification of the scale of events causing displacement.

2.3.3 Erosion of the dike slope

In the MC simulation, the erosion resistance of the turf on the levee slope was treated as a random variable ($u_{\text{Critical}_{\text{MCn}}}$). This approach accounts for the heterogeneity present in the condition of the turf. The erosion resistance of the turf was calculated using Eq. (13). The critical velocity (u_{Critical}) for erosion occurrence and its variance ($\sigma^2_{u_{\text{Critical}}}$) were set based on Table 1 under conditions of good turf growth, as established by Dean et al. (2010).

$$u_{\text{Critical}_{\text{MCn}}} = u \sim N\left(u_{\text{Critical}}, \sigma^2_{u_{\text{Critical}}}\right) \tag{13}$$

2.4 Integration of dike failure probabilities

This study uses a method incorporating point dependence to calculate the DFP for the entire levee section. Each point is termed a "segment", and segments with similar inundation are integrated into a "section". Complete dependence is assumed within a section, while partial dependence is considered between sections. This accommodates varying dependence based on the longitudinal levee scale in fast-flowing rivers. For adjacent points, higher DFP for an upstream section reduces downstream DFP due to decreased river flow. However, for segments further apart, the dependence impact diminishes, as the upstream breach may return flood flow to the river.

Initially, under the assumption of dependence between segments, we computed the DFPs for each section at various discharge conditions using Eq. (14), where dike failure occurs at the point with the maximum DFP within the section, and no failure occurs at other points:

$$P_{f,\text{section}} \left| q = \max_{i=1}^{n}(P_{f,\text{segment},i} \left| q\right.)\right. \tag{14}$$

here, $P_{f,\text{section}}$ represents the DFP in a specific section, and $|q$ denotes the conditional probability at discharge q. $P_{f,\text{segment},i}$ stands for the DFP in segment i. The parameter n indicates the number of segments included within the section.

Next, assuming that no dike failure occurs in downstream sections if a failure occurs in an upstream section (partial dependence), we calculated the DFP for the entire levee. Specifically, according to Eq. (15), the DFP for the uppermost section remains unaffected by other breach points since it is not influenced by upstream dike failures. For downstream sections, as per Eqs. (16) to (18), the DFP is determined by multiplying the non-DFP of the upstream section by the DFP in the respective downstream section.

$$P_{f,\text{dep},1} \,|q = P_{f,\text{section},1} \,|\, q \tag{15}$$

$$P_{f,\text{dep},2} \,|q = P_{f,\text{section},2} \,|\, q \cdot \left[1 - \left(P_{f,\text{dep},1} |q\right)\right] \tag{16}$$

$$P_{f,\text{dep},3} \,|q = P_{f,\text{section},3} \,|\, q \cdot \left[1 - \left(P_{f,\text{dep},1} \,|q + P_{f,\text{dep},2} \,|q\right)\right] \tag{17}$$

$$\vdots$$

$$P_{f,\text{dep},n} \,|q = P_{f,\text{section},n} \,|\, q \cdot \left[1 - \sum\nolimits_i^{n-1} \left(P_{f,\text{dep},n} |q\right)\right] \tag{18}$$

here, $P_{f,\text{dep},n}$ represents the DFP considering dependence, $|q$ denotes the conditional probability at discharge q, and n indicates the number of sections. The DFP for the entire levee

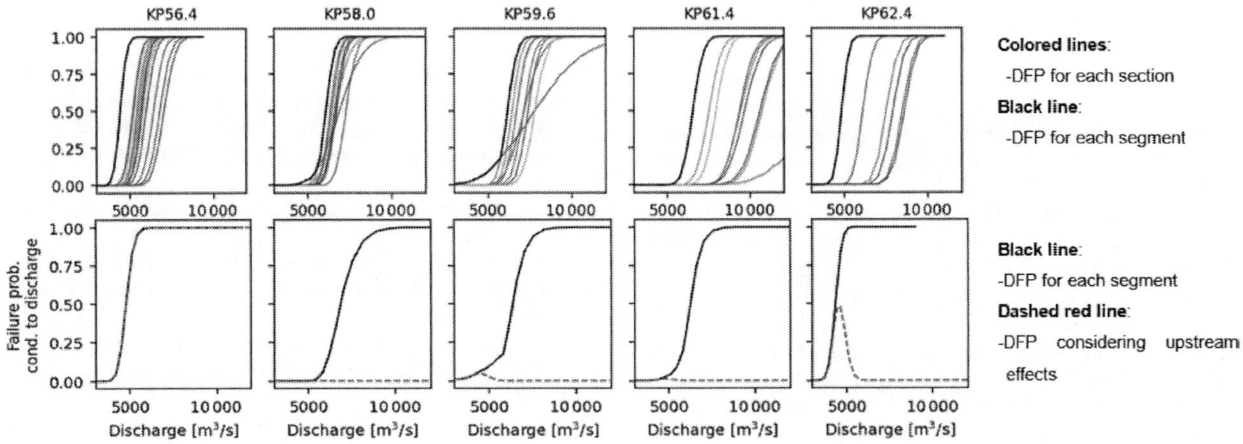

Figure 6. Dike failure probability of each section.

section under a specific discharge condition ($P_{f,\mathrm{dep,all}}|q$) is calculated according to Eq. (18) as the sum of the DFPs considering dependence for each section.

$$P_{f,\mathrm{dep,all}} = \sum_{i}^{n} \left(P_{f,\mathrm{dep},n}|q\right) \tag{19}$$

3 Results

Figure 5 depicts the segment DFP along the Tokachi River (10^{-5} to 10^{-2}) across its length. At $7000\,\mathrm{m^3\,s^{-1}}$ discharge, near crest height, a correlation is observed between increased freeboard and reduced failure probability. KP60.8, with higher conveyance capacity, exhibits the lowest DFP.

The DFP in Fig. 5 are grouped into section DFP (see Sect. 2.4). Figure 6, upper row, presents the DFP for all segments (colored lines), with the section's DFP derived from the maximum values (solid black line). The bottom graphs shows the adjusted DFP by deducting the upstream DFP, resulting in the dashed red lines. In the uppermost segment, under flow conditions exceeding $5000\,\mathrm{m^3\,s^{-1}}$, the DFP reaches 100 %. Consequently, downstream sections under the same discharge have a DFP of 0 %.

Table 2 shows DFP in the entire target area under varying dependence conditions in each section. The chosen dependence model notably impacts DFP values. Evaluating DFP along floodplains with downstream flooding, like the studied river, requires considering interdependence in breach occurrences.

4 Discussion

The fact that the DFP for each section corresponded to flow capacity indicates that the constructed model represents breach probabilities aligned with susceptibility to overflow. Moving forward, it is essential to verify elements, probability distributions, and levels of impact that should be incorporated into the model as uncertainties.

In this study, mathematical methods represented the reduction in downstream DFP from upstream breaches. Future work will incorporate hydrological approaches for downstream discharge reduction due to flooding, ensuring the consideration of upstream breaches on downstream flow in DFP assessment.

5 Conclusions

This study aims to quantify flood risk in the context of flash flood conditions in steep-gradient river. It involves the construction of a model to assess levee breach conditions as probabilities and proposes a method for calculating breach probabilities considering upstream levee breaches. The research outcomes are as follows:

- Developed a breach probability evaluation model that considers uncertainties associated with overtopping events.
- Highlighted the necessity of considering interdependence between breach points in breach events.

Data availability. The datasets generated and/or analyzed during the current study are available from the corresponding author on reasonable request.

Author contributions. FU planned, conducted, and analyzed the research, and wrote the manuscript. GR performed analysis and interpretation. SM and TY contributed to research conceptualization. TJY supervised the project. All authors approved the final manuscript for publication.

Competing interests. The contact author has declared that none of the authors has any competing interests.

Special issue statement. This article is part of the special issue "ICFM9 – River Basin Disaster Resilience and Sustainability by All". It is a result of The 9th International Conference on Flood Management, Tsukuba, Japan, 18–22 February 2023.

Acknowledgements. The study received additional support from the Rijkswaterstaat and the MLIT's Hokkaido Regional Development Bureau for providing data. The Embassy of the Kingdom of the Netherlands in Japan assisted in reporting and promoting the project's results.

Financial support. This research has been supported by the Rijksdienst voor Ondernemend Nederland (grant no. PVW4S19009).

Review statement. This paper was edited by Daisuke HARADA and reviewed by Masafumi YAMADA and one anonymous referee.

References

Dean, R. G., Rosati, J. D., Walton, T. L., and Edge, B. L.: Erosional equivalences of levees: Steady and intermittent wave overtopping, Ocean Eng., 37, 104–113, https://doi.org/10.1016/j.oceaneng.2009.07.016, 2010.

Ministry of Land, Infrastructure, Transport and Tourism: Technical Study Committee Report on River Dikes of the Damage Caused by Typhoon No. 19, Hagibis, https://www.mlit.go.jp/river/shinngikai_blog/gijutsu_kentoukai/dai03kai/pdf/houkokusyo.pdf (last access: 30 January 2024), 2020.

Vergouwe, R.: The national flood risk analysis for the Netherlands: final report, Rijkswaterstaat VNK Project Office, OCLC 952522344, 2016.

Tabata, K., Fukuoka, S., and Sezaki, T.: Study on evaluation method for probability of levee failures during large scale flood, Journal of Japan Society of Civil Engineers, Ser. B1 (Hydraulic Engineering), 71, I_1273–I_1278, https://doi.org/10.2208/jscejhe.71.I_1273, 2015.

Proc. IAHS, 386, 75–79, 2024
https://doi.org/10.5194/piahs-386-75-2024

Open Access
PIAHS

Uncertainty in simulated streamflow using runoff driven by the outputs of a high-resolution regional climate model

Aulia Febianda Anwar Tinumbang[1], Kazuaki Yorozu[1], Yasuto Tachikawa[1], Yutaka Ichikawa[1], Hidetaka Sasaki[2], and Tosiyuki Nakaegawa[3]

[1]Graduate School of Engineering, Kyoto University, Kyoto, 615-8540, Japan
[2]National Institute for Environmental Studies, Tsukuba, 305-8506, Japan
[3]Meteorological Research Institute, Tsukuba, 305-0052, Japan

Correspondence: Aulia Febianda Anwar Tinumbang (tinumbang.auliafebiandaanwar.6c@kyoto-u.ac.jp)

Received: 20 July 2023 – Revised: 10 October 2023 – Accepted: 24 February 2024 – Published: 19 April 2024

Abstract. Estimating river discharge using climate model output can aid in analyzing the potential impacts of climate change on water-related disasters. This study aimed to explore the uncertainty in simulated streamflow using the non-hydrostatic regional climate model (NHRCM) outputs in Thailand. The NHRCM was simulated at 5- and 2 km resolutions. To estimate runoff, two land surface models (LSMs) were employed: the Meteorological Research Institute–Simple Biosphere Model (MRI-SiB) in NHRCM and the Simple Biosphere including Urban Canopy (SiBUC). The NHRCM rainfalls captured the seasonal pattern of rainfall in the upper Ping River Basin, although they were underestimated. The 2 km NHRCM had less rainfall, but it captured the local topography better than the 5 km model. This difference in rainfall affected the simulated streamflow. Furthermore, the uncertainty of the simulated streamflow was influenced by the different runoff schemes used by the LSMs. For instance, MRI-SiB incorporates a direct infiltration structure from the surface to the second soil layer and estimates subsurface runoff using hydraulic diffusion and gravitational flow, while SiBUC uses a gravitational-only subsurface runoff approach. These variations led to significant disparities in surface and subsurface runoff. Future work should enhance the accuracy of rainfall from climate models and runoff from LSMs for assessing the potential impacts of climate change on water-related disasters.

1 Introduction

As global warming continues, water cycle variability and extremes will increase, causing more frequent and severe water-related disasters such as floods and droughts (IPCC, 2021). The impact of climate change is expected to vary by region, making local projections crucial. Climate models, such as general circulation models (GCMs) and regional climate models (RCMs), are the primary tools for assessing the potential impacts of climate change. RCMs are often used to investigate the local impacts of global warming owing to their high spatial resolution, which enables the capture of regional climatic features.

To mitigate and develop adaptation strategies against water-related disasters, estimating the extent to which river discharge will change owing to global warming is crucial. Much research has utilized runoff from climate models output to predict changes in future river discharge (e.g., Hirabayashi et al., 2021). The runoff can also be estimated by land surface models (LSMs) using meteorological outputs from climate models.

Nevertheless, estimated streamflow from climate models could be uncertain because of, for example, biases in the meteorological data and parameterization and structures of LSMs. Understanding the reasons for uncertainties in the model will aid in improving streamflow estimation. Past studies suggest that evapotranspiration (e.g., Haddeland et al., 2011), and snow accumulation and melt schemes (Beck et al., 2017) in the LSMs are major uncertain factors in simulated streamflow. However, the impacts of runoff schemes in

LSMs as a source of uncertainties are still not well studied. This study aimed to explore to what extent runoff process representations in LSMs impact streamflow estimation.

2 Methodology

2.1 Regional climate model output datasets

This study used the output of a non-hydrostatic regional climate model (NHRCM; Sasaki et al., 2008) with enhanced spatial resolutions of 5 and 2 km (RCM05 and RCM02, respectively) simulated for Thailand. RCM05 covered all of Thailand, and RCM02 included only the northern region, as shown in Fig. 1a. The 5 km model was nested in the general circulation model MRI-AGCM 3.2S (Mizuta et al., 2012), and the 2 km model was nested in the 5 km model. The output of the historical climate simulation from 1990 to 1999 was employed here.

2.2 Study area

The target area of this study is the upper part of the Ping River Basin in northern Thailand, as shown in Fig. 1b. It has a total catchment area of approximately 26 100 km^2. The main dam in this basin is the Bhumibol Dam. The Ping River is a major tributary of the Chao Phraya River, Thailand's most important river basin.

2.3 Land surface models

Here, the outputs of two LSMs, the Meteorological Research Institute-Simple Biosphere model (MRI-SiB, Hirai et al., 2007) and the Simple Biosphere model including Urban Canopy (SiBUC; Tanaka et al., 2005) were analyzed. MRI-SiB was incorporated into the NHRCM. Both LSMs, which were developed based on the Simple Biosphere model, have common structures, including the discretization of soil into three layers and the use of the Richards equation to estimate soil moisture. However, their details vary, such as, SiBUC estimates subsurface runoff solely on gravitational flow, whereas MRI-SiB also incorporates hydraulic diffusion, and MRI-SiB has a direct infiltration structure from the ground surface to the second soil layer (Tinumbang et al., 2023).

2.4 Flow routing model

This study used a flow routing model 1K-FRM, developed based on the 1-D kinematic wave theory, to convert total runoff (surface and subsurface runoff) into river discharge. A coupled SiBUC and 1K-FRM (Yorozu and Tachikawa, 2015) was employed here. The 1K-FRM was also used to convert the MRI-SiB runoff into river discharge.

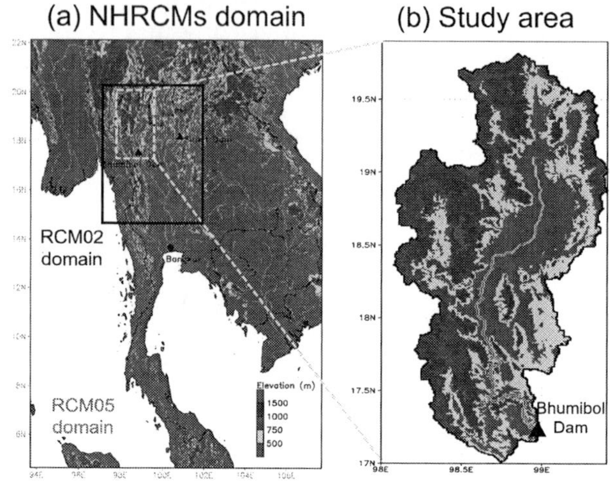

Figure 1. **(a)** Domain of NHRCM 5 and 2 km and **(b)** study area in the upper part of Ping River Basin, Thailand.

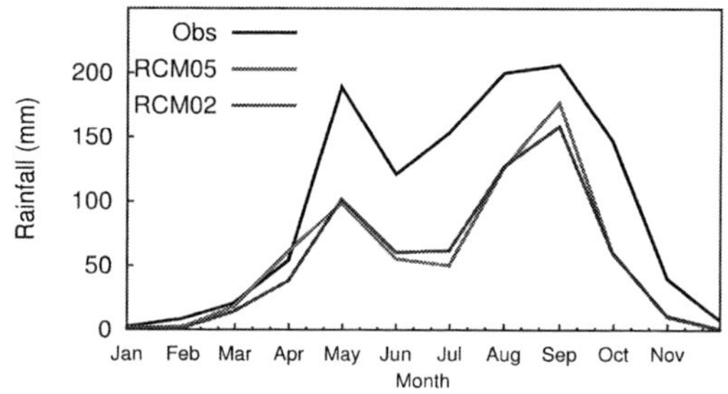

Figure 2. Reproducibility of climatological mean of monthly rainfall.

2.5 Observation data

The observed rainfall from the Climate Hazards Group InfraRed Precipitation with Station data (CHIRPS; Funk et al., 2015) was used here. The observed daily inflow at the Bhumibol Dam obtained from the Electricity Generating Authority of Thailand was also employed.

2.6 Research framework

This research is divided into three sections: evaluation of RCMs rainfall, comparison of the water budget estimated by the two LSMs driven by the RCMs output, and evaluation of the streamflow estimated by LSMs runoff. The water budget estimated by MRI-SiB was obtained from the RCMs output dataset. While the water budget by SiBUC was simulated here using precipitation, air temperature, specific humidity, downward short-wave and long-wave radiation, air pressure, and wind speed data from the RCMs output. The analysis of the water budget and streamflow driven by the RCM05 output was published in our previous study (Tinumbang et al., 2019).

 https://doi.org/10.5194/piahs-386-75-2024

Table 1. Ten-years-mean annual water budget. P, ET, and ROF indicate rainfall, evapotranspiration, and runoff, respectively.

Datasets	LSMs	Annual amount (mm)		
		P	ET	ROF
RCM05	MRI-SiB	663	649	47
	SiBUC	663	610	67
RCM02	MRI-SiB	638	625	33
	SiBUC	638	541	105

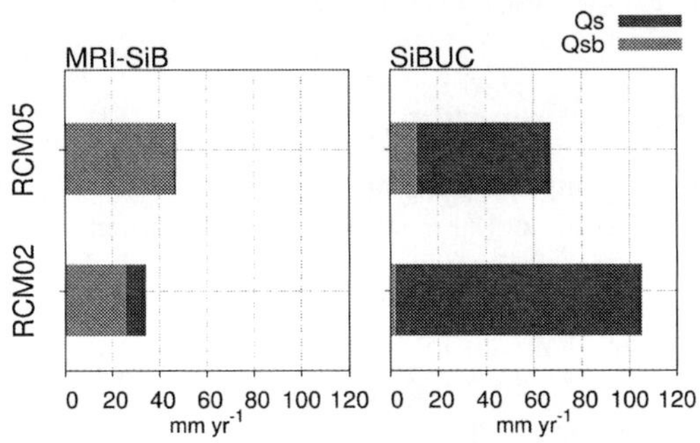

Figure 3. Annual mean of runoff components estimated by MRI-SiB and SiBUC. Qs and Qsb represent surface runoff and subsurface runoff, respectively.

3 Results and discussion

3.1 Rainfall evaluation

A comparison of the ten-year-mean monthly rainfall from observation and the RCMs is shown in Fig. 2. Overall, the RCMs rainfall could reproduce the seasonal cycle of observation in this basin well, with most rainfall occurring during the rainy season (May–October) and very low rainfall during the dry season (November–April). However, their amount was underestimated. The annual mean of observed rainfall in this basin is more than 1100 mm. In contrast, the rainfall from the RCM05 and RCM02 was 663 and 638 mm, respectively, as shown in Table 1.

3.2 Water budget comparison

A comparison of the water budgets estimated by the LSMs driven by the outputs of RCM05 and RCM02 is shown in Table 1. MRI-SiB estimated higher evapotranspiration and subsequently, lower runoff, than SiBUC. The MRI-SiB runoff driven by the RCM02 output was lower than that generated by RCM05. This was likely because RCM02 had less rainfall. The SiBUC runoff, however, exhibited the opposite trend. This could be due to the differences in runoff characteristics between the two LSMs, as shown in Fig. 3.

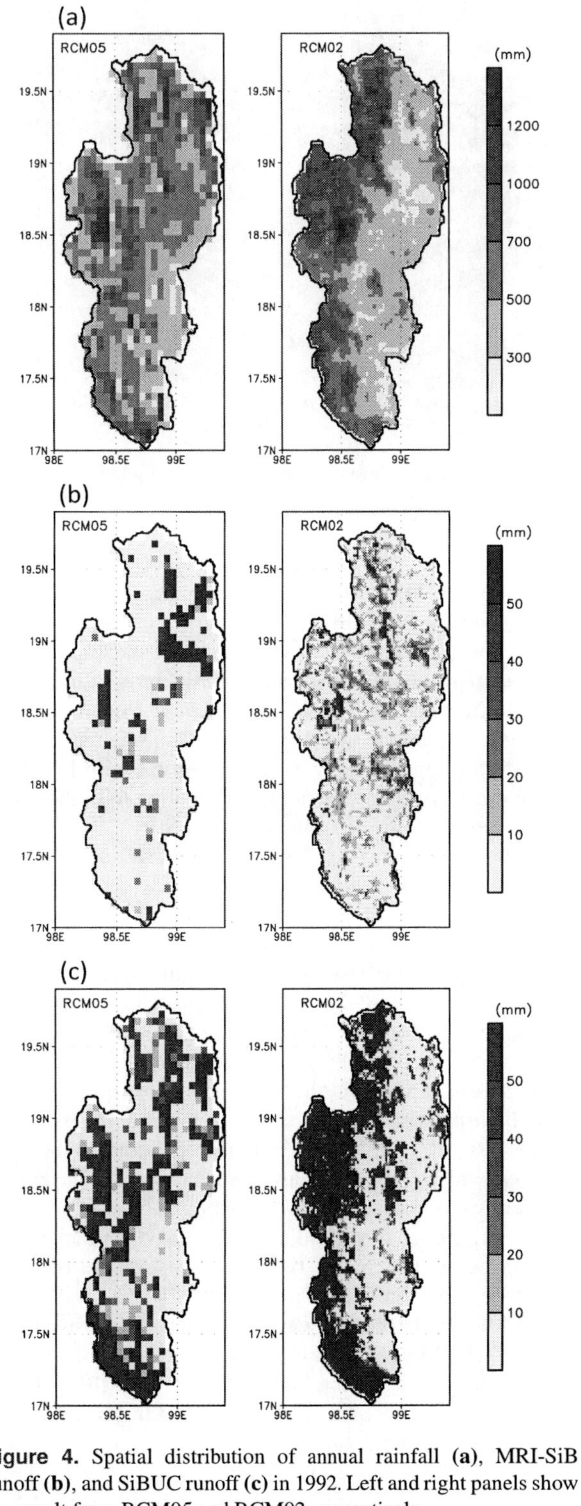

Figure 4. Spatial distribution of annual rainfall **(a)**, MRI-SiB runoff **(b)**, and SiBUC runoff **(c)** in 1992. Left and right panels show the result from RCM05 and RCM02, respectively.

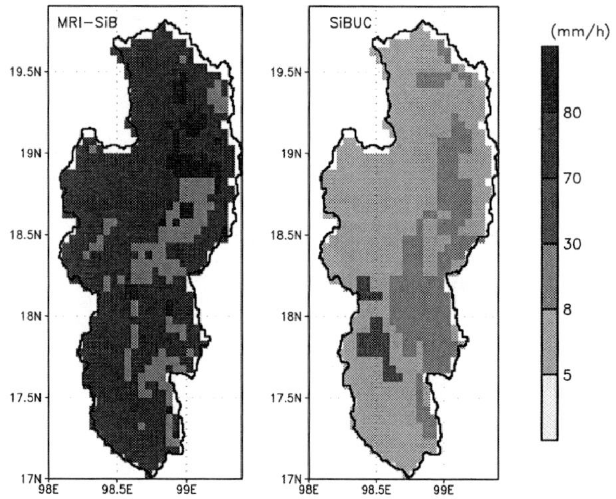

Figure 5. Comparison of saturated hydraulic conductivity between MRI-SiB and SiBUC.

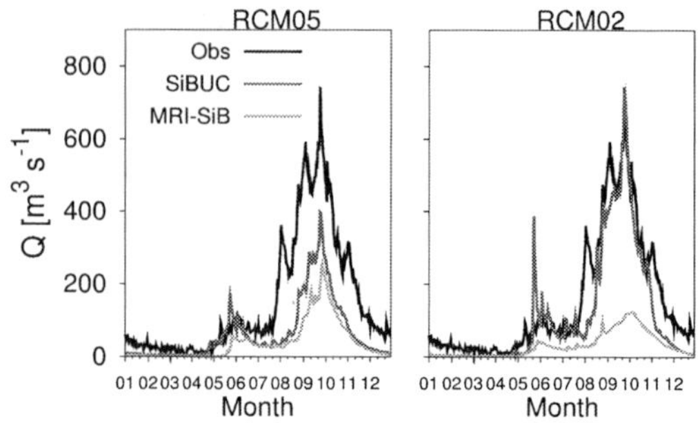

Figure 6. Comparison of annual mean daily river discharge (Q). Values indicate mean discharge.

Most of the MRI-SiB runoff was generated as subsurface runoff, whereas the SiBUC runoff was dominated by surface runoff. These variations in runoff characteristics suggest a disparity in runoff schemes, particularly in infiltration capacity between the two LSMs. Because MRI-SiB generated more subsurface runoff than SiBUC, it seemed to have a higher infiltration capacity. As more rainwater infiltrates the soil, more rainwater evaporates, contributing to higher evapotranspiration from MRI-SiB than from SiBUC.

To determine why the SiBUC runoff driven by RCM02 was higher than that driven by RCM05, the spatial patterns of rainfall and runoff were analyzed in detail. Results from 1992 were selected for the analysis. This year, RCM05 and RCM02 had similar basin-average rainfall values of 597 and 581 mm, respectively. However, as shown in Fig. 4a, their spatial distribution varied significantly. The RCM05 rainfall which was close to the basin-average value, shown in orange color, spread throughout the basin. In contrast, the RCM02 rainfall varied significantly between the western and eastern regions. RCM02 rainfall was significantly higher in the western area, which is the high-elevation zone, than in the eastern area, corresponding to the low-elevation zone.

This difference in rainfall appears to have a more significant impact on the SiBUC runoff than that on the MRI-SiB runoff, as shown in Fig. 4b and c. The SiBUC runoff driven by RCM05 and RCM02 was 52 and 86 mm, respectively, whereas the MRI-SiB runoff was 25 and 18 mm, respectively. The grids showing high value by the SiBUC runoff correspond to high rainfall. In contrast, the grids showing high MRI-SiB runoff, such as in the northern region, closely matched the high saturated hydraulic conductivity, as shown in Fig. 5. Since subsurface runoff predominates in MRI-SiB, greater values of saturated hydraulic conductivity indicate a higher infiltration rate and increased runoff.

3.3 Streamflow evaluation

Figure 6 compares the mean daily streamflow between the observation and simulations. Overall, all simulated discharges were underestimated owing to the underpredicted RCMs rainfall. The estimated discharge by MRI-SiB and SiBUC runoff driven by RCM05 was 74 % and 57 % less than the observation, and that driven by RCM02 was 75 % and 25 % lower.

The streamflow simulated by the MRI-SiB runoff using RCM02 was lower than that driven by RCM05 because the RCM02 rainfall was smaller. However, the streamflow simulated by the SiBUC runoff showed the opposite response because the SiBUC runoff driven by RCM02 was higher than that driven by RCM05. This is because more surface runoff was generated owing to more heavy rainfall in RCM02, as discussed in the previous subsection. The discharge from SiBUC runoff was nearly identical to the observed flow, although the RCM02 rainfall was less than the observation. This suggests that the amount of surface runoff that dominates SiBUC runoff may have been overestimated.

4 Conclusions

In this study, the uncertainty in the simulated streamflow using LSMs runoff driven by NHRCMs output was explored. The NHRCMs rainfall could reproduce the seasonal changes in observed rainfall in the upper part of the Ping River Basin although they were underestimated.

Both NHRCMs had similar total rainfall, though their spatial patterns differed, with the 2 km model being better at capturing the local topography than the 5 km model.

The uncertainty in simulated streamflow was found to be affected by this difference in rainfall patterns and the disparity in runoff characteristics among LSMs. Both simulated streamflow by MRI-SiB runoff, dominated by subsurface runoff, underestimated the observed discharge. In contrast, the estimated streamflow by SiBUC runoff, primarily from surface runoff, was underestimated when it was driven by RCM05, while the one driven by RCM02 was close to the observed discharge. This suggests SiBUC seemed to produce too much runoff because RCM02 rainfall underestimated the observation.

This analysis revealed that the runoff scheme in LSMs could be a major source of uncertainties in streamflow simulation, unlike previous studies which suggested evapotranspiration and snow representations were the primary uncertain factors. It is recommended that the accuracy of rainfall from the climate model and the runoff from the LSMs should be enhanced to improve the reliability of assessing the potential impacts of climate change on water-related disasters.

Data availability. Derived data supporting the findings of this study are available on request from the corresponding author.

Author contributions. AFAT and KY contributed to conceptualization and methodology; AFAT performed the investigation and wrote the original draft, supervised by KY, YT, YI, and TN; HS and TN provided the NHRCM datasets and MRI-SiB parameters.

Competing interests. The contact author has declared that none of the authors has any competing interests.

Special issue statement. This article is part of the special issue "ICFM9 – River Basin Disaster Resilience and Sustainability by All". It is a result of The 9th International Conference on Flood Management, Tsukuba, Japan, 18–22 February 2023.

Financial support. This research has been supported by the Ministry of Education, Culture, Sports, Science and Technology (MEXT) Program for the Advanced Studies of Climate Change Projection (grant no. JPMXD022680734) and Integrated Research Program for Advancing Climate Models Theme C and D (grant nos. JPMXD0717935561 and JPMXD0717935498), and the Japan Science and Technology "Moonshot R&D" (grant no. JPMJMS2283).

Review statement. This paper was edited by Mohamed Rasmy and reviewed by two anonymous referees.

References

Beck, H. E., van Dijk, A. I. J. M., de Roo, A., Dutra, E., Fink, G., Orth, R., and Schellekens, J.: Global evaluation of runoff from 10 state-of-the-art hydrological models, Hydrol. Earth Syst. Sci., 21, 2881–2903, https://doi.org/10.5194/hess-21-2881-2017, 2017.

Funk, C., Peterson, P., Landsfeld, M., Pedreros, D., Verdin, J., Shukla, S., Husak, G., Rowland, J., Harrison, L., Hoell, A., and Michaelsen, J.: The climate hazards infrared precipitation with stations-a new environmental record for monitoring extremes, Sci. Data, 2, 1–21, https://doi.org/10.1038/sdata.2015.66, 2015.

Haddeland, I., Clark, D. B., Franssen, W., Ludwig, F., Voß, F., Arnell, N. W., Bertrand, N., Best, M., Folwell, S., Gerten, D., Gomes, S., Gosling, S. N., Hagemann, S., Hanasaki, N., Harding, R., Heinke, J., Kabat, P., Koirala, S., Oki, T., Polcher, J., Stacke, T., Viterbo, P., Weedon, G. P., and Yeh, P.: Multimodel estimate of the global terrestrial water balance: setup and first results. J. Hydrometeor., 12, 869–884, https://doi.org/10.1175/2011JHM1324.1, 2011.

Hirabayashi, Y., Tanoue, M., Sasaki, Xudong, Z., and Yamazaki, D.: Global exposure to flooding from the new CMIP6 climate model projections, Sci. Rep., 11, 3740, https://doi.org/10.1038/s41598-021-83279-w, 2021.

Hirai, M., Sakashita, T., Kitagawa, H., Tsuyuki, T., Hosaka, M., and Oh'izumi, M.: Development and Validation of a New Land Surface Model for JMA's Operational Global Model Using the CEOP Observation Dataset, J. Meteor. Soc. Jpn., 85A, 1–24, https://doi.org/10.2151/jmsj.85A.1, 2007.

IPCC, Intergovernmental Panel on Climate Change Climate Change: The Physical Science Basis. Contribution of Working Group I to the Sixth Assessment Report of the Intergovernmental Panel on Climate Change, Cambridge University Press, https://doi.org/10.1017/9781009157896, 2021.

Sasaki, H., Murata, A., Hanafusa, M., Oh'izumi, M., and Kurihara, K.: Reproducibility of present climate in a non-hydrostatic regional climate model nested within an atmosphere general circulation model, SOLA, 7, 173–176, https://doi.org/10.2151/sola.2011-044, 2011.

Mizuta, R., Yoshimura, H., Murakami, H., Matsueda, M., Endo, H., Ose, T., Kamiguchi, K., Hosaka, M., Sugi, M., Yukimoto, S., Kusunoki, S., and Kitoh, A.: Climate simulations using MRI-AGCM with 20 km grid, J. Meteor. Soc. Jpn., 90A, 235–260, https://doi.org/10.2151/jmsj.2012-A12, 2012.

Tanaka, K.: Development of the new land surface scheme SiBUC commonly applicable to basin water management and numerical weather prediction model, doctoral dissertation, Kyoto University, http://hdl.handle.net/2433/145376 (last access: 5 March 2024), 2005.

Tinumbang, A. F. A, Yorozu, K., Tachikawa, Y., Ichikawa, Y., Sasaki, H., and Nakaegawa, T.: Analysis of runoff characteristics generated by land surface models and their impacts on river discharge, J. Jpn. Soc. Civ. Eng, 75, I_271–I_276, https://doi.org/10.2208/jscejhe.75.2_I_271, 2019.

Tinumbang, A. F. A., Yorozu, K., Tachikawa, Y., and Ichikawa, Y.: Developing a methodology for model intercomparison and its application to improve simulated streamflow by land surface models, J. Hydrometeor., 24, 817–833, https://doi.org/10.1175/JHM-D-22-0183.1, 2023.

Yorozu, K. and Tachikawa, Y.: The effect on river discharge estimation by considering an interaction between land surface process and river routing process, Proc. Int. Assoc. Hydrol. Sci, 369, 81–86, https://doi.org/10.5194/piahs-369-81-2015, 2015.

Proc. IAHS, 386, 81–86, 2024
https://doi.org/10.5194/piahs-386-81-2024

Artificial neural networks applied for flood forecasting in ungauged basin – the Paranaíba river study case

Abderraman R. A. Brandão[1], **Frederico C. M. de Menezes Filho**[2], **Paulo T. S. Oliveira**[1,3], and **Maria C. Fava**[4]

[1]Department of Hydraulics and Sanitation, University of São Paulo, São Carlos, SP, Brazil
[2]Institute of Exact and Technological Sciences, Federal University of Viçosa, Rio Paranaíba, MG, Brazil
[3]Faculty of Engineering, Architecture and Urbanism, and Geography, Federal University of Mato Grosso do Sul, Campo Grande, MS, Brazil
[4]Civil Engineering Department, Federal University of São Carlos, São Carlos, SP, Brazil

Correspondence: Abderraman R. A. Brandão (abderraman.brandao@usp.br)

Received: 11 July 2023 – Revised: 5 October 2023 – Accepted: 24 February 2024 – Published: 19 April 2024

Abstract. Flow simulation using artificial neural networks (ANNs) in the modelling has been widely applied and has gained prominence in regions lacking data. The hydrological variables are subject to the influence of morphological characteristics and urbanization in the watershed. Statistical models, such as ANNs, need to be able to identify the relationship between the hydrological inputs and outputs of the model, without explicitly considering the other relationships involved in physical processes. This work aimed to apply a Multilayer Perceptron (MLP) neural network for predicting flows in an urban basin subject to recurrent floods, using precipitation and flow data from previous periods as inputs. After model calibration and validation for the current state of the basin (2018–2019), its responses were analysed using input data before the basin urbanization (1985–1986) to identify the error behaviour at the output as a proxy for the basin changes effect. Its efficiency was evaluated using hydrographs, showing satisfactory results in both periods. In the urbanization period, there is more dispersion for maximum flows. For the day 4 steps back in the current forecast, NSE $=0.59$ was observed, whereas in the other period, NSE $=0.70$. The evaluation of the models for the current period of basin urbanization showed that the model could capture the basin's physical dynamics within the established static relationship. Also, the result found in the statistical relationships for the inputs showed once again the impact of urbanization on the basin.

1 Introduction

Hydrological modelling has proven to be a reliable solution for many problems. However, in traditional models, the representation of hydrological processes translates into high complexity, due to the large number of input parameters required for model setup. In parallel with the development of traditional models that aim to represent the physical processes involved in the hydrological cycle, since the 1950s, scholars in the area have been developing data-driven models for the description of complex hydrological processes without the need for several input parameters, given the difficulty in obtaining these data (Fatichi et al., 2016).

According to Kermani et al. (2020), given the computational efficiency and flexibility of applying machine learning, this approach has been applied to solve several challenges in the field of hydrological sciences and has great potential to provide more accurate and reliable predictions when compared to traditional statistical models, stochastic methods, and empirical formulations.

Artificial Neural Networks (ANNs) are part of the set of machine learning techniques based on artificial neurons. Currently, the ANN Multilayer Perceptron (MLP) architecture is widely used. Its array is divided into input neurons, which store the input vectors, and output vectors, which receive the processing response. There is a hidden layer between the input and output vectors, which may contain one or more divisions where data processing takes place, with the entire network being connected (Dazzi et al., 2021).

Considering the multilayer perceptron application perspectives, the objective of this work was to evaluate the performance of the artificial neural network to simulate average daily flows in a basin in Minas Gerais State in Brazil with input data under the effect of changes caused by the urbanization of the basin, evaluating its effectiveness according to statistical criteria. The case study was chosen due to the recurrence of Paranaíba river overflow in the rainy season, causing damage to the municipality of Patos de Minas and its population. Another factor to be highlighted is the evidence of the importance of the river for the entire Alto Paranaíba region since it supplies cities and local agricultural activities. The forecast of flows in the basin is paramount for productive environmental zoning policies, the safety of the population living in the neighbourhoods surrounding the floodplain area. According to Nogueira (2017), the area is highly susceptible to floods, while Amaral (2021) describes that floods periodically reach the urban perimeter.

2 Case Study and Problem Statement

The study was carried out in a sub-basin of the Paranaíba River, one of the main tributaries of the Paraná River basin, with the outlet defined at the fluviometric station (Code 6001100) located in the urban perimeter of Patos de Minas – MG. The drainage area of the basin is 3791.83 km^2 (Fig. 1), corresponding to 11.02 % of the entire basin of the Paranaíba River, with an extension of the main watercourse of 106.6 km.

The predominant land use and occupation class is agriculture and livestock, but there is also an urbanized area. The region has a low slope in the central region and is mountainous at the extremes and the predominant soil class is Red Latosol, according to Santos et al. (2018). The time of concentration is about 12.93 h, calculated by the U.S. Army Corps of Engineers equation (Collischown and Dornelles, 2021).

According to the Köppen-Geiger climate classification, the type of climate present in the basin is mostly of the Aw type, evidencing a tropical climate with temperatures above 18 °C, a dry season, and high annual precipitation.

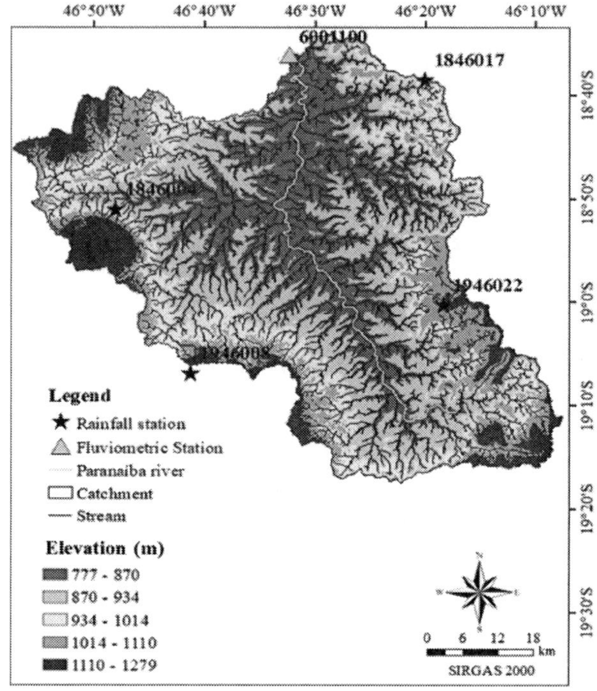

Figure 1. Location of Paranaíba River sub-catchment.

3 Material and Methods

3.1 Data description and preparation of inputs

The precipitation series were corrected using the Simple Linear Regression method, data consistency was performed using the Double Mass method developed by the Geological Survey (USA), and the average rainfall in the basin was calculated using Thiessen Polygons, all as described by Collischonn and Dornelles (2021).

The basin underwent intense urbanization from the 2000s onwards (Bessa and Soares, 2002). Thus, the rainfall and flow series for training the model before urbanization were from 1976 to 1984 (3287 daily data), and the series from 1985 to 1986 were used as a set for the test (729 daily data). For the post-urbanization, the training data selected were from 1 January 2008 to 31 December 2016 (3287), and data from 2018 to 2019 were used as a set for the test (729 daily data).

The values of past rainfall and flow to be included in the model took into account the physical characteristics of the basin to find the reaction time. We used correlation analysis to examine the interdependence between rainfall and flow for both periods. Meanwhile, autocorrelation helped identify past flow values influencing the current flow, as depicted in Fig. 3.

From the correlation graph, it can be seen that for the current period of the basin the degree of correlation between rainfall and flow has dropped in relation to the other period, requiring a smaller number of rain delays when compared to the period prior to urbanization, this being still one of the indications that the anthropological changes in the basin influenced its time of concentration. The same happens when comparing the influence of previous flows with current flows using autocorrelation with a confidence level of 99 %; there is a greater need for past flows for the current period than for the previous one, in addition to in both cases demonstrating the strong correlation of bottom recharge to the river.

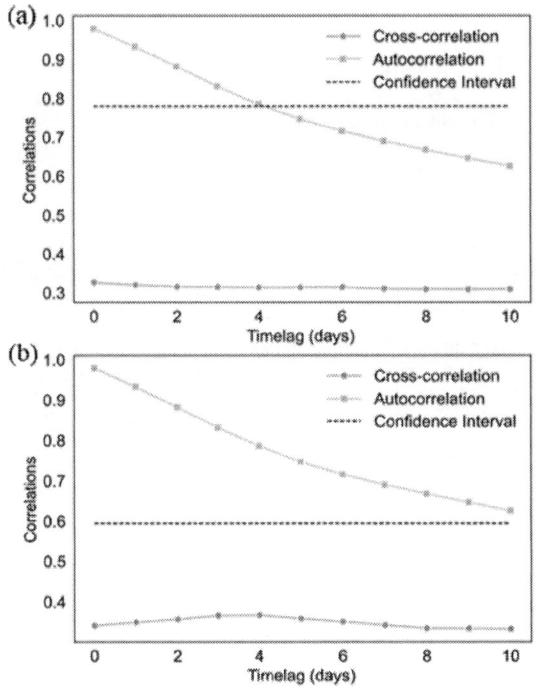

Figure 2. Correlation analysis to period **(a)** pre-urbanization and **(b)** post-urbanization.

The positive and insignificant correlations between rainfall and flow for the current (0.32) and previous period (0.36) suggest the influence of other variables on the observed flows, something already expected, under the concepts of water balance. In addition, the type of soil and its occupation, combined with the low slopes in the central area favor the recharge of the groundwater and groundwater flows; both directly influence the flow (Mendonça et al., 2021).

3.2 Model Structure definition

The training of ANNs can be supervised or unsupervised (Raschka, 2015). According to Cristaldo (2020), supervised training is the most used in hydrological forecasting problems. After training, it is expected that the network can generalize data and finds an answer in vectors different from the training. The propagation capacity of the network is related to its topology.

This study trained and validated ANNs using the Waikato Environment for Knowledge Analysis (WEKA) software, which uses a General Public License (GPL) written in Java, and has a collection of machine learning algorithms, including the MLP architecture. The software uses neurons in the hidden layer equal to half the sum of arrays and classes with the sigmoid function.

Using as neurons of the hidden layer equal to half the sum of attributes and classes, several epochs equal to 1000, a momentum of 0.2, and a learning rate of 0.1. The software library's perceptron model works with backpropagation, providing supervised training. In order to avoid the underfitting and overfitting phenomena, cross-validation k-fold type was applied ($k = 4$). A summary of the trained ANNs is presented in Table 1.

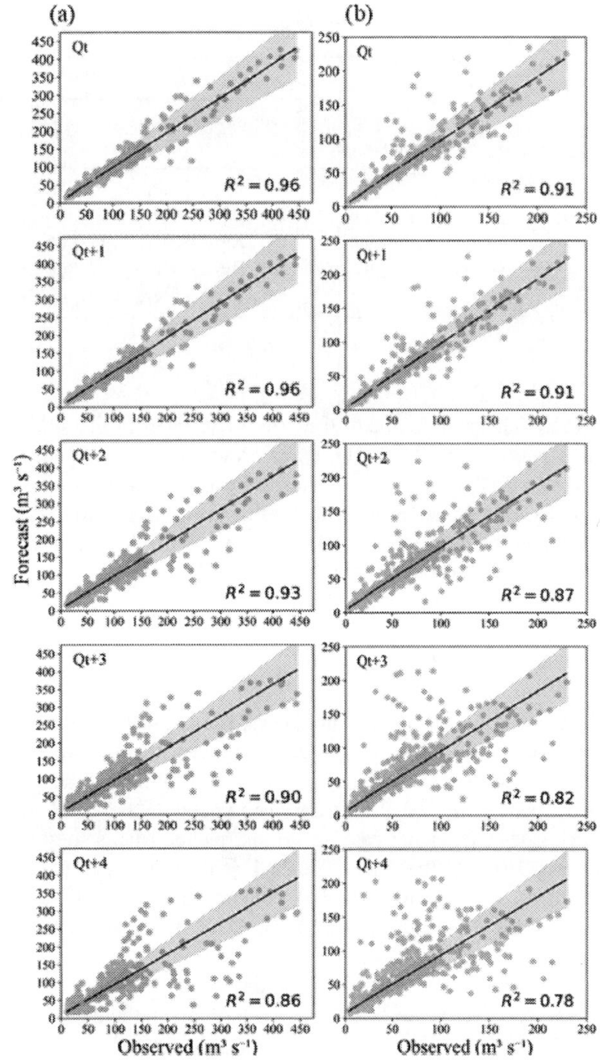

Figure 3. Observed vs. forecasted flow in Paranaíba River for **(a)** 1984 and **(b)** 2019. Comparisons include current time and 1–4 d lags. Shaded area: ±20 % error.

3.3 Metrics for Model Evaluation

The performance of the model evaluation was assessed using the determination coefficient (R^2), Root Mean Square Error (RMSE), Mean Absolute Error (MAE), Correlation Coefficient (CC) and the Nash-Sutcliffe Coefficient of Efficiency

Table 1. ANN experiment summary.

	1984		2019	
Out variable	Input variables	Hidden nodes	Input variables	Hidden nodes
Q_t	Re_t, Re_{t-1}, Re_{t-2}, Re_{t-3}, Q_{t-1}	3	Re_t, Re_{t-1}, Q_{t-1}	2
Q_{t+1}	Re_t, Re_{t-1}, Re_{t-2}, Q_t, Q_{t-1}	3	Re_t, Re_{t-1}, Q_t, Q_{t-1}	2
Q_{t+2}	Re_t, Re_{t-1}, Re_{t-2}, Q_t, Q_{t-1}	2	Re_t, Re_{t-1}, Q_t, Q_{t-1}	2
Q_{t+3}	Re_t, Re_{t-1}, Re_{t-2}, Q_t, Q_{t-1}	2	Re_t, Re_{t-1}, Q_t, Q_{t-1}	2
Q_{t+4}	Re_t, Re_{t-1}, Re_{t-2}, Q_t, Q_{t-1}	2	Re_t, Re_{t-1}, Q_t, Q_{t-1}	2

Re_{t-i} Effective rainfall (mm), i = lag time (0, 1, 2,..., day). Q_{t-k} Discharge at time ($m^3 s^{-1}$), $t-k$, $k = 0, \ldots, 4$.

(NSE). These are some of the criteria adopted in the studies of Mendonça et al. (2021), Dazzi, Vancondio and Mignosa (2021). RMSE and MAE criteria are better as closer to 0. The CC can range from −1 to 1, with values closer to the extremes indicating stronger correlations, and NSE should be closer to 1 for the best fit (Dazzi et al., 2021).

4 Results and discussion

At the outputs of the two networks, the evaluation parameters are similar, showing the decay of forecasting efficiency over the horizon. The best values were found in the pre-urbanization period (Table 2). As expected, the COE and NSE coefficients indicate an almost perfect fit as they approach 1 for the first days. The NSE still serves as an indicator of credibility for the model. According to the criteria established by Moriasi et al. (2015), the NSE values obtained for the two periods range from "very good" for the initial days to "Satisfactory" for the last day.

Despite the existence of discordant points in the dispersion results (Fig. 3), Pearson's coefficient was greater than 0.86 and 0.78 for the first and second periods respectively in the less accurate result and as expected were found for the most distant forecast day, while the other performance criteria CC, RMSE, MAE and NSE also resulted in plausible answers.

ANNs significantly influenced by the quantity and quality of the datasets that are trained. In Fig. 4, the data predominantly concentrates on lower flow values, and the model appropriately replicates these. However, the higher variability observed for intermediate values can be attributed to a lack of information within this range, leading the model to struggle in learning and accurately representing these values. Solomatine et al. (2003) support this observation, noting similar challenges in their studies. Moreover, hydrological variables, like flow, are substantially affected by anthropogenic activities (Wu et al., 2015). The replacement of native forests with agricultural practices, for instance, can intensify flood events (Housspanossian et al., 2023). In the current period, specifically in 2014/2015, a severe drought event occurred (Marengo et al., 2015). Hydrologic signatures also affect the

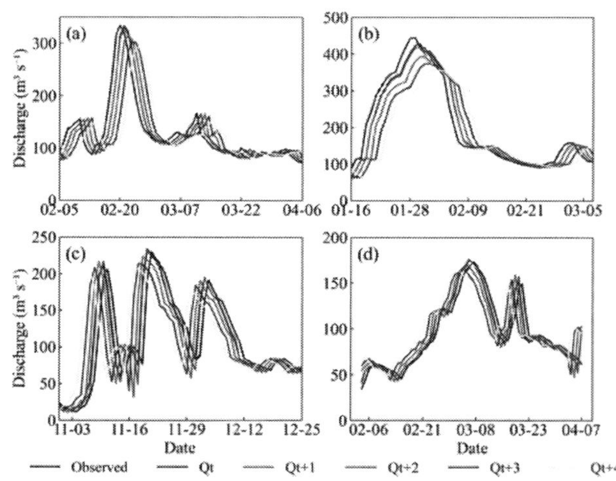

Figure 4. Flow hydrographs estimated by the MLP-ANN. **(a, b)** pre-urbanization period (1986–1985) and **(c, d)** After urbanization (2019–2019).

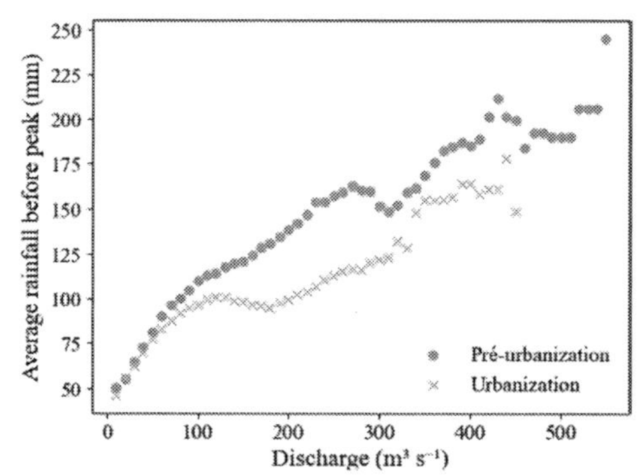

Figure 5. Average rainfall before peak of discharge (10 d) for both periods.

models (McMillan et al., 2023), and other natural mechanisms influence floods (Sharma, 2018).

These results, consequently, were also reflected in the hydrographs, where deviations between the simulated and ob-

Table 2. Comparison of errors for the verification dataset.

Prediction	Test – 1984				Test – 2019			
	CC	MAE ($m^3 s^{-1}$)	RMSE ($m^3 s^{-1}$)	NSE	CC	MAE ($m^3 s^{-1}$)	RMSE ($m^3 s^{-1}$)	NSE
Q_t	0.97	6.89	14.81	0.95	0.95	6.04	13.28	0.90
Q_{t+1}	0.97	6.77	14.72	0.95	0.95	5.26	12.23	0.91
Q_{t+2}	0.93	11.55	24.05	0.88	0.88	9.38	20.89	0.76
Q_{t+3}	0.88	15.76	31.81	0.79	0.81	12.53	25.57	0.64
Q_{t+4}	0.83	18.66	37.92	0.70	0.78	14.23	27.49	0.59

served curves are almost not perceptible (Fig. 4), demonstrating the coherence of the model and the respective numerical capture of the basin bias.

Analyzing past flow peaks (Fig. 5) revealed that a higher average of previous precipitation was necessary to trigger the flow peak compared to current conditions. This discrepancy in peak flow time is partly due to urbanization process. Soil waterproofing reduces rainwater infiltration and accelerates surface runoff. Thus, the water reaches the outlet faster in urbanized areas, bringing forward the peak flow. The results indicate that urbanization has sped up the flood arrival time, as already demonstrated in other studies (e.g., Lu et al., 2023; Nardi et al., 2018).

5 Conclusions

The study effectively simulated average daily flows using ANN, producing results aligned with existing literature. Statistical relationships in input data were crucial for selecting influential input variables and eliminating less impactful ones. The research also points out that anthropic change in the basin, altering its natural dynamics and showing better results in the pre-urbanization period. Moreover, it can aid decision-making in extreme events due to its versatile methodology, offering simpler replication than traditional models with their intensive data and parameter needs.

Data availability. Relevant data can be provided upon request to the corresponding author. The data used are open and available at the Agência Nacional de Águas e Saneamento Básico (ANA), and can be accessed at: https://www.snirh.gov.br/hidroweb/serieshistoricas (ANA, 2023).

Author contributions. ARAB conceived and presented idea, wrote the paper in consultation with MCF, FCMdMF, PTSO. All authors provided critical feedback and helped shape the research, analysis and manuscript.

Competing interests. The contact author has declared that none of the authors has any competing interests.

Disclaimer. Publisher's note: Copernicus Publications remains neutral with regard to jurisdictional claims made in the text, published maps, institutional affiliations, or any other geographical representation in this paper. While Copernicus Publications makes every effort to include appropriate place names, the final responsibility lies with the authors.

Special issue statement. This article is part of the special issue "ICFM9 – River Basin Disaster Resilience and Sustainability by All". It is a result of The 9th International Conference on Flood Management, Tsukuba, Japan, 18–22 February 2023.

Acknowledgements. The authors acknowledge the Conselho Nacional de Desenvolvimento Científico e Tecnológico (CNPq) for the scientific support, Grant/Award Numbers: 130247/2023-5 and 409527/2021-1.

Review statement. This paper was edited by Daisuke Harada and reviewed by two anonymous referees.

References

Agência Nacional de Águas e Saneamento Básico (ANA): HIDROWEB v3.3.7413.0, ANA [data set], https://www.snirh.gov.br/hidroweb/serieshistoricas, last access: 10 February 2023.

Amaral, E. A., Nascimento, A. R. T., Silva, C. R., Oliveira, A. P., and Silva, G. R.: Avaliação de impactos ambientais na APP do Rio Paranaíba e inferências para mitigação, Revista Ibero Americana de Ciências Ambientais, 12, 572–584, ISSN 2179-6858, 2021.

Collischonn, W. and Dornelles, F.: Hidrologia para engenharia e ciências ambientais, Porto Alegre: Associação Brasileira de Recursos Hídricos, ISBN 978-85-8868-634-2, 2021.

Bessa, K. C. F. O. and Soares, B. R.: Considerações Sobre A Dinâmica Demográfica Na Região Do Triângulo Mineiro/Alto Paranaíba, Caminhos de Geografia, Uberlândia, MG, 3, 22–45, https://doi.org/10.14393/RCG3615293, 2002.

Cristaldo, M. F., Jesus, L., Oliveira, P. T., Souza, C. C., Viganó, H. H. G, and Padovi, C. R.: Redes Neurais Artificiais Aplicadas À Previsão De Enchentes Para Região Do Pantanal No Mato Grosso Do Sul, Geociências, 39, 191–201, https://doi.org/10.5016/geociencias.v39i1.13644, 2020.

Dazzi, S., Vacondio, R., and Mignosa, P.: Flood Stage Forecasting Using Machine-Learning Methods: A Case Study on the Parma River (Italy), Revista Water, 13, p. 1612, https://doi.org/10.3390/w13121612, 2021.

Fatichi, S., Vivoni, E. R., Ogden, F. L., Ivanov, V. Y., Mirus, B., Gochis, D., Downer, C. W., Camporese, M., Davison, J. H., Ebel, B., Jones, N., Kim, J., Mascaro, G., Niswonger, R., Restrepo, P., Rigon, R., Shen, C., Sulis, M., and Tarboton, D.: An overview of current applications, challenges, and future trends in distributed process-based models in hydrology, J. Hydrol., 537, 45–60, https://doi.org/10.1016/j.jhydrol.2016.03.026, 2016.

Houspanossian, J., Giménez, R., Whitworth-Hulse, J. I., Nosetto, M. D., Tych, W., Atkinson, P. M., Rufino, M. C., and Jobbágy, E. G.: Agricultural Expansion Raises Groundwater and Increases Flooding in the South American Plains, Science, 380, 1344–1348, https://doi.org/10.1126/science.add5462, 2023.

Kermani, M. Z, Matta, E., Cominola, A., Xia, X., Zhang, Q., Liang, Q., and Hinkelman, R.: Neurocomputing in surface water hydrology and hydraulics: A review of two decades retrospective, current status and future prospects, J. Hydrol., 588, 125085, https://doi.org/10.1016/j.jhydrol.2020.125085, 2020.

Marengo, J. A., Nobre, C. A., Seluchi, M. E., Cuartas, A., Alves, L. M., Mendiondo, E. M., Obregón, G., and Sampaio, G.: A Seca e a Crise Hídrica de 2014–2015 em São Paulo, Rev. USP, 106, 31–44, https://doi.org/10.11606/issn.2316-9036.v0i106p31-44, 2015.

McMillan, H., Coxon, G., Araki, R., Salwey, S., Kelleher, C., Zheng, Y., Knoben, W., Gnann, S., Seibert, J., and Bolotin, L.: When good signatures go bad: Applying hydrologic signatures in large sample studies, Hydrol. Process., 37, e14987, https://doi.org/10.1002/hyp.14987, 2023.

Mendonça, L. M., Gomide, I. S., Sousa, J. V., and Blanco, C. J. C.: Modelagem chuva-vazão via redes neurais artificiais para simulação de vazões de uma bacia hidrográfica da Amazônia, Revista de Gestão de Água da América Latina, Porto Alegre, 18, 2021, https://doi.org/10.21168/rega.v18e2, 2021.

Moriasi, D. N., Gitau, M. W., Pai, N., and Daggupati, P.: Hydrologic and water quality models: Performance measures and evaluation criteria, T. ASABE, 58, 1763–1785, https://doi.org/10.13031/trans.58.10715, 2015.

Lu, M., Yu, Z., Hua, J., Kang, C., and Lin, Z.: Spatial dependence of floods shaped by extreme rainfall under the influence of urbanization, Sci. Total Environ., 857, 159134, https://doi.org/10.1016/j.scitotenv.2022.159134, 2023.

Nogueira, T. P. N.: Mapeamento Da Suscetibilidade À Inundação Na Bacia Hidrográfica Do Ribeirão Da Fábrica, Município De Patos De Minas – MG. 123 f., Master's Thesis, Environmental and Environmental Quality, Federal University of Uberlândia, Uberlândia, https://doi.org/10.14393/ufu.di.2017.304, 2017.

Nardi, F., Annis, A., and Biscarini, C.: On the impact of urbanization on flood hydrology of small ungauged basins: The case study of the Tiber river tributary network within the city of Rome, J. Flood Risk Manag., 11, S594–S603, https://doi.org/10.1111/jfr3.12186, 2018.

Raschka, S. and Mirjalili, V.: Python Machine Learning, 2 edn., Packt, Birmingham, ISBN 9781787125933, 2015.

Santos, H. G. dos, Jacomine, P. K. T., Anjos, L. H. C. dos, Oliveira, V. A. de, Lumbreras, J. F., Coelho, M. R., Almeida, J. A. de, Araujo Filho, J. C. de, Oliveira, J. B. de, and Cunha, T. J. F.: Brazilian System of Soil Classification, 5th edn. rev. and expanded, Brasília, DF, Embrapa, http://www.infoteca.cnptia.embrapa.br/infoteca/handle/doc/1094003 (last access: 15 February 2023), 2018.

Sharma, A., Wasko, C., and Lettenmaier, D. P.: If precipitation extremes are increasing, why aren't floods?, Water Resour. Res., 54, 8545–8551, 2018.

Solomatine, D. P. and Dulal, K. N.: Model trees as an alternative to neural networks in rainfall-runoff modelling, Hydrolog. Sci. J., 48, 399–411, https://doi.org/10.1623/hysj.48.3.399.45291, 2003.

Wu, J., Yin, J., Hao, Y., Liu, Y., Fan, Y., Huo, X., Liu, Y., and Yeh, T. C. J.: The Role of Anthropogenic Activities in Karst Spring Discharge Volatility, Hydrol. Process. 29, 2855–2866, https://doi.org/10.1002/hyp.10407, 2015.

Proc. IAHS, 386, 87–93, 2024
https://doi.org/10.5194/piahs-386-87-2024

Investing in Resilience: Estimating Financial Needs and Benefits of Flood Protection in Developing Asian Countries

Mikio Ishiwatari[1,2] **and Daisuke Sasaki**[3]

[1]Japan International Cooperation Agency, Tokyo, Japan
[2]Graduate School of Frontier Sciences, The University of Tokyo, Kashiwa 2778568, Japan
[3]International Research Institute of Disaster Science, Tohoku University, Sendai, 9808572, Japan

Correspondence: Mikio Ishiwatari (ishiwatari.mikio@jica.go.jp)

Received: 3 July 2023 – Revised: 14 December 2023 – Accepted: 24 February 2024 – Published: 19 April 2024

Abstract. Investment in disaster risk reduction, which the Sendai Framework for Disaster Risk Reduction highlights as a priority action, is crucial in mitigating disaster damage. This paper aims to estimate the funding gap and the benefits of investments in flood protection infrastructure in Asia. Budget and damage data were collected from major flood-prone economies in the region. Currently, nine economies invested USD 33.6 billion, or 0.21 % of GDP, in flood protection. Regression analysis indicates that the annual demand for flood protection infrastructure in developing economies will be USD 94.5 billion by 2030, or USD 98.4 billion considering impacts by climate change. Investment in flood protection is cost-effective at the regional level. If these developing economies could invest this amount, the net benefit (benefits minus costs) accumulation is projected to reach USD 263 billion by 2030.

1 Introduction

Investment is a critical factor in decreasing damages and losses and reducing disaster risks. International arenas have consistently recommended an increase in investment for disaster risk reduction (DRR). The Sendai Framework for DRR, adopted by UN Member States in 2015, highlights investment as a priority action to effectively reduce disaster risks and losses (UNISDR, 2015). The Copenhagen Accord, established in 2009 at the Conference of the Parties to the United Nations Framework Convention on Climate Change, set a target of USD 100 billion yr^{-1} by 2020 for developed countries to support mitigation and adaptation measures. Despite efforts like the establishment of the Green Climate Fund, this goal remains unmet. The Kumamoto Declaration, adopted at the 4th Asia-Pacific Water Summit in 2022, emphasized the significance of investment and the need to double investment efforts to address water-related challenges.[1]

Despite these recommendations, increasing investment levels has posed challenges, necessitating a closer examination of investment policies, plans, and financial arrangements. Estimating financial gaps and evaluating investment benefits at regional and national levels in DRR is a complex task compounded by limited availability of budget data and uncertainties associated with climate and social change. Consequently, academic literature in this field is scarce, hindering the development of well-informed investment strategies.

This paper aims to address this research gap by focusing on estimating the financial gaps in flood protection infrastructure and analysing the benefits associated with such investments. It proposes methodologies for estimating the demand for flood protection infrastructure using actual budget data and assessing generated benefits. By bridging the knowledge gap in estimating financial gaps and benefits, this study strives to facilitate evidence-based decision-making in the field of DRR.

[1]This research is a part of the project "Research on Demand Estimate on Infrastructure in Asia" by the JICA Ogata Research Institute. The contents of this paper were released in JICA Ogata Research Institute working papers (Ishiwatari and Sasaki, 2021, 2020).

Improving our understanding of financial gaps and investment benefits will enable more effective investment policies, plans, and financial arrangements in DRR. Ultimately, this research endeavour aims to enhance the resilience of communities and infrastructure in the face of disasters by informing strategic investment decisions.

2 Estimate of demand for flood protection

Several studies have attempted to calculate the theoretical costs required for adequate flood protection to ensure a specific level of safety. However, these studies have yielded highly diverse results, differing by one to two orders of magnitude (Table 1). The methodologies employed in these studies play a significant role in shaping the outcomes, leading to substantial differences in cost estimations. Also, the scenarios considered, such as different flood magnitudes or varying levels of protection, contribute to the wide range of results.

It is important to note that these studies focus solely on estimating the necessary costs without considering the actual investment that has been made in flood protection. This omission highlights the need to bridge the gap between theoretical cost calculations and real-world investment practices, which this study covers.

3 Methods and data

3.1 Data collection

Obtaining comprehensive and consistent data on investment in flood protection presents several challenges, as there are no standardized datasets available, and budget data is often not publicly accessible. The definition of flood protection infrastructure may vary across different economies, leading to inconsistencies in the data. It is important to acknowledge these limitations in this study.

To gather data for this research, expert teams visited government offices and examined publicly available data in nine flood-prone economies in developing Asia: Bangladesh, People's Republic of China (PRC), India, Indonesia, Malaysia, Pakistan, Philippines, Thailand, and Vietnam. These economies were selected due to their significant exposure to flood risks. Total population and GDP of the economies account for over 90 % of these total amounts for developing economies in Asia. Budget data for Myanmar were not available for inclusion in this study.

The expert teams collected data on flood-related damage and death tolls from public statistics and disaster management offices. Typically, the economic damage data captured physical damages caused by floods. In cases where government data were unavailable, supplementary datasets were utilized, including SIGMA developed by Swiss Re Institute, NatCatSERVICE developed by Munich RE, and the EM-DAT: Emergency Events Database developed by the Centre for Research on the Epidemiology of Disasters.

All economic data were adjusted to 2015 prices using the GDP deflators to ensure consistency and comparability across the selected years. Population data were sourced from the UN World Population Prospects. Economic indicators such as GDP, per capita GDP, and deflators were obtained from the World Economic Outlook Database.

It is important to acknowledge that this study has certain limitations due to the lack of common datasets on investment in flood protection and the availability of budget data. The absence of standardized reporting frameworks and definitions for flood protection infrastructure across economies further complicates the analysis. For instance, the inclusion of landslide protection and coastal flooding within the category of flood protection varies among different economies, leading to inconsistencies in the data.

3.2 Methods for estimating investments in flood protection

The estimation of demand for flood protection infrastructure employed a regression model utilizing budget data. Equation (1) captures the relationship between various factors:

$$B_{i,t} = \alpha_1 B_{i,t-1} + \alpha_2 y_{i,t} + \alpha_3 \mathrm{Popden}_{i,t} + F_i + \epsilon_{i,t} \tag{1}$$

Where $B_{i,t}$ is the budget per capita of flood protection in economy ith at time t; $y_{i,t}$ is Gross Domestic Product per capita of economy ith at time t; $\mathrm{Popden}_{i,t}$ is the population density of economy ith at time t; F_i is a fixed effect of economy ith, and $\epsilon_{i,t}$ is an error term.

To estimate the regression values, a dataset was constructed for a 10-year period from 2006 to 2015, encompassing five economies: PRC, India, Indonesia, Pakistan, and the Philippines. The dataset included variables such as budget, income per capita, shares of agriculture and manufacturing value-added to GDP, urbanization, and population density.

Considering that flood damage is projected to increase in the Asian region due to climate change (Takakura et al., 2019), the demand for flood protection infrastructure is expected to rise as well. Since Eq. (1) does not include damage coefficients, a simplified method was used to estimate the impact of climate change. Equation (2) is an estimate of the average annual budget increase resulting from climate change effects from 2015 to 2030:

$$A_i = \beta I_i B_i \tag{2}$$

Where A_i is the average annual budget increase from climate change effects during 2015–2030 in economy ith; β is the budget increase rate per unit of damage; I_i is the increased ratio of economic damage in the economy ith; and B_i is the average demand of flood protection infrastructure per year in economy ith during 2015–2030 that disregards climate change effects.

The budget increase rate per unit of damage (β) was calculated as 0.5 based on the correlation between budget and

Table 1. Estimate of necessary annual costs in flood protection.

Literature	Cost (USD billion yr^{-1})	Target year	disaster		area	
			Flood	Coast flood	Asia	World
Rozenberg and Fay (2019)	19.1 to 76.0 to 110.1	2030	✓		✓	
Ward et al. (2010)	1.74–3.21	2050	✓		✓	
Hinkel et al. (2014)	12–31 to 27–71	2100		✓		✓
Hallegatte et al. (2013)	50	2050		✓		136 cities

Source: Modified after Ishiwatari (2021).

damage in the PRC and the Philippines from 2001 until 2015, where both disaster damage and budgets have been increasing in recent years. Alfieri et al. (2017) provided estimates of economic damage caused by river floods globally, incorporating each country's increased ratio of economic damage (Ii). For example, damage in China is estimated to increase by an average of 6 % between 2015 and 2030, while damage in Bangladesh is estimated to increase by 17 %. This estimate does not consider the effect of socioeconomic changes such as population projections, GDP growth, or land use changes.

3.3 Methods for estimating benefits

To estimate the benefits of future investment in flood protection in Asia, this study utilized the actual economic rate of return obtained from flood protection projects supported by the Japan International Cooperation Agency (JICA). JICA, as an implementing agency for Japanese Official Development Assistance, has conducted post-project evaluations for 30 flood protection projects supported by Japanese yen loans. These evaluations include the calculation of an economic internal rate of return for each project, with an average estimated rate of 15.0 % (Ishiwatari, 2016).

The economic internal rate of return represents the discount rate at which the stream of net economic returns, derived by subtracting costs from economic benefits, equals a present value of zero. By analysing the economic internal rate of return, the economic viability and benefits of projects can be assessed. Annual average damage reduction is cumulated based on the flood flow volume. This calculation involved multiplying the damage reduction at a specific flood flow by the probability of its occurrence. The benefits evaluated encompass both direct damage reduction to assets such as houses and factories, as well as indirect damage reduction, including business interruption.

Table 2. Results of the estimated regression.

Variables	Coefficient
Lagged value of budget	0.679203 (4.51818)***
GDP per capita	0.0020472 (2.77589)***
Population density	0.038973 (1.82801)*
Observations	45
R-squared	0.9595

*** $P < 0.01$, ** $P < 0.05$, * $P < 0.1$, Source: Ishiwatari and Sasaki (2020).

4 Results

4.1 Investment estimation

The results of the regression analysis estimating the demand for flood protection infrastructure in nine developing economies in Asia are presented in Table 2. The high R-squared value of 0.9595 indicates that the demand increases with the growth of the economy and the population, adding to the previous year's budget. P-values for previous year budget and GDP per capita are at 1 % level of significance, while one for population density is at the 10 % significance level.

The analysis indicates that the total demand for flood protection infrastructure from 2016 to 2030 is projected to reach USD 1417 billion, with an average annual investment requirement of USD 94.5 billion (Table 3).

The share of investment in flood protection as a percentage of GDP is expected to increase from 0.21 % in 2015 to an average of 0.36 % during the period from 2016 to 2030.

The analysis reveals that the demand increased for flood protection infrastructure, considering the increased effects of climate change, is estimated to average USD 3.9 billion yr^{-1} for the period 2016–2030. This signifies a 4 % increase in demand specifically attributed to the impacts of climate change (Table 3).

Table 3. Demand estimate in developing Asia (USD billion yr^{-1}, 2015 price).

	Investment in flood protection in 2015	Protection (ave. between 2016–2030)	Climate Change (ave. between 2016–2030)	Total (ave. between 2016–2030)
PRC	29.9	68.7	2.2	70.9
India	1.5	13.4	1.1	14.5
Philippines	1.1	3.0	0.1	3.1
Indonesia	0.5	3.1	0.2	3.3
Thailand	0.5	1.6	0.1	1.7
Vietnam	0.1	0.6	0.03	0.63
Malaysia	0.03	1.0	0.02	1.02
Bangladesh	0.007	2.1	0.2	2.2
Pakistan	0.009	0.9	0.01	0.91
Total	33.6	94.5	3.9	98.4

Source: Modified after Ishiwatari and Sasaki (2020).

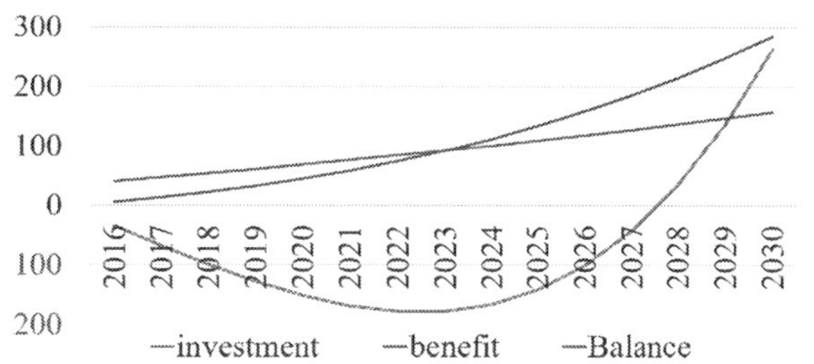

Figure 1. Benefit and investment in flood protection in Asia. Source: Ishiwatari and Sasaki (2021).

4.2 Benefits

The analysis demonstrates that the benefits of investing in flood protection infrastructure exhibit a positive trend over time, with the net benefit accumulation becoming favourable in the later stages of investment (Fig. 1). While the initial costs of implementing flood protection measures may outweigh the immediate benefits, the long-term outcomes show a favourable return on investment. This highlights the potential economic gains and improved resilience that can be achieved by investing in flood protection measures. The analysis reveals that the net benefit, calculated as the difference between the accumulated benefits and costs, turns positive in 2028. By the year 2030, the net benefit accumulation is estimated to reach a substantial value of USD 263 billion.

5 Discussion

The financing gap between future demands and current investment levels is around USD 61 billion annually or USD 65 billion when accounting for climate change effects or around 0.24 % of GDP in developing Asia. This amount is similar in magnitude to the World Bank's estimate by Rozenberg and Fay (2019). When the economies invest the amounts estimated, the net benefit accumulation is predicted to reach USD 263 billion in 2030. These findings can provide governments in Asia with evidence to strengthen their investment policies of flood protection. The projected annual demand for flood protection infrastructure, taking climate change into account, emphasizes the critical need for proactive measures to address the evolving risks associated with climate variability and extreme weather events. A significant return on investment affirms the economic viability and value of implementing flood protection measures. While upfront costs may pose initial challenges, the analysis highlights the potential for substantial economic benefits and improved disaster resilience in the future.

To ensure the steady decrease in damages caused by floods, it is crucial for governments to formulate policies that secure the budgets for flood protection in the long term. Integrating DRR into national development plans and formulating multiyear sectoral plans, as observed in the cases of PRC, the Philippines, and Japan, can help secure investment for flood protection (Ishiwatari and Surjan, 2019). Once the governments have confirmed these plans, budgets for flood protection can be allocated steadily every year, although each country's fiscal situation may fluctuate (Ishiwatari, 2021). Establishing a budget line dedicated to DRR is also useful. The Philippines has implemented DRR funds, which offer a dedicated financial resource that both national and local governments can utilize for both preparedness and recovery programs (Government of the Philippines, 2010). This innovative mechanism enables a sustained and continuous investment in DRR initiatives at a predetermined level. By providing a dedicated funding stream, the mechanism empowers governments to allocate resources effectively and consistently to enhance their disaster preparedness and response capabilities.

Innovative approaches, such as nature-based solutions, are needed to decrease costs and achieve sustainability in flood protection. Nature-based solutions, when combined with conventional engineered approaches, have shown to be more cost-effective and provide additional social and environmental benefits. An illustrative example can be found in New York City, where a synergistic blend of nature-based and conventional engineered strategies proves to be highly effective in reducing the cost of flood protection. Research conducted by Browder et al. (2019) demonstrates that this combined approach has the potential to save a substantial USD 1.5 billion, equivalent to a 22 % reduction compared to relying solely on engineered solutions. Beyond cost savings, this approach also yields a multitude of social and environmental benefits, encompassing areas such as fisheries, forestry, ecosystem protection, and recreation, as highlighted by the Global Commission on Adaptation (2019). By emphasizing research and development efforts, governments can further harness the power of innovation to drive down costs associated with DRR initiatives.

Engaging the private sector in financing flood protection is crucial, despite the fact that DRR measures may not yield immediate profits for private companies. Governments need

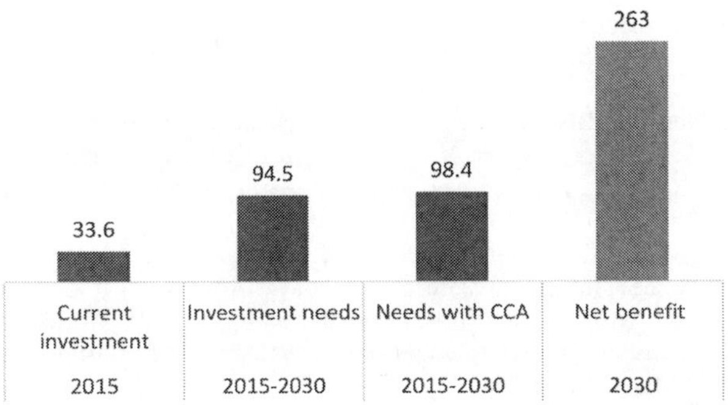

Figure 2. Key figures (USD billion). Source: Author's elaboration.

to establish mechanisms that incentivize private sector participation in flood protection, such as incorporating flood mitigation requirements in development projects or requesting private companies to contribute to infrastructure construction. Successful examples, such as the construction of retardation basins in Japan through collaboration with private companies, demonstrate the potential for public-private partnerships in achieving effective flood protection. The Tsurumigawa river basin, located in the Tokyo Metropolitan area, experienced rapid urbanization and the development of residential areas during a period of high economic growth in the 1950s and 60s. To mitigate the risk of flooding in this region, 4700 basins with a total capacity of over 3 million m^3 were constructed. These basins serve as essential flood retention structures, effectively managing and controlling the flow of water during heavy rainfall events. As a direct result of these flood protection measures, the number of houses affected by flooding has significantly decreased over the years. In the 1970s, several thousand houses were vulnerable to inundation, whereas in the 1990s and beyond, the number has been reduced to less than one hundred (Ishiwatari, 2016).

It is important to acknowledge the limitations of the methods employed in this study. The demand projection model did not consider non-economic factors such as geographical background and climate conditions, which may influence the demand for flood protection infrastructure. Additionally, the current methods of estimating benefits may not encompass all the benefits associated with flood protection programs, making it challenging to fully evaluate the saving of human lives and the mitigation of long-term indirect impacts.

Budget increases due to climate change impacts have been estimated in a simplified manner. More research is needed on the increase in damage due to climate change and the relationship between the increase in damage and the required budget.

6 Conclusions

This study estimated investment demand for flood protection infrastructure and its benefits in developing economies in Asia (Fig. 2). The results of the investment estimation analysis demonstrate the scale of investment required to effectively address the demand for flood protection infrastructure. The projected annual investment figures and the percentage of GDP allocated to flood protection underscore the significance of prioritizing and allocating resources towards DRR efforts. The positive net benefit indicates the potential to mitigate damages, enhance asset protection, and strengthen economic stability in the face of flood events.

The demand for flood protection infrastructure in nine developing economies was estimated using a methodology based on multiple regression analysis and time series data. The average annual demand for flood protection in these economies is projected to be USD 94.5 billion for the period 2016–2030. Taking into account the effects of climate change, this amount increases to USD 98.4 billion, representing a 4 % increase. The share of GDP of investment is expected to rise from 0.21 % in 2015 to an average of 0.36 % during the period 2016–2030. There exists a financing gap between the projected future demand and the current investment levels. The estimated annual shortfall is approximately USD 61 billion, or USD 65 billion when considering the impact of climate change. This financing gap represents around 0.24 % of the region's GDP.

The analysis demonstrates that these investments in flood protection are effective. The projected net benefit, calculated as the difference between the benefits and costs, is predicted to accumulate to USD 263 billion by 2030.

These findings highlight the substantial financial commitment required to address the demand for flood protection infrastructure in the region. The estimated investment levels provide valuable insights for policymakers and stakeholders involved in DRR and infrastructure planning, emphasizing the need for substantial financial resources to ensure the resilience and safety of communities in the face of increasing flood risks. By allocating resources effectively and implementing comprehensive flood protection strategies, policymakers and stakeholders can secure substantial economic returns, reduce vulnerability, and enhance the resilience of communities and economies in the face of flood events.

Further research activities should focus on the refinement and enhancement of the models used in this study. It is important to conduct additional studies that revisit the analysis of the benefits derived from investments in flood protection, taking into account the financial and socioeconomic impacts caused by the COVID-19 pandemic. The pandemic has introduced new challenges and uncertainties that may have implications for the effectiveness and economic viability of flood protection measures. The post-pandemic period presents an opportunity to reassess and strengthen investment strategies to ensure long-term resilience against flood risks.

Future research endeavours should cover developing standardized reporting mechanisms of budgets for flood protection and definitions for flood protection infrastructure across economies. This would enable the collection of more robust and comparable data, facilitating more accurate assessments of investment needs and outcomes. Collaboration among governments, international organizations, and research institutions is essential in establishing common frameworks and data-sharing protocols to enhance the accuracy and reliability of future analyses.

Data availability. The datasets generated and analysed during the current study are available from the corresponding author on reasonable request.

Author contributions. MI and DS substantially contributed to the study conceptualization. MI significantly contributed to data analysis and interpretation, and the manuscript drafting. Both authors critically reviewed and revised the manuscript draft and approved the final version for submission.

Competing interests. The contact author has declared that neither of the authors has any competing interests.

Disclaimer. Publisher's note: Copernicus Publications remains neutral with regard to jurisdictional claims made in the text, published maps, institutional affiliations, or any other geographical representation in this paper. While Copernicus Publications makes every effort to include appropriate place names, the final responsibility lies with the authors.

Special issue statement. This article is part of the special issue "ICFM9 – River Basin Disaster Resilience and Sustainability by All". It is a result of The 9th International Conference on Flood Management, Tsukuba, Japan, 18–22 February 2023.

Acknowledgements. During the preparation of this work the authors used ChatGPT in order to edit English. After using this tool/service, the authors reviewed and edited the content as needed and take full responsibility for the content of the publication.

Review statement. This paper was edited by Christophe Cudennec and reviewed by two anonymous referees.

References

Alfieri, L., Bisselink, B., Dottori, F., Naumann, G., de Roo, A., Salamon, P., Wyser, K., and Feyen, L.: Global projections of river flood risk in a warmer world, Earth's Future, 5, 171–182, https://doi.org/10.1002/2016EF000485, 2017.

Browder, G., Ozment, S., Bescos, I. R., Gartner, T., and Lange, G. M.: Integrating Green and Gray: Creating Next Generation Infrastructure, Washington DC: World Bank, World Resource Institute, http://hdl.handle.net/10986/31430 (last access: 4 March 2024), 2019.

Global Commission on Adaptation (GCA): Adapt Now: A Global Call for Leadership on Climate Resilience, Rotterdam, GCA, https://gca.org/reports/adapt-now-a-global-call-for-leadership-on-climate-resilience/ (last access: 4 March 2024), 2019.

Government of the Philippines: Philippine Disaster Risk Reduction and Management Act, Republic Act No. 10121, https://www.officialgazette.gov.ph/2010/05/27/republic-act-no-10121/ (last access: 4 March 2024), 2010.

Hallegatte, S., Green, C., Nicholls, R. J., and Corfee-Morlot, J.: Future flood losses in major coastal cities, Nat. Clim. Change, 3, 802–806, 2013.

Hinkel, J., Lincke, D., Vafeidis, A. T., Perrette, M., Nicollos, R. J., Tol, R. S. J., Marzeion, B., Fettweis, X., Ionescu, C., and Levermann, A.: Coastal flood damage and adaptation costs under 21st century sea-level rise, P. Natl. Acad. Sci. USA, 111, 3292–3297, 2014.

Ishiwatari, M.: How Disaster Resilience Can Contribute to Quality Growth Challenges Facing Disaster Risk Reduction Assistance, Tokyo, JICA Research Institute, https://jicari.repo.nii.ac.jp/records/912 (last access: 4 March 2024), 2016.

Ishiwatari, M.: Disaster Risk Reduction, in: Handbook of Climate Change Mitigation and Adaptation, edited by: Lackner, M., Sajjadi, B., and Chen, W. Y., New York, Springer, 2021.

Ishiwatari, M. and Sasaki, D.: Bridging the Gaps in Infrastructure Investment for Flood Protection in Asia, JICA Ogata Research Institute, https://www.jica.go.jp/Resource/jica-ri/publication/workingpaper/l75nbg0000199ldy-att/JICA-RI_WP_No.202.pdf (last access: 4 March 2024), 2020

Ishiwatari, M. and Sasaki, D.: Investments in flood protection: Trends in flood damage and protection in growing Asian economies, JICA Ogata Research Institute, https://www.jica.go.jp/Resource/jica-ri/publication/workingpaper/l75nbg000019v5g3-att/JICA_Ogata_Research_Institute_WP_No221.pdf (last access: 4 March 2024), 2021.

Ishiwatari, M. and Surjan, A.: Good Enough Today is not Enough Tomorrow: Challenges of Increasing Investments in Disaster Risk Reduction and Climate Change Adaptation, Prog. Disaster Sci., 1, 100007, https://doi.org/10.1016/j.pdisas.2019.100007, 2019.

Rozenberg, J. and Fey, M. (Eds.): Beyond the Gap: How Countries Can Afford the Infrastructure They Need while Protecting the Planet, Sustainable Infrastructure Series, Washington, DC, World Bank, http://documents.worldbank.org/curated/en/189471550755819133/Beyond-the-Gap-How-Countries-Can-Afford-the-Infrastructure-They-Need-while-Protecting-the-Planet (last access: 4 March 2024), 2019.

Takakura, J., Fujimori, S., Hanasaki, N., Hasegawa, T., Hirabayashi, Y., Honda, Y., Iizumi, To, Kumano, N., Park, C., Shen, Z., Takahashi, K., Tamura, M., Tanoue, M., Tsuchida, K., Yokoki, H., Zhou, Q., Oki, T., and Hijioka, Y.: Dependence of economic impacts of climate change on anthropogenically directed pathways, Nat. Clim. Change, 9, 737–741, https://doi.org/10.1038/s41558-019-0578-6, 2019.

United Nations International Strategy for Disaster Reduction (ISDR): Sendai Framework for Disaster Risk Reduction 2015–2030, Geneva, UNISDR, https://www.preventionweb.net/files/43291_sendaiframeworkfordrren.pdf (last access: 4 March 2024), 2015.

Ward, P. J., Pauw, P., Brander, L. M., Aerts, J. C. J. H., and Strzepek, K. M.: Costs of Adaptation Related to Industrial and Municipal Water Supply and Riverine Flood Protection, Climate change discussion paper No. 6, Washington DC, World Bank, http://documents.worldbank.org/curated/en/313221468326686038/Costs-of-adaptation-related-to-industrial-and-municipal-water-supply-and-riverine-flood-protection (last access: 4 March 2024), 2010.

Proc. IAHS, 386, 95–100, 2024
https://doi.org/10.5194/piahs-386-95-2024

Open Access PIAHS

Hydrodynamic Investigation of Laguna Lake, Philippines for Water Security and Flood Risk Management of Metro Manila

Eugene C. Herrera[1] **and Kensuke Naito**[2]

[1]Institute of Civil Engineering, University of the Philippines, Diliman Quezon City, 1101, Philippines
[2]International Center for Water Hazard and Risk Management – ICHARM, Public Works Research Institute (PWRI), Tsukuba, Ibaraki, Japan

Correspondence: Eugene C. Herrera (echerrera1@up.edu.ph)

Received: 20 July 2023 – Revised: 7 November 2023 – Accepted: 24 November 2023 – Published: 19 April 2024

Abstract. Laguna Lake, strategically located at the center of urban development, Metro Manila, is the focal point of national development efforts for economic activities, including flood disaster operation. It is the most stressed inland water body at the same time however, with competing and conflicting water-users, and continued environmental degradation from anthropogenic-based stressors. The use of the lake as a source of raw water for domestic supply have become more significant and critical over the past recent years with the aggravating condition of water scarcity in Metro Manila. It also serves as an important flood-detention storage for Metro Manila during the rainy season, with flow diversions from upstream Marikina river basin. This study presents a three-dimensional hydrodynamic study of Laguna Lake to understand the physical processes that govern and influence lake hydraulics for various water uses. A hydrodynamic model was set-up incorporating spatial and temporal variations in wind forcing and watershed river discharges, and dynamic sea interaction. Simulation results reveal a significant increase in lake salinity concentration for incremental increases in lake water abstraction, particularly during the dry season. A numerical experiment on bathymetric evolution also showed various Pasig River hydraulic discharge features and an increase in hydraulic residence time of the lake. Modelling analysis on the lake's flood detention function on the other hand revealed different shore-land flood inundation areas and recession durations for varying lake-flood detention levels that warrants the need for a more comprehensive and sustainable flood management operation in the metropolis. Results from this study substantiate scientific information for outlining management approaches and conservation strategies especially for fast developing environments.

1 Introduction

Laguna Lake with substantial economic value for water resource development is the most important natural resource in the Philippines. Situated adjacent to a thriving mega city, Metro Manila, it is the focal point of national development efforts not only in the agriculture and fishery, water supply and energy sectors, but in the regional development program as well (Santos-Borja and Nepomuceno, 2003). However, it is the most stressed inland water body at the same time with competing and conflicting water-users, and continued environmental degradation from anthropogenic-based stressors. The lakes' dynamic interaction with Manila Bay through Pasig River makes it ideal for fishery and aquaculture with the entry of salt-water, but at the same time discharges nutrient-rich polluted waters of Metro Manila and surrounding coastal provinces to the lake (Tabios III, 2020). The lake has also been traditionally utilized as a water resource for irrigation, power-plant cooling, hydropower generation, recreation, and transport route for people and products. It also serves as flood-detention storage for Metro Manila during the rainy season. More recently, it has been tapped as a source of raw water for domestic supply. Discussions on maximizing the use of the lake for this function have become more significant and critical over the past recent years with the aggravating condition of water scarcity in Metro Manila and the vicinity. With the lake's invaluable water resource, competition and conflicts among its stakeholders are common occurrences. Approximately twenty million inhabitants depend on the lake for environmental goods and services.

Flooding occurs when lakes, riverbeds, soil and vegetation cannot absorb all the water. In the case of Laguna Lake, it serves as a detention storage of floodwaters that drains the Pasig-Marikina-Laguna Lake basins during intense rainfall events. Mangahan Floodway is opened for flood diversion to Laguna Lake through the Rosario Weir when Sto. Nino (station along Marikina River) water elevation reaches 4.5 m a.s.l. (Fig. 1). By design, the Mangahan Floodway is capable of handling $2400\,m^3\,s^{-1}$ of water flow. To complement the floodway, the Napindan Hydraulic Control Structure (NHCS) was built in 1983 at the confluence of the Marikina River and the Napindan channel of the Pasig River to regulate the storage of flood waters in the lake during extreme rainfall events, and at the same time the tidal flow of saline water from Manila Bay to prevent the intrusion of polluted water into the lake during the dry season. It has a fully gated diversion dam at its head and was designed with a width of 260 m. Over 40 000 households are situated along the floodway's banks and these shoreline slums have reduced its effective width to 220 m. The flood-control system's operational rule, that is based on flood depths and channel discharge capacities, has not been updated since its first operation in 1986. With the competing and conflicting use of the lake waters, management of this flood-detention function has not been carried-out effectively. NHCS which was to close its gates after diverting floodwaters from Marikina River (through Mangahan Floodway) to detain water temporarily in the lake as part of flood operation were almost never closed to allow saline water to enter the lake since it benefits the aquaculture industry and fisheries sector. Thus, the NHCS was practically never operated for purposes of tidal flow regulation or pollution control.

By carrying flood waters to Laguna Lake, the Mangahan Floodway lessens flood conditions in Metro Manila but contributes to flooding of the coastal areas of Taguig, Taytay, and other towns in Laguna and Rizal along the lake. Incidents of severe floods became more frequent and lasted longer in these areas since its construction. An unusual large flood occurred in October/November 1986, lasting for 2 months and resulting in high mortality and morbidity rates due to gastroenteritis and other water-borne diseases (Jorgensen and Vollenweider, 1988).

Furthermore, pollution and sediments carried by the floodway has jeopardized the existing and potential uses of the lake. The sedimentation rate of the lake is estimated at 1.5 million $m^3\,yr^{-1}$ with the Marikina River as a major contributor of silt to the lake through the Mangahan Floodway (Santos-Borja and Nepomuceno, 2003). Additional pollution and siltation come from the shoreline settlers, living in slums up to 5 rows deep, whose waste goes directly into the floodway (JICA, 2018).

The pressures of both growing user-demand and declining water quality laid ground the value of scientific knowledge on lake environmental-functioning for the optimal and sustainable use of the lake water resources. This study investigates the hydrodynamic features of Laguna Lake with the use of high-resolution numerical modelling analyses. The study aims to clarify the circulation and transport features of the lake and provide information on the hydraulics of the lake for different lake-water use and assess critical lake ecosystem conditions.

2 Methodology

This paper presents a three-dimensional hydrodynamic study of Laguna Lake to properly understand the physical processes that govern and influence the circulation and mass transport characteristics of the lake for various water uses. The model was set-up to take into account the physical effects of natural climatic and hydrologic factors such as discharges from rivers, direct lake rainfall and evaporation, lake groundwater interaction, wind induced circulation, and tidal fluctuation at Manila Bay. Fish pen and fish cage structures used for aquaculture, affecting lake circulation were also incorporated into the model, in addition to different lake water withdrawals. The hydrodynamic component of Delft3D, FLOW was utilized to simulate the physics of the lake. Delft3D-FLOW solves the Navier Stokes equations for an incompressible fluid, under the shallow water and the Boussinesq assumptions. In the vertical momentum equation, the vertical accelerations are neglected, which leads to the hydrostatic pressure equation. In 3D models, the vertical velocities are computed from the continuity equation. The set of partial differential equations in combination with an appropriate set of initial and boundary conditions is solved on a finite difference grid. Reference is made to the technical reference manual (Delft Hydraulics, 2007) for these formulations. The unsteady shallow-water equations were solved in 3D with a vertical sigma-coordinate system laid-out on a horizontal orthogonal curvilinear grid. Flow is forced by tide at the open boundary, wind stress at the free surface, and pressure gradients due to free surface (barotropic) and density (baroclinic) gradients. Sediment transport and morphologic changes were not included in the simulations. The hydrodynamic model was calibrated and validated using hydrographic data from field surveys and long-term monitoring information (Herrera et al., 2015; Japan International Cooperation Agency, 2018).

3 Results and discussion

3.1 Flood detention operation

Laguna Lake serves as a flood-detention storage for Metro Manila (Fig. 1). Excess runoff discharge from Marikina

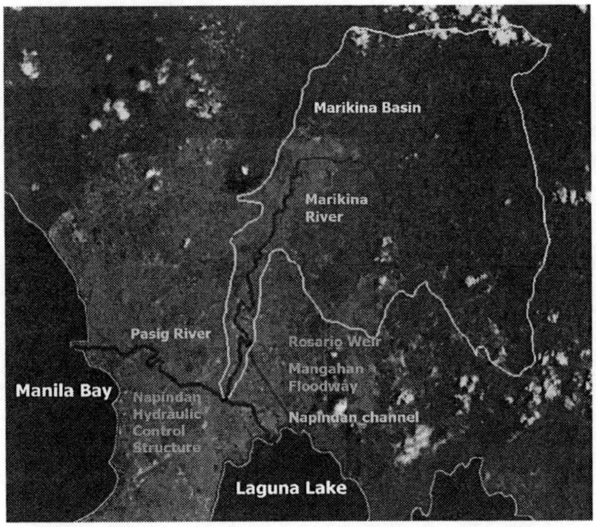

Figure 1. Metro Manila flood mitigation infrastructures and surrounding watersheds and water bodies.

Basin is diverted to the lake for temporary storage until flood waters from the watershed recedes. In 26 September 2009, typhoon Ketsana (local name Ondoy) dumped about 45 cm of rain in 12 h that resulted to widespread flooding. A 1.2 m increase in lake level that occurred due to the flood detention operation took months to recede causing flood waters in shoreland areas to last until January of the following year (Figs. 2 and 3). Although, the volume of water generated by Ketsana would have resulted to flooding nonetheless even with a fully operational flood management and control system, there are measures that can be taken to reduce the impact of such extreme event to a predetermined safety level. Among these include effective regulation of both Mangahan Floodway and NHCS operation using updated hydraulic information, enhancement of the monitoring and early warning system within the Pasig-Marikina-Laguna Lake basin, increasing the flow capacity of Napindan channel, and others. Scenario simulations revealed different shore-land flood inundation areas and recession durations for varying lake flood detention levels (Fig. 3 and Table 1). Pasig takes the longest time to recede back to normal water levels (10.5 m based on Laguna Lake Development Authority datum) at roughly 219 d. It is followed by Cainta at 194–204 d, then Napindan at 169 d, Taytay at 141 d, Pateros at 119 d and Taguig at 40–84 d. Other flooding hydraulic details from the hydrodynamic modelling conducted are presented in Table 1.

Although the operation of these existing flood control infrastructure (Rosario Weir, Mangahan Floodway, and NHCS) during typhoon Ketsana lessened the flood conditions in Metro Manila, it significantly contributed to flooding of the coastal areas around Laguna Lake. Based on the results of the numerical simulation (Fig. 3 and Table 1), there is an apparent need to effectively manage the operation of the lake for flood detention so as not to cause flooding to shoreland communities. The results also call for immediate action on updating the operational rule of the flood control system and improving the discharge capacities of the critical channels in the system, particularly the Mangahan Floodway, Napindan channel, Pasig River and lower Marikina River.

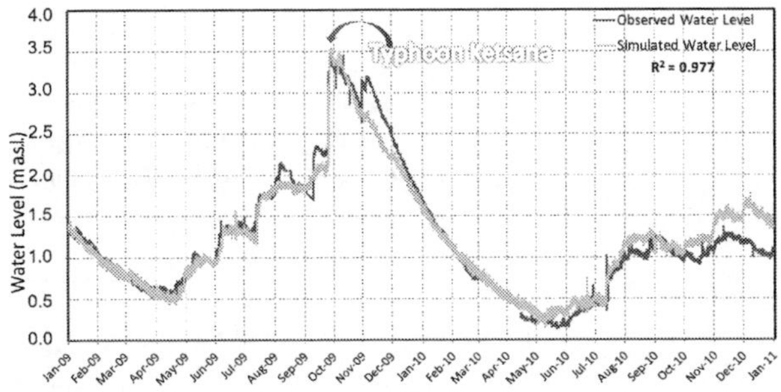

Figure 2. Laguna Lake water level simulation and validation of the Typhoon Ketsana event.

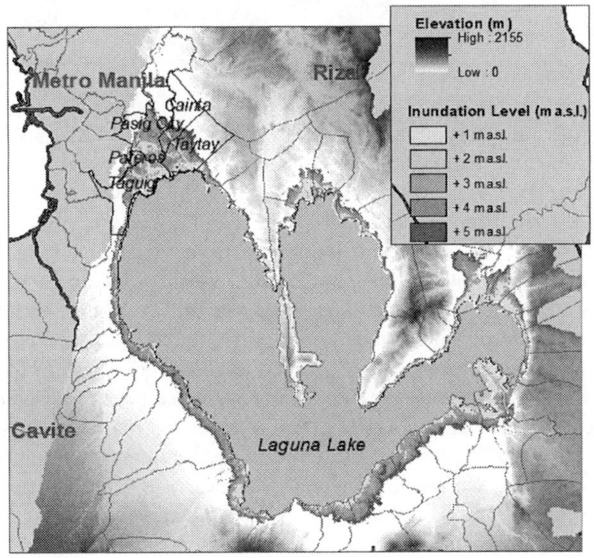

Figure 3. Laguna Lake shoreland flooding at different inundation levels.

3.2 Bathymetric change and eutrophic vulnerability

The watershed of the lake has undergone significant changes from population growth, urbanization, industrialization, and land use conversion, resulting to increased siltation, eutrophication and pollution (Nauta et al., 2003). Rivers emptying to the lake unload wastes, organic matter, and sediments that settle and accumulate on the lakebed. A hydrodynamic model experiment for different lake bathymetric configuration (1963, 1983 and 2008) showed various Pasig River

Table 1. Hydrodynamic matrix of Laguna Lake during flood detention operation.

Water Level (m a.s.l.)	Lake Area ($\times 10^6$ m^2)	Lake Volume ($\times 10^6$ m^3)	Change in Elevation (m)	Change in Flood Area ($\times 10^6$ m^2)	Pasig River Outflow ($\times 10^6$ $m^3\,d^{-1}$)	Time to Recede to normal (d)	Flooded Municipalities
3.5	1101.7	5573.1	–	–	32.3	219	Pasig
3.3	1087.7	5354.4	0.2	14.0	31.5	204	Cainta
3.0	1067.6	5031.0	0.3	20.1	29.4	194	Cainta
2.5	1033.6	4506.3	0.5	34.0	25.3	169	Napindan
2.0	998.0	3998.0	0.5	35.6	22.5	141	Taytay
1.5	960.9	3508.0	0.5	37.0	19.5	119	Pateros
1.0	923.3	3037.1	0.5	37.6	11.9	84	Taguig
0.5	891.3	2582.9	0.5	31.9	5.6	40	Taguig
0.0	849.1	2146.8	0.5	42.2	2.1	–	–

hydraulic discharge features and an increase in hydraulic residence time of the lake in terms of salinity distribution (Fig. 4). The historical change in lake bathymetry from years of deposition, erosion and reworking of sediments brought forth the changes in circulation pattern that led to the longer residence time of the lake. Erosion and deepening in the northern west and east lobes of the lake in particular could have retarded lake outflow-directed circulation. More polluted seawater was also observed to be intruded during the dry season due to the longer lake residence time. Seawater entry demonstrated larger extent in intrusion plume as a result which further established the increase in eutrophic vulnerability and higher potential for adverse response to eutrophication of Laguna Lake in recent years. Thus, it is revealed from the numerical experiment that eutrophic vulnerability in Laguna de Bay is not a simple function of anthropogenic waste loading but watershed sediment input as well with the accompanying changes in circulation patterns. Results from this numerical scenario analysis warrants the need for a better management of the sediment and pollution loading of rivers to the lake.

3.3 Increasing lake abstractions and lake hydrodynamics

Scenario simulations were also conducted to study the impact of increased lake water abstractions on the lake hydrodynamics, particularly the entry of polluted salt water from Manila Bay through the Pasig River. Simulation results reveal a significant increase in lake salinity concentration for incremental increases in lake water abstraction, particularly during the dry season. Different abstraction scenarios corresponded to different current movement patterns and concentration distributions of the lake saltwater plume (Figs. 5 and 6). With the drop in lake water level by about 1.5–2.0 cm for every 400 million $L\,d^{-1}$ lake water abstraction during the dry season, Pasig River backflow (Manila Bay to Laguna Lake) is induced by the intensified hydrodynamic imbalance between the lake and the bay resulting to higher incidence of lake salinity intrusion. With the drop in annual average Pasig River outflow (Laguna Lake to Manila Bay), the eventual exit of saline water from the lake (the lake hydraulic residence time), will take a longer period compared to the hydrodynamic condition without abstraction, increasing Laguna Lake's eutrophic vulnerability. In spite of the sheer volume of the lake waters, and its high potential for raw water abstraction and use, a scientific-based permitting guideline needs to be in place to sustainably manage requests for water abstraction and prevent any deleterious effects on the lake's environmental functioning.

4 Conclusions

A 3-dimensional hydrodynamic model was set-up for Laguna Lake, Philippines using Delft-3D to properly understand the physical and environmental processes that govern and influence the lake dynamics. The water balance of the lake was analyzed by taking into account the physical effects of all natural climatic and hydrologic factors, as well as man-made interventions. Field measurement data on various water quality and hydrodynamic parameters were used for the calibration and validation of the model. Observed lake parameters were reproduced by the model within acceptable limits of accuracy.

In general, in spite of the data limitations, the hydrodynamic set-up and simulation results showed that the hydrodynamic model can describe the water balance and transport processes in Laguna Lake within sound degree of accuracy. The validated model was able to provide useful information for understanding the dynamic condition of the lake and for predicting the impacts of future man-made interventions, such as temporary flood detention and water abstraction for domestic use. Baseline information has been generated to form the scientific foundation of probable protective

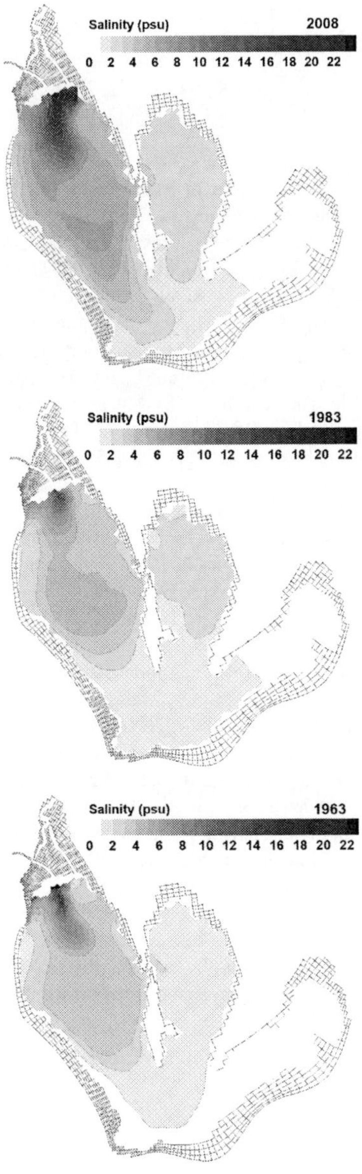

Figure 4. Hydrodynamic simulation for seawater intrusion plume of the 2008, 1983 and 1963 lake bathymetric scenarios.

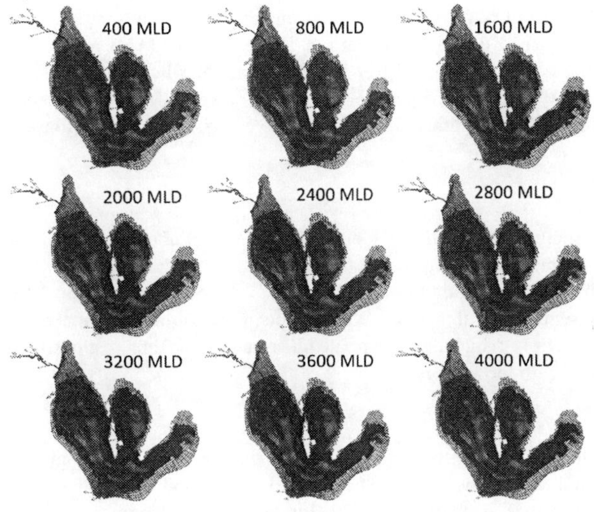

Figure 5. Laguna Lake circulation patterns for various abstraction rates.

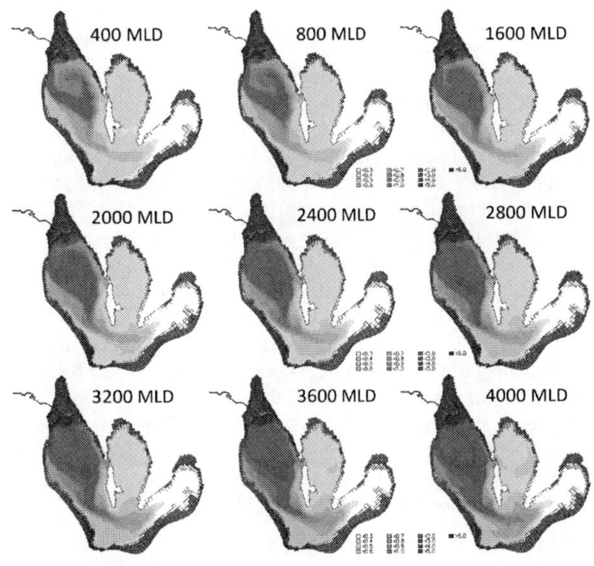

Figure 6. Laguna Lake dry season salinity plume in parts salinity units for various abstraction rates.

regulations and policies for Laguna Lake. Considering Laguna Lake is a multi-purpose reservoir, water allocation and flood-detention function require greater analysis and understanding of the dynamics of its environment for it to be managed sustainably. Thus, deeper and better understanding of the lake dynamics generated can serve as a sound basis for lake management options and for further research undertakings, essential for the lake's optimal and sustainable use as a multi-purpose resource, while preserving its environmental integrity at the same time.

Data availability. Only processed data directly used for the final version of the model are available upon request. Raw data are to be gathered from the agencies from which the authors collected such datasets.

Author contributions. ECH conceptualized, set up and simulated the models, conducted the extraction, analysis, and documentation of model results. KN reviewed the abstract and supported the overall study for its submission to the ICFM9 conference.

Competing interests. At least one of the (co-)authors is a guest member of the editorial board of *Proceedings of IAHS* for the special issue "ICFM9 – River Basin Disaster Resilience and Sustainability by All". The peer-review process was guided by an independent editor, and the authors also have no other competing interests to declare.

Special issue statement. This article is part of the special issue "ICFM9 – River Basin Disaster Resilience and Sustainability by All". It is a result of The 9th International Conference on Flood Management, Tsukuba, Japan, 18–22 February 2023.

Acknowledgements. The authors would like to express gratitude to Laguna Lake Development Authority, to their former Integrated Water Resources Management Division for all the assistance and logistical support during the field surveys, as well as the maintenance of the IMSWES monitoring platform. Appreciation is also given to the Department of Science and Technology (DOST) for facilitating the entry of instruments. This research was supported by The Japan Society for the Promotion of Science (JSPS) Grant-in-Aid for Scientific Research [No. 18254003, No. 20121007 and No. 21254002]; JSPS Core University Exchange Program; Asia-Pacific Network (APN) Grant [ARCP2006-08NMYNadaoka]; Science and Technology Research Partnership for Sustainable Development (SATREPS) in collaboration between Japan Science and Technology Agency (JST, JPMJSA1909) and Japan International Cooperation Agency (JICA); and the Department of Science and Technology (DOST) and DOST Philippine Council for Industry, Energy and Emerging Technology Research and Development (DOST-PCIEERD) through the project Eco-System Modelling and Material Transport Analysis for the Rehabilitation of Manila Bay (e-SMART) under the IM4ManilaBay Program.

Financial support. This research has been supported by the Science and Technology Research Partnership for Sustainable Development (grant no. JPMJSA1909) and the Department of Science and Technology, Philippines (grant no. Eco-System Modelling and Material Transport Analysis for the Rehabilitation of Manila Bay (e-SMART) Project).

Review statement. This paper was edited by Shinji Egashira and reviewed by two anonymous referees.

References

Delft Hydraulics: Simulation of multi-dimensional hydrodynamic flows and transport phenomena, Delft3D-FLOW User Manual Version 3.14, WL/ Delft Hydraulics, the Netherlands, 642 pp., https://content.oss.deltares.nl/delft3d4/Delft3D-FLOW_User_Manual.pdf (last access: 4 March 2024), 2007.

Herrera, E. C., Nadaoka, K., Blanco, A. C., and Hernandez, E. C.: Hydrodynamic Investigation of a Shallow Tropical Lake Environment (Laguna Lake, Philippines) and Associated Implications for Eutrophic Vulnerability, ASEAN Engineering Journal, 4-1, 48–62, https://doi.org/10.11113/aej.v4.15432, 2015.

Japan International Cooperation Agency: Resettlement Action Plan for Mangahan Floodway, Pasig-Marikina River Channel Improvement Project Phase IV Report, 1-1–1-3, September 2018.

Jorgensen, S. E. and Vollenweider, R. A. (Eds.): Principles of Lake Management, in: Guidelines of Lake Management, International Lake Environment Committee Foundation, United Nations Environment Programme, 191 pp., ISBN: 4-906356-27-3, 1988.

Nauta T. A., Bongco, A. E., and Santos-Borja, A. C.: Set-up of a decision support system to support sustainable development of the Laguna de Bay, Philippines, Mar. Pollut. Bull., 47, 211–218, https://doi.org/10.1016/S0025-326X(02)00407-1, 2003.

Santos-Borja, A. and Nepomuceno, D.: Experience and Lessons Learned Brief for Laguna de Bay, Philippines, Presented in the Lake Basin Management Initiative ILEC/LakeNet Regional Workshop for Asia: Sharing Experience and Lessons Learned in Lake Basin Management, Manila, Philippines, 1–4 September 2003, 225–258, https://iwlearn.net/documents/6001 (last access: 4 March 2024), 2003.

Tabios III, G. Q.: Water Resources Systems of the Philippines: Modeling Studies, 4, 403 pp., ISBN 978-3-030-25400-1, Springer Nature, Switzerland, https://doi.org/10.1007/978-3-030-25401-8, 2020.

Proc. IAHS, 386, 101–107, 2024
https://doi.org/10.5194/piahs-386-101-2024

A Method for Municipalities to Enhance Disaster Resilience for Potential Future Floods Using a Climate Projection Database

Akihiro Suzuki[1], **Fumihiko Uemura**[1,2], **Tsuyoshi Hoshino**[3], **Taro Yamamoto**[4], **and Tomohito J. Yamada**[1]

[1]Faculity of Engineering, Hokkaido University, Sapporo, 060-8628, Sapporo, Japan
[2]Docon. Co. Ltd., 060-0042, Sapporo, Japan
[3]The Civil Engineering Research Institute for Cold Region, Sapporo, 062-7602, Sapporo, Japan
[4]Hokkaido River Center, 060-0807, Sapporo, Japan

Correspondence: Akihiro Suzuki (suzuki1122@eng.hokudai.ac.jp)

Received: 31 July 2023 – Revised: 20 September 2023 – Accepted: 24 February – Published: 19 April 2024

Abstract. Issuing appropriate evacuation information is a daunting task for municipalities during unprecedented heavy rainfalls. Due to their limited experience, they will encounter greater challenges in responding appropriately in future climate conditions, where the scale of floods will be greater. In this study, we proposed a method to make flood scenarios based on a climate projection database. We also developed an exercise method for municipalities to consider issuing evacuation information using the scenarios. Through a demonstration experiment in Hokkaido, Japan, we conducted a simulation exercise based on the method for a municipality. As a result, they considered specific response and provided valuable insights, thereby enhancing preparedness for potential future flood disaster.

1 Introduction

In recent years, record-breaking heavy rainfalls frequently occurred in Japan. Over Hokkaido, northern-most island of Japan, severe flooding occurred due to four consecutive typhoons, especially Typhoon Lionrock in 2016. After the disaster, the central and local government established a committee to discuss flood-control measures based on specific flood-risk assessments, such as the inundation probability in past and future climate conditions based on climate projections (Yamada, 2019). Also, the Ministry of Land, Infrastructure, Transport, and Tourism of Japan (MLIT) launched a project to promote the concept of "River Basin Disaster Resilience and Sustainability by All" for all class-A river basins in Japan. The project aims to mitigate flood risks with three type of measures: flood prevention (e.g., river levees), exposure reduction (e.g., moving out of high risk area), and disaster resilience (e.g., evacuation), in collaboration with all stakeholders of the basin (Koike, 2021). If flood prevention and exposure reduction measures are insufficient (e.g., dike failure), disaster resilience measures become necessary. Municipalities issue evacuation information, based on the lead-time necessary for evacuation, first for vulnerable residents (Evacuation of the Elderly, etc.), then Evacuation Instruction is issued for every residents. However, during unprecedented heavy rainfalls, municipalities are unable to evaluate the situation correctly, thereby limiting evacuation information (Sato et al., 2011).

We discussed emergency response with municipalities using flood scenarios targeted at big rivers under future climate conditions, as calculated by Yamada (2019). As a result, it was suggested that municipalities are also concerned about flooding of small rivers, and that to consider evacuation information at specific times and areas, disaster-related information, such as warnings, is necessary (Suzuki et al., 2022). The objective of this study is to develop a disaster resilience method for potential future flood disasters. We proposed a method for making flood scenarios, which encompass small river flooding, based on climate projections. Additionally, we developed an exercise method for municipalities to contemplate the issuance of evacuation information, utilizing these scenarios to enhance their preparedness and gain

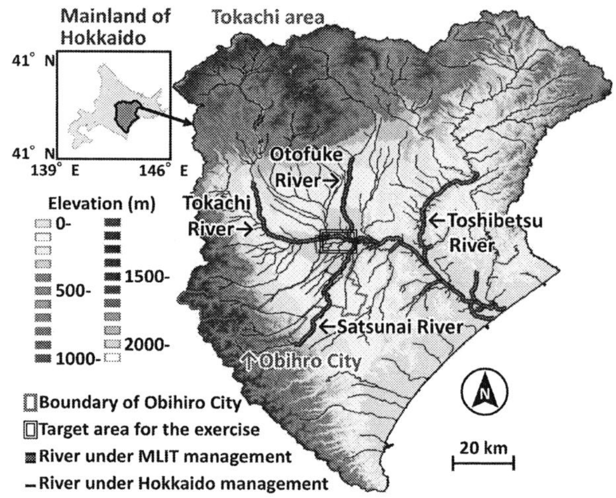

Figure 1. Elevation of Tokachi Subprefecture, river channels, and the location of the target area for the exercise method. (The elevation and river line data was provided by the Geospatial Information Authority of Japan).

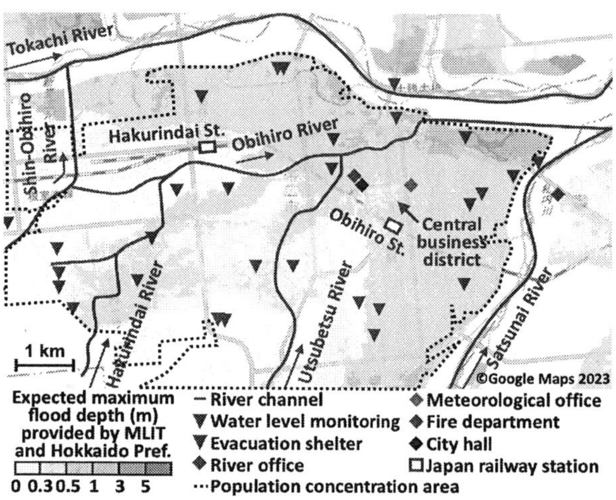

Figure 2. The target area includes information on maximum expected flood depths, river channels, water level monitoring points, flood evacuation shelters, key public facilities, and areas with high population concentration.

practical experience. The detailed procedures for conducting such exercises based on this method have been outlined in Suzuki et al. (2023). Furthermore, we conducted a simulation exercise using this method in a demonstration experiment in Hokkaido, Japan. In this paper, besides explaining these methods, we elaborate on the results of the simulation exercise and discuss the outcomes achieved, along with the challenges and further utilization ideas.

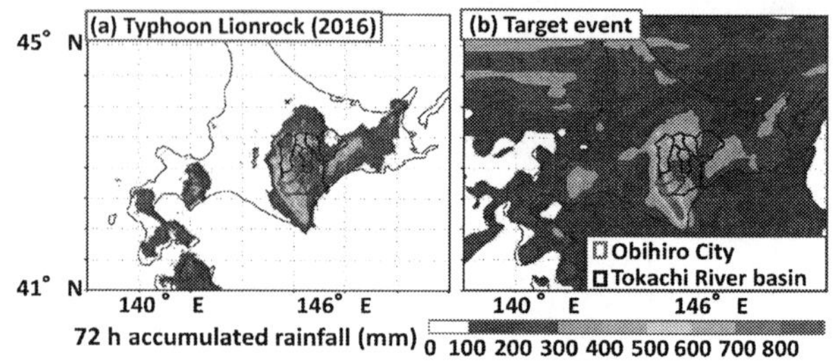

Figure 3. Peak rainfall amount over Tokachi River basin for 72 h in the target event and rainfall caused by Typhoon Lionrock (2016), and the location of the basin and Obihro City.

2 Methods

2.1 Target area

The target area for the simulation exercises was Obihiro City, in Tokachi Subprefecture, Hokkaido, and with the largest population in the subprefecture. Figure 1 shows the elevation and class A river channels, along with the target area. The city is located between 2 big rivers the Tokachi River, flowing on the north side, and the Satsunai River, flowing on the east side. Figure 2 shows the target area along with the maximum expected flood depth simulated by MLIT. It also marks river channels, water level monitoring point, flood evacuation shelters, key public facilities, and the population concentration areas. The population concentration areas, defined by the Ministry of Internal Affairs and Communications, have a population density of over 4000 people per square kilometer, including adjacent areas, with a combined population exceeding 5000. Due to the high risk of significant flooding in the city's core, there is substantial flooding risk

2.2 Data

To make potential future flood scenarios we used 5 km spatial resolution rainfall data under future climate conditions, with an average global temperature increase of 4° compared to pre-1850, from the Database for Policy Decision Making for Future Climate Change (d4PDF) (Kawase, 2023). This meteorological dataset contains thousands of years of past and future climate conditions, enabling the statistical inference of the intensity and frequency of extreme weather events (Mizuta et al., 2017; Yamada et al., 2021).

We selected a rainfall event with a route similar to that of Typhoon Lionrock (2016) and a spatial distribution of rainfall, but the scale of rainfall is greater. Figure 3 shows the peak rainfall amount over the basin for 72 h in the target rainfall event and rainfall caused by Typhoon Lionrock. The peak rainfall amount for the selected event is almost 1.5 times greater than that of Typhoon Lionrock. In Fig. 4, the route of the typhoon in the target rainfall event and the Typhoon Lionrock of 2016. In the d4PDF dataset, histori-

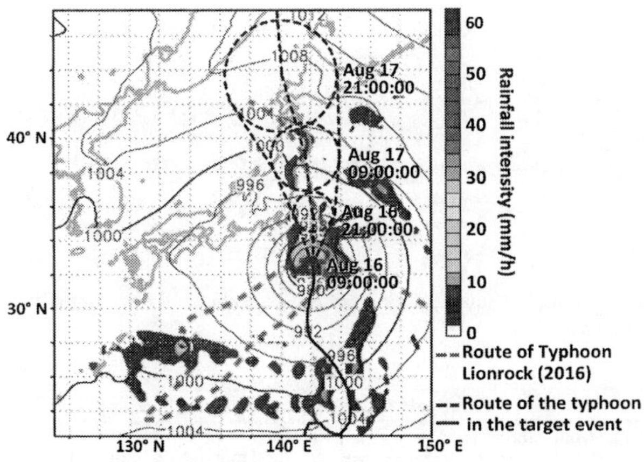

Figure 4. The routes of Typhoon Lionrock (2016) and the typhoon in the target rainfall event and sea-surface pressure in a weather report provided at start of the exercise method.

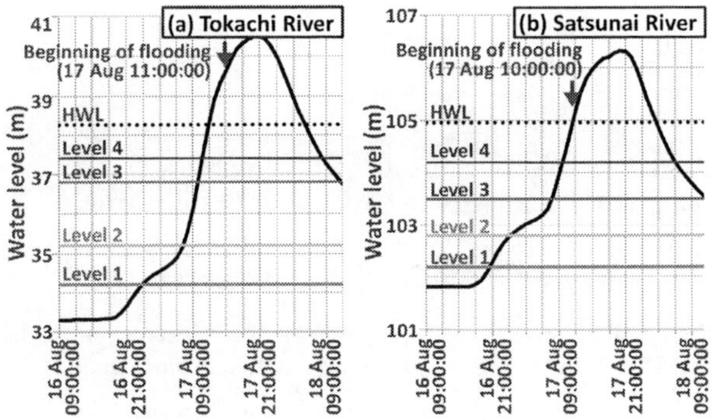

Figure 5. Water levels, and the water-level criteria of Tokachi and Satsunai Rivers at the monitoring point.

cal experiments spanning 3000 years identified 37 typhoons have similar route to Typhoon Lionrock, and in future experiments covering 5400 years, 64 case were found. This "similar route" is defined, following Hoshino et al. (2020), as typhoon with path around the Tohoku region (35–40° N) deviating by no more than 1° in longitude from Typhoon Lionrock's route. Typhoons approaching and making landfall in Hokkaido from the Pacific Ocean generally exhibit less pressure drop, increasing the likelihood of significant rainfall, a concern by municipal officials (Kitano et al., 2017).

2.3 Potential future flood scenario

The flood scenario is composed of rainfall intensity, river water level, and flood depth data at a 1 h resolution. The target rivers include large rivers managed by MLIT (the Tokachi, Satsunai Rivers) as well as relativity small rivers managed by Hokkaido Prefecture (the Hakurindai, Utsubetsu, Obihiro, and Shin-Obihiro Rivers), as shown in Fig. 2. For the rivers managed by MLIT, where survey and observation are well-developed, we calculated water level and flood depth based on the storage function method, kinematic wave model, and two-dimensional plane flow analysis (Yamada, 2019). For the rivers managed by Hokkaido, we calculated flood depth based on the two-dimensional plane flow analysis using the Nays2DFlood solver of the International River Interface Cooperative (iRIC) software, which allows for the unified analysis of river channels and floodplains (Shimizu, 2014). These two methods are employed for rivers managed by MLIT and Hokkaido for calculating the expected maximum flood depth, which is utilized in the creation of flood hazard maps for municipalities. The calculation model was provided by the MLIT, and it was developed in accordance with a manual for calculating the expected maximum flood depth (National Institute for Land and Infrastructure Management, 2017). The flood depth for the flood scenario is obtained by superimposing the maximum flooding depth of the target rivers at the same time.

Please note that all times listed in this paper are in Japan Standard Time (JST). Also, in the d4PDF for future climate conditions, the year number has no meaning and is not described. Figure 5 shows the river water level at the monitoring points as shown in Fig. 2 ,of the Tokachi and Satsunai Rivers. As shown in Fig. 5, the water level of the two rivers began to rise around 16 August at 18:00:00. Additionally, water-level criteria are shown in Fig. 5, that is equivalent to Alert Level that is a reference for issuing evacuation information as bellow. Alert Level 1: Criterion for Flood-Fighting Standby; Level 2: Advisory Water Level; Level 3: Evacuation-Warning Water Level (reference criterion for Evacuation of the Elderly, etc.); Level 4: Hazardous Water Level (reference criterion for Evacuation Instruction); High Water Level (HWL): Water levels increase to this level when the designed-maximum flow rate occurs. It is an index of flooding, although river levees are generally designed higher than this level.

Figure 6 illustrates flood depths in the target area, corresponding to Fig. 2. In this scenario, around 06:00 on 17 August, small to medium-sized rivers (Hakurinndai, Utsubetsu, Obihiro Rivers) flowing through the city's core rapidly overflowed, followed by the larger Satsunai River. The inundation area resembles the expected maximum flood shown in Fig. 2. However, as the expected maximum flood is a superposition of potential inundation areas, the scenario's inundation area is relatively smaller.

2.4 Exercise method

Figure 7 shows the flowchart of the proposed method. We summarized the method for implementing simulation exercises for municipality to consider issuing evacuation information for the flood scenarios that will potentially occur in future climate conditions. In preparing for the simulation exercise, we analyzed and created the necessary information to

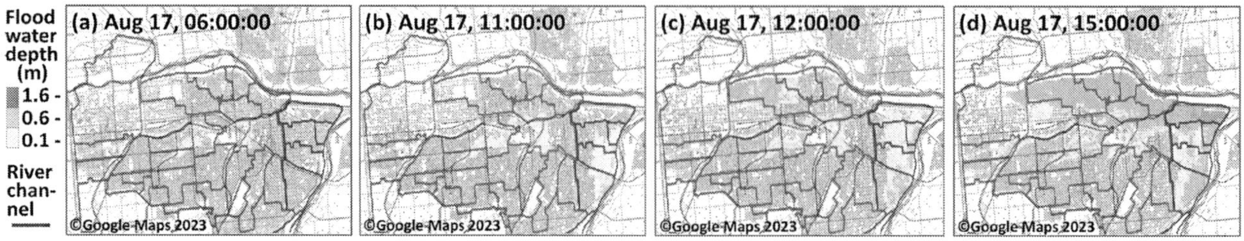

Figure 6. Flood depth at the onset of flooding of each river and the areas around the Satsunai (purple lines) and Utsubetsu Rivers (pink lines) considered for issuing evacuation information in the demonstration experiment.

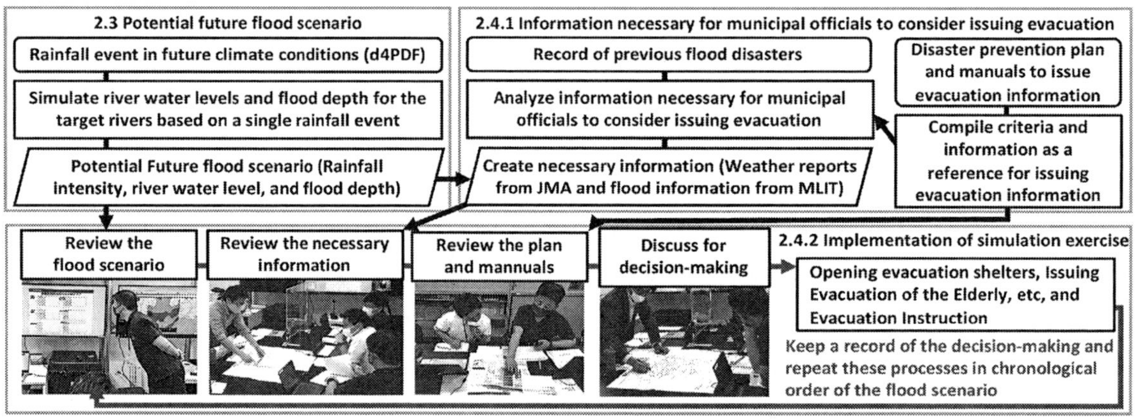

Figure 7. Flowchart of the proposed method.

assist the officials in imaging the situation and making decisions on issuing evacuation information for an unprecedented scale and pattern of flood disaster (namely, using the potential future flood scenarios).

2.4.1 Information necessary for decision-making

First, we compiled the municipal disaster-prevention plan and the manuals for issuing evacuation information to determine the criteria used as a reference for decision-making on issuing evacuation information. The criteria are detailed below: (1) Heavy Rain Advisory and Warning announced by the Japan Meteorological Agency (JMA), (2) Water level criteria of the rivers, and (3) Flood Advisory and Warning announced by MLIT, which are determined based on observations and forecasts of the water levels. These advisories are equivalent to Level 2 of Alert Level, and the warnings are equivalent to Level 3.

Next, using a method for organizing events and information among a disaster chronologically as Seki et al. (2016), we compiled a chronological table containing all the disaster-related information provided to Obihiro City and decision-making of the municipality during Typhoon Lionrock (2016), based on the disaster reports provided by MLIT. In consultation with the municipality and the river office of MLIT using the chronological table, we identify the information necessary for decision-making on issuing evacuation information. The necessary information is as follows: (1) Weather reports provided by JMA, which include sea-surface pressure, typhoon route, accumulated rainfall at observation points from the typhoon approached, forecasted daily rainfall over the region for the next 1–2 d, and Heavy Rain Advisory and Warning based on forecasted rainfall up to 6 h in advance. (2) Water level information that reaches the criteria, as well as Flood Advisory and Warning based on forecasted river water levels up to 6 h in advance, and are provided by MLIT.

Finally, we created the necessary information based on the flood scenario for the simulation exercise.

2.4.2 Implementation of simulation exercise

We set the period of the flood scenario for the simulation exercise from 09:00:00 on 16 August, a few days before the typhoon reached Hokkaido, to 21:00:00 on 17 August, when the inundation area is at its maximum. The figure suggests that the typhoon is strong, with a sea-surface pressure of 970 hPa at its center, and it is forecasted to reach Hokkaido in a few days. Figure 4 shows the sea-surface pressure and the typhoon's route in the weather report provided at the beginning of the simulation exercise.

In this exercise method, the scenario time is advanced by the hour. The flood scenario and the information necessary for decision-making on issuing evacuation information are provided to the officials corresponding to the scenario time. Then, the officials consult the disaster prevention plan and manuals for issuing evacuation information. They decide on specific areas and the level of evacuation information, which may include emergency response measures such as opening evacuation shelters and deploying patrols. Furthermore, they discuss the rationale for these decisions. Finally, they repeat all the processes in chronological order of the scenario. The flood scenario is visualized in the same format as during an actual disaster. Rainfall intensity is provided as the forecast by JMA up to 6 h in advance, river water levels are provided as the observation by MLIT, and flood depth is provided as the forecast by the Integrated Disaster Response Support System for Municipalities (IDR4M) up to 6 h in advance. IDR4M is a system that forecasts flooding risk based on machine learning and physical forecasts (Mitani, 2021). At Obihiro City, IDR4M provides trial forecasts of flooding for the demonstration experiment.

3 Results

The results of a simulation exercise based on the proposed method with the municipal officials of Obihiro City are presented as follows. The decision-making and the rationale on issuing evacuation information are detailed.

(1) On 16 August at 09:00:00, the prevention officials based on forecasted typhoon route, approaching Hokkaido, and large amount of daily rainfall in the first weather report notified all relevant agencies about the possibility of patrols deployment and opening evacuation shelters. (2) At 15:00:00, the officials initiated the opening evacuation shelters throughout the city to avoid opening the shelters at night. Voluntary Evacuation Advisory was issued for the areas around the Satsunai River bounded by purple lines in Fig. 6 after opening the shelters, because the daily rainfall forecast was a potentially record-breaking amount, according to a weather report. (3) At 18:00:00, the officials deployed patrols because the water levels of the Satsunai River exceeded Level 1 (criterion (2) in Sect. 2.3.1). (4) On 17 August at 00:00:00, the officials issued an Evacuation of Elderly, etc. for the area around the Hakurindai, Utsubetsu, and Obihiro Rivers bounded by pink lines in Fig. 4, because Heavy Rainfall Warning (criterion (1) in Sect. 2.3.1) was announced and flooding was forecasted in 6 h by IDR4M. The area for Evacuation of Elderly, etc. was expanded because of Flood Warning (criterion (3) in Sect. 2.3.1) for the Tokachi River and a heavy rainfall was predicted in the upstream part of the river based on rainfall data. Moreover, flooding within a few hours was a subject of concern Evacuation of Elderly, etc. was issued for the area bounded by purple lines in Fig. 4. (5) At 05:00:00, the officials issued Evacuation Instruction for the area bounded by pink lines in Fig. 4 because the river water levels of the Hakurindai and Utsubetsu Rivers exceeded Level 4. Moreover, Evacuation Instruction was issued for the area bounded by purple lines in Fig. 4 because the water level was rising rapidly and Flood Warning was announced.

4 Discussion

Based on the post-simulation exercise review, we describe achievements of this method. Furthermore, we will discuss the challenges and further utilization ideas.

4.1 Achievements

In the simulation exercise, prevention officials gained valuable experience and insight. Initially, officials believed that Satsunai River would experience flooding based on their past experiences and the forecasted daily rainfall. However, sudden overflows occurred in smaller rivers like Hakurindai River in this flood scenario, giving them new insights into these flooding patterns. During the simulation exercise, the flood scenario and necessary information were presented in the same format and were as comprehensive as in an actual disaster, enabling officials to compare it with established criteria and manuals. This facilitated decision-making regarding the issuance of evacuation information for specific areas and times. Additionally, longer-term forecast information, such as weather reports, allowed for considerations not only in the immediate decision-making before a flood but also for the early opening of evacuation shelters and flood-fighting activities, including the deployment of patrols.

4.2 Challenges

4.2.1 Enhancement of model accuracy

Current flood analysis models, similar to the models used in the creation of maximum expected inundation depth maps employed in the development of flood hazard maps, do not incorporate the influence of elevated water levels at confluence points downstream. Furthermore, the effects of inland flooding, including runoff from sewage systems, have not been considered. These factors may result in accuracy issues in the calculation results.

4.2.2 Multi-hazard scenario incorporation

In actual emergencies during heavy rainfall, municipalities must respond not only to river-based floods but also to sediment disasters. However, the current scenarios only consider river-based floods and do not encompass other types of hazards. Therefore, to avoid underestimating the necessary responses during emergencies, it is necessary to encompass the various types of hazards caused by rainfall.

4.2.3 Approximation to actual disaster conditions

Municipalities rely on consultations with the JMA observatory and MLIT river offices, as well as reports from citizens, as key sources of information for emergency decision-making. To better reflect real conditions, additional situational information and participation of related organizations are effective.

4.2.4 Scenario selection and prioritization

Comprehensive preparation for future floods requires exercises involving numerous heavy rainfall scenarios. However, conducting these exercises is impractical due to extensive procedures, including scenario creation, coordination with municipalities for simulation training, and prioritization of representative scenarios. Addressing this challenge involves selecting a subset of critical scenarios that must be experienced at least and evaluating the effective order for simulation exercises.

4.3 Further utilization ideas

4.3.1 Expanding for residents and companies

In the "River Basin Disaster Resilience and Sustainability for All" project, it is crucial that all stakeholders within the basin contribute to mitigation efforts. To enhance disaster resilience, residents and companies must actively prepare on a daily basis and respond appropriately in emergency. Our method has the potential to support them in creating emergency plans, gaining practical experience, and enhancing their preparedness. However, it is necessary to understand that key actions for them during emergency response (e.g., issuance of evacuation information by municipalities) and the necessary information for decision-making (e.g., weather and river warnings for municipalities) are essential. Furthermore, developing methods of exercises to simulate as actual emergency for them is equally important.

4.3.2 Testing new forecast and disaster information

Testing new forecasting and support information for municipal decision-making during the trial phase usually involves assuming past disaster events. In our method, there is the potential to assess the impact of this new information on future floods within the context of real emergency situations, including currently available forecasts and disaster-related information. Additionally, it can contribute to exploring the optimal integration of new information with existing data and determining when to utilize it in emergency decision-making, all from the perspective of information recipients.

5 Conclusions

In this study, to enhance disaster resilience against potential future flood disasters, we proposed a method to make flood scenarios based on climate projections, considering the risk of flooding in small rivers. Focusing on the information necessary for decision-making on issuing evacuation information, we also developed an exercise method for municipalities using the flood scenarios. Moreover, we conducted a simulation exercise with municipal officials in Hokkaido using this method. As a result, issuing evacuation information for specific areas and times and providing valuable insights, including knowledge of new flooding patterns and disaster prevention issues, were achieved, thereby enhancing preparedness for potential future flood disasters. Furthermore, we discussed the challenges in applying this method to municipal disaster prevention strategies and explored additional ways to utilize it in the future.

Code and data availability. The software code and research data supporting the findings of this study were developed through collaborative efforts with local municipalities and administrations. Due to the nature of this partnership, public disclosure poses significant challenges. As a result, the software code and research data are not publicly available at this time. However, in cases where there is a recognized need and an agreement has been reached with relevant stakeholders, access to the software code and research data can be facilitated by the corresponding author, Akihiro Suzuki.

Author contributions. AS: Conceptualization; data curation; formal analysis; investigation; methodology; project administration; software; validation; visualization; writing – original draft; writing-review & editing. FU: Data curation; formal analysis; investigation; software; visualization. TH: Data curation; formal analysis; investigation; software; visualization. TY: Project administration. TJY: Conceptualization; data curation; funding acquisition; methodology; project administration; resources; software; supervision; validation; writing – original draft; writing-review & editing.

Competing interests. The contact author has declared that none of the authors has any competing interests.

Special issue statement. This article is part of the special issue "ICFM9 – River Basin Disaster Resilience and Sustainability by

All". It is a result of The 9th International Conference on Flood Management, Tsukuba, Japan, 18–22 February 2023.

Acknowledgements. This study is supported by the Council for Science, Technology and Innovation (CSTI), Cross-ministerial Strategic Innovation Promotion Program (SIP), Development of a Resilient Smart Network System against Natural Disasters, with funding provided by the National Research Institute for Earth Science and Disaster Resilience (NIED), and The Advanced Studies of Climate Change Projection (SENTAN) Grant Number JPMXD0722680734 from the Ministry of Education, Culture, Sports, Science, and Technology (MEXT), Japan.

The utilization of d4PDF in this study was made possible through collaboration with the Earth Simulator and its associated science programs (SOUSEI, TOUGOU, SI-CAT, DIAS) of MEXT, Japan.

The Crisis Management Section of the City of Obihiro, Hokkaido, provided valuable cooperation during the empirical experiments in this study.

The Ministry of Land, Infrastructure, Transport and Tourism, Japan, supported the development of rainfall runoff and inundation calculation models for flood scenarios by providing observation data and software code.

Flood analysis in small rivers was conducted using the Nays2DFlood solver of the International River Interface Cooperative (iRIC) software.

Financial support. This research has been supported by the Japan Society for the Promotion of Science (grant no. 22H01594) and the Ministry of Education, Culture, Sports, Science and Technology (grant no. JPMXD0722680734).

Review statement. This paper was edited by Daisuke HARADA and reviewed by two anonymous referees.

References

Hoshino, T., Okachi, H., Takehara, Y., and Yamada, T. J.: Evaluation of heavy rainfall risk of typhoon hagibis (2019) associated with typhoon track, J. Jpn. Soc. Civ. Eng. Ser. B1, 76, 414–423, https://doi.org/10.2208/journalofjsce.10.1_534, 2020.

Kawase, H.: Dynamical downscaling ensemble data covering the whole Japan from d4PDF with 5-km nonhydrostatic regional climate model (NHRCM), Data Integration and Analysis System (DIAS) [data set], https://doi.org/10.20783/DIAS.657, 2023.

Kitano, Y., Yamamoto, T., Kobayashi, A., and Yamada, T. J.: Statistical analysis of typhoon related events in Hokkaido and surroundings in the last 56 years including the 2016 heavy rainfall, J. Jpn. Soc. Civ. Eng. Ser. B1, 73, 4, 1231–1236, https://doi.org/10.2208/jscejhe.73.I_1231, 2017 (in Japanese).

Koike, T.: Evolution of Japan's flood control planning and policy in response to climate change risks and social changes, Water Policy, 23, 77–84, https://doi.org/10.2166/wp.2021.287, 2021.

Mitani, Y.: Development of innovative technology for new disaster prevention information system, J. Struct. Mater. Civ. Eng., 37, 21–27, 2021 (in Japanese).

Mizuta, R., Murata, A., Ishii, M., Shiogama, H., Hibino, K., Mori, N., Arakawa, O., Imada, Y., Yoshida, K., Aoyagi, T., Kawase, H., Mori, M., Okada, Y., Shimura, T., Nagatomo, T., Ikeda, M., Endo, H., Nosaka, M., Arai, M., Takahashi, C., Tanaka, K., Takemi, T., Tachikawa, Y., Temur, K., Kamae, Y., Watanabe, M., Sasaki, H., Kitoh, A., Takayabu, I., Nakakita, E., and Kimoto, M.: Over 5,000 years of ensemble future climate simulations by 60-km global and 20-km regional atmospheric models, Bull. Amer. Meteor., 98, 1383–1398, https://doi.org/10.1175/BAMS-D-16-0099.1, 2017.

National Institute for Land and Infrastructure Management: Manual for Creating Flood Inundation Assumed Area. 4th edn., 2017.

Sato, S., Hayashi, H. Tamura, K., and Urata, Y.: Lessons and Issues Learned from Emergency Response of the 2009 HeavyRain – TRENDREADER Analysis of Open Ended Answer Data in "Survey of Municipal Governments Published Evacuation Advisory and Order", J. Jpn. Soc. Nat. Disaster Sci., 30, 123–145, 2011 (in Japanese).

Seki, K., Wakigawa, K., and Okayasu, T.: Study on the decision-making system for effective evacuation during flood events, Adv. River Eng., 22, 362–366, https://doi.org/10.11532/river.22.0_361, 2016 (in Japanese).

Shimizu, Y.: iRIC Software Nays-2D Flood Solver Manual, https://i-ric.org/solvers/nays2dflood/ (last access: 5 November 2020), 2014 (in Japanese).

Suzuki, A., Yamada, T. J., Yamamoto, T., and Uemura, F.: Efforts toward evacuation decision making and prior learning using ensemble climate information in Tokachi River basin, Hokkaido, Proc Jpn. Soc. Civ. Eng., Hokkaido Blanch 2022 Annual Meeting, 2022 (in Japanese).

Suzuki, A., Uemura, F., Hoshino, T., Ishihara, M., Yamamoto, T., Hashimoto, S., and Yamada, T. J.: A method for municipality to plan issuing evacuation information for flood risk considering impact of the climate change, J. Jpn. Soc. Civ. Eng. Ser. G, 79, 23-27045, https://doi.org/10.2208/jscejj.23-27045, 2023 (in Japanese).

Yamada, T. J.: A Case Study: Adaptation Measures for Extreme Floods in Japan, Using Huge Ensemble of High-resolution Climate Model Simulation, United Nations Framework Convention on Climate Change, Bonn, Germany, 28–30, https://unfccc.int/documents/196569 (last access: 1 December 2023), 2019.

Yamada, T. J., Hoshino, T., and Suzuki, A.: Using a massive high-resolution ensemble climate data set to examine dynamic and thermodynamic aspects of heavy precipitation change, Atmos. Sci. Lett., 22, e1065, https://doi.org/10.1002/asl.1065, 2021.

Proc. IAHS, 386, 109–114, 2024
https://doi.org/10.5194/piahs-386-109-2024

Open Access
PIAHS

Sediment transport processes in the Sangu River basin using a rainfall-sediment runoff model for sustainable river management

Md. Majadur Rahman[1,2], **Daisuke Harada**[2], **and Shinji Egashira**[2]

[1]Disaster management program, National Graduate Institute for Policy Studies (GRIPS), 7-22-1 Roppongi, Minato-ku, Tokyo 106-8677, Japan

[2]International Centre for Water Hazard and Risk Management (ICHARM) under the Auspices of UNESCO, Public Works Research Institute (PWRI), 1–6 Minamihara, Tsukuba, Ibaraki 305-8516, Japan

Correspondence: Md. Majadur Rahman (mimrahman102@gmail.com)

Received: 20 July 2023 – Revised: 29 September 2023 – Accepted: 24 February 2024 – Published: 19 April 2024

Abstract. The Sangu River, whose basin area is about 3600 km^2, conveys a huge volume of sediment laden water during the monsoon, which decreases the river's capacity, causing flooding, channel change, and navigability problems. In this study, a basin-scale rainfall-sediment runoff (RSR) model was used to analyse the sediment transport in the Sangu River. Simulations were conducted for the year 2011, and the results were validated in terms of suspended sediment concentration. The simulation results show deposition trends particularly in the lower and middle reaches, and that nearly 7.5×10^5 m^3 of sediment annually discharges to the Bay of Bengal. The simulation results also show the tributaries actively produce sediment, which results in sediment deposition in downstream channels. The analyses also find that the implementation of check dams in the upstream area significantly reduces sediment deposition in the downstream area. These findings offer valuable insights for the development of sustainable sediment management strategies.

1 Introduction

The Sangu River is one of the five river systems in the southeastern hilly area of Bangladesh, which primarily consists of hilly terrain and a long strip of low-lying coastal plains along the Bay of Bengal. This morphological feature makes the river vulnerable to disasters. The continued practice of deforestation and topsoil cutting in the upstream hilly region are driven by the need to fulfil various human requirements such as food production, human settlements, transportation infrastructure, brick manufacturing, and agricultural land preparation. During the monsoon season, floods due to heavy rainfall carry sediment from the upstream hilly area to the downstream low-lying area. As a result, the downstream part of the Sangu River faces siltation, which reduces river conveyance, increases flooding during the monsoon season, and decreases water availability during the dry season, resulting in navigability problems and scarcity of water uses that have adverse impacts on the natural environment (BWDB, 2021). Thus, basin-scale assessment tools can be useful to estimate erosion and sediment yield for proper planning and management of the river basin (Owens, 2005; Battarai and Dutta, 2007).

In order to address various issues and promote sustainable management, this study investigates sediment transport processes in the Sangu River. Initially, a rainfall-sediment runoff (RSR) model was developed for the purpose of evaluating sediment transport processes. Subsequently, the RSR model was revised and updated to include structural countermeasures for sediment management within the river basin. Several suitable sites were identified for the installation of check dams with the aim of promoting sustainable management of the river.

2 Methodology

The RSR model is an extension of the rainfall-runoff-inundation (RRI) model coupled with a sediment-runoff model based on unit channel concept (Rahman et al., 2022).

The sediment continuity equation in unit channel is:

$$\frac{\partial z}{\partial t} = \sum_i \frac{1}{(1-\lambda)} \left[\frac{1}{B_j L_j} \left\{ Q_{\text{b}i}\left(x_j\right) + Q_{\text{b}i}\left(y_j\right) - Q_{\text{b}i}\left(x_{j+1}\right) \right\} - E_i + D_i \right] \quad (1)$$

where z is the river bed elevation, λ is the porosity of bed materials, B_j and L_j are the width and length of unit channel j, $Q_{\text{b}i}(x_j)$ and $Q_{\text{b}i}(y_j)$ are the incoming bed load transport rates to unit channel j for sediment class i, $Q_{\text{b}i}(x_{j+1})$ is the outflow bed load transport rate from unit channel j for sediment size class i, E_i and D_i are the erosion and deposition rates of the suspended sediment for sediment class i.

Egashira et al. (1997) proposed the following formula for the bed load transport rate:

$$q_{\text{b}*i} = \frac{4}{15} p_i \frac{K_1^2 K_2}{\sqrt{f_\text{d} + f_\text{f}}} \tau_{*i}^{5/2} \quad (2)$$

where $q_{\text{b}*i}$ is the non-dimensional bed load transport rate, p_i is the fraction of particle size i, τ_{*i} is the non-dimensional shear stress, and K_1, K_2, f_d, and f_f are the parameters specified in Egashira et al. (1997).

The convection equation for suspended sediment transport in each unit channel is described as:

$$\frac{\partial h c_i}{\partial t} = \frac{1}{B_j L_j} \left\{ c_{\text{s}i}\left(x_j\right) Q\left(x_j\right) + c_{\text{s}i}\left(y_j\right) Q\left(y_j\right) - c_{\text{s}i}\left(x_{j+1}\right) Q\left(x_{j+1}\right) \right\} + E_i - D_i \quad (3)$$

where c_i is the suspended sediment concentration for sediment class i, $c_{\text{s}i}\left(x_j\right)$ and $c_{\text{s}i}\left(y_j\right)$ are the incoming and $c_{\text{s}i}\left(x_{j+1}\right)$ is the outgoing sediment concentrations from unit channel j for sediment class i. $Q\left(x_j\right)$ and $Q\left(y_j\right)$ are the incoming flow discharges to unit channel j, and $Q\left(x_{j+1}\right)$ is the outgoing flow discharge from unit channel j.

The erosion rate for suspended sediment can be evaluated using the Harada et al. (2022) formula:

$$E_i = p_i W_\text{e} c_\text{s} \quad (4)$$

where c_s is the sediment concentration of the bed load layer, and W_e is the entrainment velocity, which can be evaluated by the density stratified flow formula proposed by Egashira and Ashida (1980).

The deposition rate is calculated using following equation:

$$D_i = w_{oi} c_i \quad (5)$$

where w_{oi} is the fall velocity of sediment particle size class i

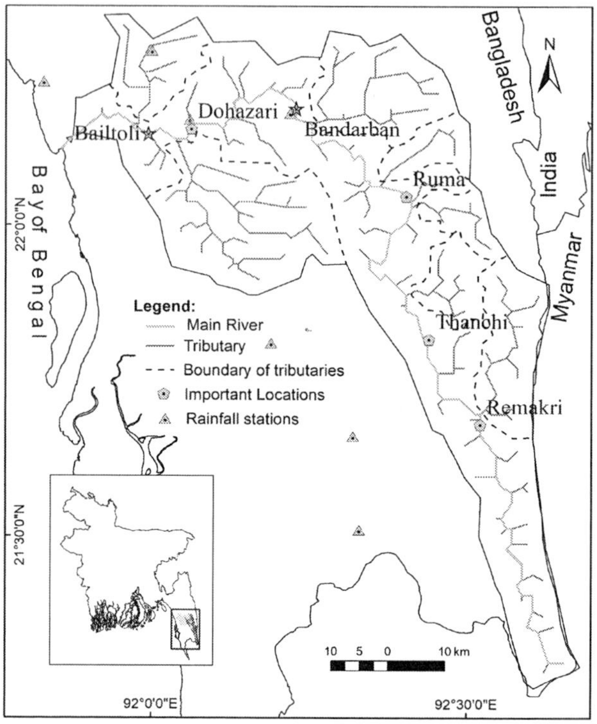

Figure 1. Sangu River basin in Bangladesh.

3 Study area and calculation conditions

The Sangu River originates from the Arakan Hills and flows north to the Bay of Bengal (Fig. 1). The total basin area is $3600\,\text{km}^2$, of which about 80 % is in hilly terrain and about 20 % is along coastal plain land. The Sangu River is a meandering river with an average longitudinal slope of 1/6000 from the river mouth to Thanchi, after which the bed slope becomes steep with 1/800. The basin mostly consists of brown hilly soil and is mostly covered by clay loam. The river bed material is predominantly sandy in the downstream part, gravels in the upstream part, and boulders in the most upstream part. The climate is tropical, with stormy monsoon seasons in the summer and dry winters. The flow in the Sangu River is only from rainfall runoff within the basin, keeping the basin's small watercourses (charas) and waterfalls alive during the dry season (BWDB, 2021).

The RSR model was prepared for the Sangu River basin using SRTM 30S data. The computation was conducted using observed rainfall data collected in 2011. Model calibration and validation were conducted using the observed discharge at the Bandarban station. The Nash-Sutcliffe efficiency (NSE) was 0.72 for 2011 and 0.68 for 2007 for calibration and validation, respectively (Rahman et al., 2022). As a result of the model calibration, Manning's roughness for river, Manning's roughness for slope, soil depth, porosity, and hydraulic conductivity were determined to be 0.03,

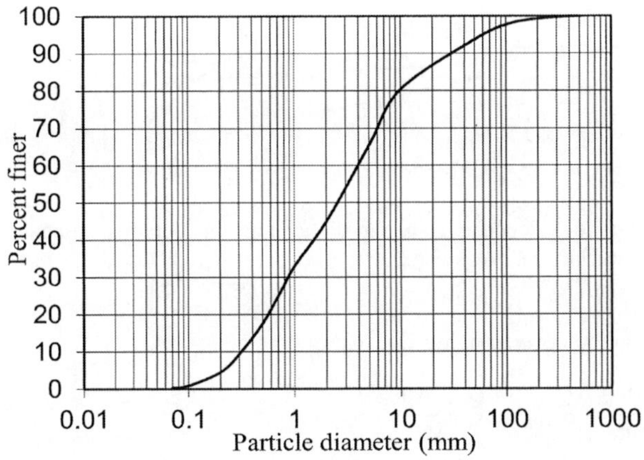

Figure 2. Initial sediment size distribution of bed materials.

0.40, 3 m, 0.50, 0.001 m s^{-1}, respectively. Sensitivity analysis was performed to create an initial sediment particle size distribution consisting of 13 size classes ranging from 70 µm to 250 mm in diameter. Figure 2 shows the initial sediment particle size distribution of the bed materials. The exchange layer thickness for non-uniform sediment computation was set to 0.10 m. In this study, we neglected sediment supplies from slopes.

4 Results and Discussions

4.1 Sediment transport and sediment yield

Figure 3 compares simulated and observed suspended sediment discharges (Qs). The field-measured data were collected in 2018 and 2020 at Bandarban and Bailtoli, respectively. The measurement was done from a boat, in which ADCP discharge measurements and water sampling were conducted simulatniously. Several water samples were collected at each site, with variations in depth. Observations were made every two weeks during the flood season at the Bandarban site. At the Bailtoli site, which is tidally influenced, 21 surveys were made within three days. Figure 3 shows the validity of the RSR model in terms of suspended sediment transport.

Figure 4 shows the total sediment yeild (combination of bed load and suspended load) and the specific sediment yeild at different locations of the main river. The total sediment yeild increases from upstream to downstream because of sediment supply from the tributaries. It decreases near the outlet as the sediment transport capacity decreases due to the wider river width around the river mouth. On the other hand, the specific sediment yeild decreases from upstream to downstream because of a higher sediment supply rate in the upstream area due to the steeper bed slopes. According to the model simulation results, about 7.5 × 10^5 m^3 of sediment discharges to the Bay of Bengal annually.

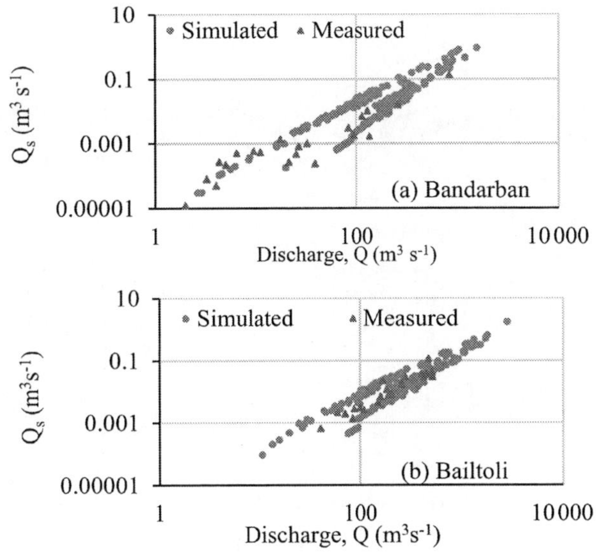

Figure 3. Comparison of simulated suspended sediment discharge with field measurements at **(a)** Bandarban and **(b)** Bailtoli (Rahman et al., 2022).

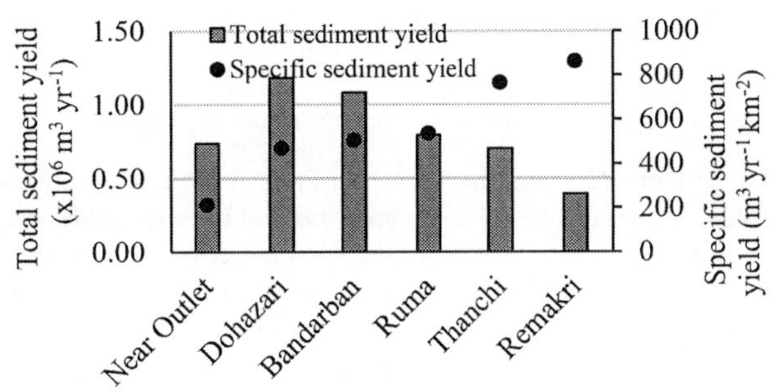

Figure 4. Total sediment yield and specific sediment yield.

4.2 Bed deformation

Figure 5 shows the simulated bed deformation within one year. The tributaries actively supply sediment to the main river. Most of the sediment in the Sangu River comes from the hilly area through the tributaries and deposited in the downstream part of the main river. The maximum erosion at the upstream part is 0.36 m, and the maximum deposition is 0.74 m at the downstream part. Since the river is suspended load dominant, the bed deformation is mainly caused by suspended sediment. According to the results of the model computation, a considerable amount of sediment is transported from the upstream part and subsequently deposited in the lower and middle stream reaches, where sand bars form as a result of the phenomina. These characteristics of the river are the main causes of flooding, bank erosion, and navigability problems.

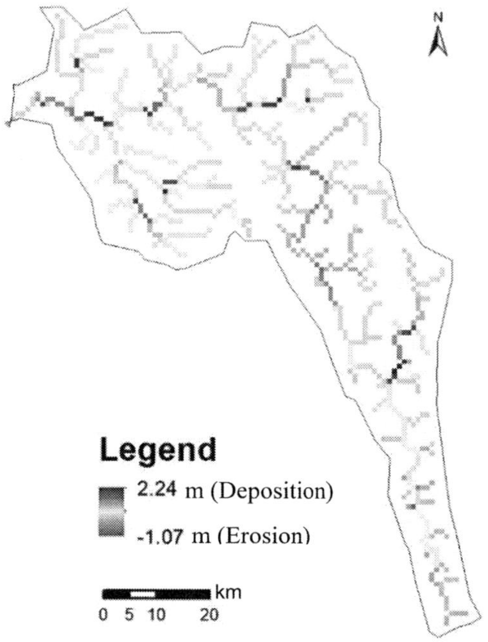

Figure 5. Bed deformation within one year.

4.3 Countermeasures for sediment management

For proper sediment management of the river basin, countermeasures should be taken considering the river morphology, as well as environmental factors, such as ecosystems. Countermeasures should improve, or at least maintain, the current state of waterways. Excess inflow sediment that is produced by the tributaries can be controlled by taking soft and hard measures. Check dams are considered to be one of the effective and widely utilized soil conservation measures for sediment trapping, erosion mitigation, and watershed sedimentation control (Abbasi et al., 2019; Zhao et al., 2017). Though preventive soft measures for sediment yield reduction, such as land use regulation, are preferred, structural measures may be necessary for improving the current situations with significant sediment trapping. Since the river is dominated by suspended load (Rahman et al., 2022), it is imperative to derive a formula that can effectively estimate the quantity of sediment that can be trapped by check dams.

The temporal change in sediment concentration in the water stored at a check dam (Fig. 6) can be defined as:

$$\frac{\mathrm{d}c\overline{V}}{\mathrm{d}t} = c_{\text{in}}Q - cQ - w_o cA \tag{6}$$

Where $\overline{V}$ is the volume of the dam storage, c is the suspended sediment concentration inside the dam storage, c_{in} is the incoming suspended sediment concentration into the dam storage, Q is the flow discharge, w_o is the particle fall velocity, and A is the area of the dam storage. The following equation

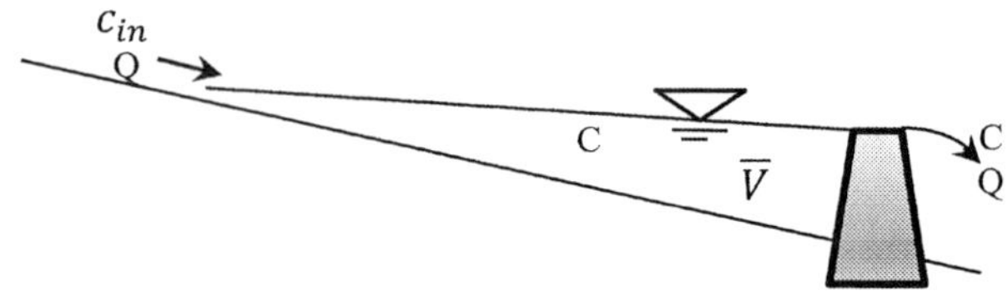

Figure 6. Schematic diagram of incoming and outgoing suspended sediment at a check dam.

can be derived using Eq. (6):

$$\frac{\mathrm{d}c}{\mathrm{d}t} = \frac{Q}{\overline{V}}c_{\text{in}} - \frac{Q + Aw_o}{\overline{V}}c \tag{7}$$

$$\text{Let, } a = \frac{Q}{\overline{V}}c_{\text{in}}, b = \frac{Q + Aw_o}{\overline{V}} = \frac{1 + \frac{A}{Q}w_o}{\frac{\overline{V}}{Q}}$$

$$\text{Then, } \frac{a}{b} = \frac{Qc_{\text{in}}}{Q + Aw_o} = \frac{1}{1 + \frac{A}{Q}w_o}c_{\text{in}} \tag{8}$$

Equation (7) can be normalized as:

$$\frac{\mathrm{d}c}{\mathrm{d}t} = a - bc \tag{9}$$

By integrating Eq. (9):

$$c = \frac{1}{b}(a - ae^{-bt} + bc_o e^{-bt}) \tag{10}$$

Where c_o is the initial concentration. When $t = 0, c_o = 0$ Eq. (10) can be expressed as follows:

$$c(t) = \frac{a}{b}(1 - e^{-bt}) \tag{11}$$

When Eq. (11) is solved for $t = \infty$:

$$c = \frac{a}{b} = \frac{1}{1 + \frac{A}{Q}w_o}c_{\text{in}} \tag{12}$$

For the sediment discharge trapped by the check dam, $Q_{\text{s trap}}$ can be determined by the following equation:

$$Q_{\text{s trap}} = Qc_{\text{in}} - Qc = (c_{\text{in}} - c)\,Q \tag{13}$$

By replacing c in Eq. (13) using Eq. (12):

$$Q_{\text{s trap}} = \frac{1}{1 + \frac{Q}{Aw_o}}c_{\text{in}}Q \tag{14}$$

As per Eq. (14), the RSR model was updated to incorporate check dams as countermeasures.

Investigation was conducted by installing seven check dams at different tributaries in the upstream hilly area to control sediment load. In this study, the height of the check dam was assumed to be 10 m. Figure 7 shows the model-simulated bed deformations with check dams experimentally installed

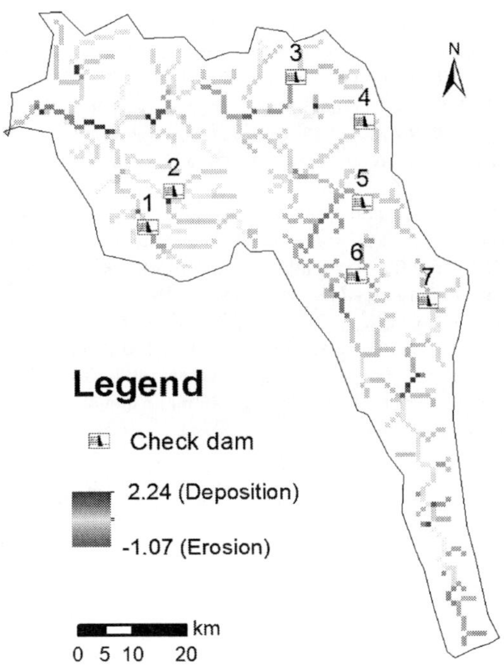

Figure 7. Bed deformation with check dams.

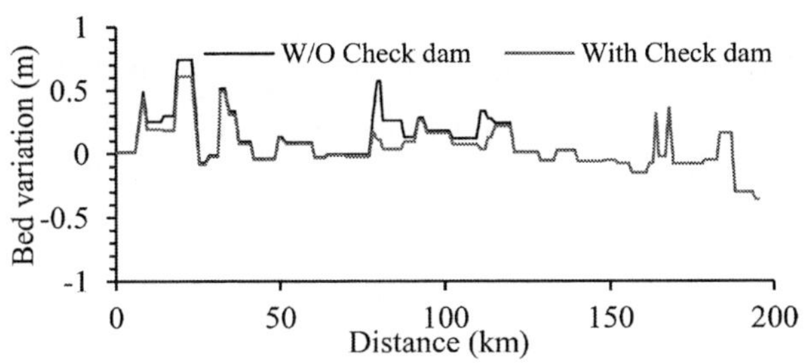

Figure 8. Bed variation along the main river at the end of simulation.

on different tributaries. Figure 8 compares the bed deformations along the main river with and without (W/O) check dams.

Figures 7 and 8 show that deposition is improved along the main river and its tributaries after installing the check-dams. In general, a significant improvement in the middle stream, as evidenced by a reduction in maximum depositions from 0.57 to 0.11 m and a corresponding decrease in volume by about 50 %. Similarly, the downstream also shows a decrease in the maximum deposition from 0.74 to 0.61 m, with a corresponding volume decrease of about 20 % and annual amount of sediment goes to the Bay of Bengal reduced from 7.5×10^5 to 6.5×10^5 m^3.

The findings suggest that check dams can serve as effective measures for controlling sediment. However, it is worth noting that sediment build-up in check dams may lead to significant operational challenges and failures due to their limited lifespan (Zhao et al., 2017). Hence, it is imperative to take sustainable measures for sediment management of check dams; for example, sediment removal is the most suitable strategy. In order to determine the longevity of a check dam, a computational model was employed for a period of two consecutive years, specifically 2011 and 2012. The longevity of check dams 1, 2, and 3 can be considered significant. Check dams 4, 5, 6, and 7 will reach their maximum sediment capacity within a period of four to five years. The demand for sediment is significantly high in Bangladesh because, for any development activity, land filling measures are necessary since Bangladesh is a deltaic region with low-lying terrain. The potential removal of trapped sediment presents an opportunity to generate revenue for the government through its sale, given the significant demand for sediment in the region.

5 Conclusion

In this paper, we conducted basin-scale numerical simulations for assessing sediment transport processes for sustainable river sediment management considering structural measures in the Sangu River. The simulation results show that sediment deposition occurs in most areas of the lower and middle streams. In addition, the computation results with seven check dams in the tributaries show the effectiveness of such countermeasures. It should be noted that thorough feasibility analysis needs to be conducted prior to the implementation of any structure.

Data availability. The data are available from the corresponding author on reasonable request.

Author contributions. MMR, DH and SE conceptualized the study; MMR performed the numerical simulations; MMR wrote the original draft of the manuscript; DH and SE reviewed and edited the manuscript.

Competing interests. At least one of the (co-)authors is a guest member of the editorial board of *Proceedings of IAHS* for the special issue "ICFM9 –River Basin Disaster Resilience and Sustainability by All". The peer-review process was guided by an independent editor, and the authors also have no other competing interests to declare.

Disclaimer. Publisher's note: Copernicus Publications remains neutral with regard to jurisdictional claims made in the text, published maps, institutional affiliations, or any other geographical representation in this paper. While Copernicus Publications makes every effort to include appropriate place names, the final responsibility lies with the authors.

Special issue statement. This article is part of the special issue "ICFM9 – River Basin Disaster Resilience and Sustainability by All". It is a result of The 9th International Conference on Flood Management, Tsukuba, Japan, 18–22 February 2023.

Acknowledgements. The authors would like to express their gratitude to Bangladesh Water Development Board (BWDB) and JICA for supporting this study. The authors also wish to thank the editor and two anonymous reviewers for their constructive comments and suggestions, which helped to improve the quality of the paper.

Review statement. This paper was edited by Kensuke Naito and reviewed by two anonymous referees.

References

Abbasi, N. A., Xu, X., Lucas-Borja, M. E., Dang, W., and Liu, B.: The use of check dams in watershed management projects: Examples from around the world, Sci. Total Environ., 676, 683–691, https://doi.org/10.1016/j.scitotenv.2019.04.249, 2019

Bhattarai, R. and Dutta, D.: Estimation of soil erosion and sediment yield using gis at catchment scale, Water Resour. Manag. J., 21, 1635–1647, 2007.

BWDB: Feasibility Study for Restoration of Sangu and Matamuhuri River Basin. Final report, IWM, 2021.

Egashira, S. and Ashida, K.: Studies on the Structures of Density Stratified Flows, Bulletin of the Disaster Prevention Research Institute, Kyoto University, 29, 165–198, 1980.

Egashira, S., Miyamoto, K., and Itoh, T.: Bed load rate in view of two phase flow dynamics, Annual Journal of Hydraulic Engineering JSCE, 41, 789–794, 1997.

Harada, D., Egashira, S., Ahmed, T. S., and Ito, H.: Entrainment of bed sediment composed of very fine material, Earth Surf. Proc. Land., 47, 3051–3061, https://doi.org/10.1002/esp.5442, 2022.

Owens, P. N.: Conceptual Models and Budgets for Sediment Management at the River Basin Scale, J. Soils Sediments, 5, 201–212, https://doi.org/10.1065/jss2005.05.133, 2005

Rahman, M. M., Harada, D., and Egashira, S.: Basin scale sediment transport processes using rainfall sediment runoff model – a case study of Sangu river basin, Journal of JSCE, Ser. B1 (Hydraulic Engineering), 78, I_1099-I_1104, https://doi.org/10.2208/jscejhe.78.2_I_1099, 2022.

Zhao, G., Kondolf, G. M., Mu, X., Han, M., He, Z., Rubin, Z., Wang, F., Gao, P., and Sun, W.: Sediment yield reduction associated with land use changes and check dams in a catchment of the Loess Plateau, China, Catena, 148, 126–137, https://doi.org/10.1016/j.catena.2016.05.010, 2017.

Proc. IAHS, 386, 115–120, 2024
https://doi.org/10.5194/piahs-386-115-2024

Flood Simulation Studies with Rainfall-Runoff-Inundation (RRI) Model, Over Yangon City, Myanmar

Pyae Pyae Phyoe[1] **and Tatsuhiko Uchida**[2]

[1]Department of Advanced Science and Engineering Science, Graduate School of Advanced Science and Engineering Science, Hiroshima University, 1-4-1 Kagamiyama, Higashi-Hiroshima, Hiroshima 739-8527, Japan

[2]Graduate School of Advanced Science and Engineering Sciences, Hiroshima University, 1-4-1 Kagamiyama, Higashi-Hiroshima, Hiroshima 739-8527, Japan

Correspondence: Pyae Pyae Phyoe (pyaepyaephyoe6@gmail.com)

Received: 6 August 2023 – Revised: 15 November 2023 – Accepted: 24 February 2024 – Published: 19 April 2024

Abstract. Flood risks are increasing due to climate change and human activities in river basins. Flooding is one of the most common disasters in Myanmar, as well as in many other countries. The floods occur during the southwest monsoon season, which is from May to October. Yangon City is at risk of flooding when heavy monsoon rainfall coincides with high tides. In this study, the Yangon area was used as a case study, and the Rainfall-Runoff-Inundation (RRI) model was applied. The RRI model is one of the models that has been applied in various river basins to estimate river discharges and flood inundation from rainfall data. The model is a two-dimensional model capable of simulating rainfall-runoff and flood inundation simultaneously. This study aimed to clarify the characteristics of the model's parameters, such as lateral saturated hydraulic conductivity k_a and soil depth h_d, for a flat river basin in Yangon. Eighty numerical simulations were conducted for four flood events in 2011, 2012, 2013, and 2014, with six soil depths h_d and five different k_a values in the Bago River. The discharge hydrographs with different k_a and soil depths h_d values for this river basin were compared. The results show that k_a and soil depth h_d values significantly differ between Myanmar and other countries due to the geological characteristics and climate zones of the countries.

1 Introduction

According to a global disaster database, flood and storm are the primary causes of disasters in Myanmar (World Meteorological Organization, 2011). Flooding is one of the most devastating natural phenomena in the world. It occurs when the watershed system experiences heavy or continuous rainfall events, causing the water flow rate to exceed the channel's capacity. There are three types of flood characteristics: pluvial floods, fluvial floods, and coastal floods (storm surges). Yangon City, located in the center of Lower Myanmar, which encompasses the coastal lowlands and the confluence of the Bago River and Hlaing River, is prone to fluvial and pluvial floods. These frequently occur during each rainy season. Fluvial floods in Yangon happen when the water levels in the surrounding rivers or canals are higher than those in areas within the city. Pluvial floods can also occur even at low tides in some locations due to blockage of the drainage system by solid waste. Flood inundations, with over 0.5 m and a duration of 3 to 5 h, occurs 6 to 10 times per year. Based on records from 1947 to 2008, there is a 3.2 % annual probability that Yangon City has been affected by a tropical cyclone and a 6.7 % probability that a storm surge has occurred in nearby coastal areas (Asian Development Bank, 2016). Severe flood events occurred in the city in 1973, 1974, 1976, 1988, 1991, 1997, 2002, 2004, 2007, and 2014. The flood event in Yangon in 2014 was different from previous floods, and this happened in the Yangon area in June and July due to continuous heavy rainfall. Nearly all the rivers and creeks overflowed, and the inundation lasted for more than seven days. The heavy rainfall in 2014 led to flooding in 14 townships in Yangon City, causing significant damages to infrastructure and economic losses for the residents.

There is a strong need for an available model that simulates rainfall-runoff, and flood inundation models have been developed and successfully applied to river basins in many parts of the world (Bates and De Roo, 2000; Leandro et al., 2014; Liu et al., 2015). A one-dimensional analysis method is generally used for predicting flood flows through the rivers in rainfall runoff analysis (Collection of Hydraulic Formulas, 2019). To consider the rainfall distribution and topography, a distributed runoff model can be used. Recently, the Rainfall-Runoff Inundation (RRI) model was developed to analyze the processes of rainfall runoff, flooding, and inundation in an integrated manner using rainfall as input conditions (Sayama, 2014). The RRI model has also been developed and applied to evaluate river discharges and flood inundation depths in flood-prone areas (Shimizu et al., 2021a, b). In this paper, we selected part of the Bago River Basin, specifically the downstream section that includes Yangon City, as our case study area to investigate the rainfall and runoff characteristics of floods in Yangon City. In the Bago River basin, other researchers have previously applied the RRI model to both the upper and lower Bago River (Bhagabati and Kawasaki, 2017). They analyzed the applicability of the RRI model to a flat river basin and confirmed the model's performance by comparing the results with satellite-based data from UNOSAT (United Nations Satellite Centre). However, the basin area in this study differs from the previous research. Furthermore, the effect of parameters in the model on the discharge hydrograph has not been revealed for this basin. This paper discusses the applicability of the RRI model to a flat river basin to understand the runoff characteristics of floods in Yangon for future research considerations. The objectives of this study include understanding the rainfall and runoff characteristics of the 2014 flood in Yangon and clarifying the RRI parameters, specifically lateral saturated hydraulic conductivity k_a values and soil depth h_d for the large, flat river basin in Yangon.

2 Background of the Study Area

Yangon City is located in the eastern part of the Ayeyarwady Delta, with a patchwork of rivers, creeks, and canal drains around the area. Yangon is $794.43\,\text{km}^2$ in land area, and is situated at approximately 96.141762° longitude and 17.132582° latitude. Yangon has a tropical climate. The average annual temperature is around 26 °C, and the average annual rainfall is 93.5 in. (237.49 cm) (Asian Development Bank, 2016).

The area of the Yangon model basin is $22\,700\,\text{km}^2$, with approximately 1100 grid cells. The upper boundary of the catchment area is formed by the Hlaing River and Bago River, while the lower boundary is defined by the Yangon River and the coastal line. The location Map of the Yangon Basin Model is shown in Fig. 1. This map was generated 70 using ArcGIS Pro 3.2 (https://pro.arcgis.com/en/pro-app/latest/getstarted/download-arcgis-pro.htm, last access: 6 March 2024).

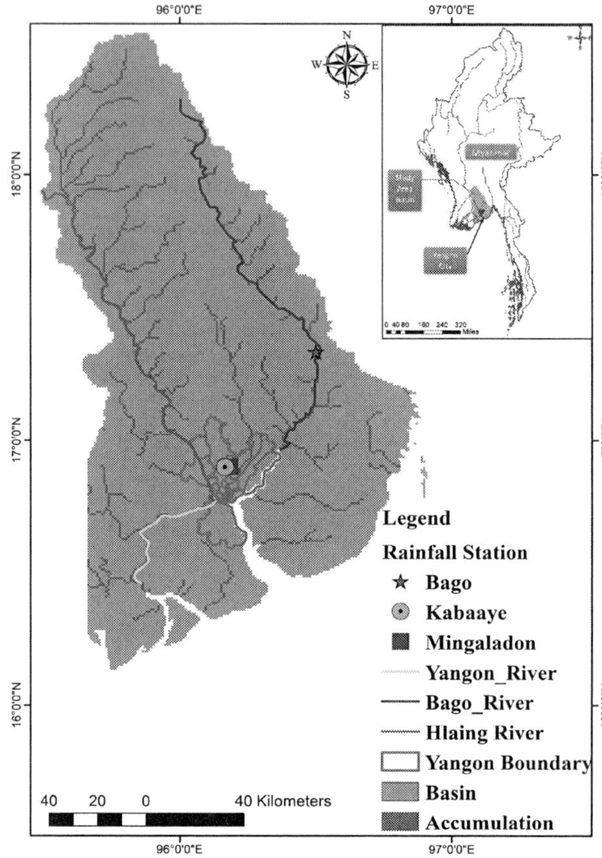

Figure 1. Location Map of the Yangon Basin Model.

This basin area comprises of eleven townships: Yangon, Kungyangon, Kyauktan, Thongwa, Hlegu, Hmawbi, Htantabin, Taikkyi, Kawkhmu, Hinthada, and Bago. Among these, Yangon City is the largest.

3 Methodology

Rainfall-Runoff-Inundation (RRI) model

The RRI model was developed to simulate the water flow mechanism between slope grid cells and river grid cells (Sayama, 2014). The continuity equation (Eq. 1) is solved to calculate the temporal variation in water volume within the grid of watershed slopes.

$$\frac{\partial z_s}{\partial t} + \frac{\partial q_x}{\partial x} + \frac{\partial q_y}{\partial y} = r - f \qquad (1)$$

where, z_s = water level, q_x, q_y, = unit width flow rate in the x, y direction, r = rainfall intensity, f = vertical penetration intensity. The unit width flow rates are divided into surface

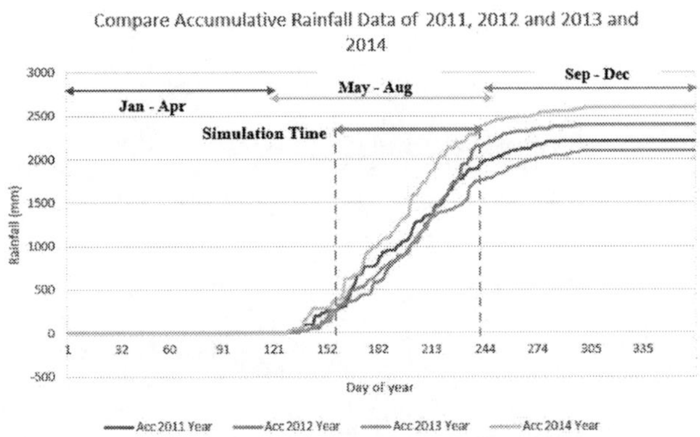

Figure 2. Comparison between the Accumulative Rainfall data of 2011, 2012, 2013 and 2014.

flow and penetrating flow. Generally, the unit width flow rate in the x and y directions is described by the shallow water equation as:

$$\begin{cases} \frac{\partial uh}{\partial t} + \frac{\partial u^2 h}{\partial x} + \frac{\partial uvh}{\partial y} = -gh\frac{\partial z_s}{\partial x} - \frac{\tau_{0x}}{\rho} \\ \frac{\partial vh}{\partial t} + \frac{\partial uvh}{\partial x} + \frac{\partial v^2 h}{\partial y} = -gh\frac{\partial z_s}{\partial y} - \frac{\tau_{0y}}{\rho} \end{cases} \tag{2}$$

where, $g =$ gravitational acceleration, $\tau_{0x}, \tau_{0y} =$ shear stress terms in x, y directions which can include other flow resistance such as vegetation. In the RRI model, a diffusion wave is assumed for surface flow, in which the convection terms in Eq. (2) are ignored. In this study, Eq. (3) was used to calculate the surface flow flux (Shimizu et al., 2021a, b).

$$u = \frac{1}{n}h^{\frac{2}{3}}\frac{\partial z_s}{\partial x}S^{-\frac{1}{2}}, \quad v = \frac{1}{n}h^{\frac{2}{3}}\frac{\partial z_s}{\partial y}S^{-\frac{1}{2}},$$

$$S^2 = \left(\frac{\partial z_s}{\partial x}\right)^2 + \left(\frac{\partial z_s}{\partial y}\right)^2 \tag{3}$$

where, $n =$ the roughness coefficient of Manning. The flow in the ground was divided into the basin slope, the plain area, and the mountainous area. In the mountainous area, vertical seepage flow was not considered ($f = 0$), and lateral seepage flow was calculated using Darcy's law as shown in Eq. (4).

$$q_x = -k_a(h_d + h)\frac{\partial z_s}{\partial x} + uh,$$

$$q_y = -k_a(h_d + h)\frac{\partial z_s}{\partial y} + vh \tag{4}$$

where q_x and $q_y =$ unit width flow rate in the x, and y direction, respectively $k_a =$ saturated hydraulic conductivity, and $h_d =$ water depth in the soil layer, where the lateral infiltration flow is saturated and surface flow occurs.

4 RRI model Calibration and Validation

Calibration was performed using known rainfall data and observed river discharge data, adjusting selected model parameters in the model to match the model's outputs with the observed values. Generally, the calibration procedure consists of a combination of manual and automated calibrations. In this study, finding acceptable parameters was based on manual calibration through trial and error. Table 1 shows the calibrated parameters, and Table 2 shows the best Nash-Sutcliff efficiencies for each rainy season's flood event of 2011, 2012, 2013, and 2014. The simulation period, from 7 June to 23 August, was selected based on the accumulative rainfall data, as shown in Fig. 2. For the initial condition of the simulation, the water depth was set to zero after the dry season. The calibration process was conducted for different soil depths h_d and different lateral saturated hydraulic conductivity k_a values in the Bago River basin on a daily flow basis. These values were considered to depend on the soil texture and particle size distribution. The soil depth h_d values depend on how much water can be stored in the soil conditions and the size of sediment materials in the basin area. As indicated in Eq. (4), the subsurface velocity of flux varies with the gradient of the water level. If large values of k_a are used, the water can pass through the river in the earlier stages before saturating the soil. When the water depth exceeds the soil depth above an impermeable layer, such as bedrock, infiltration stops, and the surface water flows as described in Eq. (3). If the soil depth h_d is high in the calculation, a large amount of water enters the river with $k_a = 0$, and only surface water goes into the river. On the other hand, if large values are used, the soil has storage space, and then the water slowly move into the river that the discharge will be decreased. This is the effect of k_a and soil depth h_d. Therefore, the depth of soil depth h_d and lateral saturated hydraulic conductivity k_a are important parameters for fixing these calibration events.

Table 1. The calibration parameter values base on three land areas.

Calibration Parameters	Mountain	Flat Plain Area	Urban Area
Ns River	0.03	0.03	0.03
Ns slope	0.4	0.4	0.4
Soil depth h_d	1, 1.5, 2, 2.5, 3, 3.5	1	0.5
φ	0.463	0.501	0.475
k_{sv}	3.670×10^{-6}	1.000×10^{-6}	1.000×10^{-7}
Sf	8.890×10^{-2}	1.668×10^{-1}	3.163×10^{-1}
k_a	0, 0.025, 0.05, 0.1, 0.2	0	0

Table 2. The best Nash-Sutcliff efficiencies in calibration process for different event with different k_a value and soil depth h_d.

Sr	Events	k_a ($\mathrm{m\,s^{-1}}$)	Soil depth h_d (m)	Nash Sutcliffe Efficiency
1	2011	0.05	1.5	0.89
2	2012	0.05	2.5	0.93
3	2013	0	3	0.89
4	2014	0	3	0.89

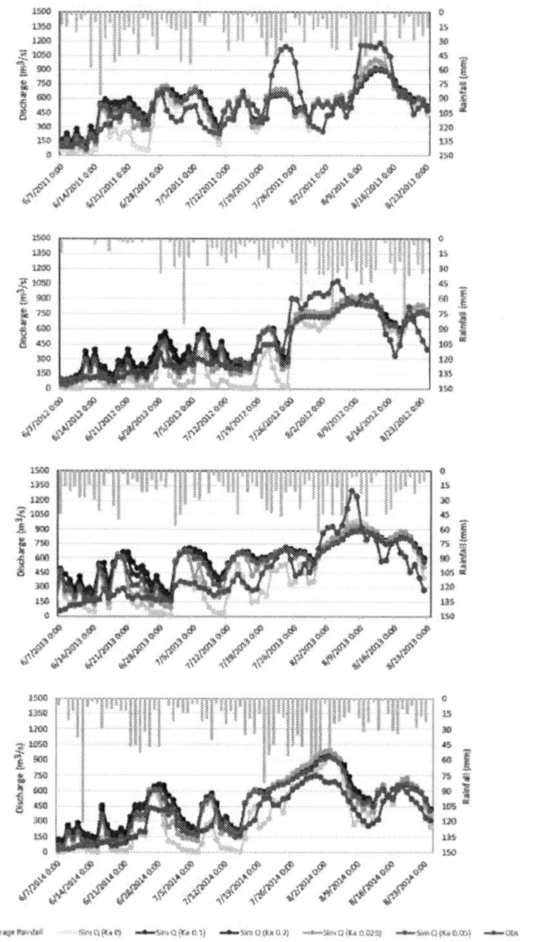

Figure 3. Simulated Hydrographs with different k_a values and calibrated soil depth h_d for each year.

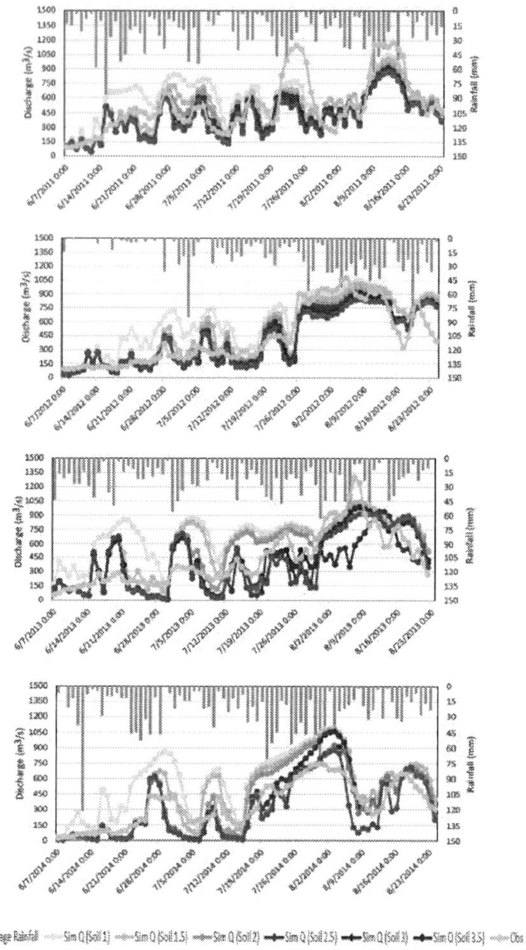

Figure 4. The Simulated Hydrographs with different soil depth h_d values and calibrated k_a for each year.

A total of 80 numerical runs were conducted with perturbations of these parameters.

5 Results and Discussion

In this study, the discharge hydrographs with different k_a values and soil depths h_d for this river basin were compared between Myanmar and other countries. The simulated hydrographs with different k_a values and calibrated soil depth h_d are shown in Fig. 3. For each year, increasing the k_a value decreases the simulation discharge in the early stage but increases it in the late season because the river basin is already saturated. As a result, the water flows through the river on the water surface flows. If large values are used, the water can pass through the river in the earlier stage before saturating the soil. However, if very small k_a values are used, the ground becomes saturated in the early rainy season.

The simulated hydrographs in different soil depth h_d values and calibrated k_a are shown in Fig. 4. When larger soil depth h_d values are used for year, the discharge becomes close to the observed discharge because the soil depth is high in the calculation, allowing a large amount of water to enter the river.

According to the simulation results, Table 3 shows the calibrated values for k_a and soil depth h_d for a mountainous area in Japan, Myanmar, and other countries. When calculated using the Nash Sutcliffe Efficiency value, which indicates a close relationship between the observed and simulated flow, the model performance is good. As a result, the best Nash Sutcliffe Efficiency values for the 2011 event were found to be soil depth $h_\mathrm{d} = 1.5$ m and $k_a = 0.05\,\mathrm{m\,s^{-1}}$, $h_\mathrm{d} = 2.5$ m and $k_a = 0.05\,\mathrm{m\,s^{-1}}$ for the 2012 event, $h_\mathrm{d} = 3$ m and $k_a = 0\,\mathrm{m\,s^{-1}}$ for the 2013 event, and $h_\mathrm{d} = 3$ m and $k_a = 0\,\mathrm{m\,s^{-1}}$ for the 2014 event, indicating a close correlation between the observed and simulated flows. In Japan's

Table 3. Comparison of the differences in k_a values and soil depths h_d among different countries.

Parameter	Japan (Subarctic & Subtropical)	Myanmar (This study) (Subtropical)	Oman (Arab Country) (Subtropical)	Thailand (Tropical)	Indonesia (Tropical)	Pakistan (Continental)
Lateral saturated hydraulic conductivity (k_a)	0.03–0.53	0–0.05	0.05	0.1	0.1	0.1–0.4
Soil Depth (h_d)	0.5–2	1.5–3	0.7–2	0.3	1.0	0.1

case, for the mountainous area, the values of lateral saturated hydraulic conductivity k_a ranged from 0.03 to 0.53 , and the soil depth h_d values ranged from 0.5 to 2 (Sayama et al., 2020; Shimizu et al., 2019; Yagi et al., 2020). In Myanmar's case, for the mountainous area, the lateral saturated hydraulic conductivity k_a values ranged from 0 to 0.05, and the values for the soil depth h_d ranged from 1.5 to 3. In Oman's case, for the mountainous area, the values of lateral saturated hydraulic conductivity k_a is 0.05, and the soil depth h_d values ranged from 0.7 to 2 (Abdel-Fattah et al., 2018).

In Thailand's case, for the mountainous area, the values of lateral saturated hydraulic conductivity k_a is 0.1, and the soil depth h_d value is 0.3 (Sayama et al., 2017; Try et al., 2020). In Indonesia's case, for the mountainous area, the values of lateral saturated hydraulic conductivity k_a is 0.1, and the soil depth h_d value is 1.0 (Nastiti et al., 2018). In Pakistan's case, for the mountainous area, the values of lateral saturated hydraulic conductivity k_a ranged from 0.1 to 0.4, and the soil depth h_d value is 0.1 (Sayama et al., 2012).

Therefore, the values of lateral saturated hydraulic conductivity k_a of Myanmar are smaller and the soil depth h_d values are larger than those of other countries.

The best parameter range values and sensitive parameters will change for different countries due to the climate zone, geographical features, flood characteristics, river morphology and sediment size of the basin area.

According to the table, the parameter values for soil depth h_d in Myanmar have a larger range than those in other countries, and the values of k_a are smaller. When the gap between the soil and the sediment materials size is large, water can easily flow into the ground layer. In Myanmar's case, the size of the sediment materials is not too large, hence; water cannot easily flow into the ground layer.

6 Conclusions

In this study, the RRI model was applied to rainfall and flooding events that happened in Myanmar in 2011, 2012, 2013, and 2014. The hydrographs were created, demonstrating acceptable performance in simulating flood events and are closely related to meteorological hydrographs observed, indicating the usefulness of the RRI model for flood inundation analysis. In summary, this study investigated the differences in values of lateral saturated hydraulic conductivity k_a and the soil depth h_d between Myanmar and other countries. However, in terms of Myanmar's river basin, the k_a value ranges from 0 to 0.05, and the soil depth h_d ranges from 1.5 to 3. In this study, considering these two parameters, the evaluation of this model's simulation focuses on the infiltration process. Infiltration parameters vary with rainfall, soil porosity, and the topography of the study area. The rate of infiltration depends on measurable soil properties, such as hydraulic conductivity, and the soil depth h_d. The effects of the parameters of lateral saturated hydraulic conductivity k_a and the soil depth h_d were investigated within the optimal range of $k_a = 0$ to 0.05 and $h_d = 1.5$ to 3, which indicate deep and compacted soil layers compared to other countries. Therefore, these two parameters were used to provide satisfactory output for measured data, and those coefficients closely match what is expected from the study basin area in Myanmar. Apart from that, the difference in these two parameters indicates the geographical features of Japan, Myanmar, and other countries where sediment deposits are deeply in large flat areas. As a result, Japan, and Oman's values range of k_a and the soil depth h_d are very close to Myanmar's case because they are in the same climate zone. More detailed investigation of the differences in parameters and geographical features is required for future research avenues.

Code availability. The RRI model is accessible on the International Centre for Water Hazard and Risk Management (ICHARM) website: https://www.pwri.go.jp/icharm/research/rri/index.html (Sayama, 2024).

Data availability. Meteorological and Hydrological data were provided by the Department of Meteorology and Hydrology (DMH) in Myanmar. The data sets are not publicly accessible.

Author contributions. PPP designed, planned, collected the data, systematized the processing and analysis of meteorological data, and wrote the original manuscript in consultation with TU. TU supervised the project, provided research tools, and edited the manuscript.

Competing interests. The contact author has declared that neither of the authors has any competing interests.

Disclaimer. Publisher's note: Copernicus Publications remains neutral with regard to jurisdictional claims made in the text, published maps, institutional affiliations, or any other geographical representation in this paper. While Copernicus Publications makes every effort to include appropriate place names, the final responsibility lies with the authors.

Special issue statement. This article is part of the special issue "ICFM9 – River Basin Disaster Resilience and Sustainability by All". It is a result of The 9th International Conference on Flood Management, Tsukuba, Japan, 18–22 February 2023.

Acknowledgements. We acknowledge the Department of Meteorology and Hydrology (DMH) in Myanmar for providing the data of the Bago River.

Financial support. This research has been supported by the Program for Japanese Grant Aid for Human Resource Development Scholarship (JDS) (Fellowship number B0012019MMR012).

Review statement. This paper was edited by Mohamed Rasmy and reviewed by two anonymous referees.

References

Abdel-Fattah, M., Kantoush, S. A., Saber, M., and Sumi, T.: Rainfall-Runoff modelling for extreme flash floods in Wadi Samail, Oman, J. JSCE Ser. B, 74, I_691–I_696, https://doi.org/10.2208/jscejhe.74.5_I_691, 2018.

Asian Development Bank: Republic of the Union of Myanmar, Transformation of urban management – flood management component final report part 2 (part 2 of 7), 1–57, Project Number: 46496-001, 2016.

Bates, P. D. and De Roo, A. P. J.: A simple raster-based model for flood inundation simulation, J. Hydrol., 236, 54–77, https://doi.org/10.1016/S0022-1694(00)00278-X, 2000.

Bhagabati, S. S. and Kawasaki, A.: Consideration of the Rainfall-Runoff-Inundation (RRI) Model for flood mapping in a deltaic area of Myanmar, Hydrol. Res. Lett., 11, 155–160, https://doi.org/10.3178/hrl.11.155, 2017.

Collection of Hydraulic Formulas: 2018 Edition, JSCE, Maruzen, Japan, 2019.

Leandro, J., Chen, A. S., and Schumann, A.: A 2D parallel diffusive wave model for floodplain inundation with variable time step (P-DWave), J. Hydrol., 517, 250–259, https://doi.org/10.1016/j.jhydrol.2014.05.020, 2014.

Liu, L., Liu, Y., Wang, X., Yu, D., Liu, K., Huang, H., and Hu, G.: Developing an effective 2-D urban flood inundation model for city emergency management based on cellular automata, Nat. Hazards Earth Syst. Sci., 15, 381–391, https://doi.org/10.5194/nhess-15-381-2015, 2015.

Nastiti, K. D., An, H., Kim, Y., and Jung, K.: Large-scale Rainfall–Runoff–Inundation modeling for upper Citarum river watershed, Indonesia, Environ. Earth Sci., 77, 640, https://doi.org/10.1007/s12665-018-7803-x, 2018.

Sayama, T.: Rainfall-Runoff-Inundation (RRI) Model technical manual, Tsukuba, International Center for Water Hazard and Risk Management (ICHARM), ISSN 0386-5878, 2014.

Sayama, T.: Rainfall-Runoff-Inundation (RRI) Model, Tsukuba, International Center for Water Hazard and Risk Management (ICHARM), https://www.pwri.go.jp/icharm/research/rri/index.html, last access: 6 March 2024.

Sayama, T., Ozawa, G., Kawakami, T., Nabesaka, S., and Fukami, K.: Rainfall–Runoff–Inundation analysis of the 2010 Pakistan flood in the Kabul river basin, Hydrolog. Sci. J., 57, 298–312, https://doi.org/10.1080/02626667.2011.644245, 2012.

Sayama, T., Tatebe, Y., and Tanaka, S.: An emergency response-type Rainfall-Runoff-Inundation simulation for 2011 Thailand floods, J. Flood Risk Manage., 10, 65–78, https://doi.org/10.1111/jfr3.12147, 2017.

Sayama, T., Yamada, M., Sugawara, Y., and Yamazaki, D.: Ensemble flash flood predictions using a high-resolution nationwide distributed Rainfall-Runoff Model: case study of the heavy rain event of July 2018 and Typhoon Hagibis in 2019, Prog. Earth Planet Sci., 7, 75, https://doi.org/10.1186/s40645-020-00391-7, 2020.

Shimizu, R., Uchida, T., and Kawahara, Y.: Flood analysis in the Nuta river basin during the Western Japan heavy rain in July 2018, 38th IAHR World Congress, International Association for Hydro-Environment Engineering and Research (IAHR), 1614–1620, https://doi.org/10.3850/38WC092019-1332, 2019.

Shimizu, R., Kaida, K., Matsuda, M., Uchida, T., Sayama, T., and Kawahara, Y.: Enhanced Rainfall-Runoff flood simulation with RRI Model incorporating river vegetation resistance based on 2D numerical calculation results and slope direction surface flow flux, J. JSCE Ser. B, 77, 84–91, https://doi.org/10.2208/jscejhe.77.1_84, 2021a.

Shimizu, R., Uchida, T., Kawahara, Y., Yamashita, A., and Fukuda, T.: High-accuracy prediction of water level in Rainfall-Runoff analysis coupled with unsteady 2-D calculation results, Advan. River Eng., 27, 241–246, https://doi.org/10.11532/river.27.0_241, 2021b.

Try, S., Tanaka, S., Tanaka, K., Sayama, T., Oeurng, C., Uk, S., Takara, K., Hu, M., and Han, D.: Comparison of gridded precipitation datasets for Rainfall-Runoff and Inundation modeling in the Mekong river basin, PLoS ONE, 15, e0226814, https://doi.org/10.1371/journal.pone.0226814, 2020.

World Meteorological Organization.: Manual on flood forecasting and warning (WMO-No.1072), 1–142, https://library.wmo.int/idurl/4/35881 (last access: 6 March 2024), 2011.

Yagi, F., Uchida, T., and Kawahara, Y.: Numerical investigation of three-dimensional flow structures and turbulence energy distributions in bank erosion reaches during large flood events, J. JSCE Ser. B, 10, 136–144, https://doi.org/10.2208/journalofjsce.10.1_136, 2022.

Proc. IAHS, 386, 121–126, 2024
https://doi.org/10.5194/piahs-386-121-2024

Open Access PIAHS

Hydraulic Analysis of the Marikina River Floodplain During Typhoon Vamco using Numerical Modelling

Jonathan S. Serrano[1], **Eugene C. Herrera**[1], **and Kensuke Naito**[2]

[1]Institute of Civil Engineering, University of the Philippines, Diliman, Quezon City, 1101, Philippines
[2]International Centre for Water Hazard and Risk Management, Tsukuba-shi, 305-8516, Japan

Correspondence: Jonathan S. Serrano (jsserrano@up.edu.ph)

Received: 22 July 2023 – Revised: 5 December 2023 – Accepted: 24 February 2024 – Published: 19 April 2024

Abstract. Typhoon Vamco in 2020 costed the Philippines over PHP 19 billion in damages. One of the heavily affected floodplains is the Marikina River floodplain. Discharges of up to $2582\,m^3\,s^{-1}$ along Marikina River were noted. This is the strongest flashflood incident in the area after Typhoon Ketsana in 2009. This study aims to investigate the hydraulics of flood development within the Marikina River floodplain during Typhoon Vamco using numerical modeling. The hydraulic model in HEC-RAS utilized a 1 m LIDAR DTM, calibrated discharges from a hydrologic model and water levels as upstream and downstream boundary conditions, respectively, and a rain-on-grid input. Results showed a max inundation area of $\sim 75\,km^2$ and flood depths of up to 4 m which accounts for the damages in some municipalities and cities in Metro Manila and Rizal Province. Model validation showed the simulated peak level (11.31 m a.s.l.) to be only 1.0 % lower than the peak observed water level (11.43 m a.s.l.). Model results can be used for measures to minimize risk and damages.

1 Introduction

Around 80 % of the natural disasters occurring in the Philippines are attributed to hydrometeorological events like typhoons and floods (Jha et al., 2018). These events cause a lot of socioeconomic damages. About twenty typhoons enter the Philippine Area of Responsibility (PAR) each year, with about eight to nine of these making landfall, and about five are considered destructive based on the Asian Disaster Reduction Center (Santos, 2021). On 26 September 2009, Typhoon Ketsana or Tropical Storm Ondoy struck parts of Luzon. Around 455 mm of rains were poured in 24 h, although about 347 mm of these poured in just 6 h. Ondoy caused 464 fatalities and costed more than PHP 11 billion (NDCC, 2009) in damages. A more recent typhoon, Vamco a.k.a. Typhoon Ulysses, made landfall on 11 November 2020 and heavily struck parts of Luzon. Vamco poured 356 mm of rains in 24 h, killing 101 people and costing over PHP 19 billion (NDRRMC, 2021) in damages. Communities beside Marikina and Pasig Rivers were highly affected due to downpour over Marikina River Basin. Flood risk is extremely high in the said areas due to the size of the Marikina River Basin and due to the highly urbanized communities within Metro Manila and Rizal Province. Damages to infrastructure and agriculture brought about by Vamco in Metro Manila alone were around PHP 717 million (NDRRMC, 2021). The River Basin and its floodplain are also highly monitored, which makes it a good site for flood modeling studies. The purpose of this study is to assess the hydraulics of flood development (i.e. flood depths, inundation extents, and timing of flood propagation, etc.) within the floodplain of Marikina River Basin using numerical modeling and to know what solutions can be done to minimize the flood risk. The Hydrologic Engineering Center (HEC) tools of the US Army Corps of Engineers (USACE) were used to assess the flood development of Typhoon Vamco.

2 Materials and methods

2.1 Study site

The Marikina River Basin has a catchment area of around $546\,km^2$. It is located at the southern part of Luzon, less than 5 km northwest of Laguna Lake, and around 9 km east of Manila Bay. About 90 % of the basin is in the mountainous province of Rizal while the rest and more downstream section is in the highly urbanized Metro Manila. The whole study site is comprised of the Marikina River Basin, along with Manggahan Subbasin ($\sim 85\,\mathrm{km}^2$) and Taguig Subbasin ($\sim 43\,\mathrm{km}^2$) which both drain to Laguna Lake. The two subbasins are also part of the floodplain of Marikina River due to the complex river network in the floodplain. The Marikina River Basin has a dendritic drainage system, with its vast network of streams converging into the Upper Marikina River. The Upper Marikina River diverges into the Lower Marikina River and the Manggahan Floodway, which is a manmade channel constructed to divert flows to Laguna Lake before eventually discharging into Pasig River during storm events. The Lower Marikina River diverges into Pasig River to the west which naturally discharges to Manila Bay and into Napindan Channel to the east which leads to Laguna Lake. Along the Upper Marikina River is the Sto. Niño Water Level Monitoring Station (WLMS), which is considered a crucial point for the models in this study. Figure 1 shows the study site, where both hydrologic and hydraulic model domains are delineated; its location within Luzon, which is the largest island in the Philippines; and a closer look at the river system within the Marikina River floodplain.

2.2 Methodological framework

The flood model consists of both a hydrologic and a hydraulic model. The hydrologic model transforms the rainfall over the more upstream areas of the Marikina River Basin into runoff along the Upper Marikina River. The model's extent is set such that the outlet is set at the Sto. Niño WLMS. Discharges available at this station were used to calibrate the hydrologic model. After calibration, flows at a more upstream segment of the Upper Marikina River were extracted from the model and were used as upstream boundary condition for the 2D hydraulic model. Aside from these flows, water levels at the lake and at the confluence of Pasig River and Marikina River were used as boundary conditions. The hydraulic model results to flood maps at any timestep within the simulation period. This model was calibrated so that the simulated water levels at Sto. Niño WLMS will be as close as possible to the observed water levels at that section. Once calibrated, flood maps were generated and flood depths, inundation areas, and flood volumes were extracted. Both secondary meteorologic and topographic data were used in the hydrologic and hydraulic models. The methodological framework is also shown in Fig. 1.

2.3 Data

2.3.1 Terrain

Digital Elevation Models (DEM) were used as terrain input for both hydrologic and hydraulic models. Interferometric Synthetic Aperture Radar – Digital Terrain Model (IFSAR – DTM) with 5 m resolution (NAMRIA, Philippines, 2013) was used for the hydrologic model while Light Detection and Ranging (LIDAR) DTM with 1 m resolution was used for the hydraulic model. Bathymetry of main rivers were included in the hydraulic model for more accurate computations of water depths and velocities. Since bathymetric points were only available for the reaches of Lower Marikina River, Manggahan Floodway, and Napindan Channel, bathymetric data along the Upper Marikina River was extrapolated using the elevations at the confluence of Manggahan Floodway and Marikina River, and the slope of the water level surface along the Upper Marikina River as reflected by the LIDAR DTM (assumed equal to the bed slope).

2.3.2 Land and soil covers

The 2015 Land Cover Map from the National Mapping and Resource Information Authority (NAMRIA, Philippines, 2015) and soil data from the Food and Agriculture Organization (FAO-UNESCO, 1981) were used to determine SCS curve numbers for the hydrologic model. The same land cover map was used for Manning's N values in the hydraulic model.

2.3.3 Meteorologic data

The meteorologic data used for both models were hourly rainfall datasets. A single time-series data was used for each model, with each time-series data comprising a combination of point rainfall data that would yield the highest rainfall volume. The point rainfall datasets were from the ground-based weather gauges in the vicinities of respective model domains. Rainfall volume was maximized to attain enough flood volume both for the hydrologic and hydraulic models, since it was seen that using a spatially averaged rainfall data prepared using Thiessen weights yielded an underestimated flood volume. The data used for the hydrologic model were from the following gauges: Aries, Boso-Boso, Mt. Campana, Mt. Oro, and Nangka. For the hydraulic model, Mt. Campana was removed, while the gauges of Napindan and Science Garden were added to the list.

2.3.4 Water levels and discharges

Rating curve-derived discharges at Sto. Niño WLMS were used to calibrate the hydrologic model. Meanwhile, the water levels at the same station were used to calibrate the hydraulic model. Other secondary water level data were used as downstream boundary conditions for the hydraulic model.

2.4 Model set-up

2.4.1 Hydrologic model

HEC-HMS (Hydrologic Modeling System) was used as the hydrologic modeling platform. The model domain was first delineated into 43 subbasins and 21 reaches. The methods

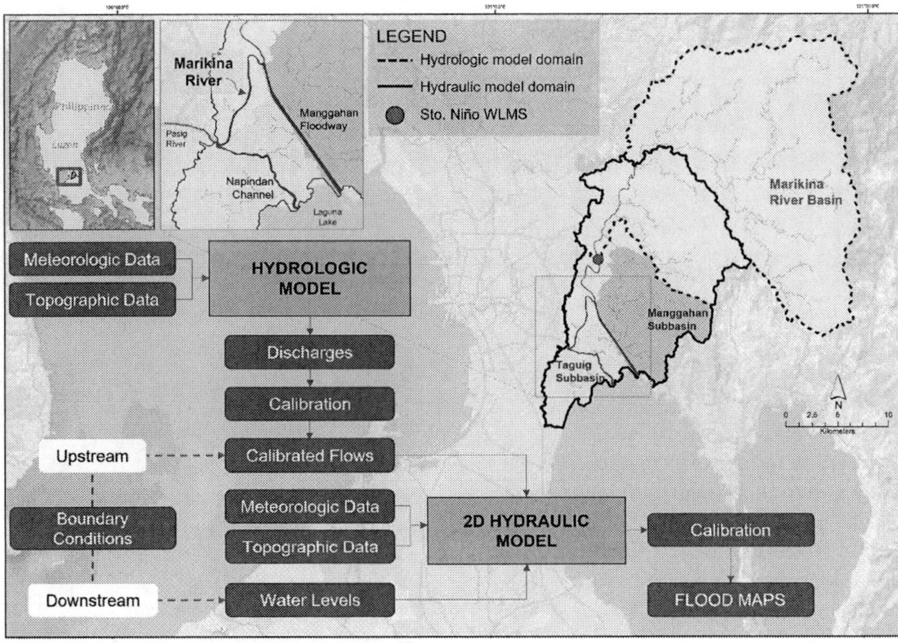

Figure 1. Study site, model domains, and methodological framework (Ocean Basemap Source: ESRI).

chosen for the model are as follows: SCS Unit Hydrograph for the rainfall-runoff transformation; SCS Curve Number for infiltration loss; Simple Canopy and Simple Surface for the stored water in canopies and surfaces respectively; Constant Monthly for the baseflow; and Kinematic Wave method for the river routing of flows. The simulation period used for the hydrologic model was from 10 November 2020, 00:00:00 LT to 13 November 2020, 23:00:00 LT. This three-day period was enough to include a warm-up for the model, as well as enough time for flood wave attenuation after peak discharge occurs at the outlet. Computation time interval was set to one minute. The main parameter adjusted for calibration was the lag time in the SCS Unit Hydrograph method. Initial values for this input were computed using watershed parameters obtained from topographic data.

2.4.2 Hydraulic model

A 40×40 m grid was set up over the 2D flow area in HEC-RAS (River Analysis System). A Manning's N map was also set over the same area. Boundary condition lines were placed in respective positions: at the upstream end of the Upper Marikina River within the model domain; at the confluence of Marikina River and Napindan Channel, which is at the west boundary of the model extent; and at the edge of the model domain adjacent to Laguna Lake at the southeast. Simulation period was set from 11 November 2020, 02:00:00 LT to 13 November 2020, 23:00:00 LT. The diffusion wave equation was utilized to maintain model computation stability at five-second computation time interval. Adjustments of the Manning's N values of channels were made until the closest fit between the observed and simulated water levels at Sto. Niño WLMS were attained.

3 Results and discussion

3.1 Flood development

The inundation area was determined for every two hours starting from the 11 November 2020, 03:00:00 LT timestep, and for the initial timestep at 11 November 2020, 02:00:00 LT to see the flooding brought about solely by the initial conditions in the model. The inflow hydrograph, shown in Fig. 2, peaks at $2126\,m^3\,s^{-1}$ on 12 November 2020, 09:00:00 LT. Maximum flood volume also occurs at this timestep. However, maximum inundation in terms of area ($\sim 75\,km^2$) was attained four hours prior to this, signifying that flood extent is maxed earlier than occurrence of peak discharge and starts to decline even before flood depths increase in more flood prone areas. Figure 2 shows the rainfall hyetograph, upstream discharges (i.e. the inflow hydrograph to the hydraulic model), the inundated area within the floodplain, along with the timing of peak values.

A notable increase in the flooded areas in km^2 can be seen from the ninth to eleventh data points (11 November 2020, 17:00:00, 19:00:00, and 21:00:00 LT; 22, 36, $54\,km^2$ respectively). The start of this period coincides with the time when inflow discharges start increasing from its baseflow value of

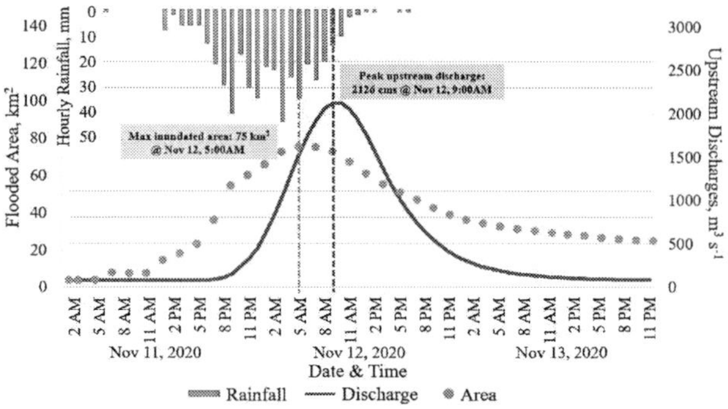

Figure 2. Flood development within the Marikina River floodplain.

$75\,m^3\,s^{-1}$. Rainfall also starts to increase significantly in the said period. Results show it took 16 h of continuous raining within the floodplain before maximum inundation area was attained (starting from 11 November 2020, 13:00:00 LT), and another 19 h for the flooded area to decrease by 50 %.

3.2 Maximum flood depths

The average maximum flood depths per land cover type as well as per barangay were extracted. Max flood depths per land cover type were noted to know the extent of damage in built-up areas and other types in which damage from floods may have economic implications such as grasslands and agriculture. Average max depths in built-up areas, agricultural lands, and grasslands were found to be 0.8, 2.1, and 1.3 m, respectively. Average max flood depth over built-up areas was the lowest among all land cover types within the floodplain since it covers most of the model domain. Rain-induced local ponding in some areas which are much shallower than riverine floods tend to decrease the average depth. The most heavily flooded barangays based on the max flood depth map were Dulong Bayan I (4.00 m), Guitnang Bayan I (2.97 m), Concepcion Uno (2.22 m), Guinayang (2.20 m), and Jesus de la Peña (2.18 m). Figure 3 below shows the resulting max flood depth map and the five most heavily flooded barangays in terms of flood depths.

3.3 Model calibration

For the hydrologic model, adjusting the lag time of the subbasins among other watershed parameters resulted to a Nash Sutcliffe efficiency (NSE) of 0.97 and a percent bias (PBIAS) of 1.7. These values are classified as having "very good" model performance rating (Moriasi et al., 2007). However, simulated peak discharge ($2854\,m^3\,s^{-1}$) was 10.5 % higher than the observed ($2582\,m^3\,s^{-1}$) at Sto. Niño. The simulated peak discharge also occurred one hour ahead (12 November 2020, 09:00:00 LT) of the observed. Further calibration and input data reassessment can minimize this gap. The hydrologic model calibration plot is shown in Fig. 4.

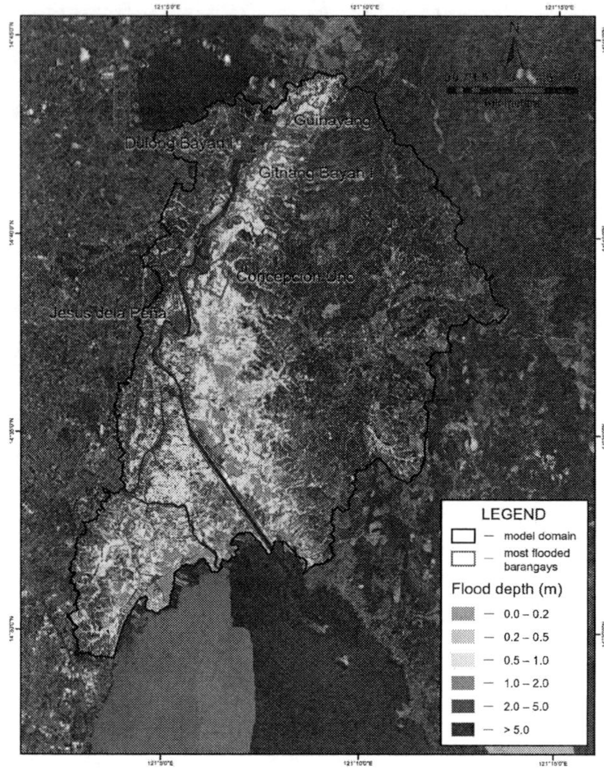

Figure 3. Maximum flood depth map and the five most flooded barangays (World Imagery Basemap Source: ESRI).

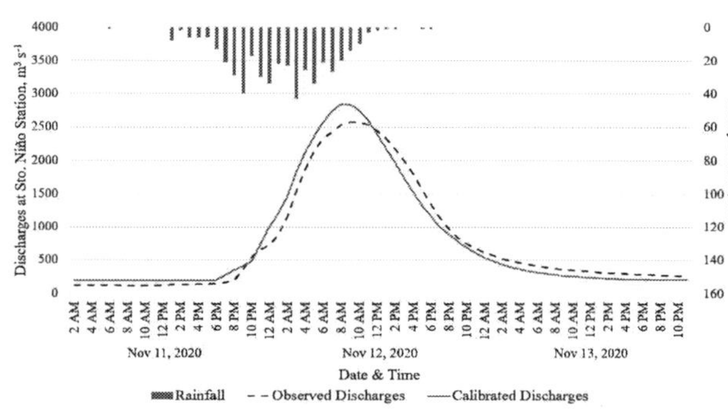

Figure 4. Hydrologic model calibration plot at Sto. Niño Station.

For the hydraulic model, adjusting the Manning's N of main channels to 0.035 yielded the nearest peak water level in terms of both depth and timing. This resulted to an NSE of 0.88, with the simulated level (∼ 11.31 m a.s.l.) being only 1 % lower than the observed (∼ 11.43 m a.s.l.). This peak occurs one hour later in the simulation (12 November 2020, 11:00:00 LT) than the observed. The hydraulic model calibration plot is shown in Fig. 5.

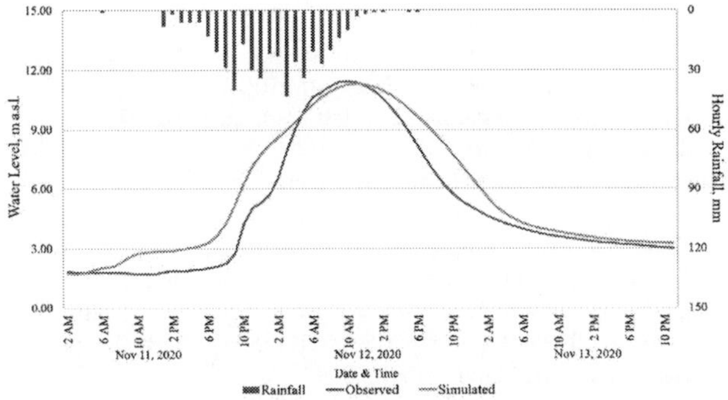

Figure 5. Hydraulic model calibration plot at Sto. Niño Station.

This figure translates to a time to peak of ten[1] hours. This is the length of time between the centroid of the rainfall hyetograph and the peak discharge. More sense or implication can be deduced upon looking at the cumulative rainfall of the data used. Results showed that 20 h of continuous rainfall that accumulates to ~ 230 mm, and/or 24 h accumulating to ~ 350 mm of rains can produce the same inundation caused by Typhoon Vamco. These values, along with information on flood extent can give lead times of ten and/or six hours respectively before peak flood volume occurs. Further development and calibration of both models can still improve the results. Continuous datasets of water levels and discharges of another storm event can also be used for validation. Such flood models can be the key to flood risk mitigating solutions.

3.4 Simulation-based recommendations

From the results, it can be recommended to authorities to shift the focus from water level monitoring to rainfall monitoring as this would give a lot more lead time for evacuation. It is also highly recommended to use updated flood maps in determining safe evacuation sites. Regularly updating flood risk maps may also be helpful since lots of factors can alter the topography which play a crucial role in the flood development during storm events.

4 Conclusions

This study showed the effectiveness of using numerical models to investigate the flood development of a strong typhoon such as Typhoon Vamco. A coupled hydrologic model (HEC-HMS) and hydraulic model (HEC-RAS) that yielded an NSE of 0.97 for discharge hydrographs and 0.88 for stage hydrographs respectively, implies the nearness of the model results to that which occurred during Typhoon Vamco. A difference of 1 % in the peak water level at Sto. Niño WLMS and the one-hour difference in time-to-peak of simulated and observed values also imply a certain level of accuracy. Average max flood depths and timing of flood extents as well as flood volumes for different timesteps were determined. This study can help assess already existing flood control plans in the area, with the overall objective of minimizing flood risk and damages in the future.

[1] considering hydrograph from the simulation

Data availability. Only processed data directly used for the final version of the model are available upon request. Raw data are to be gathered from the agencies from which the authors collected such datasets.

Author contributions. ECH conceptualized the study upon the request of a stakeholder for a research project. JSS set up and simulated the models, conducted the extraction, analysis, and documentation of model results upon the supervision of ECH. JS wrote the draft and edited it based on the review done by ECH. KN also reviewed the abstract and supported the overall study for its submission to the ICFM9 conference.

Competing interests. At least one of the (co-)authors is a guest member of the editorial board of *Proceedings of IAHS* for the special issue "ICFM9 – River Basin Disaster Resilience and Sustainability by All". The peer-review process was guided by an independent editor, and the authors also have no other competing interests to declare.

Special issue statement. This article is part of the special issue "ICFM9 – River Basin Disaster Resilience and Sustainability by All". It is a result of The 9th International Conference on Flood Management, Tsukuba, Japan, 18–22 February 2023.

Acknowledgements. This research was supported by the Project for Development of a Hybrid Water-Related Disaster Risk Assessment Technology for Sustainable Local Economic Development Policy Under Climate Change in the Philippines (HyDEPP-SATREPS) of the Japan International Cooperation Agency (JICA) and Japan Science and Technology Agency (JST); and the Department of Science and Technology (DOST) and DOST Philippine Council for Industry, Energy and Emerging Technology Research and Development (DOST-PCIEERD) through the project Eco-System Modelling and Material Transport Analysis for the Rehabilitation of Manila Bay (e-SMART) under the IM4ManilaBay Program.

Financial support. This research has been supported by the Science and Technology Research Partnership for Sustainable Development (grant no. JPMJSA1909) and the Department of Science and Technology, Philippines (grant no. Eco-System Modelling and Material Transport Analysis for the Rehabilitation of Manila Bay (e-SMART) Project).

Review statement. This paper was edited by Mohamed Rasmy and reviewed by two anonymous referees.

References

FAO-UNESCO: Digital Soil Map of the World, FAO-UNESCO [data set], https://www.fao.org/soils-portal/data-hub/soil-maps-and-databases/faounesco-soil-map-of-the-world/en (last access: 18 March 2024), 1981.

Jha, S., Martinez, A., Quising, P., Ardaniel, Z., and Wang, L.: Natural Disasters, Public Spending, and Creative Destruction: A Case Study of the Philippines, ADBI Working Paper 817, https://doi.org/10.2139/ssrn.3204166, 2018.

Moriasi, D. N., Arnold, J. G., Van Liew, M. W., Bingner, R. L., Harmel, R. D., and Veith, T. L.: Model Evaluation Guidelines for Systematic Quantification of Accuracy in Watershed Simulations, T. ASABE, 50, 885–900, https://doi.org/10.13031/2013.23153, 2007.

NAMRIA, Philippines: IFSAR DEM, Photogrammetry Division, Mapping and Geodesy Branch [data set], https://namria.gov.ph/Files/RequestforIFSARandLIDARData,Orthoimage,OrthophotoandAerialPhotographs.pdf (last access: 18 March 2024), 2013.

NAMRIA, Philippines: Land Cover Map, Resource Data Analysis Branch, 2015.

National Disaster Coordinating Council (NDCC): SitRep 48 on TS "Ondoy" and Typhoon "Pepeng", Report, https://reliefweb.int/attachments/b5da56c0-e8de-30ea-9352-f59e4948cd61/D2F7A40978978E214925766C001B455A-Full_Report.pdf (last access: 18 March 2024), 2009.

National Disaster Risk Reduction and Management Center (NDRRMC): SitRep 29 re Preparedness Measures and Effects for Typhoon "ULYSSES" (I.N. VAMCO), Report, https://reliefweb.int/attachments/c81781d2-633e-3fae-bd53-82d7517e94dd/SitRep_no_29_re_TY_Ulysses_as_of_13JAN2021.pdf (last access: 18 March 2024), 2021.

Santos, G. D.: 2020 tropical cyclones in the Philippines: A review, Tropical Cyclone Research and Review, 10, 191–199, https://doi.org/10.1016/j.tcrr.2021.09.003, 2021.

Proc. IAHS, 386, 127–132, 2024
https://doi.org/10.5194/piahs-386-127-2024

Assessment of flood damage to agricultural crops under climate change scenarios using MRI-AGCM outputs in the Solo River basin of Indonesia

Badri Bhakta Shrestha, Mohamed Rasmy, Tomoki Ushiyama, Ralph Allen Acierto, Takatoshi Kawamoto, Masakazu Fujikane, Hiroyuki Ito, and Takafumi Shinya

International Centre for Water Hazard and Risk Management,
Public Works Research Institute, 1-6 Minamihara, Tsukuba, Ibaraki, 305-8516, Japan

Correspondence: Badri Bhakta Shrestha (shrestha@icharm.org)

Received: 11 July 2023 – Revised: 6 October 2023 – Accepted: 24 February 2024 – Published: 19 April 2024

Abstract. Understanding the effects of climate change on agricultural crops is an essential part of flood-risk management and adaptation measures. However, the quantitative impact of flood damage on agricultural crops under climate change is still not clearly understood. It is thus crucial to investigate flood damage to agricultural crops under climate change scenarios to understand future crop damage better. In this study, we first explored the relationship between extreme rainfall and damaged paddy area for a historical period, and then focused on the quantitative analysis of flood impact on rice crops under climate change using MRI-AGCM3.2S climate model outputs for the past (1979–2002) and future (2075–2098) climate periods for the Solo River basin in Indonesia. We developed a quantitative damage assessment method by coupling water and energy budget-based rainfall-runoff-inundation model outputs and a flood loss model. Flood damage to rice crops was defined as a function of flood depth and duration, and depth-duration-damage curves were applied to quantify the damage. The results show that flood damage to rice crops will be more severe in the future than in the past, and the findings can be useful in establishing risk reduction and adaptation measures.

1 Introduction

Floods are the most frequently occurring natural disaster in many countries (Komolafe et al., 2019). Flood damage to crops causes significant economic losses and impacts food security and rural livelihoods, particularly in developing countries (Shrestha et al., 2016). Flood damage may be more serious in the future because of the increasing frequency and intensity of extreme weather events due to climate change. Therefore, understanding the impact of climate change on agricultural crops and quantitatively assessing flood damage to agricultural crops under climate change scenarios are essential for planning and implementing flood prevention and adaptation measures. However, the quantitative impacts of flood damage to agricultural crops and other sectors under climate change have not been clearly understood since they have spatially heterogeneous characteristics. Therefore, this study focused on the quantitative analysis of flood impact on rice crops under extreme climate change scenarios using atmospheric general circulation model (AGCM) outputs produced by the Meteorological Research Institute (MRI) (MRI-AGCM3.2S).

We selected the Solo River basin (SRB) in Indonesia for this study. First, we explored the relationship between extreme rainfall and damaged paddy area for a historical period based on linear regression analysis. Then, we formulated a flood damage assessment framework for rice crops under climate change to quantify the damage by coupling hydrologic-hydraulic model outputs and a flood damage estimation model, and assessed the rice crop damage for the past (1979–2002) and future (2075–2098) periods using MRI-AGCM3.2S climate model outputs.

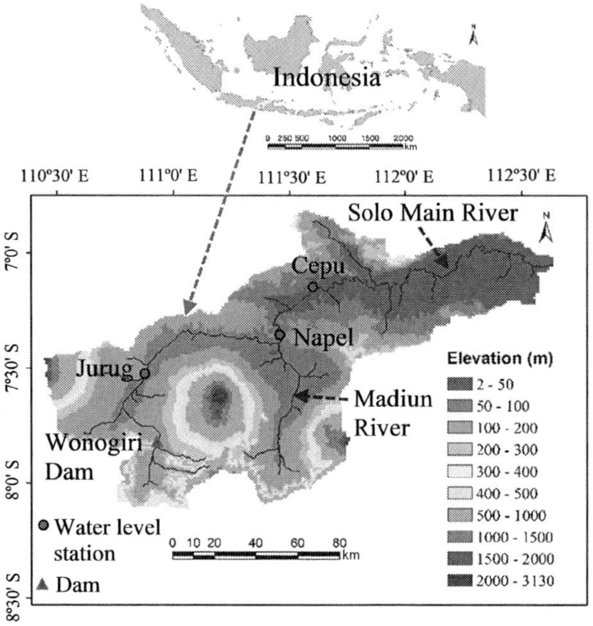

Figure 1. Location of the Solo River basin, Indonesia (Elevation data: HydroSHEDS digital elevation model).

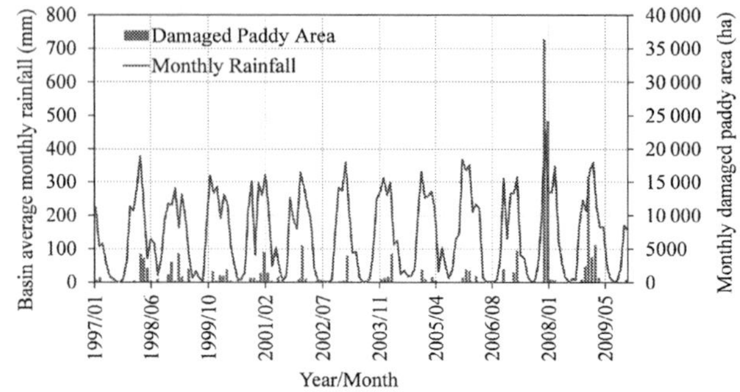

Figure 2. Monthly rainfall and monthly damaged (affected) paddy area from January 1997 to December 2009.

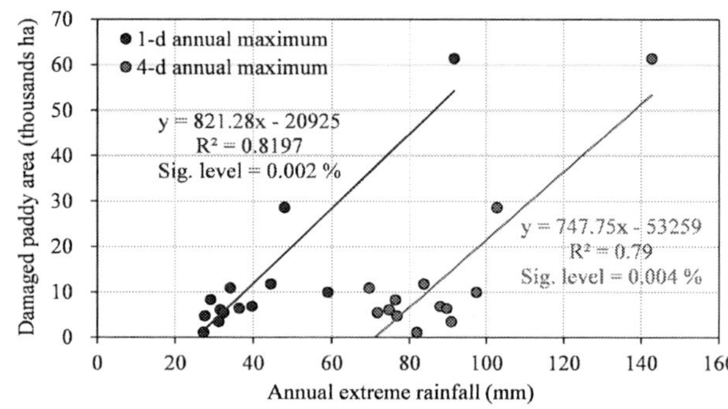

Figure 3. Plots between damaged paddy area and annual extreme rainfall (1 or 4 d annual maximum) and their linear relationships (Sig. = Significance).

2 Study area

The SRB is located in Java Island in Indonesia (Fig. 1). The catchment area of the study boundary is about 15 752 km^2, and the main Solo River is approximately 600 km long. The Wonogiri multipurpose dam, with a reservoir capacity of 730×10^6 m^3, is in the upper part of the basin. The SRB is one of the country's important river basins in terms of agriculture production. However, agricultural crops are often damaged due to flooding in the area. Therefore, the assessment of flood damage to agricultural crops under climate change in this basin is essential to implement effective preventive and adaptation measures.

3 Relationship between extreme rainfall and damaged paddy area

We collected data on damaged paddy area for 1997–2009 from the Directorate of Food Crop Protection (DFCP, 2010). The collected data were used to study the relationship between extreme rainfall and damaged paddy area based on linear regression analysis. Figure 2 shows the plot between monthly rainfall and total monthly damaged paddy area for the regencies in the study area. The figure shows that flood damage to rice crops usually occurred in December–March, when the basin average monthly rainfall was higher than 200 mm. The figure shows that a significantly large area of rice paddies was affected during the flood in December 2007–January 2008, and the basin average monthly rainfalls in these months were 455 and 270 mm, respectively.

In SRB, the monsoon period is from November to April, and floods usually occur between December and March. We thus defined the period from 1 August (beginning from the month with the least rainfall in a year) to the end of July as the hydrological year. The relationship between damaged paddy area (i.e., flooded paddy area with rice crop damage reported by local government (DFCP, 2010)) and extreme rainfall was analyzed using the annual data for the hydrological year. The 1 and 4 d maximum rainfall were considered for analysis to represent floods due to heavy rainfall over a single day or several days, based on rainfall and discharge characteristics. Figure 3 shows the linear relationship between the annual damaged paddy area and the 1 or 4 d annual maximum basin average rainfall. The damaged paddy area varies positively with the extreme rainfall, and the linear relationship between the damaged paddy area and the 1 or 4 d maximum rainfall was statistically significant, with a significance level of less than 0.005 %. The linear trends show that floods may affect rice paddies when 1 or 4 d rainfall exceeds 25 or 71 mm, respectively, although flooding may depend on the spatial distribution and the occurrence time of rainfall.

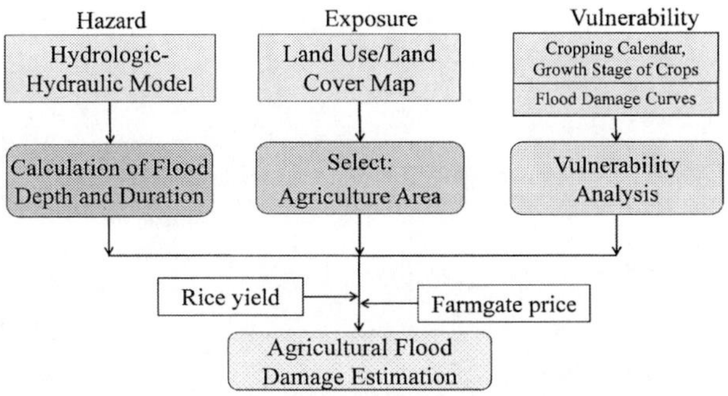

Figure 4. Process of estimating flood damage to rice crops.

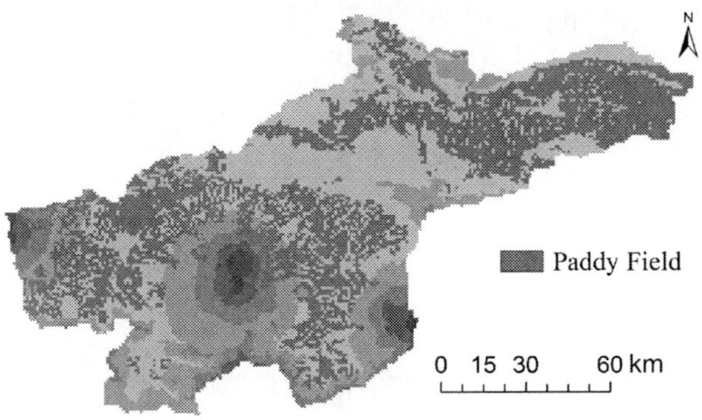

Figure 5. Paddy area based on the 2006 land cover produced by the Ministry of Environment and Forestry, Indonesia.

The 1 and 4 d maximum basin average rainfalls during the 2007–2008 flood event, the largest recorded flood event in the basin, were 91.6 and 142.7 mm, respectively.

4 Flood damage assessment under climate change

4.1 Global climate model outputs

This study used the global climate model (GCM) outputs of MRI-AGCM3.2S, which are the high-resolution climate outputs from the Climate Model Intercomparison Project-phase 5. The GCM outputs of MRI-AGCM3.2S (past period: 1979–2002, future period: 2075–2098), which were dynamically downscaled to a 5 km domain using a weather research forecasting model and statistically corrected the bias using ground-observed rainfall data for the past period, were employed to simulate flood hazard conditions under climate change scenarios. For the future period, the representative concentration pathway (RCP) 8.5 scenario, the high-end emissions scenario experiments, was used.

4.2 Flood hazard analysis

We used the water and energy budget-based rainfall-runoff-inundation (WEB-RRI) model developed by Rasmy et al. (2019) for flood hazard simulation. The WEB-RRI model takes into account all the natural hydrological processes in a catchment and simulates evapotranspiration, the rainfall-runoff process, infiltration, and flood inundation. We used the digital elevation model data obtained from the HydroSHEDS. First, the parameters of the WEB-RRI model were calibrated and validated by comparing calculated discharges with observed discharges for past flood events. Outflows from the Wonogiri dam were also considered for the flood events of calibration and validation. Then, flood simulations were conducted using the calibrated and validated WEB-RRI model with MRI-AGCM3.2S outputs for the past and future climate periods. The WEB-RRI model was set up using bias-corrected rainfall data and other forcing data, such as temperature and radiation, obtained from downscaled MRI-AGCM for the past and future climates. The details on other required data for WEB-RRI model and their sources can be found in Rasmy et al. (2019). Natural flow conditions (i.e., no dam conditions) were considered in flood simulations for the past and future climate periods. Inundation may occur due to fluvial flooding when water overflows the river banks or pluvial flooding when extreme rainfall accumulates in the area.

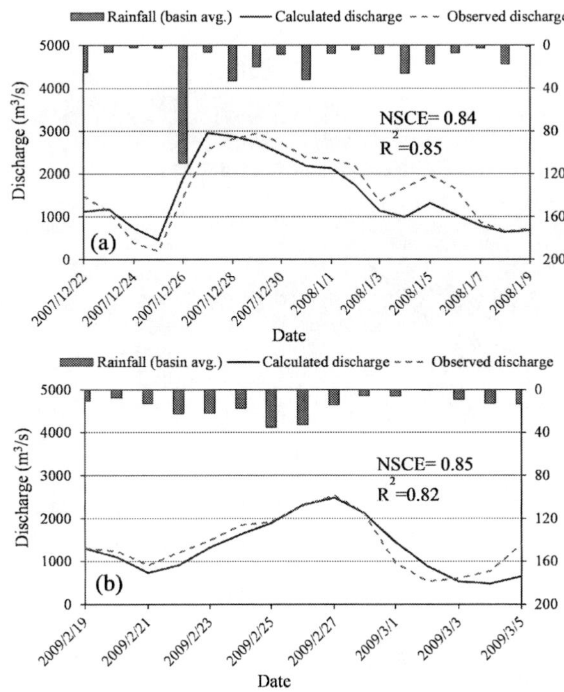

Figure 6. Comparison of calculated and observed discharges at the Cepu station for flood events in 2007/2008 and 2009.

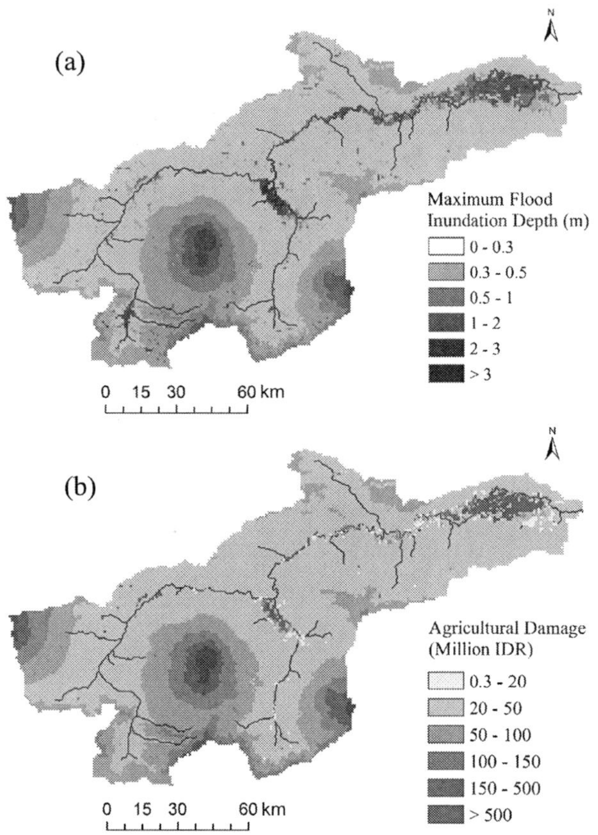

Figure 7. **(a)** Calculated flood inundation depth and extent and **(b)** estimated rice-crop damage, during 2007/2008 flood.

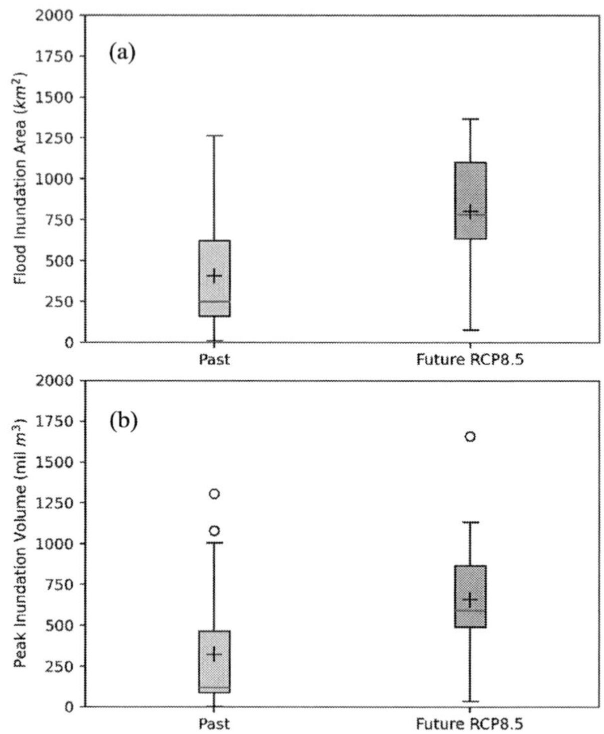

Figure 8. Boxplots of **(a)** flood inundation area and **(b)** peak inundation volume for the past and future climate periods (plus mark: average value).

4.3 Flood damage assessment

Figure 4 shows the process of estimating flood damage to rice crops. The flood damage to rice crops in terms of economic losses was calculated as the product of the value of the exposed rice crops in hazardous areas and flood vulnerability. Flood depth and duration were considered as key flood parameters that affect rice crops during flooding. The paddy fields in the study area were extracted from the land cover map for 2006 produced by the Ministry of Environment and Forestry, Indonesia (Fig. 5).

The flood damage curves (FDCs) for rice crops presented by Shrestha et al. (2016) were applied to quantify the damage, which can be applied to the study area because major rice crops grown in Southeast Asia are similar types (Shrestha et al., 2021). FDCs were defined as the function of flood depth, duration, and crop growth stage. The applicability of FDCs or the damage assessment method was also validated by comparing the calculated damage with reported data for past floods. When flooding occurs during the vegetative stages, farmers normally replant rice crops, and in such a case, the value of rice crop damage was estimated as losses of the cost of input. However, when flooding occurs during the reproductive and maturity stages of rice plants, the value of rice crop damage was estimated as the volume of production losses with farmgate price. The volume of rice production losses was estimated as the product of rice yield, damaged paddy area, and yield loss (Shrestha et al., 2016). The volume and value of rice crop loss were estimated using the following values: a rice yield of 5230 kg ha^{-1} (Panuju et al., 2013), a farm gate price of IDR 4650 per kg (Wright and Meylinah, 2014), and a cost of input of IDR 1 970 414 per ha (Zakaria et al., 2004). The growth stage of the rice plant during the past flood events was identified with reference to the duration of each growth stage of the rice plant (Shrestha et al., 2016) and the cropping calendar published by DSRI (2001). After the validation of the method, flood damage to rice crops was assessed for the past (1979–2002) and future (2075–2098) periods under climate change scenarios using MRI-AGCM3.2S outputs with scenarios of different rice growth stages.

5 Results and discussion: flood damage assessment

Figure 6 compares the calculated discharge using the WEB-RRI model with the observed discharge at the Cepu for the

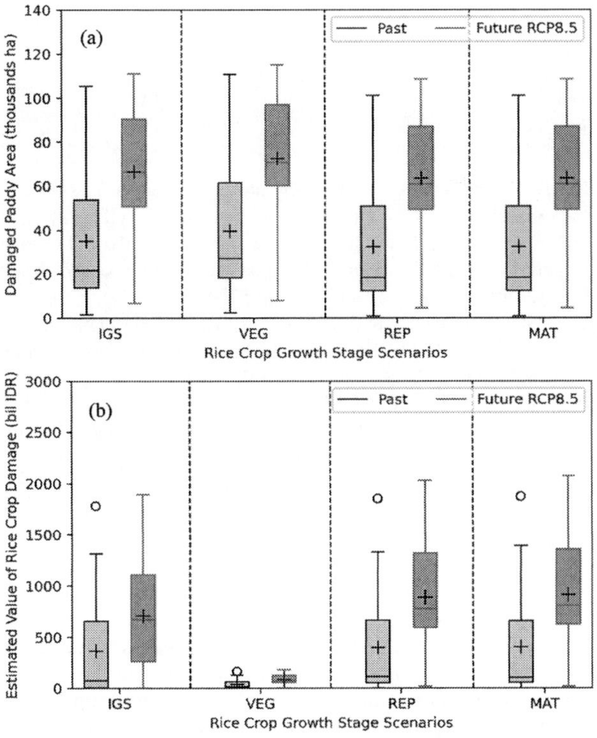

Figure 9. Boxplots of **(a)** calculated damaged paddy area and **(b)** estimated rice-crop damage for the past and future climate periods (plus mark: average value).

2007/2008 and 2009 floods. The results show good model performance with high values of the Nash-Sutcliffe coefficient of efficiency (NSCE) and R-squared. Figure 7 shows the calculated flood inundation depth and the estimated rice-crop damage for the 2007/2008 flood. Based on the cropping calendar and the duration at each rice growth stage, the rice plant during the 2007/2008 flood was at the vegetative and maturity stages in the downstream and upstream parts of the basin, respectively. The estimated value and area of rice crop damage during the 2007/2008 flood were 89.1 billion IDR and 73 324 ha, respectively, while their reported values were 93.3 billion IDR (Hidayat et al., 2008) and 60 630 ha (DFCP, 2010).

Figure 8 shows the boxplots of the flood inundation area and the peak inundation volume for the past and future periods. The figure shows that the first quartile values of the flood inundation area and the peak inundation volume for the future climate period were comparatively higher than the third quartile values of those for the past period, which indicates that floods with a larger inundation area and a higher flood depth may frequently occur in the future.

Figure 9 shows the boxplots of calculated damaged paddy area and estimated rice-crop damage for the past and future periods with different scenarios of rice growth stages,

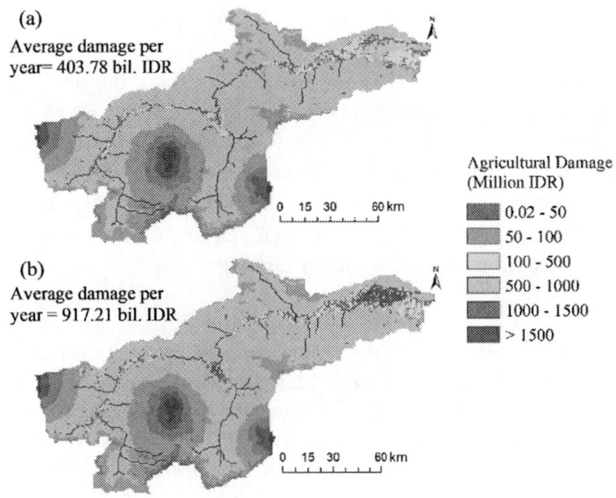

Figure 10. Spatial distribution of average rice crop damage at the maturity stage case for **(a)** past and **(b)** future periods (in million rupiah per unit grid at approximately 920 m).

(i.e., identified approximate rice growth stage during flooding based on cropping calendar (IGS) and also assumed if floods occur during the vegetative stage (VEG) or reproductive stage (REP) or maturity stage (MAT)). The figure shows that the first or third quartile values of the calculated damaged paddy area and the estimated rice-crop damage for the future period (RCP8.5) were comparatively higher than the first or third quartile of those for the past period. The results indicate that rice paddies will be frequently affected by highly damaging floods in the future. The average rice crop damage per year is approximately two times higher at all growth stages for the future period than for the past period, suggesting that rice crop damage can be more serious in the future due to climate change.

Figure 10 shows the comparison of the spatial distribution of the average rice crop damage (in million rupiah per unit grid at approximately 920 m) at the maturity stage case for the past and future periods. The results show that the rice crop damage in the future will be more severe in the areas farthest downstream of the basin and immediately upstream of the Madiun River confluence point.

6 Conclusions

We explored the relationship between extreme rainfall and damaged paddy area and assessed flood damage to rice crops under climate change scenarios. The findings show a strong correlation between extreme rainfall and damaged paddy area in the historical period. The results also show that flood inundation and rice crop damage might be more severe due to climate change in the future under the RCP8.5 scenario than in the historical period. The findings indicate that floods with a larger inundation area and a higher flood depth may frequently occur in the future. The rice paddies might be frequently affected by floods, suffering severe crop damage. To reduce such rice crop damage in the future, the development and implementation of adaptation measures and land-use plans are necessary. Using flood submergence-tolerant rice varieties in flood-prone areas or changing the timing of rice cultivation could also help to reduce rice crop damage in the future. The findings of this study can be useful in changing agricultural practices under climate change.

Data availability. The DEM data are available at https://www.hydrosheds.org/hydrosheds-core-downloads (Lehner et al., 2008). The damaged paddy area data are available from the DFCP (2010), and the hydro-meteorological data from the Ministry of Public Works and Housing, Indonesia. The data is not available directly.

Author contributions. BBS, MR, and TU conceived and designed the research. All authors contributed to research, activities, and paper writing.

Competing interests. At least one of the (co-)authors is a guest member of the editorial board of *Proceedings of IAHS* for the special issue "ICFM9 – River Basin Disaster Resilience and Sustainability by All". The peer-review process was guided by an independent editor, and the authors also have no other competing interests to declare.

Disclaimer. Publisher's note: Copernicus Publications remains neutral with regard to jurisdictional claims made in the text, published maps, institutional affiliations, or any other geographical representation in this paper. While Copernicus Publications makes every effort to include appropriate place names, the final responsibility lies with the authors.

Special issue statement. This article is part of the special issue "ICFM9 – River Basin Disaster Resilience and Sustainability by All". It is a result of The 9th International Conference on Flood Management, Tsukuba, Japan, 18–22 February 2023.

Acknowledgements. The authors would like to thank the Ministry of Public Works and Housing, Indonesia, for providing rainfall and discharge data; and the Ministry of Environment and Forestry, Indonesia, for providing land cover maps.

Financial support. This research has been supported by the Integrated Research Program for Advancing Climate Models (TOUGOU) and the Program for the Advanced Studies of Climate Change Projection (SENTAN) from the Ministry of Education, Culture, Sports, Science and Technology, Japan.

Review statement. This paper was edited by Shinji Egashira and reviewed by two anonymous referees.

References

DSRI (Department of Settlement and Regional Infrastructure): Comprehensive development and management plan study for Bengawan Solo River Basin under Lower Solo River Improvement Project, Final Report, DSRI, Directorate General of Water Resources, Indonesia, 3-1–3-73 pp., 2001.

DFCP (Directorate of Food Crop Protection): Indonesia broad flood damage in rice plant: Solo River Basin, Flood Damage Data Published by DFCP, Indonesia [data set], 2010.

Hidayat, F., Sungguh, H. M., and Harianto: Impact of climate change on floods in Bengawan Solo and Brantas River Basins, Indonesia, Proc. of 11th Int. Riversymp., 1–4 September 2008, Brisbane, Australia, http://archive.riversymposium.com/index.php?element=HIDAYAT (last access: 28 September 2020), 2008.

Komolafe, A. A., Herath, S., and Avtar, R.: Establishment of detailed loss functions for the urban flood risk assessment in Chao Phraya River basin, Thailand, Geomatics, Nat. Hazard. Risk, 10, 633–650, 2019.

Lehner, B., Verdin, K., and Jarvis, A.: New global hydrography derived from spaceborne elevation data, Eos, Transactions, American Geophysical Union [data set], 89, 93–94, https://www.hydrosheds.org/hydrosheds-core-downloads (last access: 8 June 2021), 2008.

Panuju, D. R., Mizuno, K., and Trisasongko, B. H.: The dynamics of rice production in Indonesia 1961–2009, J. Saudi Soc. Agric. Sci., 12, 27–37, 2013.

Rasmy, M., Sayama, T., and Koike, T.: Development of water and energy Budget-based Rainfall-Runoff-Inundation model (WEB-RRI) and its verification in the Kalu and Mundeni River Basins, Sri Lanka, J. Hydrol., 579, 124–163, 2019.

Shrestha, B. B., Okazumi, T., Miyamoto, M., and Sawano, H.: Flood damage assessment in the Pampanga River basin of the Philippines, J. Flood Risk Manag., 9, 355–369, 2016.

Shrestha, B. B., Kawasaki, A., and Zin, W. W.: Development of flood damage functions for agriculture crops and their applicability in regions of Aisa, J. Hydrol.-Reg. Stud., 36, 100872, https://doi.org/10.1016/j.ejrh.2021.100872, 2021.

Wright, T. and Meylinah, S.: Indonesia grain and feed annual report 2014, GAIN Report Number ID1407, https://apps.fas.usda.gov/newgainapi/api/report/downloadreportbyfilename?filename=GrainandFeedAnnual_Jakarta_Indonesia_4-23-2014.pdf (last access: 7 March 2024), 2014.

Zakaria, W. A., Aring, H. L. D., and Indriani, Y.: The impact of irrigation development on rice production in Lampung Province, Research Report, University of Lampung, Indonesia, https://citeseerx.ist.psu.edu/document?repid=rep1&type=pdf&doi=4a0ef2eddfafed4606e6c4983ab2e77417be69ba (last access: 7 March 2024), 2004.

Proc. IAHS, 386, 133–140, 2024
https://doi.org/10.5194/piahs-386-133-2024

Open Access
PIAHS

Evaluation of climate change impacts on urban flooding using high-resolution rainfall data

Hideo Amaguchi[1], **Jonas Olsson**[2], **Akira Kawamura**[1], **and Yoshiyuki Imamura**[1]

[1]Department civil and environmental engineering, Tokyo Metropolitan University, Hachioji, 192-0397, Japan
[2]Swedish Metrological and Hydrological Institute, Norrkoping, 601, Sweden

Correspondence: Hideo Amaguchi (amaguchi@tmu.ac.jp)

Received: 21 July 2023 – Revised: 2 October 2023 – Accepted: 24 February 2024 – Published: 19 April 2024

Abstract. Evaluation of the effects of climate change on urban river flow has traditionally depended on design rainfall data due to the scarcity of other data sources. However, high-resolution precipitation data, both temporally and spatially, have become accessible from regional climate models. Despite this availability, few studies have tapped into such high-resolution data to analyze urban river flow systems. This study incorporates an event-based storm runoff and inundation simulation with data from a regional climate model. This model offers a spatial resolution of 5 km and a temporal resolution of 10 min for both current and projected future climate scenarios. Rainfall analysis reveals that the 20-year probability of an hourly rainfall of 75 mm, specific to Tokyo, decreases to around 13 years for the period 2016–2035 and further to approximately 8 years for 2076–2095. This trend indicates a growing frequency of heavy rainfall events. Moreover, flood discharge analysis shows an increased downstream flow during such rainfall events if reservoir capacities remain unchanged. Flood risk assessment underscores a significant threat: areas inundated to depths of 25 cm or more may increase by 1.25 times for 2016–2035 and by 1.55 times for the period 2076–2095 when compared to current conditions.

1 Introduction

In small to medium-sized urban river watersheds, flood damage often results from pluvial and fluvial flooding, especially when the development of flood control facilities lags and heavy rainfall surpasses facility capacities. According to the Fifth Assessment Report of the Intergovernmental Panel on Climate Change (IPCC, 2013), Japan may experience a rise in the intensity and frequency of short-duration rainfall by the century's end due to an increase in the global average surface temperature.

Many studies have assessed the climate change impact on Class A river systems using the database for policy decision-making regarding future climate change, known as d4PDF (Mizuta et al., 2017). The d4PDF originates from the global model MRI-AGCM3.2, which is the global atmospheric model of the Meteorological Research Institute of Japan with a resolution of 60 km. This has been further refined to a 20 km resolution using a regional model, NHRCM, which covers the Japan area. Flood risk research utilizing the d4PDF includes investigations into future shifts in extreme river flow for the Arakawa, Shonai, and Yodo River watersheds (Tachikawa et al., 2017) and an evaluation of the correlation between cumulative rainfall and peak flood discharge in the Omono River watersheds (Abe et al., 2019). Flood risk assessments due to climate change traditionally target watersheds spanning thousands of square kilometers, constrained by the temporal and spatial resolution of precipitation data. Few assessments focus on urban river watersheds covering tens to hundreds of square kilometers. For urban rivers, the one-hour precipitation data from d4PDF proves inadequate. Accurate flood runoff and inundation risk assessment necessitate finer resolution, such as 10 min rainfall data, where, for instance, the maximum 10 min average rainfall intensity aligns with peak river water levels (Takakura et al., 2016). Climate forecasts suggest a 1.3-fold increase in short-duration heavy rainfall events in Tokyo by the century's end (Social Infrastructure Development Council, 2022). Evaluating runoff analyses based on these future rainfall projec-

Published by Copernicus Publications on behalf of the International Association of Hydrological Sciences.

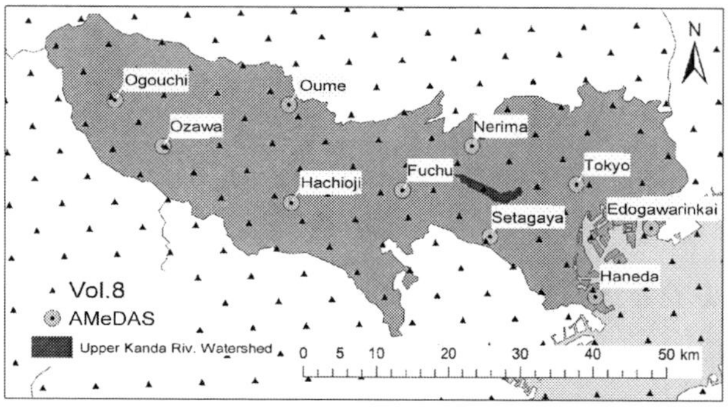

Figure 1. NHRCM 5 km analysis grid and AMeDAS stations.

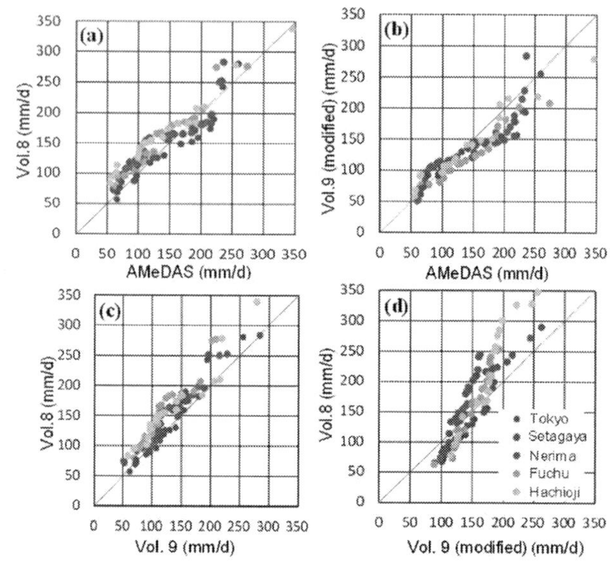

Figure 2. Comparison of annual maximum daily rainfall: **(a)** and **(b)** cover the P1980 period, **(c)** and **(d)** cover the P2076 period.

tions becomes crucial for determining necessary flood control measures.

A suitable dataset for assessing urban watershed flood risks due to climate change is the NHRCM 5 km precipitation data, found in Global Warming Prediction Information Volume 8 (hereafter "Vol.8"). This data is computed by the non-hydrostatic regional climate model from the Japan Meteorological Agency Meteorological Research Institute (JMA/MRI). It offers a spatial resolution of 5 km and a temporal resolution of 10 min (Japan Meteorological Agency, 2013). This dataset encompasses three timeframes: the past (1980–1999), the near future (2016–2035), and the distant future (2076–2095). These values are experimental, based on the SRES A1B scenario (IPCC, 2007).

This study utilized design rainfall with a 20-year probability spanning three periods: 1980–1999 (P1980), 2016–2035 (P2016), and 2076–2095 (P2076). The data was sourced from the NHRCM 5 km precipitation in Vol. 8. Flood runoff and inundation risks in the upper Kanda River watershed due to climate change were assessed using the Tokyo Storm Runoff (TSR) model.

2 Design rainfall based on NHRCM 5 km rainfall data

2.1 Comparison with forecast scenario RCP 8.5

Global Warming Projection Information Vol.9 (Japan Meteorological Agency, 2017) provides NHRCM 5 km precipitation data, computed under the RCP8.5 projection scenario. This data is accessible via DIAS with a daily temporal resolution for the P1980 members. Data from the four members of P2076 is also available. This study utilized daily precipitation data from Vol. 9 to assess the 10 min rainfall data in Vol. 8. Figure 1 illustrates the NHRCM 5 km analysis grid alongside the layout of the Automated Meteorological Data Acquisition System (AMeDAS) stations in the vicinity of Tokyo.

Figure 2 presents the rank statistics of annual maximum daily precipitation for five AMeDAS stations in Tokyo, compared with the analytical grid data from Vol. 8 and 9. The data for Vol. 8 was derived from the 10 min intervals, while bias-corrected data was used for Vol. 9. Figure 2a and b represent the P1980 period, whereas Fig. 2c and d depict the P2076 period. In comparison with the current climate data, the projected values from Vol. 8 surpass those from AMeDAS (Fig. 2a), while the estimates in Vol. 9 are generally lower than AMeDAS data (Fig. 2b). The predicted annual maximum daily precipitation for P1980 in Vol. 8 exceeded the values in Vol. 9 for stations St. Hachioji and St. Fuchu, both situated in western Tokyo. However, the predictions for St Tokyo remain largely consistent (Fig. 2c). For the P2076 timeframe, Vol. 8's predictions tend to be higher than those of Vol. 9 when the annual maximum daily precipitation surpasses approximately 100–150 mm d^{-1}. The annual maximum daily precipitation of A1B scenario's annual maximum daily precipitation was consistently higher than that of the RCP8.5 climate scenario.

2.2 Preparation of design rainfall

For flood runoff analyses in urban watersheds, the integration of short-duration rainfall intensity proves crucial. Several methods have been proposed for this objective. However, no singular approach has achieved broad consensus as the most effective (Ioannis, 2022). In Japan, a prevalent strategy when factoring in climate change impacts for flood runoff analysis involves applying a multiplier (a.k.a. climate factor) to the projected rainfall values based on current design rainfall (Arnbjerg-Nielsen et al., 2013; Toda et al., 2020; National Institute for Land and Infrastructure Management, 2017).

Given that the 10 min data in Vol. 8 is not bias-corrected, the study emphasizes the need to evaluate the reproducibility of short-term rainfall. To address potential underestimations crucial to urban flood runoff analysis, this research adopted a method to adjust rainfall intensity, utilizing climate factors influenced by climate change-induced precipitation changes. A coefficient (α_t) was obtained by assuming a linear relationship between the probability of rainfall in the ground rain gauge (G_{t_T}), the probability of rainfall in the present climate (D_{t_T}), and by the following formula (Kojima et al., 2018).

$$G_{t_T} = \alpha_t D_{t_T} \quad (1)$$

where t denotes the rainfall duration(min); and T denotes return period (year). Specifically, grid data at 69 locations were extracted from the NHRCM 5 km time-series precipitation data of P1980 for Tokyo, and annual maxima of 10, 20, 30, 60, 120, and 180 min were generated for each grid (Fig. 1). Time series of precipitation data were generated, and 20-year rainfall probabilities were calculated using a generalized extreme value distribution. Similar processing was applied to ground-based rainfall observation data, and the coefficient α, determined as the average value of all grid data, yielded a value of 1.71, 1.34, 1.27, 0.97, 0.85 and 0.80 respectively. Despite having a temporal resolution of 10 min intervals, the rainfall intensity multiplier α displays pronounced bias for short-duration rainfall, with values of 1.27 for 30 min, 1.34 for 20 min, and 1.71 for 10 min. Given the absence of ground-based rainfall data for P2016 and P2076, the coefficients α_t for these periods were adjusted using the coefficients derived from P1980.

The Cleveland type adopted in the Tokyo Metropolitan Government's (TMG) River Plan (Tokyo Metropolitan Government Bureau of Construction, 2012) was used to develop a rainfall intensity equation. The constants were set to 10, 20, 30, 60, 120, and 180 min for the corrected 20-year rainfall probabilities of P1980, P2016, and P2076.

$$I = \frac{a}{t^n + b} \quad (2)$$

where I denotes the rainfall intensity (mm h^{-1}); t denotes the rainfall duration (min); and a, b, and n denote constants.

Figure 3 shows the rainfall intensity curve with 20-year return period, (a) before and (b) after correction. The NHRCM 5 km rainfall tended to underestimate rainfall shorter than 60 min and overestimate rainfall longer than 60 min. For the annual maximum 60 min rainfall, the climate factor is 1.10 for P2016 and 1.22 for P2076. This factor is less than the change factor of 1.3 for RCP8.5 as determined by d4PDF but exceeds the change factor of 1.1 for RCP2.6.

A 60 min probability rainfall intensity curve was developed for the 60 min rainfall to assess the future return period of a 20-year 60 min rainfall of 75 mm in a present climate, which is the target of the Tokyo Metropolitan Government's river plan (Tokyo Metropolitan Government Bureau of Construction, 2012) (Fig. 4). The reproducibility of the annual

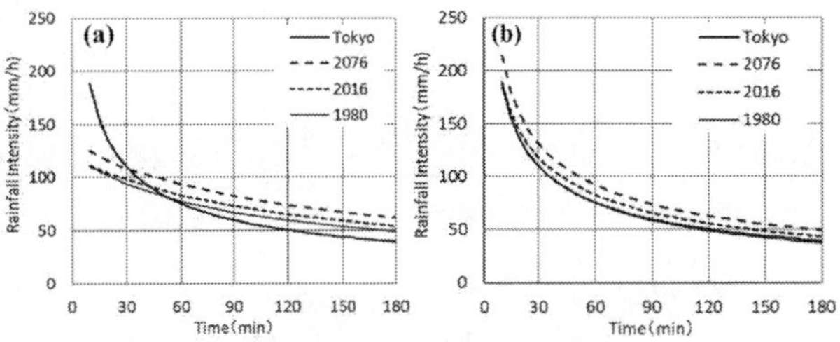

Figure 3. Rainfall intensity curve with 20-year return period, **(a)** before modification, **(b)** after modification.

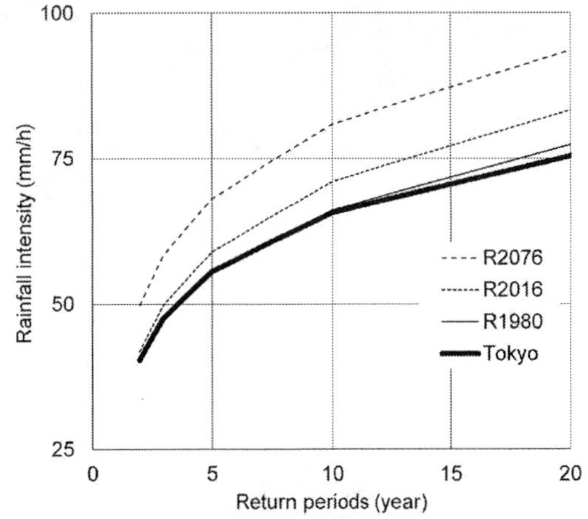

Figure 4. 60 min intensity-frequency curve.

maximum 60 min rainfall in the present climate was very high as the 60 min probability intensity curves for the Tokyo District Meteorological Observatory and NHRCM 5 km almost overlapped. Therefore, no corrections were made to the rainfall probabilities of P2016 and P2076. The return period of 75 mm hourly rainfall will decrease to 13 years in the near future (P2016) and 8 years in the future (P2076), suggesting an increasing trend in heavy rainfall frequency.

3 Study area and flood modelling

3.1 Outline of the study area

The Kanda River, a quintessential urban river in Tokyo, starts its journey at Inogashira Pond and merges into the Sumida River via the confluence with the Zenpukuji and Myoshoji Rivers. This study focuses on the upper Kanda River segment, spanning from Inogashira Pond to the Zenpukuji River confluence (with a watershed area of 11.8 km^2 and a channel length of 9 km). Within this watershed, sewage systems are integrated, channeling waste to the sewage treatment facility through a primary sewer line. Figure 5 illustrates the upper Kanda River watershed's urban landscape GIS, as detailed by Koga et al. (2016), derived from a 1 : 2500 scale topographic

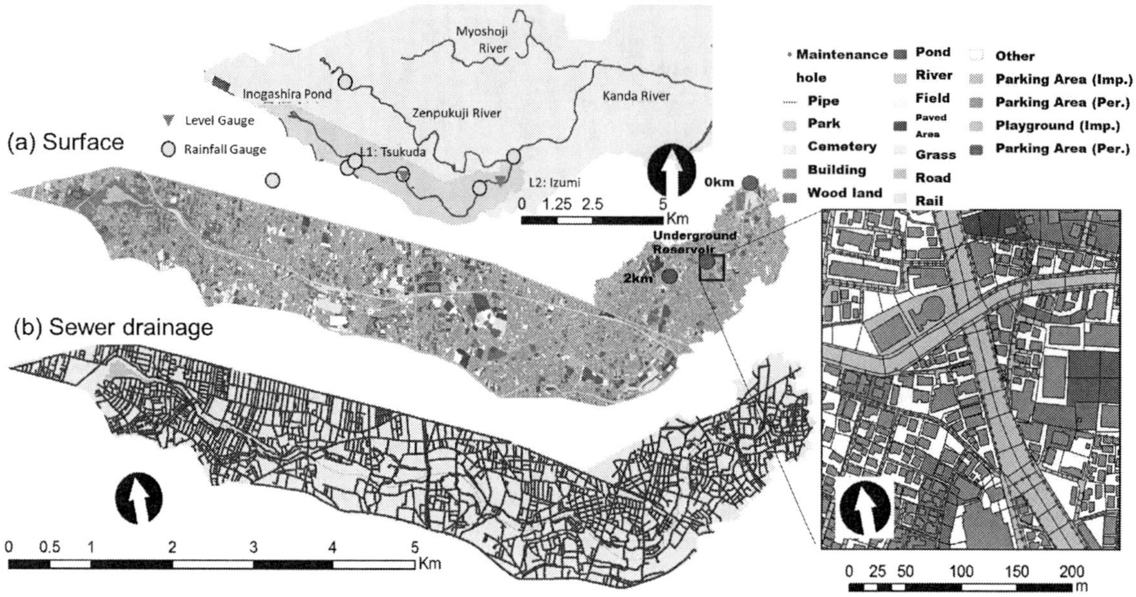

Figure 5. Urban landscape GIS delineation of the Upper Kanda River watershed, **(a)** overland, **(b)** sewer.

map of Tokyo. The GIS revealed a total of 104 342 surface polygons, with area distribution percentages as follows: buildings (29 %), roads (16 %), green spaces (9 %), paved areas (3 %), private premises excluding buildings (30 %), and other zones (13 %). As for the stormwater pipe network GIS data, it identified manholes with points and pipelines with polylines, recording 9632 manholes and 9904 pipelines.

3.2 Overview of the TSR model

The Tokyo Storm Runoff (TSR) model (Amaguchi et al., 2012) served as the primary tool for flood runoff and inundation analysis in urban watersheds for this study. Figure 6 illustrates a conventional rainfall-runoff process captured by the model. Rainfall distribution within the watershed corresponded to each land use element, derived from the urban landscape GIS. Within each urban block, direct runoff from impervious zones and rainfall surpassing the permeability of infiltrating zones was computed based on land use characteristics for infiltration and impermeability. Rainwater from structures was directed to adjacent road sections. In road elements, rainwater either entered storm and sewer pipelines, where manholes existed or continued its surface flow in their absence. Consequently, water entering storm and sewer lines navigated through the pipeline network, ultimately discharging into the river and exiting the watershed.

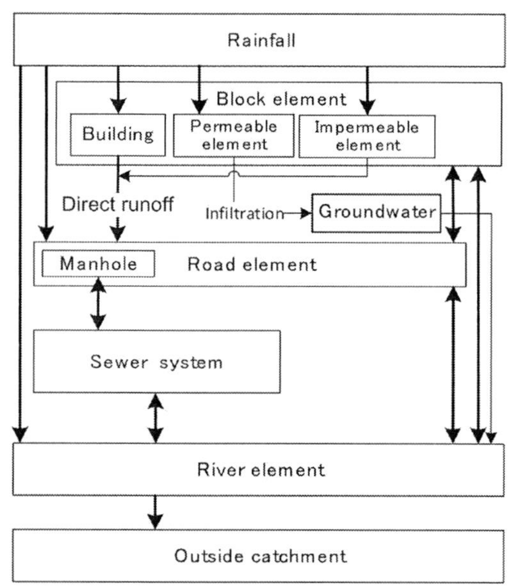

Figure 6. Schematic of the rainfall-runoff process.

Parameters in the TSR model, such as initial loss, initial infiltration capacity, final infiltration capacity, infiltration capacity attenuation coefficient, and equivalent roughness coefficient of land-use elements, were set according to values from an earlier study (Amaguchi et al., 2013). The model addressed the significant flood event on 4 September 2000, within the Kanda watershed. Despite no flood damage in the target watershed, largely due to effective underground control basins, the model simulated the event as if these control basins were non-operational. Model validation occurred by contrasting its outputs with river water levels measured at upstream sites, unaffected by underground control basins.

This validation confirmed the model's ability to accurately reproduce the observed river water levels across the board.

3.3 Conditions of application

The underground control reservoir of Loop Route 7, currently connected to the northern part of the Zempukuji River, has a total storage capacity of 540 000 m^3. In this study, the storage capacity was set at 240 000 m^3, which corresponds to the section between the Kanda Rivers. To model the underground reservoirs, the stormwater and sewerage pipeline modelling method in the TSR model was applied, and the overflow height and width were used as connection information between the river channel and reservoirs.

Three types of watershed measures in the river plan were covered: rainwater penetration, infiltration trenches, and permeable pavements installed by the year 2000 (Araki et al., 2012). Tabulated data enabled the attribution of the quantity of rainwater infiltration facilities per building, infiltration trench length per building (in meters), and paved area (in square meters) to the Kanda River Watershed's land-use GIS data based on provided addresses and additional details. The watershed contained approximately 10 000 rainwater penetration inlets, an infiltration trench length totaling 40 km, and permeable pavement covering 220 000 m^2.

The unit infiltration capacities of rainwater penetration and trenches were 0.234 $m^3\,pcs^{-1}\,h^{-1}$ and 0.220 $m^3\,m^{-1}\,h^{-1}$, respectively (Tokyo Metropolitan Council for Comprehensive Flood Control Measures, 2019). For the permeable pavement, a thickness of 0.20 m and porosity of 0.1 was used for the permeable pavement with a storage volume of 20 mm (Tokyo Metropolitan Council for Comprehensive Flood Control Measures, 2019). The TSR model was calculated using different building infiltration capacities and storage volumes for each road.

4 Results and discussion

4.1 Flooding risk assessment using design rainfall

Employing design rainfall based on a future 20-year return period, variations in channel flow and peak inundated areas within the study region were compared across periods to assess flood risk. Channel flow changes are examined at the confluence depicted in Fig. 7 and at a location 2 km away from the confluence. An underground control reservoir is located 1.3 km from this confluence.

Due to the elongated shape of the watershed in the downstream direction, the peak flow from the 2 km point is already in a downstream state. Therefore, the maximum flow is larger at the 2 km point for all periods, and the flow decreases as it reaches the confluence point. In the case of P2076, a swift rise in discharge occurred post the peak rainfall at the confluence location. This surge results from the reservoir hitting its capacity, preventing water transfer to the underground reservoir.

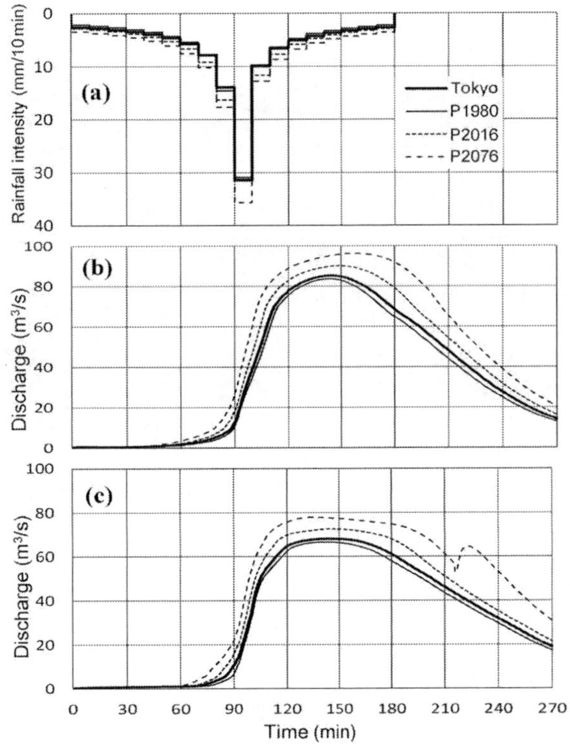

Figure 7. Results of flood discharge analysis, **(a)** rainfall intensity, **(b)** discharge at 2 km, and **(c)** discharge at confluence point flow rate at the confluence.

The peak flow variation was more pronounced 2 km away than at the confluence for all durations. When comparing peak flow multipliers relative to P1980, P2016 increased by 1.08 times and P2076 by 1.15 times at the 2 km location. Meanwhile, for the confluence location, P2016 increased by 1.09 times and P2076 by 1.17 times. This suggests that the risk of flooding seemed to amplify further downstream from the control reservoir.

Figure 8 shows the maximum inundation map for depths of 25 cm or greater, and Table 1 lists the corresponding inundation areas for roads, city blocks, and houses. The maximum inundation area with a depth exceeding 25 cm is approximately 0.56 km^2 in the present climate (P1980), approximately 0.70 km^2 in the P2016 period, and around 0.87 km^2 in the future P2076. When compared to the present climate, the inundated area was approximately 1.25 times larger in P2016 and approximately 1.55 times larger in the future (P2076).

4.2 Flood risk assessment using rainstorm events

In flood runoff analyses based on projected rainfall, assessing river flow variations due to rainfall events, including factors, such as rainfall magnitude and duration, is difficult. To evaluate the effects of rainfall patterns on the NHRCM 5 km, flood

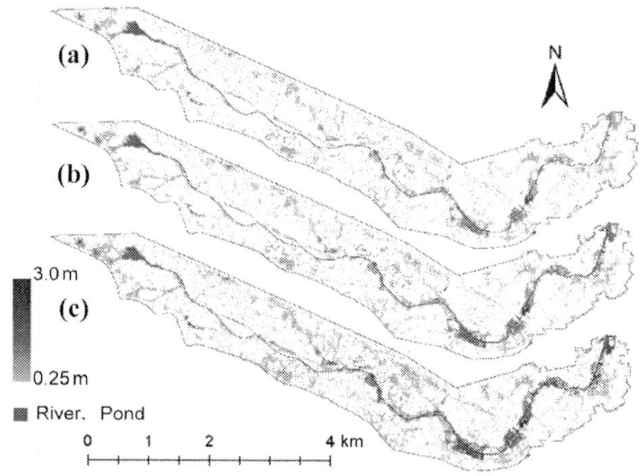

Figure 8. Maximum inundated area above 25 cm depth, **(a)** P1980, **(b)** P2016, and **(c)** P2076.

Table 1. Area inundated depth of more than 25 cm (km^2).

	Tokyo	P1980	P2016	P2076
Road	0.15	0.14	0.18	0.22
Block	0.20	0.18	0.25	0.34
Building	0.24	0.23	0.27	0.30
Sum	0.59	0.56	0.70	0.87

runoff characteristics due to rainstorm events were estimated with a maximum hourly rainfall exceeding 20 mm. As mentioned above, rainfall with a duration of more than 60 min can be overestimated. However, it is important to consider rainstorm events in terms of flood frequency, which cannot be captured by analyzing the designed rainfall. Figure 9 shows the selected heavy rainfall events, where the vertical axis represents the 60 min maximum rainfall and horizontal axis represents the total rainfall. In the near future, P2016 will have a high frequency, and P2076 will have many events with a 60 min maximum rainfall exceeding 60 mm.

Using the selected rainstorm events, changes in river water levels accumulated over time, and the risk of river channel flooding were assessed. For the upstream and downstream sides of the regulating reservoir, the cumulative time at which the river water levels exceeded the bank height and bank height −1 m was calculated.

Figure 10 displays the accumulated time of exceedance based on the criteria previously mentioned. When comparing the accumulated time beyond both bank elevation and bank elevation −1 m between the two sites for all periods, the upstream river is found to be flooded more than the downstream river. This suggests that the flooding risk in terms of flood frequency is higher on the upstream side of the underground reservoir. During the P2016 period, although 24 events registered rainfall surpassing 20 mm h^{-1}, the peak was around

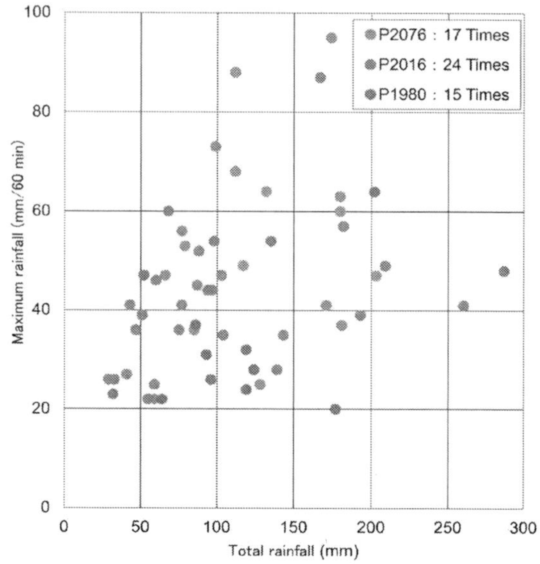

Figure 9. Characteristics of selected rainstorm events.

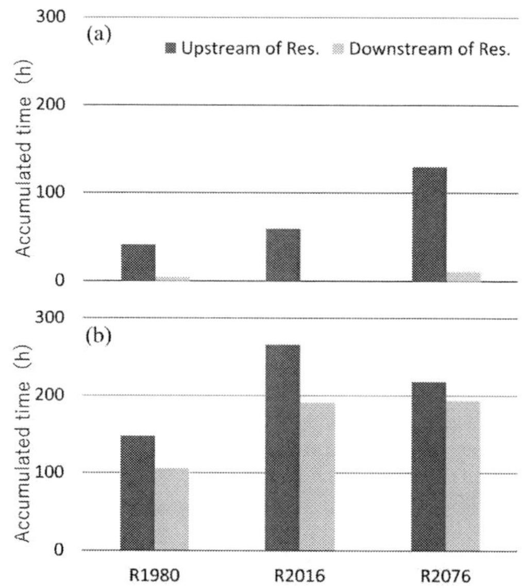

Figure 10. Accumulated time of exceedance of the reference level at points upstream and downstream of the underground reservoir: **(a)** bank-level and **(b)** bank-level −1 m.

60 mm, leading to more time at bank elevation −1 m. In the P2076 period, the number of events, 17, mirrored that of the P1980 period. However, six of these events had a maximum hourly rainfall exceeding 60 mm, resulting in the water level surpassing the bank height for approximately three times the hours observed in P1980.

5 Summary and conclusion

In this study, the NHRCM 5 km 10 min precipitation data were used to generate a design rainfall with a 20-year return period for 1980–1999, 2016–2035, and 2076–2095. Furthermore, flood runoff and inundation analyses were conducted for the upper Kanda River watershed using the Tokyo Storm Runoff (TSR) model to examine the risk of inundation due to climate change.

The rainfall analysis indicated that the future return period of today's 1 h 20-year rainfall (75 mm), which is the river plan for the Tokyo metropolitan area, is about 13 years for 2016–2035 and about 8 years for 2076–2095, suggesting an increasing trend in the frequency of heavy rainfalls. It was confirmed by the flood discharge analysis that the flow rate during rainfall increased, and a more pronounced flow rate was observed downstream than in the middle part of the river when the reservoir's capacity was assumed to remain unchanged from its current state. Through the flood risk assessment, it was determined that the maximum inundated area with depths of 25 cm or greater would experience an increase of approximately 1.25 times for the 2016–2035 period and approximately 1.55 times for the 2076–2095 period when compared to the present conditions.

To conclude, we have shown how historical rainfall statistics and future high-resolution climate projections can be combined in order to estimate future short-duration rainfall extremes. These data may then be used for detailed evaluation of climate adaptation measures to reduce urban flood risk, as demonstrated here. The development of high-resolution convection-permitting RCMs that well represent observed small-scale rainfall extremes is highly encouraging (e.g. Olsson et al., 2021). Still only a limited amount of future projections are available because of the high computational demand, but much research and development is currently devoted to this field. It is our hope that this will improve the accuracy and reliability of urban flood risk assessment and, in turn, lead to better-informed adaptation measures being taken.

Data availability. The rainfall datasets used were collected and provided under the Data Integration and Analysis System (DIAS), which was developed and operated by the Ministry of Education, Culture, Sports, Science and Technology of Japan, available at: https://search.diasjp.net/ja/dataset/JMA_GWP (last access: 10 March 2023).

Author contributions. All the authors HA, JO, AK, and YI participated in conceptualizing the study, collecting data, analyzing data, and drafting the manuscript.

Competing interests. The contact author has declared that none of the authors has any competing interests.

Disclaimer. Publisher's note: Copernicus Publications remains neutral with regard to jurisdictional claims made in the text, published maps, institutional affiliations, or any other geographical representation in this paper. While Copernicus Publications makes every effort to include appropriate place names, the final responsibility lies with the authors.

Special issue statement. This article is part of the special issue "ICFM9 – River Basin Disaster Resilience and Sustainability by All". It is a result of The 9th International Conference on Flood Management, Tsukuba, Japan, 18–22 February 2023.

Acknowledgements. This study used data produced with the Earth Simulator by the Program for Risk Information on Climate Change (SOUSEI) and Integrated Research Program for Advancing Climate Models (TOUGOU) from the Ministry of Education, Culture, Sports, Science and Technology (MEXT), Japan. We thank the reviewers for constructive comments on the original manuscript.

Financial support. This research has been supported by the Tokyo Metropolitan Advanced Research (gran no. R4-2).

Review statement. This paper was edited by Daisuke HARADA and reviewed by two anonymous referees.

References

Abe, S., Watanabe, S., Yamada, M., Kotsuki, S., and Watanuki, A.: Effects of observed precipitation and sea surface temperature patterns on inundation analysis using large-scale climate prediction information, J. Jpn. Soc. Civ. Eng. Ser. B1, 75, I_1081–I_1086, https://doi.org/10.2208/jscejhe.75.2_I_1081, 2019 (in Japanese with English abstract).

Amaguchi, H., Kawamura, A., Olsson, J., and Takasaki, T.: Development and testing of a distributed urban storm runoff event model with a vector-based catchment delineation, J. Hydrol., 420/421, 205–215. https://doi.org/10.1016/j.jhydrol.2011.12.003, 2012.

Amaguchi, H., Nagasaka, T., Kawamura, A., Takasaki, T., and Nakagawa, N.: A proposal of storm runoff model considering process of building inundation for an urban catchment, Adv. River Eng., 19, 211–216, 2013 (in Japanese with English abstract).

Araki, T., Amaguchi, H., Kawamura, A., and Takasaki, T.: Development of a groundwater recharge model for an urban catchment using urban landscape GIS delineation and its simulation for actual catchment, J. Jpn. Soc. Civ. Eng. Ser. B1, 68, 109–124, 2012 (in Japanese with English abstract).

Arnbjerg-Nielsen, K., Willems, P., Olsson, J., Beecham, S., Pathirana, I., Bülow Gregersen, I., Madsen, H., and Nguyen, V.-T.-V.: Impacts of Climate Change on Rainfall Extremes and Urban Drainage Systems: a review, Water Sci. Technol., 68, 16–28, https://doi.org/10.2166/wst.2013.251, 2013.

Social Infrastructure Development Council (SIDC): Countermeasures of Water-related Disasters due to Global Warming, available at: https://www.mlit.go.jp/river/shinngikai_blog/chisui_kentoukai/index.html (last access: 10 November 2022), 2020 (in Japanese).

Ioannis, M. K. and Vassilios, A. T.: Update of intensity-duration-frequency (IDF) curves under climate change: a review, Water Supply, 22, 4951–4974, https://doi.org/10.2166/ws.2022.152, 2022.

IPCC (Intergovernmental Panel on Climate Change): Climate change 2007: The Physical Science Basis, Contribution of Working Group I to the Fourth Assessment Report of the Intergovernmental Panel on Climate Change, edited by: Solomon, S., Qin, D., Manning, M., Chen, Z., Marquis, M., Averyt, K. B., Tignor, M., and Miller, H. L., Cambridge University Press, Cambridge, United Kingdom and New York, NY, USA, 996 pp., ISBN: 978-0521-88009-1, 2007.

IPCC (Intergovernmental Panel on Climate Change): Climate change 2013: The Physical Science Basis, Contribution of Working Group I to the Fifth Assessment Report of the Intergovernmental Panel on Climate Change, edited by: Stocker, T. F., Qin, D., Plattner, G. K., Tignor, M., Allen, S. K., Boschung, J., Nauels, A., Xia, Y., Vex, V., and Midgley, P. M., Cambridge University Press, Cambridge, UK, and New York, USA, 1535 pp., ISBN: 9781107415324, 2013.

Japan Meteorological Agency (JMA): Global Warming Prediction Information, Vol. 8, https://www.data.jma.go.jp/cpdinfo/GWP/Vol8/index.html (last access: 10 March 2023) 2013 (in Japanese).

Japan Meteorological Agency (JMA): Global Warming Prediction Information, Vol. 9, https://www.data.jma.go.jp/cpdinfo/GWP/Vol9/index.html (last access: 10 March 2023) 2017 (in Japanese).

Koga, T., Kawamura, A., Amaguchi, H., and Tanouchi, H.: Assessing impervious area ratios of grid-based land-use classifications on an example of an urban watershed, Hydrol. Sci. J., 61, 1728–1739, https://doi.org/10.1080/02626667.2015.1133909, 2016.

Kojima, T., Maruya, Y., and Harada, M.: Proposal of the correction formula of probability rainfall produced by 20 km d4PDF for Gifu Prefecture, J. Jpn. Soc. Civ. Eng. Ser. B1, 74, I_133–I_138, https://doi.org/10.2208/jscejhe.74.5_I_133, 2018 (in Japanese with English abstract).

Mizuta, R., Murata, A., Ishii, M., Shiogama, H., Hibino, K., Mori, N., Arakawa, O., Imada, Y., Yoshida, K., Aoyagi, T., Kawase, H., Mori, M., Okada, Y., Shimura, T., Nagatomo, T., Ikeda, M., Endo, H., Nosaka, M., Arai, M., and Kimoto, M.: Over 5,000 Years of Ensemble Future Climate Simulations by 60 km Global and 20 km Regional Atmospheric Models, Bull. Am. Meteorol. Soc., 98, 1383–1398, https://doi.org/10.1175/BAMS-D-16-0099.1, 2017.

National Institute for Land and Infrastructure Management (NILIM): Research on Climate Change Adaptation in River and Coastal Management – With focusing on the results of Project Research titled "Development of base technology to support forming a portfolio of management alternatives for large-scale water-related disasters under climate change", Project research report, NILIM, 56, ISSN 1880-0114, 2017 (in Japanese).

Olsson, J., Du, Y., An, D., Uvo, C.B., Sörensen, J., Toivonen, E., Belušić, D., and Dobler, A.: An Analysis of (Sub-)Hourly Rainfall in Convection-Permitting Climate Simulations Over Southern Sweden From a User's Perspective, Front. Earth Sci., 9, 681312, https://doi.org/10.3389/feart.2021.681312, 2021.

Tachikawa, Y., Miyawaki, K., Tanaka, T., Yorozu, K., Kato, M., Ichikawa, Y., and Kim S.: Analysis of future changes in extreme river flows using super multiple ensemble climate prediction experimental data, J. Jpn. Soc. Civ. Eng. Ser. B1, 73, 77–90, https://doi.org/10.2208/jscejhe.73.77, 2017 (in Japanese with English abstract).

Takakura, N. and Ikeda, N.: Prediction analysis of water level change of the city river caused by local torrential Rain, Tokyu Constr. Tech. Inst. Bull., 42, 75–80, 2016 (in Japanese with English abstract).

Toda, A., Ozaki, T., Hashimoto, A., and Toda, K. : Estimation of future probable precipitation using d4PDF and its impact on water drainage planning – case study on Kinki region, J. Jpn. Soc. Civ. Eng. Ser. G, 76, II_227–II_235, https://doi.org/10.2208/jscejer.76.6_II_227, 2020 (in Japanese with English abstract).

Tokyo Metropolitan Government Bureau of Construction: Future Maintenance of Small and Medium-sized Rivers in Tokyo, https://www.kensetsu.metro.tokyo.lg.jp/jigyo/river/chusho_seibi/chusho_arikata/index_chusho_arikata.html (last access: 31 March 2020) 2012 (in Japanese).

Tokyo Metropolitan Council for Comprehensive Flood Control Measures: Tokyo Metropolitan Government Technical Guidelines for Rainwater Storage and Infiltration Facilities, http://www.tokyo-sougou-chisui.jp/shishin/index.html (last access: 10 March 2023) 2019 (in Japanese).

Proc. IAHS, 386, 141–146, 2024
https://doi.org/10.5194/piahs-386-141-2024

Dual-Stage Attention-Based LSTM Network for Multiple Time Steps Flood Forecasting

Fan Wang[1,2,3], **Weiqi Wang**[1,2], **Wuxia Bi**[1,2], **Wenqing Lin**[1,2], **and Dawei Zhang**[1,2]

[1]State Key Laboratory of Simulation and Regulation of Water Cycle in River Basin, Beijing, 100038, P. R. China
[2]China Institute of Water Resources and Hydropower Research, Research Center on Flood & Drought Disaster Reduction of the Ministry of Water Resources, Beijing, 100038, P. R. China
[3]National Key Laboratory of Water Disaster Prevention, Hohai University, Nanjing, 210098, P. R. China

Correspondence: Fan Wang (wangfan@iwhr.com)

Received: 21 July 2023 – Revised: 7 October 2023 – Accepted: 24 February 2024 – Published: 19 April 2024

Abstract. Flood forecasting plays a crucial role in supporting decision-making for flood management. In addition to conceptual and physical-based models, the data-driven models have garnered increasing attention in recent years. The proposed model in this study employs LSTM networks, Encoder-Decoder framework, as well as feedback and attention mechanism to effectively utilize diverse observed data and future rainfall as inputs for multiple time steps flood forecasting. The accuracy and reliability of the model have been validated across case studies in multiple watersheds in China. The results demonstrate the high performance of the LSTM-based flood forecasting model. Meanwhile, the efficacy of both the feedback mechanism and attention mechanism has been validated in the domain of flood prediction.

1 Introduction

The artificial neural network (ANN) stands out as the most widely adopted algorithm among data-driven flood forecasting models, showcasing exceptional performance and remarkable stability. After years of development, the simple feedforward network has evolved into sophisticated architectures such as radial basis function (RBF) network and recurrent neural network (RNN) (Xiang et al., 2020; Kao et al., 2020), among which, the RNN with gating mechanism, such as the LSTM network (Hochreiter and Schmidhuber, 1997), has shown great performance in the field of series prediction. The structure of RNN can effectively utilize the information inherent in time series data, making it highly suitable for predicting rainfall and runoff series characterized by significant autocorrelation and cross-correlation.

Based on the CAMELS dataset (Addor et al., 2017; Newman et al., 2015), the LSTM network has been employed to construct runoff prediction models for multiple watersheds in the United States (Kratzert et al., 2018). Thapa et al. (2020) developed an LSTM-based discharge model driven by snowmelt in a Himalayan basin, which utilized MODIS snow images and remotely sensed meteorological data as input variables. The LSTM network and seq2seq architecture have been utilized to construct flood forecasting models for Clear Creek and Upper Wapsipinicon River in the United States, demonstrating the effective application of the LSTM-seq2seq model in short-term flood forecasting (Xiang et al., 2020). The LSTM network and Encoder-Decoder framework have been adopted to achieve multi-step prediction, providing a maximum lead time of 6 h for reservoir inflow prediction (Kao et al., 2020). The LSTM-based flood forecasting model developed by Dazzi et al. (2021) utilized the observed upstream and downstream water levels as inputs to accurately predict the flood stages of critical gauge stations.

In this paper, a flood forecasting model based on the LSTM network, Encoder-Decoder framework, as well as the feedback and attention mechanism are proposed to achieve multiple time steps flood forecasting. Three representative watersheds with distinct climate and underlying surface characteristics are selected to conduct case studies to validate the performance of the proposed model.

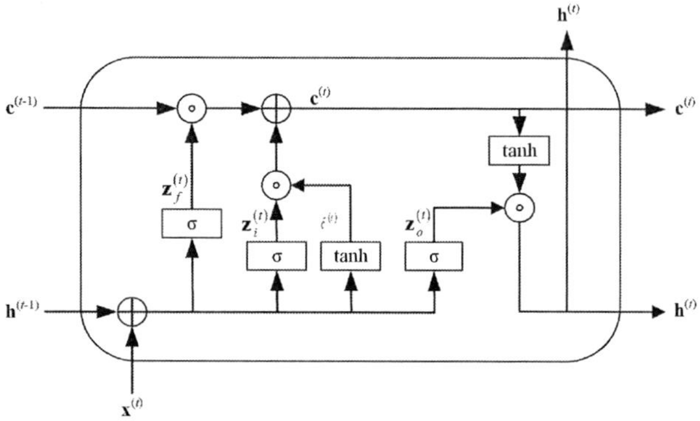

Figure 1. Internal structure of LSTM cell.

2 Methodology

2.1 LSTM-based flood forecasting model

The internal structure of the LSTM cell is illustrated in Fig. 1, encompassing three gates for regulating information: the "forget gate", "input gate", and "output gate".

The information processing in "forgetting gate" is shown in Eq. (1), which mainly controls the information to be discarded in the internal state.

$$z_{\mathrm{f}}^{(t)} = \sigma\left(\mathbf{W}_{\mathrm{f}}\left[\boldsymbol{h}^{(t-1)}; \boldsymbol{x}^{(t)}\right] + \boldsymbol{b}_{\mathrm{f}}\right) \tag{1}$$

In which, σ is sigmoid function, $\mathbf{W}_*$ is weight matrix, $\boldsymbol{b}_*$ is bias weight, $\boldsymbol{h}^{(t-1)}$ is the hidden state of previous time step, $\boldsymbol{x}^{(t)}$ is the input at time t.

The computation of vectors in the "input gate" is illustrated in Eq. (2), which determines the novel information incorporated into the internal state at each time step. The candidate states are obtained through a feedforward network layer using the "tanh" function as the activation function, as depicted in Eq. (3).

$$z_{i}^{(t)} = \sigma\left(\mathbf{W}_{i}\left[\boldsymbol{h}^{(t-1)}; \boldsymbol{x}^{(t)}\right] + \boldsymbol{b}_{i}\right) \tag{2}$$

$$\tilde{\boldsymbol{c}}^{(t)} = \tanh\left(\mathbf{W}_{\tilde{c}}\left[\boldsymbol{h}^{(t-1)}; \boldsymbol{x}^{(t)}\right] + \boldsymbol{b}_{\tilde{c}}\right) \tag{3}$$

The "output gate" process, depicted in Eq. (4), governs the transmission of information from the internal state to the hidden state at each time step. Nonlinear transformation is also essential in the output process and can be achieved through activation functions, as illustrated in Eq. (5).

$$z_{\mathrm{o}}^{(t)} = \sigma\left(\mathbf{W}_{\mathrm{o}}\left[\boldsymbol{h}^{(t-1)}; \boldsymbol{x}^{(t)}\right] + \boldsymbol{b}_{\mathrm{o}}\right) \tag{4}$$

$$\boldsymbol{h}^{(t)} = z_{\mathrm{o}}^{(t)} \odot \tanh\left(\boldsymbol{c}^{(t)}\right) \tag{5}$$

The proposed forecasting model is referred to as LSTM-DSA, with DSA representing the dual stage attention mechanism. The architectural depiction of LSTM-DSA can be observed in Fig. 2. The dual stage attention operates on both the Encoder and Decoder components individually. The Encoder part utilizes the attention mechanism to enable adaptive selection among multiple driving factors, thereby facilitating the utilization of point rainfall data as input instead of area rainfall data. In the Decoder part, the attention mechanism assigns weight coefficients to the hidden state of the encoder at each time step. Additionally, the model encompasses a feedback mechanism within the Decoder part to utilize future rainfall data as additional input.

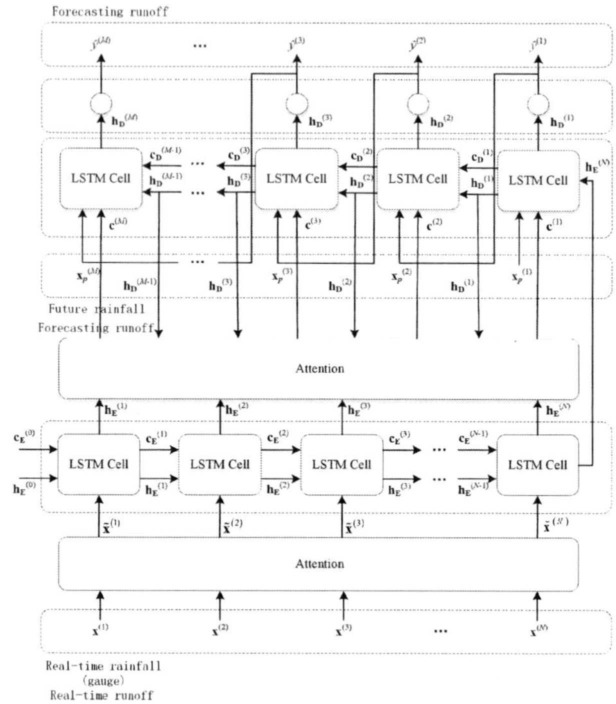

Figure 2. Architecture of the LSTM-DSA model.

2.2 Benchmark model

The forecasting model based on Encoder-Decoder framework proposed by Kao et al. (2020) is utilized as the benchmark model, named as LSTM-ED, whose architecture is illustrated in Fig. 3.

2.3 Evaluation metrics

The model performance is assessed based on the simulated peak value, peak occurrence time, and flood process. The evaluation metrics include the relative error of the peak, error in peak occurrence time, and Nash efficiency coefficient.

The relative error of the peak is calculated as follows:

$$\delta_{\mathrm{RPE}} = \frac{Q_{\max}^{\mathrm{sim}} - Q_{\max}^{\mathrm{obs}}}{Q_{\max}^{\mathrm{obs}}} \times 100\,\% \tag{6}$$

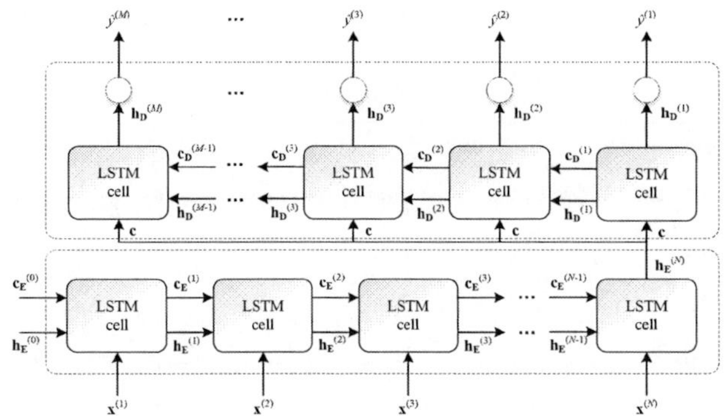

Figure 3. Architecture of the benchmark model.

In which, $Q_{\text{max}}^{\text{sim}}$ ($\text{m}^3\,\text{s}^{-1}$) and $Q_{\text{max}}^{\text{obs}}$ ($\text{m}^3\,\text{s}^{-1}$) are the peak value of the simulated and the observed sequence, respectively.

The computation of the peak time error is as follows:

$$\Delta_{\text{PTE}} = T_{\text{peak}}^{\text{sim}} - T_{\text{peak}}^{\text{obs}} \tag{7}$$

Where, $T_{\text{peak}}^{\text{sim}}$ (h) is the peak occurrence time of the simulated sequence and $T_{\text{peak}}^{\text{obs}}$ (h) is the peak occurrence time of the observed sequence.

Equation (8) shows the computation of the Nash efficiency coefficient, which quantifies the overall discrepancy between the simulated and observed sequence.

$$C_{\text{NSE}} = 1 - \frac{\sum_{t=1}^{M}\left(Q_t^{\text{obs}} - Q_t^{\text{sim}}\right)^2}{\sum_{t=1}^{M}\left(Q_t^{\text{obs}} - \bar{Q}^{\text{obs}}\right)^2} \tag{8}$$

In which, Q_t^{obs} and Q_t^{sim} represent the value of the observed and simulated sequence, respectively. $\bar{Q}^{\text{obs}}$ is the mean value of the observed sequence.

3 Study Watersheds and Data

In order to comprehensively assess the performance of the proposed model, we carefully select three watersheds as study areas, each characterized by distinct climatic and geomorphic features. The three watersheds are situated within Shaanxi Province, China, and boast well-maintained rain gauges and hydrological stations, along with an extensive collection of observational data. The basic information the study watersheds is shown in Table 1, and the locations are shown in Fig. 4.

4 Experiments

4.1 Model inputs and outputs

The input steps of the LSTM-based models for the three watersheds are determined as 3, 9 and 9 h by analysing the ACF and PACF of the rainfall and runoff data comprehensively as well as conducting preliminary modelling analysis. The inputs and outputs of each flood forecasting model are illustrated in Table 2.

Table 1. Basic information of the study watersheds.

Watershed	Area (km^2)	Rainfall stations	Data period	Flood events
Xichuan	717.43	5	1971–2013	35
Bahe	1605.72	11	1970–2015	160
Yuehe	2821.65	14	1962–2014	139

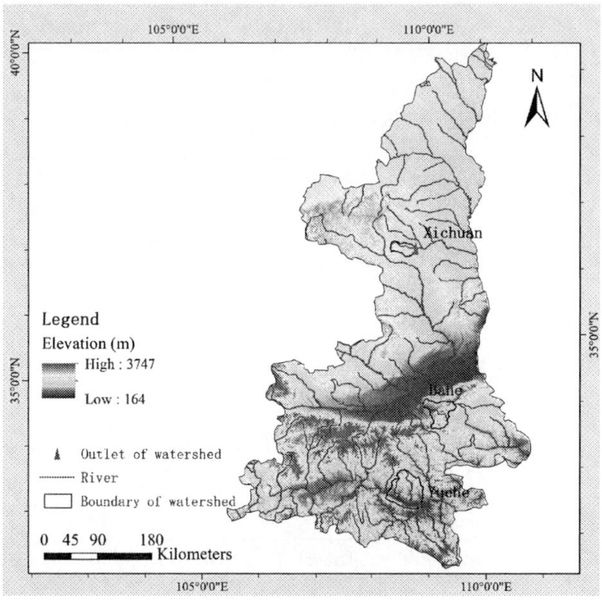

Figure 4. Locations of the study watersheds.

4.2 Data pre-processing and sample division

The linear normalization method is employed for pre-processing, as indicated in Eq. (9), to account for the disparities in dimensions and scales between rainfall and runoff data. This approach ensures improved accuracy and convergence speed of the model.

$$x_{\text{norm}} = \frac{x - x_{\text{min}}}{x_{\text{max}} - x_{\text{min}}} \tag{9}$$

The sample set is constructed using normalized rainfall and runoff data. Firstly, 20 % of all flood events are randomly selected as testing events. Secondly, the remaining 80 % of the events are sliced according to input and output steps to generate samples, from which 10 % are randomly chosen as validation set. Finally, the remaining samples constitute the training set. Table 3 shows the number of samples in each set.

Table 2. Model inputs and outputs.

Watershed	Model	Inputs	Outputs
Xichuan	LSTM-ED LSTM-DSA	Point rainfall data and runoff data from $T-2$ to T, area rainfall data from $T+1$ to $T+6$	Runoff data from $T+1$ to $T+6$
Bahe	LSTM-ED LSTM-DSA	Point rainfall data and runoff data from $T-8$ to T, area rainfall data from $T+1$ to $T+6$	
Yuehe	LSTM-ED LSTM-DSA	Point rainfall data and runoff data from $T-8$ to T, area rainfall data from $T+1$ to $T+6$	

Table 3. Sample set information.

Watershed	Total number	Training events	Testing events	Training samples	Validation samples
Xichuan	35	28	7	306	35
Bahe	160	128	32	4881	543
Yuehe	139	112	27	4770	523

Table 4. Hyperparameters of models.

Watershed	Model	Number neurons	Batch size
Xichuan	LSTM-ED	16; 16	64
	LSTM-DSA	16; 32	64
Bahe	LSTM-ED	16; 16	256
	LSTM-DSA	16; 32	512
Yuehe	LSTM-ED	16; 16	256
	LSTM-DSA	16; 32	512

4.3 Settings and training

The forecasting model is constructed based on the Keras in Python, with a single-layer LSTM utilized for both the Encoder and Decoder layer, while a full connection layer serves as output layer. The "MSE", "Adam" and "ReLU" are selected as loss function, optimizer and activation function, respectively. The other hyperparameters of the model are determined through trial calculations with the loss function as indicators, wherein the number of neurons and batch size are optimized using powers of 2, which are presented in Table 4.

5 Results and Discussion

The evaluation metric distributions of the prediction results for each model are depicted in Figs. 5–7 respectively. The predicted hydrographs for typical flood events in each watershed are illustrated in Figs. 8–10.

For Xichuan watershed, the Nash efficiency coefficient and relative error of flood peak significantly deteriorate when the prediction step exceeds 3 h. Overall, the estimations of

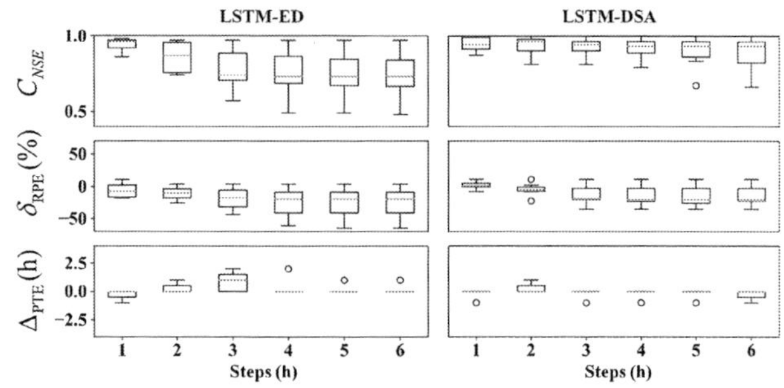

Figure 5. Metrics distribution for models in Xichuan.

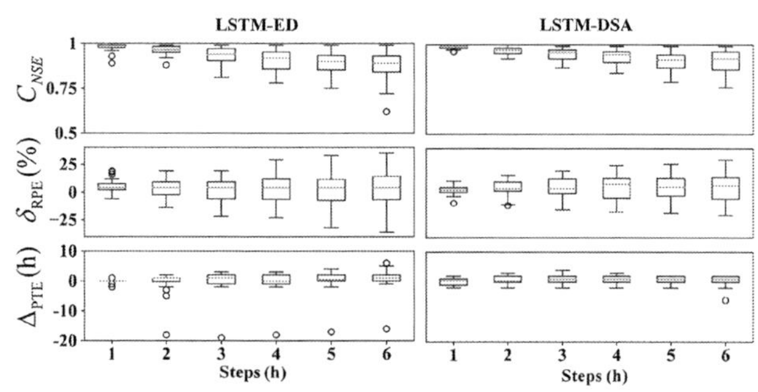

Figure 6. Metrics distribution for models in Bahe.

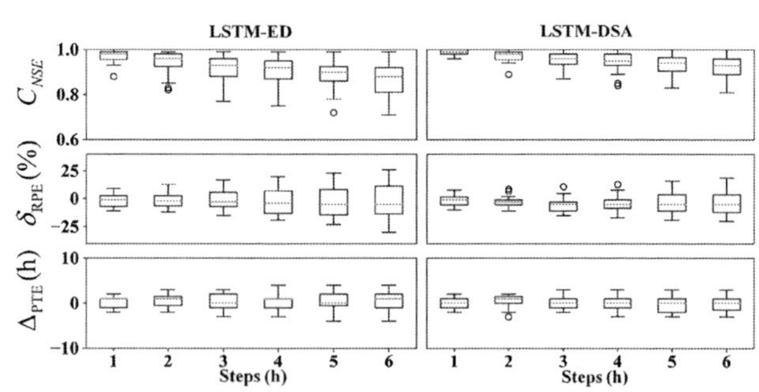

Figure 7. Metrics distribution for models in Yuehe.

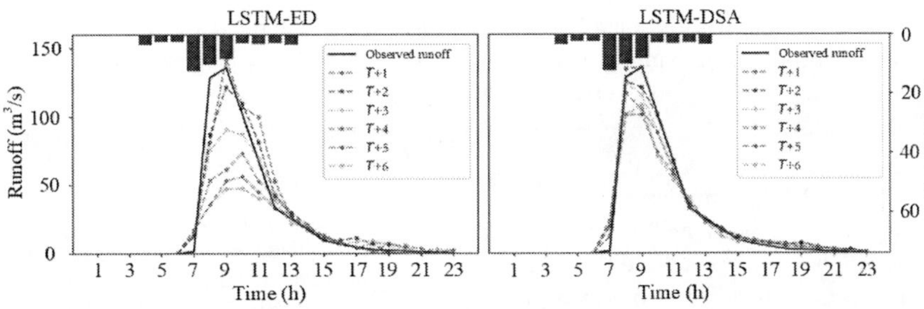

Figure 8. Results for the "19910610" event in Xichuan.

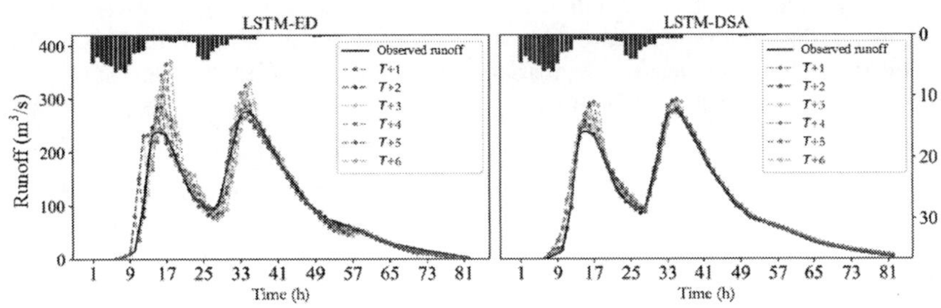

Figure 9. Results for the "19860909" event in Bahe.

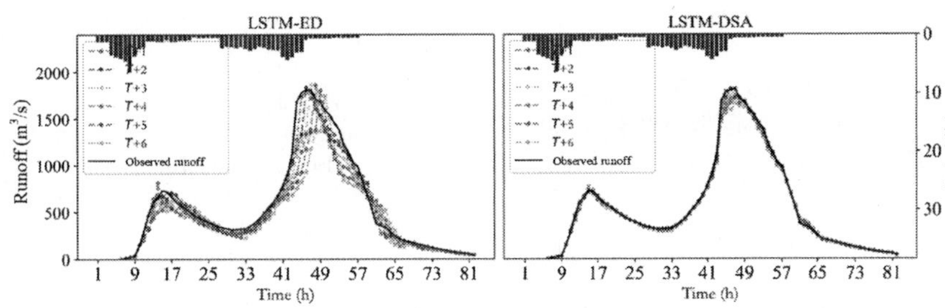

Figure 10. Results for the "20100718" event in Yuehe.

flood peaks are slightly lower than the observed value. The performance of the LSTM-DSA model, meanwhile, surpasses that of the LSTM-ED model.

For Bahe watershed, the performances of both models decrease with the growing of prediction step. For flood peaks, the predicted values are higher than the observed one. The evaluation metrics show improvement with the inclusion of future rainfall data.

The metrics distributions in Yuehe watershed exhibit a comparable phenomenon. The prediction accuracy of both models decreases as the prediction step extends. The estimations of flood peaks are lower than the observed value. Incorporating feedback and a dual-stage attention mechanism enhances the model performance.

The experiments reveal that the performance of all models has exhibited varying degrees of decline as the prediction step extends. The attenuation of information in the input data is believed to contribute to this portion of the prediction error. This phenomenon has been effectively mitigated through the implementation of the feedback mechanism. Overall, the performance of the LSTM-DSA model is greatly enhanced by incorporating feedback and a dual-stage attention mechanism, surpassing that of the LSTM-ED model. The inclusion of the attention mechanism has significantly enhanced prediction accuracy for the Bahe and Yuehe watersheds, whereas its impact on the Xichuan watershed is less pronounced, suggesting that the attention mechanism proves more effective in larger-scale watersheds.

Although we employ a sophisticated model structure, accurate predicting certain floods remains challenging. We believe this error can be attributed to two factors. Firstly, the input of each model solely relies on rainfall and runoff data without considering other relevant variables. The information contained in the input data is insufficient for precise flood prediction. Secondly, there is still room for improvement in optimizing the model, refining the loss function, adjusting batch size, and fine-tuning the hyperparameters.

6 Conclusions

The paper introduces the LSTM-DSA, a dual-stage attention-based LSTM network for multiple time steps flood forecasting. Three representative watersheds are selected as study areas to access the model performance. The experiment results are evaluated, and the sources of prediction errors are analysed. In summary, the proposed model exhibits a significantly high level of accuracy in predicting floods within distinct prediction steps, demonstrating its exceptional predictive capability. The efficacy of both the feedback mechanism and attention mechanism has also been validated in the domain of flood prediction.

Code and data availability. The code is available in Github (https://github.com/Vingttrois/Flood-foresting-model-based-on-LSTM.git, last access: 7 March 2024) and Zenodo (https://doi.org/10.5281/zenodo.10791419, Wang, 2024).

The data are available upon request (wangfan@iwhr.com).

Author contributions. FW established the forecasting model and wrote the text. WW, WB and WL took care of the data processing. DZ revised the results and the text.

Competing interests. The contact author has declared that none of the authors has any competing interests.

Disclaimer. Publisher's note: Copernicus Publications remains neutral with regard to jurisdictional claims made in the text, published maps, institutional affiliations, or any other geographical representation in this paper. While Copernicus Publications makes every effort to include appropriate place names, the final responsibility lies with the authors.

Special issue statement. This article is part of the special issue "ICFM9 – River Basin Disaster Resilience and Sustainability by All". It is a result of The 9th International Conference on Flood Management, Tsukuba, Japan, 18–22 February 2023.

Acknowledgements. This work is supported by the Belt and Road Special Foundation of the National Key Laboratory of Water Disaster Prevention (Grant No. 2021491511).

Financial support. This research has been supported by the State Key Laboratory of Hydrology-Water Resources and Hydraulic Engineering (grant no. 2021491511).

Review statement. This paper was edited by Mamoru Miyamoto and reviewed by two anonymous referees.

References

Addor, N., Newman, A. J., Mizukami, N., and Clark, M. P.: The CAMELS data set: catchment attributes and meteorology for large-sample studies, Hydrol. Earth Syst. Sci., 21, 5293–5313, https://doi.org/10.5194/hess-21-5293-2017, 2017.

Dazzi, S., Vacondio, R., and Mignosa, P.: Flood Stage Forecasting Using Machine-Learning Methods: A Case Study on the Parma River (Italy), Water, 13, 1612, https://doi.org/10.3390/w13121612, 2021.

Hochreiter, S. and Schmidhuber, J.: Long short-term memory, Neural Comput., 9, 1735–1780, https://doi.org/10.1162/neco.1997.9.8.1735, 1997.

Kao, I. F., Zhou, Y. L., Chang, L. C., and Chang, F. J.: Exploring a Long Short-Term Memory based Encoder-Decoder framework for multi-step-ahead flood forecasting, J. Hydrol., 583, 124631, https://doi.org/10.1016/j.jhydrol.2020.124631, 2020.

Kratzert, F., Klotz, D., Brenner, C., Schulz, K., and Herrnegger, M.: Rainfall–runoff modelling using Long Short-Term Memory (LSTM) networks, Hydrol. Earth Syst. Sci., 22, 6005–6022, https://doi.org/10.5194/hess-22-6005-2018, 2018.

Newman, A. J., Clark, M. P., Sampson, K., Wood, A., Hay, L. E., Bock, A., Viger, R. J., Blodgett, D., Brekke, L., Arnold, J. R., Hopson, T., and Duan, Q.: Development of a large-sample watershed-scale hydrometeorological data set for the contiguous USA: data set characteristics and assessment of regional variability in hydrologic model performance, Hydrol. Earth Syst. Sci., 19, 209–223, https://doi.org/10.5194/hess-19-209-2015, 2015.

Thapa, S., Zhao, Z. B., Li, B., Lu, L., Fu, D. L., Shi, X. F., Tang, B., and Qi, H.: Snowmelt-Driven Streamflow Prediction Using Machine Learning Techniques (LSTM, NARX, GPR, and SVR), Water, 12, 1734, https://doi.org/10.3390/w12061734, 2020.

Wang, F.: Vingttrois/Flood-foresting-model-based-on-LSTM: V1.0.0 (V1.0.0), Zenodo [code], https://doi.org/10.5281/zenodo.10791419, 2024.

Xiang, Z. R., Yan, J., and Demir, I.: A Rainfall-Runoff Model With LSTM-Based Sequence-to-Sequence Learning, Water Resour. Res., 56, e2019WR025326, https://doi.org/10.1029/2019wr025326, 2020.

Proc. IAHS, 386, 147–152, 2024
https://doi.org/10.5194/piahs-386-147-2024

Open Access
PIAHS

Development of model predictive controls for flood control and power generation of reservoir cascade in Mahaweli basin, Sri Lanka to mitigate energy and economic crisis provoked by the COVID-19 pandemic

Ranapura Dewage Thilini Kaushalya[1], **Sajana Pramudith Hemakumara**[2], **Selvarajah Hemakanth**[1], and **Seenipellage Chaminda Sugeeshwara**[1]

[1]Irrigation Department, Government of Sri Lanka, Colombo, 0700, Sri Lanka
[2]Norwegian University of Science and Technology (NTNU), Trondheim, 7034, Norway

Correspondence: Ranapura Dewage Thilini Kaushalya (kaushalyardt@gmail.com)

Received: 21 July 2023 – Revised: 13 October 2023 – Accepted: 24 February 2024 – Published: 19 April 2024

Abstract. Mahaweli basin is the largest river basin in Sri Lanka which receives high-intensity rainfall reasoned to frequent flooding. Mahaweli River passes through Victoria, Randenigala and Rantambe reservoirs which cater for drinking, irrigation, hydropower, and environmental water requirements. These reservoirs contribute to 63 % of the hydropower production of the country. The development of flood control methods for this cascade will mitigate flash floods and lower the risk of reservoir spill during peak hours. Therefore, a system dynamic model is developed using VENSIM software to analyse the complex correlation between flood storage and power flow considering short-term weather prediction. Model predictive controls are applied to analyse the system to enhance power generation through storage optimization. This brings direct economic benefits and be a sustainable solution for the energy crisis provoked by the economic loss that occurred during the COVID-19 pandemic.

1 Background

Sri Lanka is a water-rich tropical country that receives rainfall through the monsoon, convection, and depression. The mean annual rainfall of the country varies from 900 mm in the dry zone to 5000 mm in the wet zone (Department of Meteorology, 2016). The water resources system of Sri Lanka comprises groundwater sources and surface water sources such as rivers, lakes and reservoirs and small tank cascades.

Although Sri Lanka has a radial type of river network which starts at the central highlands and flows all over the island before reaching the sea, currently it faces diverse problems when addressing water issues. The frequent occurrence of high-intensity short duration rainfall cause floods in the wet climatic zone of the country and prolonged hot weather in the dry zone cause significant water deficiencies in the dry zone. Therefore, it is noted that the water management practices in both dry and wet zones need to be carefully analysed and revise appropriately.

Energy generation in Sri Lanka mostly depends on the use of hydropower and coal. The share of electricity production by hydropower, coal, crude oil, and wind energy are 45 %, 43 %, 9 %, and 3 % respectively (Ceylon Electricity Board, 2022). However, due to the impact of the Covid-19 pandemic, the country underwent several periods of lockdown to prevent the spread of the virus. Therefore, Sri Lanka could limit the death count to 16 774 (Health Promotion Bureau, 2022) which is a relatively lesser amount compared with world statistics. As a result of the Covid-19 pandemic, the economy of Sri Lanka experienced a drastic downturn and currently, the whole country is suffering due to the economic crisis as well as the energy crisis. Considering the current situation of the country, it is noted that the optimization of hydropower production will bring direct benefits to the economy. But it should be carried out carefully to avoid any disruption to domestic and irrigation water supply because the food supply also plays a key role in the economy.

Therefore, the optimization of reservoir storage to cater for additional flood storage was assessed for the reservoir cascade located in the upper watershed of the Mahaweli river basin. A System Dynamic (SD) model is developed using the VENSIM software to analyse the complex correlation between flood storage, irrigation, drinking, environmental and power flow requirements considering short-term weather prediction.

2 Study Area

Mahaweli basin is the largest river basin in Sri Lanka covering 10 327 km^2 (Survey Department, 2012) area which is about 16 % of the total land area of the country. Mahaweli river is the largest and longest river in the country which is 335 km (Survey Department, 2012) in length and passes through all three climatic zones namely wet, dry, and intermediate zones. The river originated at the Hatton Plateau of Sri Pada Mountain and flows to the sea at Koddiyar Bay in Trincomalee.

The basin receives rainfall by North-East and South-West monsoon rainfall periods. The upper catchment of the basin receives rainfall during the Southwest monsoon from May to September. The wet climatic zone of the country usually receives high-intensity rainfall during this season, and it causes extreme floods in the wet zone. However, the downstream reaches of the Mahaweli river lie within the dry zone and more water can be released to downstream for optimal flood control during this period. Therefore, the establishment of an efficient water management system in the Mahaweli river basin is vital to store a portion of flood discharge for future use.

The upper catchment of the Mahaweli river basin comprises a series of multi-purpose reservoirs which cater for the drinking, irrigation, hydropower, and environmental water requirements. At the upstream reach, the river passes through Victoria, Randenigala, and Rantambe reservoirs generating hydroelectric power at each location. These three reservoirs contribute to 63 % (Ceylon Electricity Board, 2022) of the hydropower production of the country. The cascade system supplies irrigation water for agricultural areas during both Maha and Yala seasons which are the two main cultivation seasons of Sri Lanka. The location of the selected reservoir cascade is shown in Fig. 1.

Victoria reservoir receives inflows from the Polgolla barrage and its own catchment. The outflow from Victoria reservoir flows into Randenigala reservoir which is situated immediately downstream. This reservoir acts as a regulating reservoir. Rantambe reservoir which is comparatively the smallest in this cascade receives reservoir releases from Randenigala and inflows from its own catchment. The schematic diagram of the study area is shown in Fig. 2.

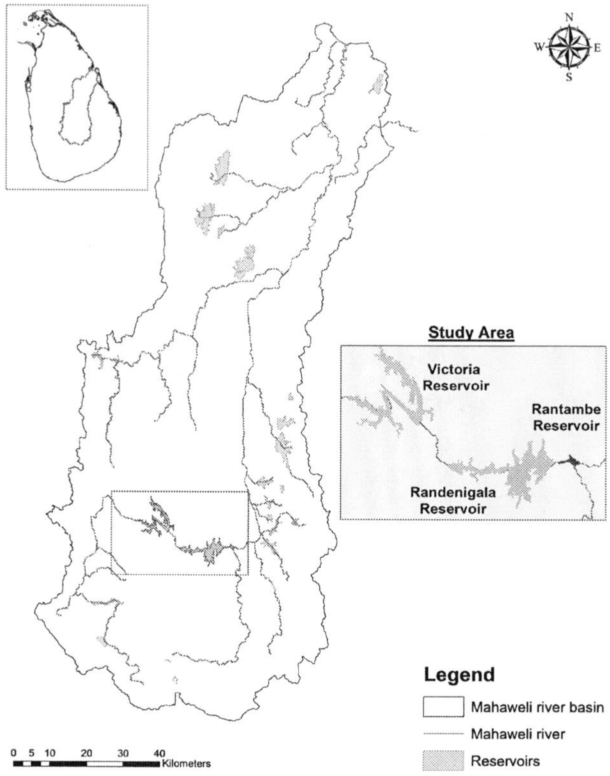

Figure 1. Study area of the research.

3 Methodology

The three main reservoirs namely, Victoria, Randenigala and Rantambe which are in the upper reaches of Mahaweli river basin were selected as the study area for this research. A system dynamic model is developed for this reservoir cascade using VENSIM software PLE 9.3.1 platform to analyse the complex interrelationship between the storage and hydropower production of this cascade system. The main purpose of developing this type of model is to check the ability to optimize power generation from these reservoirs by optimizing the reservoir storage during rainfall events as a solution to the current energy and economic crisis. The methodology of this research is depicted in Fig. 3.

3.1 Development of System Dynamic (SD) Model

The SD-based Mahaweli cascade system model comprises five sectors, namely, reservoir storage, drinking, irrigation, power production, and environmental release. The reservoir storage sector of the Mahaweli cascade system is the main component of the model. The interrelationships among the five sectors were modelled considering feedback in a dynamic environment. The model was simulated from 2000 to 2020 period on daily basis.

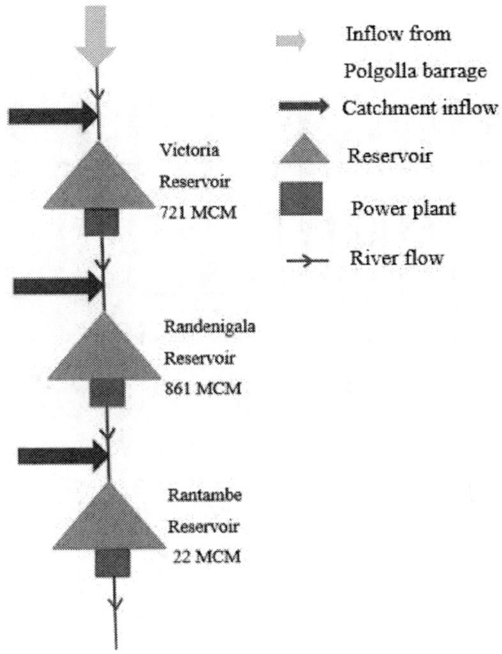

Figure 2. Schematic diagram of the study area.

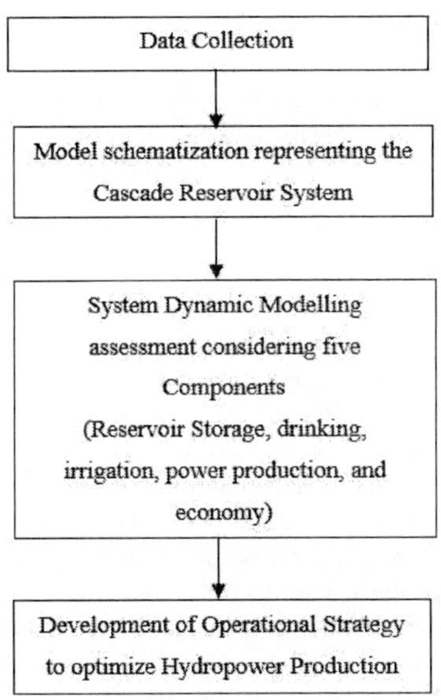

Figure 3. Methodology of the research.

3.2 Reservoir storage component of the model

The water balance assessment was carried out for three reservoirs to compute the reservoir storage. The water balance equation applied for each reservoir depicts in Eq. (1).

$$\Delta S = R + I - O - S_p - L, \tag{1}$$

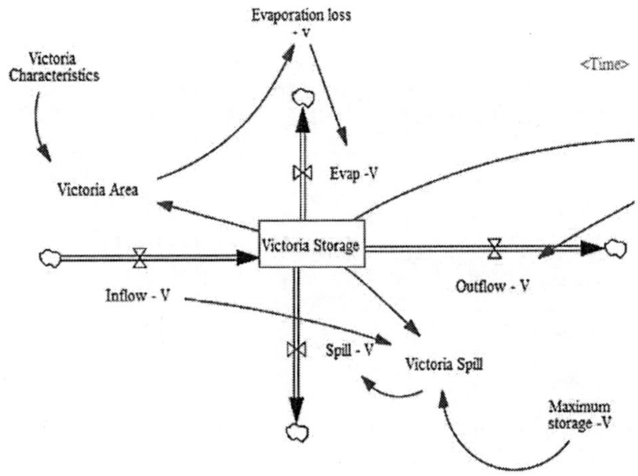

Figure 4. Schematisation of the reservoir storage component in the VENSIM graphical interface.

where ΔS is the change in reservoir storage and R, I, O, S_p and L are rainfall, inflow, outflow, spill flow and losses respectively.

The "Outflow" in the above equation represents the water release from the reservoir for hydropower, irrigation, drinking, and environmental water demands. The schematised Reservoir storage component of Victoria reservoir is illustrated in Fig. 4.

3.3 Drinking water component model

In this component of the SD model, the emphasis is placed on the population that is served by each water supply intake. It considers both the per capita water consumption and the size of the served population.

3.4 Hydropower component model

This is one of the major components in this model as the reservoir cascade is mainly generating hydropower. The power flow, power production, and power pricing were considered in this component.

3.5 Irrigation component model

The irrigation water requirements on daily basis were considered in this component of the model. Paddy is the major crop grown in this basin during the Maha season which is the main agricultural season. Various types of cash crops along with paddy are grown during the Yala season, which is the second agricultural season. The agricultural water requirements for land preparation and crop growth were modelled corresponding to each cultivation period.

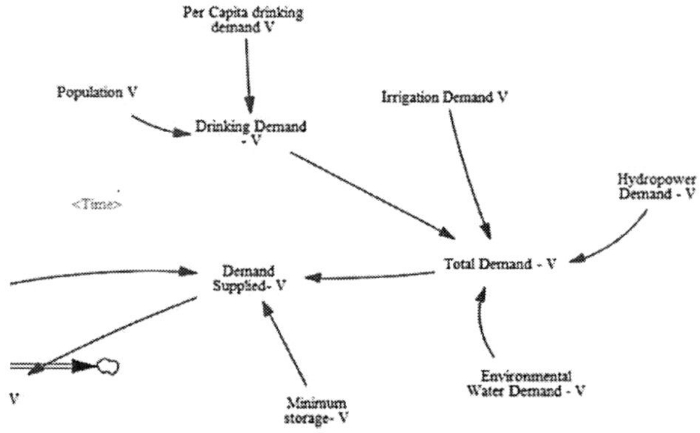

Figure 5. Schematisation of the demand components in the VENSIM graphical interface.

3.6 Environmental flow component model

The mandatory environmental releases for downstream reaches were modelled under this component.

The schematic view of the demand components of the Victoria reservoir model is illustrated in Fig. 5.

4 Analysis

Firstly, SD model was developed for the Victoria reservoir system. The aim of the research is to examine variation of the power generation of Victoria reservoir with respect to the operational scenarios of reservoir. In this study, the receipt of short-term rainfall forecast for 5 d in advance was used analyse the optimization of energy production with respect to the variations of Victoria reservoir capacity. The reservoir storage was optimized in such a way to cater for maximum power production while releasing water for other water requirements. Accordingly, the spill flow from the reservoir was minimized in this scenario. The optimized daily energy production was compared with the original daily energy production (in GWh) of the Victoria reservoir for the period from 2000 to 2020 and shown in Fig. 6.

The highest annual energy production was observed in the year 2013, where it produced 1426 and 1553 GWh for the original and modelled scenario respectively. A decreasing trend in energy production is noted for the periods 2000–2009, 2010–2012 and 2013–2020.

The difference in modelled and original annual energy production depicts the effect of the optimized reservoir storage on energy production in each specific year. This difference was presented as a percentage of annual energy increment which was evaluated with the annual maximum rainfall (mm) in Fig. 7. As annual maximum precipitation signifies extreme wet events, the variation of energy increment with the flood events is evaluated.

The increment in energy production was reflected in the years 2006, 2011–2015, 2018, and 2020. The annual maximum rainfall was higher than 100 mm in these specific years. The highest increment in energy production was observed in 2013. A decreasing trend was observed in energy increment from this year onwards as well as the maximum precipitation. On the other hand, the annual maximum rainfall is highest in 2012 whilst the energy production increment is fairly less. This maximum precipitation occurred in late December and that event couldn't progress for a few days in this particular year might be the reason for it.

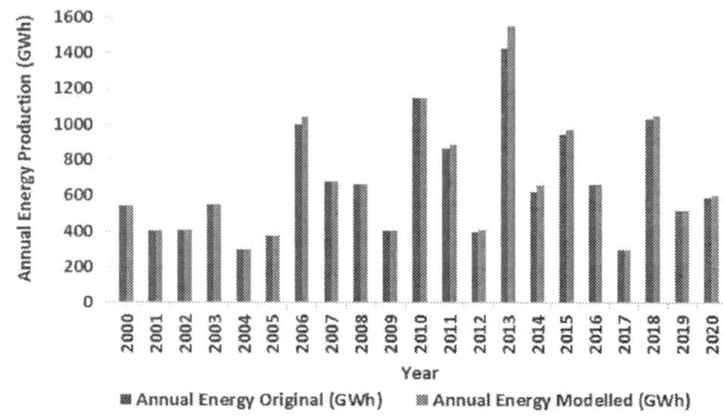

Figure 6. Comparison of the annual energy production: original (blue) and modelled (orange) scenarios.

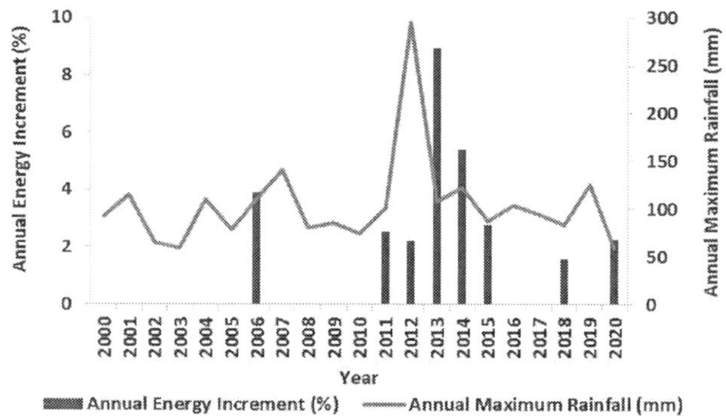

Figure 7. Annual energy increment (%) and annual maximum rainfall (mm).

As it is also mentioned previously, the Victoria reservoir was optimized in order to maximize power production by minimizing the spillage of the reservoir. The original and optimized spill volumes (in Million Cubic Meters (MCM)) were assessed with the annual rainfall (mm) in Fig. 8.

The spill volume was optimized for the years 2006, 2011–2015, 2018 and 2020. The annual rainfall was over 1500 mm for every year mentioned above except in 2020. The highest spill volume optimization can be seen in 2013 and the lowest in 2012. According to the model results, the spill volume can be utilised for power production up to 122 MCM depending on the intensity of the rainfall and the reservoir storage. The increment in hydropower production with this

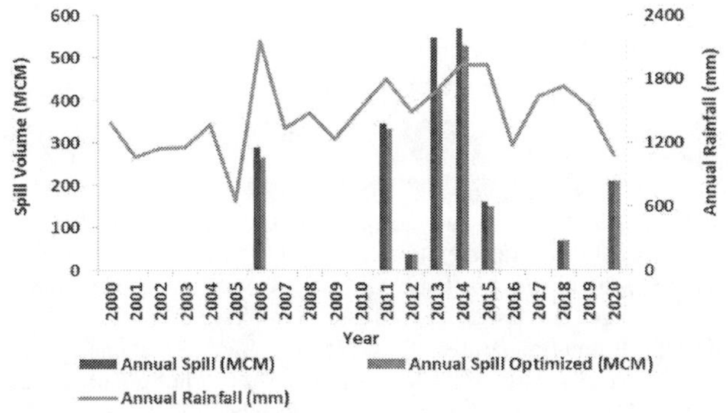

Figure 8. Annual spill volume (MCM) for original and optimized scenarios and annual maximum rainfall (mm).

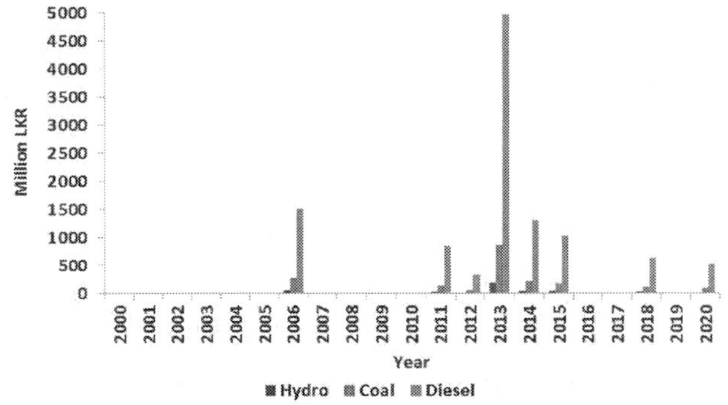

Figure 9. Comparison of electricity production costs (million LKR) for energy increment using hydro, coal, and diesel.

optimization causes additional production costs. Assuming this increment in hydropower production would have been done with thermal (coal and diesel) power, the increment in energy production costs for hydro, coal, and thermal power are calculated and illustrated in Fig. 9. The unit prices (in Sri Lankan Rupees (LKR)) for hydro and thermal (coal and diesel) power for producing 1 kWh electricity are considered as LKR 1.5, LKR 6.78, and LKR 38.88 respectively (Ceylon Electricity Board, 2022).

The energy production is highest in the year 2013 and the cost is about LKR 192 million, LKR 866 million and LKR 4964 million for hydro, coal, and diesel energy sources respectively. Therefore, it is observed that if the reservoir operated in this pattern, the energy production costs will be reduced by a significant amount as the hydropower production cost is the least out of the other two sources.

5 Discussion

The volume of the spill in Victoria reservoir was optimized for a 5 d short-term rainfall prediction in order to maximize the hydropower production without varying any other water requirements. Under the scenario developed for this research, the continuous agricultural water supply for both Maha and Yala seasons, uninterrupted supply of drinking water even during dry months and environmental water supply were considered as the first priority. The optimization of power production through the change of reservoir operations was introduced as the second priority focusing on reduction of spill flow, which will further reduce the existing downstream flood risk. The spill volume optimization is highest in the wet years. As spill volume is minimized, more water allocation can be provided for power production. The higher the minimization of the spillage, greater the hydropower is produced and lesser the downstream flood inundation. On the other hand, harnessing hydropower for electricity production is the cheapest among coal and diesel as coal and diesel power production are 5 and 26 times more costly than hydropower. Higher the increment of energy production by spill optimization, the greater the difference in production costs among the energy sources.

6 Conclusion and recommendations

The Victoria reservoir storage was optimised by minimizing its spillage to obtain maximum power production by keeping the other water requirements constant. Although power production will add production charges, hydropower is the cheapest energy source compared with coal and diesel. It is observed that maximizing hydropower production makes direct economic benefits and is a sustainable solution for the energy crisis provoked by the economic loss that occurred during the COVID-19 pandemic. The same approach can be followed for the entire cascade system considering Rantambe and Randenigala reservoirs as well. Further, it is expected to develop a few scenarios to evaluate the energy production of this cascade by optimizing irrigation water usage by introducing irrigation modernization techniques to reduce water consumption.

Data availability. The data supporting the findings of this study are available from the corresponding author on request.

Author contributions. RDTK, SPH, SH and SCS designed the methodological and conceptual frameworks of the study. SH and SCS provided hydro-meteorological data. RDTK performed the simulations, and SPH analysed the results. RDTK and SPH prepared the manuscript.

Competing interests. The contact author has declared that none of the authors has any competing interests.

Disclaimer. Publisher's note: Copernicus Publications remains neutral with regard to jurisdictional claims made in the text, published maps, institutional affiliations, or any other geographical representation in this paper. While Copernicus Publications makes every effort to include appropriate place names, the final responsibility lies with the authors.

Special issue statement. This article is part of the special issue "ICFM9 – River Basin Disaster Resilience and Sustainability by All". It is a result of The 9th International Conference on Flood Management, Tsukuba, Japan, 18–22 February 2023.

Review statement. This paper was edited by Mamoru Miyamoto and reviewed by two anonymous referees.

References

Ceylon Electricity Board: Electricity Generated, https://ceb.lk/electricity-generated/en (last access: 24 October 2022), 2022.

Department of Meteorology: Climate of Sri Lanka, http://www.meteo.gov.lk/ (last access: 24 October 2022), 2016.

Health Promotion Bureau: COVID-19: Live Situational Analysis Dashboard of Sri Lanka, https://hpb.health.gov.lk/covid19-dashboard/ (last access: 24 October 2022), 2022.

Survey Department: National Atlas, Colombo: Survey Department, Sri Lanka, ISBN 9559059041, 2012.

Proc. IAHS, 386, 153–158, 2024
https://doi.org/10.5194/piahs-386-153-2024

Open Access
PIAHS

Study on a flood damage reduction countermeasure by installing a spillway on a riverine levee for promoting the initiative of River Basin Disaster Resilience and Sustainability by All

Osamu Itagaki[1,a], **Miho Ohara**[2,b], **and Toshio Koike**[3]

[1]Upper Kiso-gawa River Office, Ministry of Land, Infrastructure, Transport and Tourism, Gifu, 500-8801, Japan
[2]Institute of Industrial Science, University of Tokyo, Tokyo, 153-0041, Japan
[3]International Center for Water Hazard and Risk Management, Public Works Research Institute, Tsukuba, 305-8516, Japan
[a]formerly at: Flood Disaster Prevention Division, River Department, National Institute for Land and Infrastructure Management, MLIT, Tsukuba, 305-0804, Japan
[b]formerly at: International Center for Water Hazard and Risk Management, Public Works Research Institute, Tsukuba, 305-8516, Japan

Correspondence: Osamu Itagaki (itagaki-o573co@pwri.go.jp)

Received: 21 July 2023 – Revised: 6 October 2023 – Accepted: 25 February 2024 – Published: 19 April 2024

Abstract. In light of the frequent occurrence of unprecedented heavy rains and floods attributable to climate change effects, the importance of the "River Basin Disaster Resilience and Sustainability by All" initiative is being emphasized and promoted strongly in Japan. For promoting discussions on a wide range of countermeasures for flood damage reduction by the initiative, this paper presents a trial estimation result of quantitative flood damage reduction effect by spillway installation on a riverine levee in a protected area where severe flood disasters have occurred in recent years. The result shows that the countermeasure considerably reduce the flooded area, the days required for drainage, and house flood depths. Results also indicate that the average annual damage to paddy fields in the area can be expected to decrease, although the inundation frequency will increase.

1 Introduction

In response to an unprecedented increase in torrential rains and floods due to the impact of climate change associated with global warming, it is necessary to accelerate the development of flood control structures, as well as, for example, land use regulation, and community resilience enhancement. It emphasizes the need for "River Basin Disaster Resilience and Sustainability by All" (Ryuuiki-chisui) initiative aimed at sustainable development through building a resilient society prepared for water disasters, such as devising ways to live, evacuating in the event of a disaster, reducing economic damage, and swift recovering and reconstructing after a disaster.

This study aims to provide material for promoting discussion on a wide range of countermeasures for flood damage reduction through the initiative by presenting a trial estimation result of quantitative flood damage reduction effect by spillway installation on a riverine levee in a protected area. This kind of damage mitigation measures include "Nokoshi", a measure with lowering a part of a levee system to avoid severe damage such as destruction of an irrigation weir downstream, which has been recorded since the Edo period (from 17 to 19 century) in Japan (Tanabe and Okuma, 2001), and the "Managed Overtopping" of the US Army Corps of Engineers (US Army Corps of Engineers, 2019). This research is conducted on a trial basis in one of the areas where severe floods occurred in recent years.

Published by Copernicus Publications on behalf of the International Association of Hydrological Sciences.

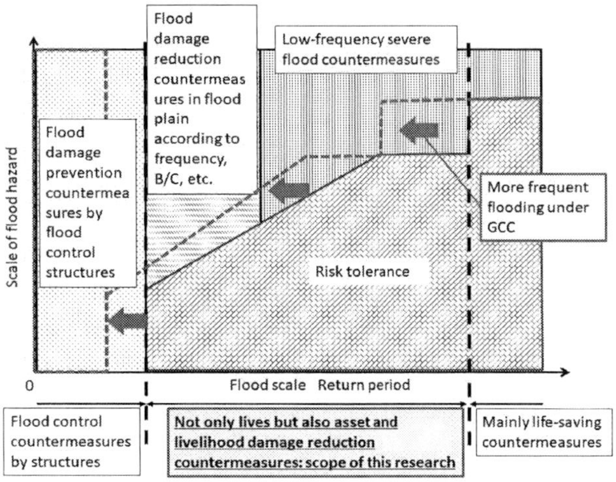

Figure 1. Classification of flood countermeasures according to flood scale and the hazard scale (modified from Itagaki et al., 2022).

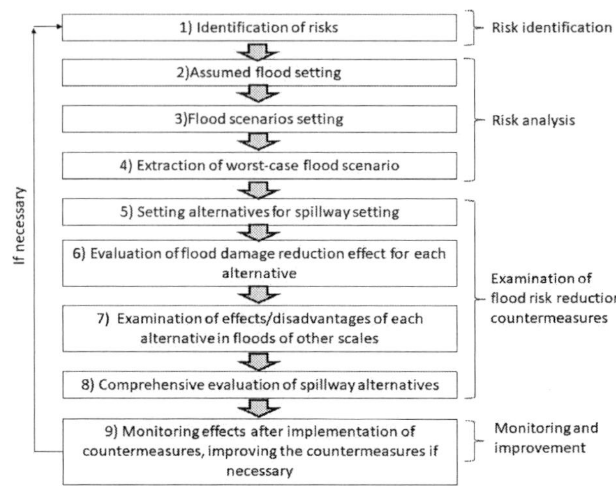

Figure 2. Procedure for examining spillway installation in riverine levee system (Itagaki et al., 2022).

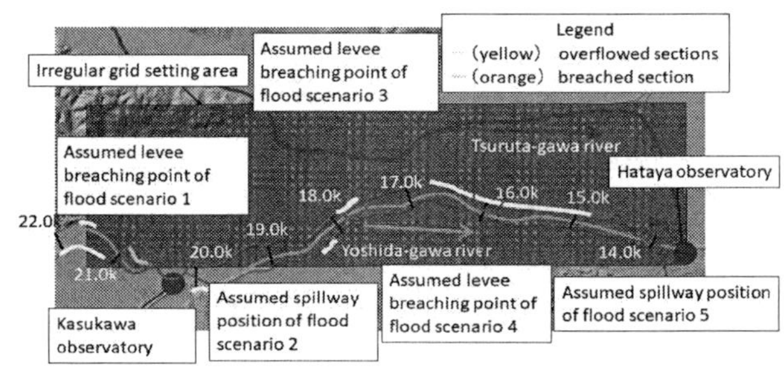

Figure 3. Trial application area (Itagaki et al., 2022). * Created by iRIC, using the Geospatial Information Authority color-coded elevation map as the background (same for Figs. 5, 6, and 7).

2 Level 1.5 flood damage reduction countermeasures

2.1 Classification of flood damage reduction countermeasures

Figure 1 shows the classification of countermeasures (including risk tolerance) based on two axes: flood scale/recurrence period and scale of hazard (flood depth at each location etc.). "Risk tolerance" refers to cases in which no countermeasures are taken based on the recurrence period of a disaster etc., and "Low-frequency severe flood countermeasures" refers to countermeasures against floods that have a long recurrence period but are expected to cause severe damage when they do occur. In the figure, the left-pointing arrows indicate that the recurrence period of heavy rainfall is shortened due to climate change, and the recurrence period of flooding is shortened.

The above-mentioned installing a spillway to reduce damage by reducing the flood water volume is considered to be one of the low-frequency severe flood countermeasures, and are considered to be countermeasure options to bridge the gap between the development of flood control structures (level 1 countermeasures) and the protection of human lives through evacuation (level 2 countermeasures). They are damage prevention and mitigation countermeasures (so to speak, level 1.5 countermeasures) that fill the gap, including not only lives but also assets and livelihoods.

2.2 Study method for countermeasures to reduce flood water volume into protected areas

Figure 2 shows the procedure for examining countermeasures to reduce the flood water volume into the protected area. In the following chapter, some items are omitted due to space limitations.

3 Experimental examination of countermeasure options to reduce flood water volume into protected areas

3.1 Overview of trial examination area

Figure 3 shows an area where trial examination is conducted. The Tsuruta-gawa River in the Takagi-gawa River system siphons across the Yoshida-gawa River in the Naruse-gawa River system.

The trial application area is about 2.1 km × 7.4 km, and the riverbed slope of the series of reaches including the reaches within the area is about 1/3000 for the Yoshida-gawa River and about 1/1000 to 1/22 000 for the Tsuruta-gawa River.

In the flood caused by Typhoon No. 19 (East Japan Typhoon) in 2019 (hereinafter referred to as "Actual Flood"), farmlands were submerged and it took 12 d for drainage, and

Table 1. Set flood scenarios (Itagaki et al., 2022).

Scenario No.	Flooding point into protected area
1	Left bank 20.9 kp (reproduction of Actual Flood)
2	Left bank 20.3 kp (spillway installation option 1)
3	Left bank 17.8 kp
4	Left bank 16.0 kp
5	Left bank 14.0 kp (spillway installation option 2)

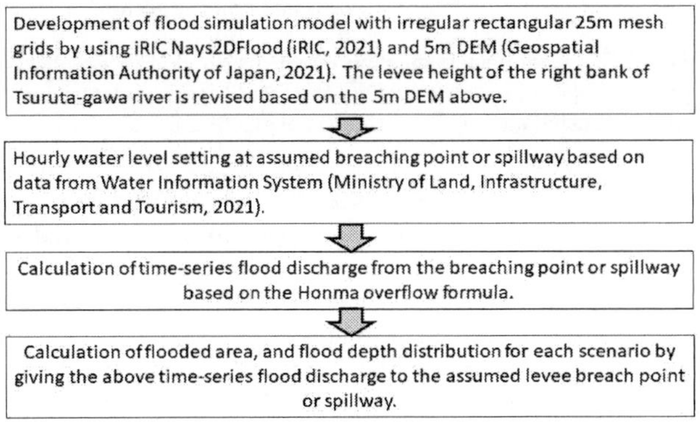

Figure 4. Flood simulation procedure (Itagaki et al., 2022).

residential houses were flooded near the ceiling of the first floor and some residents moved out after the disaster.

3.2 Setting flood scenarios

Figure 3 shows overflowed sections of the Yoshida-gawa River during Actual Flood (yellow indicates overflowed, orange indicates levee breached after overflow) (Naruse-gawa River Levee Investigation Committee, 2020). The maximum overflow depth was about 0.4 m at the levee breach point (near 20.9 kp (kilo meter post)). Based on assumed levee breach points on the left bank of the Yoshida-gawa River based on the major overflowed sections during Actual Flood, and the options for levee breach prevention measures in Sect. 3.5, a group of flood scenarios (conditions of flood into the protected area) were set as shown in Table 1. Figure 4 shows the flood simulation procedure, and Fig. 5a shows the flood simulation result of flood scenario 1 (reproduction simulation of Actual Flood).

3.3 Verification of flood simulation accuracy

A visual comparison between estimated inundation depth distribution map by Geospatial Information Authority of Japan and Fig. 5a reveals no significant difference in the extent and the maximum depth of inundation in the area between the Yoshida-gawa River and Tsuruta-gawa River. When comparing the trace inundation depth (12 points)

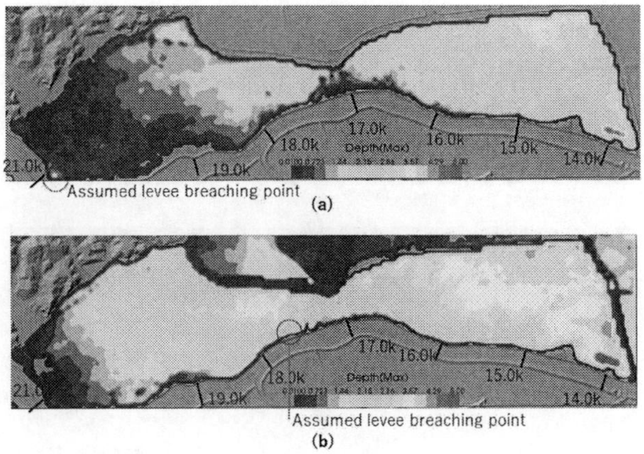

Figure 5. Maximum inundation depth distribution (**(a)** flood scenario 1, **(b)** flood scenario 3) (Itagaki et al., 2022).

(2019 Typhoon No. 19 Flood Disaster Survey Team, Hydraulic Engineering Committee, Japan Society of Civil Engineers, 2020) (Umeda, 2019) with the maximum inundation depth of the flood simulation result of scenario 1, the difference between the two was about 0.2 to 1.8 m, and the average absolute value of the difference was about 0.85 m. The difference between the two is large near embankments, near the levee breach point, and at points where the ground height is about 1 m higher than the surrounding area. The number of days required for drainage was 14 d in flood scenario 1, and the difference from the actual 12 d was 17 %. Here, the number of days required for drainage is roughly estimated by simply dividing the flood water volume in the protected area by the drainage capacity, based on the drainage capacity of the existing drainage pump station near 14.0 kp on the left bank of the Yoshida-gawa River, which is $8.5\,m^3\,s^{-1}$.

3.4 Extraction of worst case scenario

Comparing flood scenarios 1, 3 (Fig. 5b), and 4 (omitted), scenario 3 exhibits the largest flooded area, and scenario 4 the maximum flood depth. Therefore, flood scenarios 3 and 4 are the worst flood scenarios among these three scenarios, 3 for flooded area and 4 for the maximum flood depth.

3.5 Setting countermeasure options for flood water volume reduction

By installing a spillway whose crest height is equal to the design high water level of Yoshida-gawa River with water energy dissipator, part of river water will flow into the area protected by the levee, and the water level of the Yoshida-gawa River will be lowered. Countermeasures to prevent overflowing from levees other than the spillway as much as possible were examined. In this trial examination, based on local hearings, the target is set to avoid flooding above the floor of res-

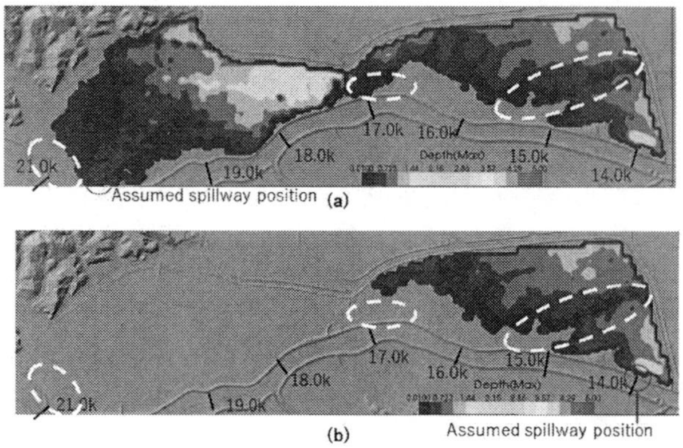

Figure 6. Maximum inundation depth distribution of spillway installation options ((**a**) option 1 (scenario 2), (**b**) option 2 (scenario 5)) (Itagaki et al., 2022).

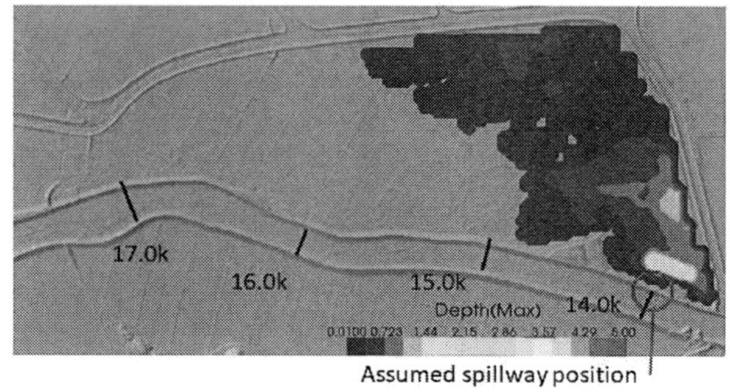

Figure 7. Maximum inundation depth distribution of flood with peak water level of 0.6 m higher than the design high water level, spillway installation option 2 (Itagaki et al., 2022).

idential houses, and reduce the number of days required for drainage to 3 d or less.

It is also assumable that human damage can be reduced by extending the time from overflow to levee breach by introducing overflow-resilient levees (crisis management-type hardware countermeasures) across the entire levee system, but there could be an argument for downstream risk transfer. In addition, since the method of evaluating the functioning of overflow-resilient levees is still under study, we did not treat it as an option in this paper. Besides, we think that it is not realistic to replace the entire levees of the Yoshida-gawa River with a length of about 8 km in the area with spillways, because a vast area will be required for installing water energy dissipators on both banks.

Option 1 was set to install a spillway directly downstream of the village on the left bank near the upper end of the Yoshida-gawa River in the area, and option 2 to install a spillway near the lower end of the Yoshida-gawa River in the area, where the lowest elevation in the area. Here, the width of the spillway (length in the downstream direction of the Yoshida-gawa River) was set at 50 m (see Itagaki et al., 2022).

Figure 6 shows the flood simulation result for each option. In both cases, the inundation area and depth were smaller than those of scenarios 1, and 3.

3.6 Comprehensive evaluation of countermeasure options for flood water volume reduction

In option 2 (in parentheses for option 1), the inundation area during Actual Flood was reduced to about 34 % (82 %) of the minimum inundation area among flood scenarios 1, 3, and 4, and the maximum flood depth was reduced to about 57 % (54 %). In addition, while the number of days required for drainage was 14 to 25 d for scenarios 1, 3 and 4, it was significantly shortened to about 2 d for option 2, approximately 86 % reduction (5 d for option 1, approximately 64 % reduction).

Comparing inundation depths of residential houses inside the white dashed line on the right side of Fig. 6b between scenario 1 and scenario 5 (spillway installation option 2), the latter decreased by about 2 m compared to the former. In other words, even in places where the flood depth was such that near the ceiling of the first floor of the house was submerged during Actual Flood, it is possible to reduce the damage to subfloor flooding level by raising the height of the residential land by about 1 m in addition to the spillway installation.

The disadvantage of this countermeasure is that the installation of the spillway increases the frequency of water overflowing into the protected area. In order to grasp this concretely, we simulated the maximum inundation depth distribution of a flood with a peak water level higher than the design high water level by 0.6 m that is the half of 1.2 m which is the difference between the design levee height and the design high water level (Fig. 7). The result showed that the number of days required for drainage was only about one day.

In addition, the annual exceedance probability for each assumed flood was roughly evaluated based on annual maximum water level data at each observatory (Ministry of Land, Infrastructure, Transport and Tourism, 2021), and the expected annual damage to crops was roughly estimated (assuming uniform paddy rice fields, without considering damage by sedimentation, damage to houses, agricultural facilities, roads, etc.). The expected annual damage was reduced to about 1/3 of the minimum expected annual damage among flood scenarios 1, 3, and 4. Finally, with using a diagram of the effect of countermeasures (the diagram shows the increase in damage due to the impact of climate change by flood return period) (Climate Change Adaptation Research Group, NILIM, MLIT, 2017), we compared the existing levee system and spillway installation option 2 on the degree of adaptation to the impact of climate change. In Fig. 8, the same scale of flood damage is assumed when the flood

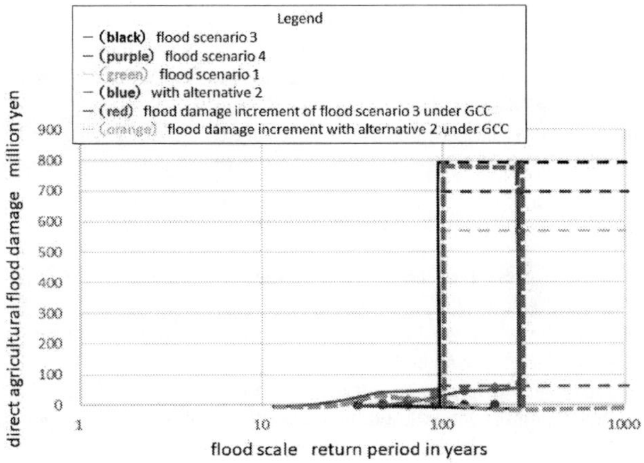

Figure 8. Comparison of adaptability to climate change (Itagaki et al., 2022).

scale is larger than Actual Flood, and is indicated by dashed lines. Due to the impact of climate change, the flood recurrence period (the recurrence period of Actual Flood based on the observed water level data at Hataya observatory station is roughly estimated to be about 265 years) will be shortened. Besides, we do not enter into the discussion of how much the recurrence period is shortened, but we assumed that the relationship curve would simply shift to the left. Figure 8 shows the case where the recurrence period of Actual Flood is shortened to about 100 years. These years correspond to the positions where the curve rises from the horizontal axis. The increase in damage in option 2 is significantly reduced compared with scenario 3, the largest damage scenario among scenarios 1, 3, and 4.

4 Conclusions

The effect of lowering the water level of the Yoshida-gawa River at the time of Actual Flood by installing a spillway (about option 2 which is more effective than option 1) in the trial application area is as follows:

1. Maximum overflowing depth is reduced by approximately 0.25 m from approximately 0.39 m (estimated from nearby water level observations) to approximately 0.14 m.
2. The length of the section where the water level exceeds the design levee crest height has been shortened by approximately 3.8 km from approximately 6 km to approximately 2.2 km.
3. The duration of overflowing was shortened by about 1 h and 50 min from about 4 h to about 2 h and 10 min.

It is estimated that the maximum overflowing depth among the sections where water overflowed during Actual Flood but not breached was about 0.36 m, and the duration of overflow was about 3 h (Naruse-gawa River Levee Investigation Committee, 2020). Therefore, by installing a spillway, both the depth and duration of overflow are reduced to below those in the above-mentioned non-breach sections. In addition, by raising the height of some residential lands by about 1 m, the target of preventing over-floor flood damage can be achieved. Besides, it is important to make the levee resilient against overflow, because great uncertainty cannot be avoided in the functioning of the levee when the river water level exceeds the design high water level.

Besides, it is essential to equalize the damage burden by supporting the raising of houses and roads to reduce flood damage in areas where the frequency of inundation increases.

Based on the results of this research, we would like to continue to promote concrete discussion on potential countermeasures for flood damage reduction by River Basin Disaster Resilience and Sustainability by All initiative.

Code availability. Code is available at https://i-ric.org/solvers/nays2dflood/ (iRIC, 2021).

Data availability. Data is available at https://fgd.gsi.go.jp/download/menu.php (Geospatial Information Authority of Japan, 2021) and http://www1.river.go.jp/ (Ministry of Land, Infrastructure, Transport and Tourism, 2021).

Author contributions. Conceptualization and validation were carried out by all three co-authors. Supervision was carried out by TK. Data curation, formal analysis, investigation, methodology, visualization, original draft preparation were carried out by OI.

Competing interests. At least one of the (co-)authors is a guest member of the editorial board of *Proceedings of IAHS* for the special issue "ICFM9 – River Basin Disaster Resilience and Sustainability by All". The peer-review process was guided by an independent editor, and the authors also have no other competing interests to declare.

Disclaimer. Publisher's note: Copernicus Publications remains neutral with regard to jurisdictional claims made in the text, published maps, institutional affiliations, or any other geographical representation in this paper. While Copernicus Publications makes every effort to include appropriate place names, the final responsibility lies with the authors.

Special issue statement. This article is part of the special issue "ICFM9 – River Basin Disaster Resilience and Sustainability by All". It is a result of The 9th International Conference on Flood Management, Tsukuba, Japan, 18–22 February 2023.

Acknowledgements. We would like to express our sincere gratitude to Makoto Umeda of Nihon University for providing the valuable field data, everyone involved in the Osaki City Flood-Resilient Urban Development Expert Council, and everyone from the Water and Disaster Management Bureau, and the National Institute for Land and Infrastructure Management of the Ministry of Land, Infrastructure, Transport and Tourism, Japan, for helpful discussions.

Review statement. This paper was edited by Shinji Egashira and reviewed by two anonymous referees.

References

2019 Typhoon No. 19 Flood Disaster Survey Team, Hydraulic Engineering Committee, Japan Society of Civil Engineers: Report of 2019 Typhoon No. 19 Flood Disaster Survey Team (Tohoku area), Hydraulic Engineering Committee, Japan Society of Civil Engineers, Japan Society of Civil Engineers, 63, https://committees.jsce.or.jp/report/system/files/taifu19.pdf (last access: 8 March 2024), 2020.

Climate Change Adaptation Research Group, National Institute for Land and Infrastructure Management, Ministry of Land, Infrastructure, Transport and Tourism: Research on climate change adaptation in river & coastal management, Project Research Report of NILIM, No. 56, 201–203, ISSN 1880-0114, https://www.nilim.go.jp/lab/bcg/siryou/kpr/prn0056.htm (last access: 8 March 2024), 2017.

Geospatial Information Authority of Japan: 5m DEM, https://fgd.gsi.go.jp/download/menu.php, last access: 29 October 2021.

iRIC: Nays2DFlood, https://i-ric.org/solvers/nays2dflood/, last access: 29 October 2021.

Itagaki, O., Ohara, M., and Koike, T.: Study of flood damage reduction by spillway installation on a riverine levee in a protected area, J. Jpn. Soc. Hydrol. Water Res., 35, 325–338, https://doi.org/10.3178/jjshwr.35.1739, 2022.

Ministry of Land, Infrastructure, Transport and Tourism: Water Information System, http://www1.river.go.jp/, last access: 29 October 2021.

Naruse-gawa River Levee Investigation Committee: Naruse-gawa River Levee Investigation Committee report, 2-11–3-16, https://www.thr.mlit.go.jp/Bumon/B00097/k00360/saigai-info/2019_taihuu19gou/2006teiboutyousa_naruse.pdf (last access: 8 March 2024), 2020.

Tanabe, T. and Okuma, T.: The study on the roles and effects of "NOKOSHI" on the Jobaru-river Basin Example of flood control that allows overflowing and its future possibility, Historical Studies in Civil Engineering, 21, 147–158, https://doi.org/10.2208/journalhs1990.21.147, 2001.

Umeda, M.: 2019 Typhoon No. 19 trace flood water level data, 2019.

US Army Corps of Engineers: Managed overtopping of levee systems, Engineering and Construction Bulletin No. 2019-8, 12, https://wbdg.org/ffc/dod/engineering-and-construction-bulletins-ecb/usace-ecb-2019-8 (last access: 8 March 2024), 2019.

Proc. IAHS, 386, 159–164, 2024
https://doi.org/10.5194/piahs-386-159-2024

Open Access
PIAHS

Methods to create hazard maps for flood disasters with sediment and driftwood

Daisuke Harada and Shinji Egashira

International Centre for Water Hazard and Risk Management (ICHARM), Public Works Research Institute (PWRI), Tsukuba, Ibaraki, 305-8516, Japan

Correspondence: Daisuke Harada (d-harada55@pwri.go.jp)

Received: 4 August 2023 – Revised: 16 October 2023 – Accepted: 24 February 2024 – Published: 19 April 2024

Abstract. As flood disasters with sediment and driftwood are becoming severe, it is required to evaluate the risk of such disasters. The present study proposes methods to create hazard maps for such type of flood disasters using a watershed model and a 2-D flood flow model, and applied the methods to the Akatani river flood disaster in 2017 to discuss their applicability and challenges. The upstream boundary conditions of 2-D flood flow model are specified for the hydrographs of flow discharge, sediment discharge and driftwood discharge by using the results obtained from their basin model. As a result of the Akatani river simulation with the proposed methods, we found that the computational results reproduce the inundation area and elevation change, and the proposed methods are effective in creating flood hazard maps with sediment and driftwood.

1 Introduction

When heavy rainfall occurs in mountainous areas, it causes landslides and debris flows at numerous locations in a watershed. The landslides and debris flows generate numerous amounts of sediment and large wood. Once these sediment and large wood are supplied to the river channel, they are transported downstream by the flood flow. In downstream areas, large wood accumulation takes place at many locations, such as bridges and sediment deposition areas and influences the flood flow. Recently, these types of flood hazards have been reported in numerous places over the world (e.g., Comiti et al., 2008; Harada and Egashira, 2018; Lucia et al., 2018).

To mitigate such flood disasters, it is necessary to predict possible hazard occurrences and develop hazard maps so that residents understand the possible hazards and evacuate appropriately. However, compared to flood hazard maps, there are not enough methods to assess and develop hazard maps for flood hazards with sediment and driftwood. To evaluate such flood hazards, it is necessary to obtain a time series of the water, sediment, and driftwood discharged from the watershed and to evaluate its effect on flood flows at downstream areas. Although several previous studies have attempted to evaluate such flood hazards (e.g., Liu et al., 2022; Huang et al., 2022), methods to evaluate flood hazards from the above perspective have not yet been established. This study proposes methods to create hazard maps for such types of flood disasters by combining a watershed model and a 2-D flood flow model, and applied the methods to the Akatani river flood disaster in 2017 to discuss their applicability and challenges.

2 Methods

2.1 Evaluation of water, sediment and driftwood runoff from a watershed

In this study, we propose an integrated method to simulate rainfall-runoff, landslide and debris flow, and sediment and driftwood transport in the river channel to obtain a time series of sediment and driftwood discharged from the watershed at an arbitrary location.

Figure 1 illustrates these processes conceptually. The occurrence of landslide, debris flow, and driftwood transport induced by a landslide on a hill slope are evaluated based on the method of Yamazaki and Egashira (2019). Slope stability analysis to evaluate the landslide occurrence is performed in the slope cell of the Rainfall-runoff-inundation (RRI) model developed by Samaya et al. (2012). The sediment transport

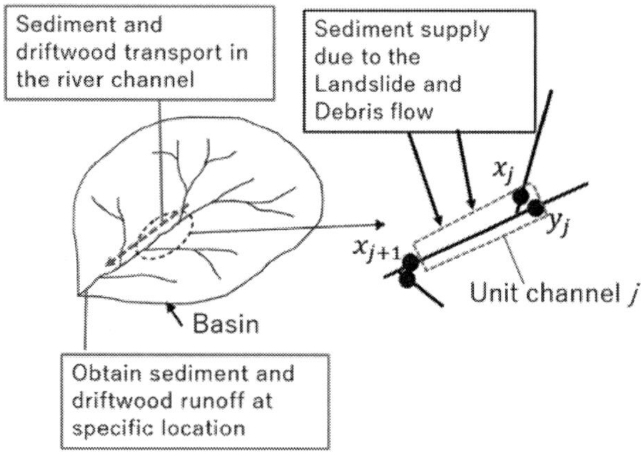

Figure 1. Conceptual illustrations of a method for evaluating water, sediment and driftwood runoff from a basin.

Table 1. Parameters employed for the watershed computations.

Item	Value
Mesh size (m)	10×10
Soil depth (m)	1.0
Saturated hydraulic conductivity (cm s^{-1})	0.5
Equivalent roughness coefficient	0.4
Soil porosity: λ	0.475
Internal friction angle (°)	35
Cohesion (kN m^{-2})	12.5
Soil density (kg m^{-3})	2650
Water density (kg m^{-3})	1000

that accompanies the landslide is analyzed using the equation of a mass system, from the point of origin to the location where the deposition occurs, and when the sediment reaches the river channel, it is treated as a sediment supply to the channel.

In the river channel, following the method proposed by Egashira and Matsuki (2000), a section that includes the upstream confluence (x_j, y_j) and excludes the downstream confluence point (x_{j+1}) is designated as the unit channel, as shown in Fig. 1, and the sediment and large wood runoff for the entire basin is predicted by allocating the unit channels in series and parallel. This modeling allows for simple and stable analysis of sediment and large wood transport in complex channel networks. Therefore, the riverbed evolution in each unit channel is described as follows;

$$\frac{\partial z_b}{\partial t} = \sum_i \frac{1}{(1-\lambda) B_j L_j} \left\{ Q_{bi}\left(x_j\right) + Q_{bi}\left(y_j\right) - Q_{bi}\left(x_{j+1}\right) + V p_{Vi} \right\} - E_i + D_i \quad (1)$$

where z_b is the bed elevation, λ is the sediemnt porosity, B_j and L_j are the width and length of the unit channel j, respectively; $Q_{bi}\left(x_j\right)$ is the bedload transport rate for grain size d_i of the unit channel j; V is the sediment volume supplied from the slope; p_{Vi} is the content ratio in the sediment supplied from the slope for the grain size d_i, E_i and D_i are the erosion and deposition rate of suspended sediment for grain size d_i.

To treat a large number of driftwood transport, we assume that the wood pieces behave as neutral buoyant particles, as this assumption allows the introduction of the convection equation (Harada and Egashira, 2018). In this method, we assume that the erosion and deposition of driftwood occurs in proportion to sediment erosion and deposition, based on the idea that when the sediment is deposited, the wood particles are also deposited on the riverbed, and when sediment is eroded, wood particles are also taken into the water. Also, assuming that large wood accumulation occurs at artificial structures such as bridges, the convection equation is coupled with the storage equation of driftwood in the channel bed.

In case of $\partial z_b / \partial t > 0$ (Sediment deposition)

$$\frac{\partial c_{drf} h}{\partial t} = \frac{1}{B_j L_j} \left\{ c_{drf}\left(x_j\right) Q\left(x_j\right) + c_{drf}\left(y_j\right) Q\left(y_j\right) - c_{drf}\left(x_{j+1}\right) Q\left(x_{j+1}\right) \right\} - c_* \frac{\partial z_b}{\partial t} c_{drf} r\left(t, x, y\right) - v_n c_{drf} p_b \delta\left(x - x_i, y - y_i\right) \quad (2)$$

$$\frac{\partial S}{\partial t} = \frac{\partial z_b}{\partial t} c_{drf} r\left(t, x, y\right) + v_n c_{drf} p_b \delta\left(x - x_i, y - y_i\right) + V_{drf} \quad (3)$$

where c_{drf} is the driftwood concentration, h is the depth, $Q\left(x_j\right)$ is the flow discharge of the unit channel j, c_* is the sediment concentration of the river bed, v_n is the is the inward velocity normal to the structure such as the bridge, p_b is the probability that large wood is captured at structures, ranging from 0 to 1 (we employ 1 in this study), V_{drf} is the volume of driftwood supplied from the slope due to the debris flow. Function $r(txy)$ is introduced to describe the phenomena that the driftwood erosion does not occur at depths shallower than a certain water depth, and large wood deposition does not occur at depths deeper than a certain water depth. Due to space limitation, please refer to Harada and Egashira (2018) for specific functions $r(txy)$ and treatment for the sediment erosion case.

2.2 Evaluation of flood flow with sediment and driftwood

In order to create flood hazard maps for these hazards, we employ the depth averaged 2-D numerical simulations. The flow field is simulated using 2-D depth-integrated governing equations, which include mass and momentum conservation equations (e.g., Shimizu et al., 2019). The upstream boundary conditions for the 2-D computation are obtained by the methods in Sect. 2.1, in which we obtain a time series of sediment and driftwood discharged from the watershed. The 2-D forms of the Eqs. (2) and (3) are employed to evaluate the driftwood behaviour.

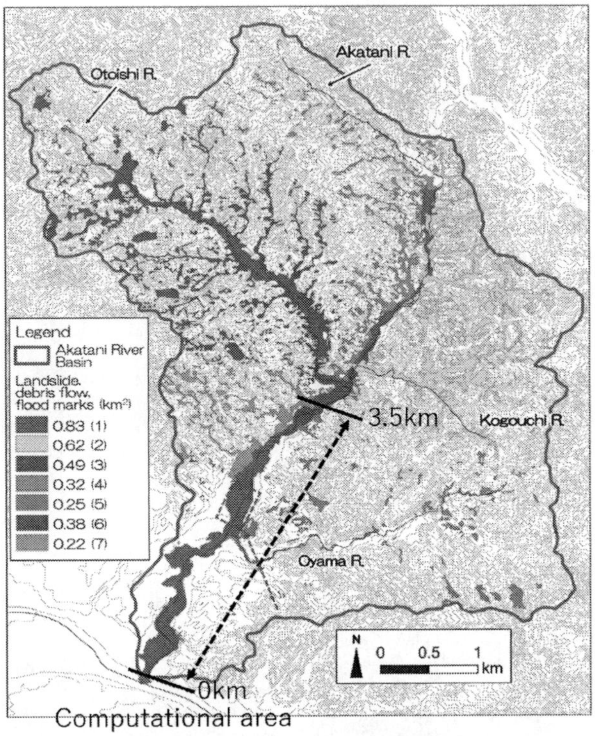

Figure 2. The Akatani River basin with debris flows and flood marks identified from aerial photos (Nagumo and Egashira, 2019; was partially modified by the authors). The background image is provided by the Geographical Information Authority of Japan.

3 Application to the Akatani river flood disaster in 2017

3.1 Computational conditions

The proposed method is applied to evaluate the flood disaster in the Akatani river, 2017, where a large amount of sediment and driftwood was produced in the mountainous area, as shown in Fig. 2. Parameters employed for the rainfall-runoff and landslide calculation is shown in Table 1. Initial sediment size distributions set in the channel network are given as shown in Fig. 3, referring to the sediment size distributions observed after the flood. The density of standing trees for the driftwood runoff simulation is set as 0.06 ($m^3\ m^{-2}$).

The 2-D flood flow simulations with sediment and driftwood are conducted within the 3.5 km section as shown in Fig. 2. For upstream conditions related to flow discharge, sediment, and driftwood, we use the watershed computation results at the 3.5 km location. Calculations are performed for the two cases, flow only (Case 1) and with sediment and driftwood (Case 2) to compare the differences in results depending on the presence of sediment and driftwood.

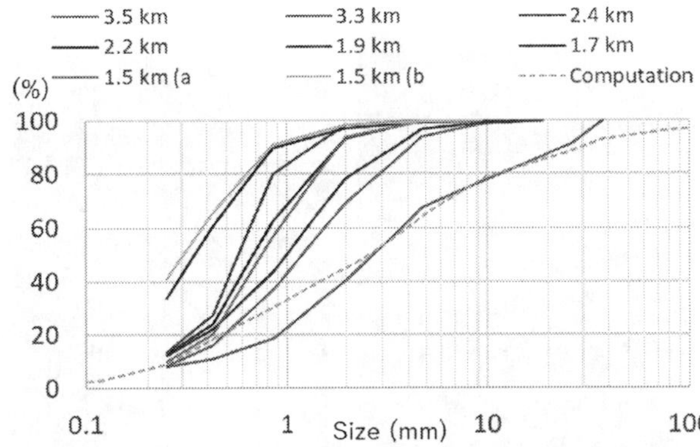

Figure 3. Sediment size distribution in the riverbed surveyed after the disaster. The (km) in the legend is the distance from the Chikugo river confluence point, as shown in Fig. 2.

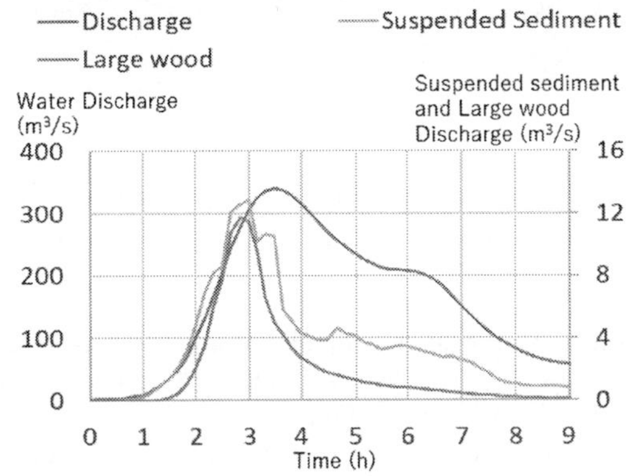

Figure 4. Computational results of the watershed model for flood water, suspended sediment, and driftwood discharge at the 3.5 km point.

3.2 Results

Figure 4 shows the results of the watershed computation at 3.5 km point; the location corresponds to the upstream boundary of the 2-D flood flow computation. According to the figure, suspended sediment and driftwood discharge are concentrated around 3 h after the start of the event. This is due to the results that the occurrence of landslide and associated sediment transport are concentrated during this time period. This result is used as the upstream end boundary condition for the 2-D flood flow analysis, and the results in Fig. 4 indicate that a large amount of sediment and driftwood are discharged to the target section before the flood discharge reaches its peak flow.

Figure 5 compares the computational results for the inundation area between Case 1 and Case 2. There are seven bridges within the 3.5 km section, and the driftwood deposition in the computation exceeds 1 m in some of the bridges, thus the flow is obstructed, causing the flow to divert around

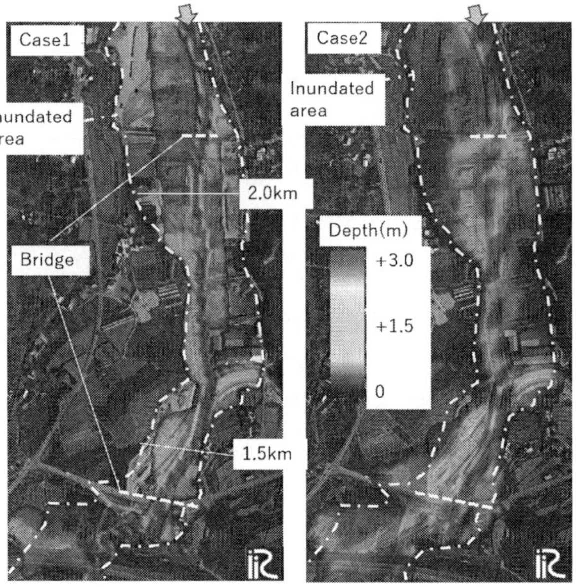

Figure 5. Computational results of the water depth at peak discharge. The comparison is between Case 1 (flow only; left) and Case 2(with sediment and driftwood; right); The background image was taken from © Google Maps.

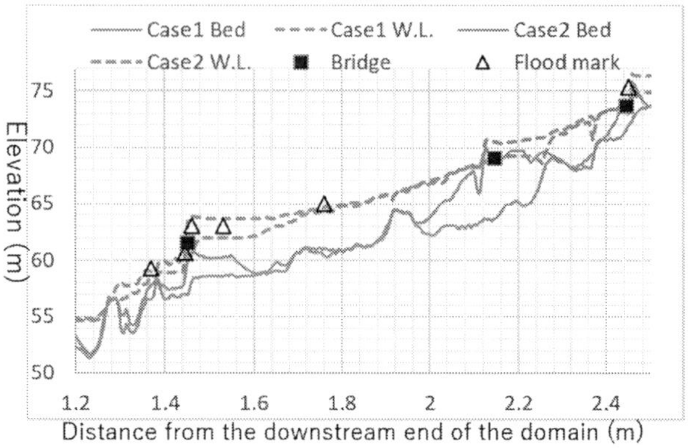

Figure 6. Comparison of the results with the water level mark and river bed elevation in the longitudinal direction of the river channel during peak flow. Bridge location shows the elevation of the bridge.

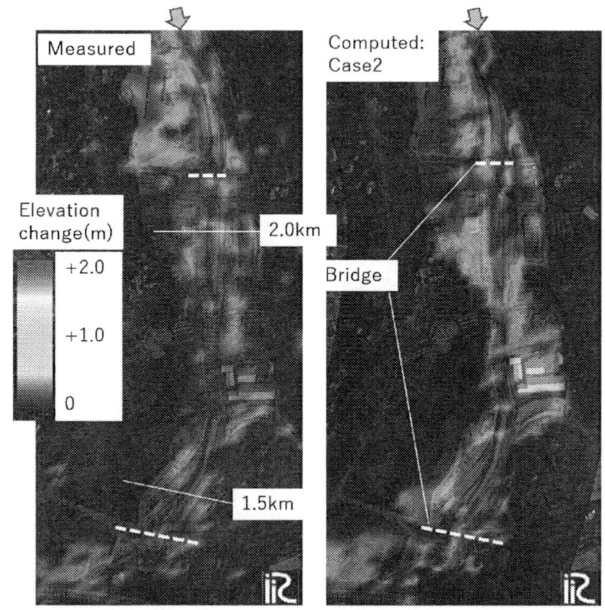

Figure 7. Comparison of elevation changes before and after the flooding measured by aerial laser survey (left) and elevation change at the end of Case 2 computation (right). The background image was taken from © Google Maps.

the bridge. As a result, the inundation area in Case 2 is larger than that in Case 1, which is closer to the actual inundated area, shown as white dotted line.

Figure 6 compares the computation results at the peak flow with the water level marks to partially show the validity of this computation. In Case 2, the channel capacity has decreased due to sediment deposition on the river channel. Comparing the water levels in Case 1 and 2 with the water level mark, for example, around the 1.5 km location, where is the upstream of the bridge, Case 1 is about 1 m below the water level mark and Case 2 is about 50 cm above the water level mark. In Case 2, the bed shear stress in the river channel is reduced at upstream of the bridge due to the driftwood accumulation at the bridge, that causing significant sediment deposition here.

Figure 7 compares the difference in sediment deposition during the flood event between the measured by an aerial laser survey (left figure) and the computational results in Case 2 (right figure). The trends in the two figures are generally consistent in terms that more than 2 m sediment deposition takes place in the river channel (red-coloured area) and that sediment is deposited several tens of centimeters thickness in the areas where inundation occurred.

4 Discussion

In the Akatani River disaster, sediment conditions in the river channel changed drastically because of the large amount of fine sediment deposited in the channel supplied by landslide and debris flows in many places. According to Fig. 4, the supplied sediment is transported to the downstream, thus in the downstream river channels, this change seems to have occurred before the water discharge peak time. Figure 6 also shows that the river channel capacity is drastically reduced even before the flood flow peak comes, and the driftwood discharged from the watershed is deposited at the bridges, which significantly affect the water surface profile and flood flow around the bridges. To evaluate such a disaster, the physics-based method proposed in this study is effective.

Compared to existing methods for treating sediment and driftwood in the numerical simulation, this method allows us to perform physics-based integrated analysis of water, sediment, and driftwood runoff from the watershed. In particular, the unit channel concept in Fig. 1 and the convection equation in Eq. (2) allow the treatment of numerous amounts of sediment and driftwood transport. On the other hand, it should be mentioned that some challenges remain in creating hazard maps using such methods. For example, validation of the computed results is required where disasters have not yet occurred, thus validation should be conducted using the existing data for the present sediment transport or past disasters. Particularly, parameters that affect the results such as the soil cohesion, and sediment size distribution that we give as conditions should be carefully investigated in future.

5 Conclusion

In this research, we proposed methods to create flood hazard maps with active sediment and driftwood transport, using a watershed model and a 2-D flood flow model, where the driftwood transport is based on the convection and storage equations, and applied the methods to the Akatani river flood disaster in 2017 to discuss their applicability and challenges. As a result, the 2-D flood flow simulation that uses the results of watershed water, sediment, and driftwood runoff simulation reproduces the disaster in terms of inundation area, water depth, and sediment deposition. The methods proposed in this study are suitable for creating flood hazard maps in cases where the sediment conditions in the river channel change drastically, as in the case of the Akatani river disaster. However, further studies will be required for the use of practical hazard map creation in terms of the validation where disasters have not yet occurred, sediment size distribution, sediment sorting processes, river bank erosion, and so on.

Data availability. The data are available from the corresponding author on reasonable request.

Author contributions. DH and SE conceptualized the study; DH performed the numerical simulations; DH wrote the manuscript draft; SE reviewed and edited the manuscript.

Competing interests. At least one of the (co-)authors is a guest member of the editorial board of *Proceedings of IAHS* for the special issue "ICFM9 – River Basin Disaster Resilience and Sustainability by All". The peer-review process was guided by an independent editor, and the authors also have no other competing interests to declare.

Disclaimer. Publisher's note: Copernicus Publications remains neutral with regard to jurisdictional claims made in the text, published maps, institutional affiliations, or any other geographical representation in this paper. While Copernicus Publications makes every effort to include appropriate place names, the final responsibility lies with the authors.

Special issue statement. This article is part of the special issue "ICFM9 – River Basin Disaster Resilience and Sustainability by All". It is a result of The 9th International Conference on Flood Management, Tsukuba, Japan, 18–22 February 2023.

Acknowledgements. The authors would like to thank Naoko Nagumo, Yousuke Nakamura, and Yusuke Yamazaki for their contribution to the field survey and data preparation for this study.

Financial support. This research has been supported by the Japan Society for the Promotion of Science (grant no. 22K14334).

Review statement. This paper was edited by Kensuke Naito and reviewed by Dimitrios Fytanidis and one anonymous referee.

References

Comiti, F., Mao, L., Preciso, E., Picco, L., Marchi, L., and Borga, M.: Large wood and flash floods: evidence from the 2007 event in the Davča basin (Slovenia), WIT Transactions on Engineering Sciences, 60, 173–182, 2008.

Egashira, S. and Matsuki, T.: A method of predicting sediment runoff caused by erosion of stream channel bed, Annual Journal of Hydraulics Engineering, Japan Society of Civil Engineers, 44, 735–740, 2000 (In Japanese).

Harada, D. and Egashira, S.: Flood flow characteristics with fine sediment supply and drift woods – analysis on the akatani river flood hazards in july, 2017, Annual Journal of Hydraulic Engineering, Japan Society of Civil Engineers, 74, I_937–I_942, 2018 (in Japanese).

Huang, R., Ni, Y., and Cao, Z.: Coupled modeling of rainfall-induced floods and sediment transport at the catchment scale, Int. J. Sed. Res., 37, 715–728, 2022.

Liu, H., Du, J., and Yi, Y.: Reconceptualising flood risk assessment by incorporating sediment supply, CATENA, 217, 106503, https://doi.org/10.1016/j.catena.2022.106503, 2022.

Lucía, A., Schwientek, M., Eberle, J., and Zarfl, C.: Planform changes and large wood dynamics in two torrents during a severe flash flood in Braunsbach, Germany 2016, Sci. Total Environ., 640, 315–326, https://doi.org/10.1016/j.scitotenv.2018.05.186, 2018.

Nagumo, N. and Egashira, S.: Flood Hazard Analysis and Locations of Damaged Houses Based on Land Classification in the Akatani River Basin Following Torrential Rainfall in Northern Kyushu, 2017, Journal of Geography (Chigaku Zasshi), 128, 835–854, 2019 (in Japanese).

Sayama, T., Ozawa, G., Kawakami, T., Nabesaka, S., and Fukami, K.: Rainfall–runoff–inundation analysis of the 2010 Pakistan flood in the Kabul River basin, Hydrol. Sci. J., 57, 298–312, 2012.

Shimizu, Y., Nelson, J., Arnez Ferrel, K., Asahi, K., Giri, S., Inoue, T., Iwasaki, T., Jang, C.L., Kang, T., Kimura, I., and Kyuka, T.: Advances in Computational Morphodynamics Using the International River Interface Cooperative (iRIC) Software, Earth Surf. Process. Landf., 45, 11–37, https://doi.org/10.1002/esp.4653, 2019.

Yamazaki, Y. and Egashira, S.: Run out processes of sediment and woody debris resulting from landslides and debris flow, Association of Environmental and Engineering Geologists, special publication 28, 2019.

Proc. IAHS, 386, 165–172, 2024
https://doi.org/10.5194/piahs-386-165-2024

Investment decision support for business sustainability improvement based on flood inundation simulations with a variety of extreme rainfall scenarios

Saki Nonaka, Kazuki Suzuki, Nobuyuki Iwamae, Daisuke Nohara, and Yoriyuki Yamada

KAJIMA Technical Research Institute, Tokyo, 182-0036, Japan

Correspondence: Saki Nonaka (nonakasa@kajima.com)

Received: 29 July 2023 – Revised: 26 October 2023 – Accepted: 24 February 2024 – Published: 19 April 2024

Abstract. In recent years, heavy rainfall disasters have occurred frequently in Japan, which increases the importance for private companies to take flood countermeasures to prepare improved business continuity plans (BCPs). Although river authorities have published flood hazard maps for design rainfall and probable maximum rainfall in Japan, hazard information for floods with other return periods is also required for the target business facilities to make cost-effective plans and investment decisions based on flood risk assessment. In order to overcome the situation where public flood hazard maps do not provide enough information for risk-based business continuity planning, we developed a method to provide tailored flood risk information for a business site based on a series of flood inundation simulations for rainfall scenarios with various return periods in the expected inundation area of the Tama River, Tokyo, Japan. Calculated results showed that the inundation characteristics highly depend on the topographic features of the flood plain, suggesting the importance of assessment at a respective business site. We also present an example of the application of the proposed method of simulation results in designing the BCP.

1 Introduction

In recent years, extreme rainfall has become more frequent and severe. Japan Meteorological Agency (JMA) reported that the frequency of heavy rainfall ($> 50\,\mathrm{mm\,h^{-1}}$) has increased at a rate of 27.5 times per decade in the past 45 years (JMA, 2021). Corresponding to intensified heavy rain, severe flood inundation disasters have occurred and caused enormous damages in Japanese river basins in recent years, namely, Kinu River basin in 2015 (Nagumo et al., 2016), Oda River basin in 2018 (Nohara et al., 2020), Chikuma River basin in 2019 (Fan et al., 2020), Kuma River basin in 2020 (Ministry of Land, Infrastructure, Transport and Tourism (MLIT), 2021), and Rokkaku River basin in 2021 (Yamamoto et al., 2022). In some of those cases, the magnitude of rainfall exceeds that for the river planning (design rainfall). In response to these backgrounds, the Flood Fighting Act was amended by the Japanese national government in May 2015 to oblige local governments to publish expected inundation areas (flood hazard maps) for probable maximum rainfall (PMR), in addition to that for design rainfall (MLIT, 2015a). Accordingly, information on the expected flood inundation area or depth is now available both for design rainfall and PMR. The design rainfall generally has a return period of 100 to 200 years for most major river systems in Japan, while the return period of PMR is typically around or over 1000 years (MLIT, 2015b).

In designing flood countermeasures such as construction of flood walls at business facilities, the maximum inundation depth is often used as hazard information at the site. Feasible height of flood walls is usually up to three meters due to constraint of cost and social reasons (land use, landscape, etc.). On the other hand, expected inundation depth of PMR is often much greater than the feasible height of walls in floodplains. This does not mean that construction of flood walls is useless for extreme floods, because it can be effective against extreme floods with smaller return periods (but greater than that of design rainfall) where expected maximum inundation depth can be smaller than the feasible wall height.

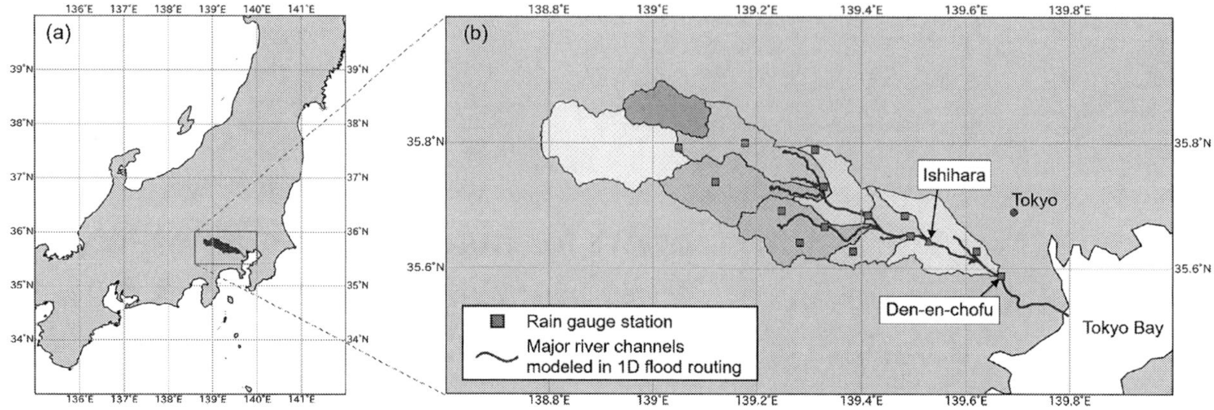

Figure 1. (a) Location of the Tama River basin (filled in blue). **(b)** Watershed subdivision map for the upstream of Den-en-chofu used in the runoff analysis.

To take cost-effective measures and investment decisions based on flood risk assessment, it is therefore important to consider possible floods with return periods between the designed level and probable maximum, which occurs more frequently than the probable maximum. However, hazard maps for floods with an intermediate magnitude between the design rainfall and PMR have not been provided to the public. Business entities or companies, therefore, need to prepare such intermediate-scale hazard information to design measures for improved business continuity against floods.

In order to overcome the problem described above, we developed a method to estimate comprehensive flood risk information based on a series of flood inundation simulations for rainfall scenarios with a wide range of magnitudes and spatiotemporal distributions for a business site located in the expected inundation area of the Tama River, Tokyo, Japan. First, we examined the effect of the difference in rainfall scenario on the simulated river discharge and inundation depths in the study area. A risk analysis for examining optimal flood protection measures was then applied to an individual site located in the floodplain of the Tama River. Finally, we discussed how the proposed framework could provide useful information for decision-making regarding flood countermeasures.

2 Outline

2.1 Study area

The Tama River is located in the western part of Tokyo, Japan, and flows down from west to east (Fig. 1). The basin area is $1240\,km^2$, and the estimated basin population is 4.14 million as of 2020 (MLIT, 2023a). The upper stream area is mountainous, and the urban area spreads out within the expected inundation area from the midstream to the downstream.

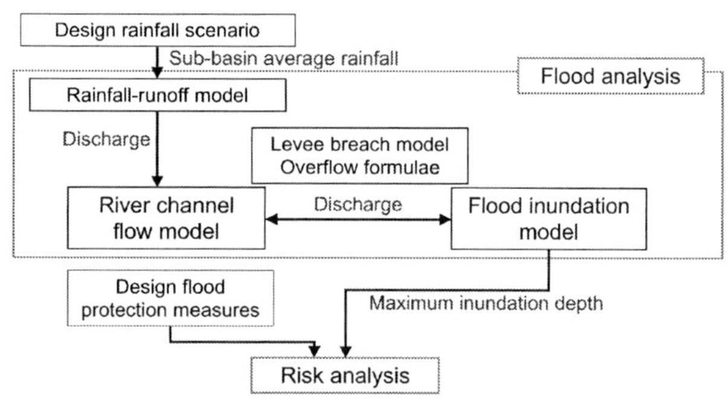

Figure 2. Framework for investment decision support.

2.2 Framework

Figure 2 shows the framework for decision support employed in this study, which consists of two parts: a flood analysis and a risk analysis.

In the flood analysis, the distribution of inundation depth on floodplain is obtained by using rainfall-runoff model and flood inundation model. The rainfall-runoff model is based on the storage function method with two parameters and lag time proposed by Kimura (1961). The flood inundation model consists of a one-dimensional (1D) unsteady flow model to simulate river channel flow and a two-dimensional (2D) unsteady flow model for inundation flow on land. In the integrated model, the water exchange between land and river is determined by overflow formulae when the water level exceeds the levee top. The risk analysis was conducted based on the maximum inundation depth at the target site obtained from the flood analysis and the designed flood protection plan.

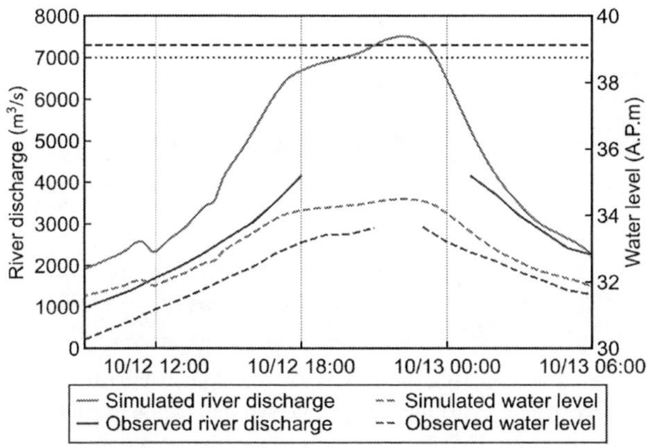

Figure 3. Observed and simulated river discharge and water level at the Ishihara station for typhoon No. 19 event in 2019 (black dashed horizontal line is the estimated peak river discharge without overflow and dam operation, and black dotted horizontal line is the estimated peak river discharge with overflow and dam operation).

2.3 Simulation condition

In the rainfall-runoff simulation, the Tama River basin (upstream of Den-en-chofu) was divided into eight sub-basins (Fig. 1b). The runoff discharge calculated by the rainfall-runoff model for each sub-basin was used as the boundary condition (discharge at the upper-end or lateral inflow of the target river channel section) of the river channel flow model. The calculation domain of the river channel flow model was from 0.0 to 61.8 km upstream from the river outlet and parts of tributaries as shown by the blue lines in Fig. 1b, and discretized only in the longitudinal direction (1D model) with 200 m spacing. The reservoir operation and river upstream overflow are not considered when simulating river channel flow.

For the flood inundation simulation, only inundation from the river channels was considered in this study, assuming that no inundation occurs from the sewage systems or rain water in the calculation domain of the flood inundation model. The calculation domain of the flood inundation model was set large enough so that the flooded water did not reach the boundary. Grid spacing is 25 m, uniform in the whole calculation domain. The roughness coefficients are set according to land use. The effects of buildings and other land structures preventing flood flow are considered as porosity and transmissivity. The river levee breach can also be considered.

2.4 Validation of the analysis model

The validation of the analysis model is described in this section. Since there was no large-scale flood inundation due to river overflow in this study area, the rainfall-runoff model and the river flow model were validated against the river water level and discharge at Ishihara station for the heaviest rainfall recorded by typhoon No. 19 event in 2019. The observed data are derived from the Water Information System managed by MLIT. The simulated and observed data at the Ishihara station (see Fig. 1b for its location) is shown in Fig. 3. The sub-basin averaged rainfall input for the rainfall-runoff model was calculated from the radar-AMeDAS composite rainfall of JMA in the simulation. Observation of river water level was missing around the peak at Ishihara station in this flood event, so we also referred to water level data observed by the simplified water level gauge (denoted by SWLG hereafter) installed nearby Ishihara station as a backup (MLIT, 2022). It was seen in Fig. 3 that although the simulated water level tends to be excessive throughout, it generally shows the same trend as the observed water level. The time series of simulated and observed water levels are also consistent in that both of them have two peaks with the second peak being larger.

The observed discharge was also missing during flooding, and the peak discharge estimated from the observation data by the SWLG was 7000 $m^3\,s^{-1}$ (black dotted horizontal line in Fig. 3, MLIT, 2022). According to the hypothetical reanalysis, assuming there was no reservoir operation or river upstream overflow, the peak river discharge was estimated to be 7300 $m^3\,s^{-1}$ in that condition (black dashed horizontal line in Fig. 3, MLIT, 2023b). As described in Sect. 2.3, reservoir operation and river upstream overflow are not considered in this study when simulating river flow in the target area. When comparing the peak river discharge, which is a key variable for maximum inundation depth in the flood plain, the simulated peak river discharge (7500 $m^3\,s^{-1}$) showed good agreement with the reanalyzed peak river discharge when overflow and dam operation was neglected in the upstream (7300 $m^3\,s^{-1}$), with overestimation by 2.7 %. The difference between the simulated peak river discharge and that estimated from the observed water level (7000 $m^3\,s^{-1}$) is also not very large, which also shows that the simulated results by the model are not far from the actual condition even though the model neglected the effects of dam operation and upstream overflow.

The flood inundation model used in this study was constructed in accordance with the guidelines for making expected inundation area maps published by MLIT (2015c). Since there is no record of large-scale inundation in the Tama River in recent years, this model was verified using the expected inundation area published by MLIT (2015d). Comparing the simulated maximum inundation depth with published data when the same levee breach point was selected for the design rainfall, it was confirmed that the distribution of inundated water was similar (Fig. 4). Therefore, this series of analysis models employed in this study is considered acceptable for estimating hazard of large-scale flooding for the Tama River.

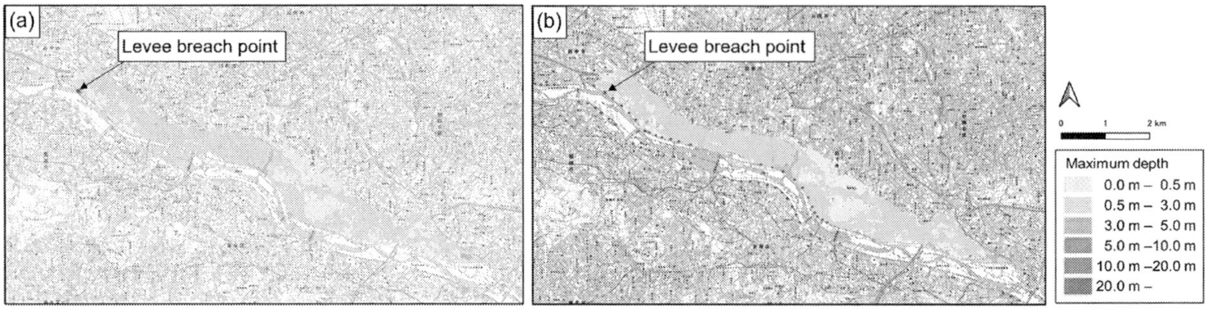

Figure 4. **(a)** Simulated maximum inundation depth. Background map is sourced from GSI Tiles (https://maps.gsi.go.jp/development/ichiran.html, last access: 10 March 2024) created by Geospatial Information Authority of Japan. **(b)** Published expected inundation depth (MLIT, 2015d).

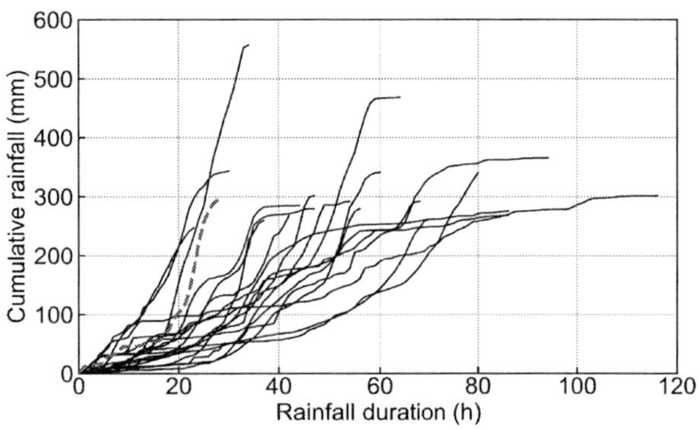

Figure 5. Selected 20 rainfall events waveform (the red dashed line is the event used in the case of Fig. 6).

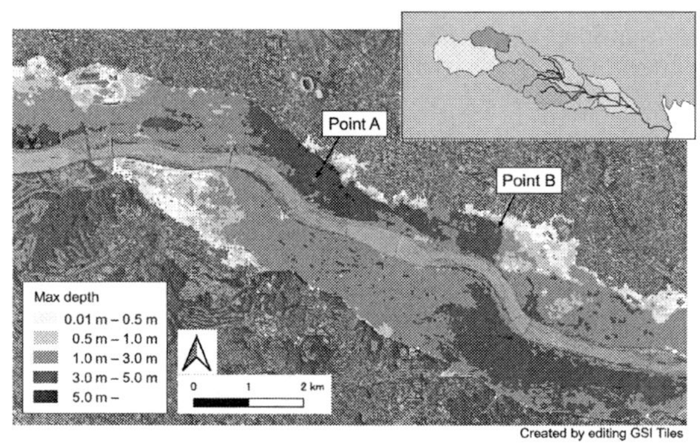

Figure 6. Model calculated maximum inundation depth distribution. Background aerial photo is sourced from GSI Tiles (https://maps.gsi.go.jp/development/ichiran.html, last access: 10 March 2024) created by Geospatial Information Authority of Japan. The main map shows the area within the red square in the upper right.

3 Assessment of flood inundation characteristics in the target river basin

3.1 Rainfall scenarios

This study used hourly rainfall data from 14 stations: 11 rainfall stations from the Water Information System managed by MLIT and three Automated Meteorological Data Acquisition System (AMeDAS) rainfall stations provided by JMA. The period of obtained rainfall data was for 62 years, from 1958 to 2019. The top 20 rainfall events were selected in terms of total basin average rainfall (Fig. 5). The selected events have a wide range of intensity and duration.

The design rainfall for the Tama River basin is 457 mm in 48 h rainfall with return periods of 200 years (MLIT, 2016a), and the PMR is 588 mm in 48 h rainfall with return periods of over 1000 years (MLIT, 2016b). However, no information on estimated inundation has been published for an intermediate magnitude of floods between the two. In order to identify estimated inundation processes of such floods, each of 20 rainfall events was stretched so that maximum 48 h rainfall was identical to 457, 460–580 (10 mm intervals), and 588 mm, respectively. By this operation, 300 rainfall scenarios with different scales and patterns were generated (20 rainfall waveforms from different events $\times 15$ different stretching factors). The Thiessen method was used to generate time series of sub-basin average rainfall as input for the rainfall-runoff model.

3.2 Inundation scale

In this section, the effect of difference in rainfall scenario on estimated river discharge and inundation depths in the study area was investigated. Only overflow from the levee top was considered as the cause of flood inundation, assuming that no levee breach occurs. As an example of the result from the flood inundation simulation, Fig. 6 shows the distribution of the maximum inundation depth in the midstream estimated by using a rainfall scenario, which was derived by stretching the one shown by the red dashed line in Fig. 5 to be 588 mm in 48 h. Overflow occurred at several locations on both sides of the river, resulting in widespread inundation.

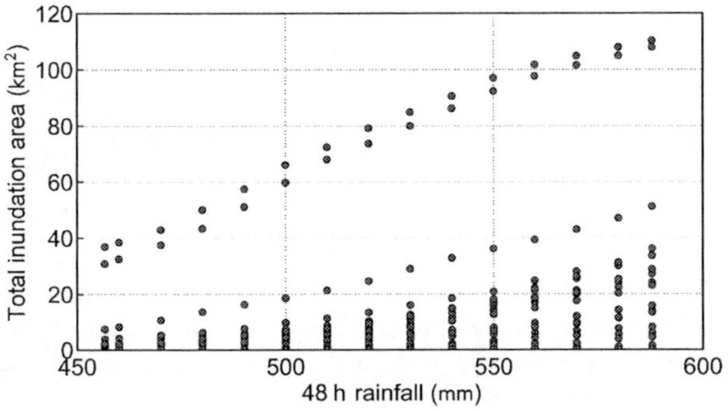

Figure 7. Relationship between 48 h rainfall and total inundation area.

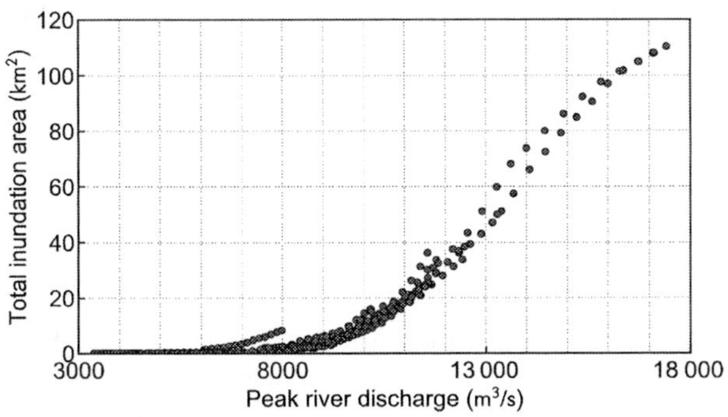

Figure 8. Relationship between peak river discharge and total inundation area.

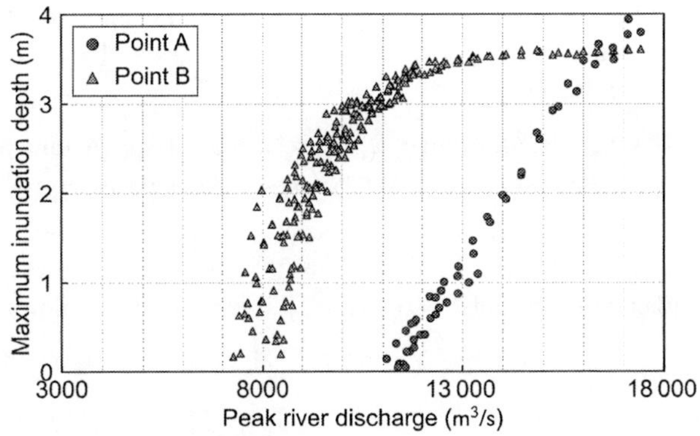

Figure 9. Relationship between peak flow rate and maximum inundation depth at Point A and B (see Fig. 4 for location).

Figure 7 shows the relationship between 48 h rainfall and total inundation area. Even though rainfall scenarios with maximum 48 h rainfall with the same return period were applied, total inundation area varies widely because of their different rainfall pattern and peak intensity.

Figure 8 shows the relationship between the peak river discharge at Den-en-chofu station (see Fig. 1 for its location) and the total inundation area. Here, the values of peak river discharge were obtained from the rainfall-runoff analysis where overflow was not permitted to see the maximum potential river discharge at the river section, while the river bank overtop process in the target area was considered to calculate the inundation area. There is a correlation between the peak discharge and inundation area, regardless of the rainfall waveform.

3.3 Inundation characteristics at individual site

Effects of peak discharge on flood inundation depth at the individual site were then analyzed. Figure 9 shows the relationship between peak discharge and maximum inundation depth at two sites, Points A and B (Fig. 6). We examined it at all inundated points and selected two locations with different features under the condition that the maximum inundation depth of 300 cases is the same level (3.5–4.0 m). At Point A, the maximum inundation depth increases linearly, with the peak discharge exceeding 11 000 $m^3\,s^{-1}$. On the other hand, at Point B, the maximum inundation depth rapidly increases as the peak discharge exceeds 7000 $m^3\,s^{-1}$, and it then approaches a constant depth. This difference in characteristics of increase in the maximum inundation depth can attributed to the difference in topography around the two sites. Point A is located in the middle of floodplain, where elevation gradually decreases along the river and inundated water smoothly flows down to the adjacent areas with lower elevation. On the other hand, since Point B is located in the lowest end of the floodplain surrounded by the terraces, inundated water tends to concentrate around there.

The tendency of increase in the maximum inundation depth with increasing peak river flow rate differs greatly for each location even in the same floodplain. It is therefore important to conduct flood simulations with various extreme rainfall scenarios to assess the vulnerability and risk of inundation at an individual business site. The measures to be taken against flood inundation can differ between sites where the inundation depth increases rapidly when the peak flow rate exceeds a certain level and those where the inundation depth increases linearly with the increase in the peak flow rate.

4 Risk analysis to determine on-site structural measures

4.1 Procedure of flood risk analysis

Flood risk analysis was conducted based on the method proposed by Morita (2008). Construction of flood walls surrounding the target business site was considered as a flood protection measure. For simplification, the following assumptions were made regarding the relationship between inundation depth and the damage cost in this work:

- No damage cost occurs when the inundation depth around the target site is less than the design height of the flood wall.
- When the inundation depth exceeds the design height, a constant amount of damage (α JPY) occurs regardless of flood depth.

Under the above assumptions, the damage cost for floods with a return period T, $D(T)$ [JPY], is expressed as follows:

$$D(T) = \begin{cases} 0\left(T < T_\mathrm{p}\right) \\ \alpha\left(T \geq T_\mathrm{p}\right) \end{cases} \tag{1}$$

where T_p is the flood protection level, which is identical to the return period of floods that can be protected by the designed protection measure (flood walls in this work). In this case study, damage cost is used in which damage reaches its maximum value as soon as inundation occurs, assuming an ideal site where there are highly expensive facilities vulnerable to submergence located on the ground or underground. In this case, flood damage largely depends on whether or not inundation occurs at the target site, no matter how deep the site is inundated. This characteristic of flood damage curve is often the case in the business site where heavy experimental or factory equipment is installed on the ground or underground. However, This assumption is not appropriate for estimating flood damage for the business site with facilities where the vertical distribution of assets is not concentrated below or on the ground. In that case, damage cost would vary depending on the flood depth in accordance with the vertical distribution of assets. The relationship between the maximum inundation depth and return period if floods can be obtained by the assessment of flood inundation characteristics described in the Sect. 2.

The expected annual damage (EAD) can be calculated by the following equation:

$$\mathrm{EAD}_{T_\mathrm{p}} = \int_0^\infty D(t) \cdot p(t)\,\mathrm{d}t = \int_{T_\mathrm{p}}^\infty \alpha \cdot p(t)\,\mathrm{d}t = \frac{\alpha}{T_\mathrm{p}} \tag{2}$$

where $p(t)$ is the probability density of flood occurrence at return period t. As an indicator of the effectiveness of flood protection measures, the degree of risk reduction (RR) by construction of flood walls with the flood protection level T_p can be defined as the following equation:

$$\mathrm{RR}_{T_\mathrm{p}} = \mathrm{EAD}_{T_0} - \mathrm{EAD}_{T_\mathrm{p}} \tag{3}$$

where EAD_{T_0} is the EAD expected with no protection measure taken, and $\mathrm{EAD}_{T_\mathrm{p}}$ is that expected when a measure with protection level T_p is taken.

The construction cost is assumed to be proportional to the wall height. This assumption can be applied when the wall height is not too high, and the construction costs do not increase significantly with respect to the wall height. If the wall height exceeds several meters, the design of the foundation or temporary structure might be changed, so it is necessary to consider the detailed structure. Based on the above assumption, the construction cost C_{T_p} is expressed as the following equation:

$$C_{T_\mathrm{p}} = \beta h_{T_\mathrm{p}} \tag{4}$$

where β is a construction cost for a unit height of the walls, and h is the wall height, respectively. Note that in actual situations, fixed costs need to be included, and variable costs are not necessarily proportional to wall height.

Since $\mathrm{RR}_{T_\mathrm{p}}$ can be considered as the benefit of the protection measure, the optimal plan for the structural measure (height of flood walls in this work) can be obtained by maximizing the benefit-cost ratio, which is given by the following equation:

$$\text{maximize} \frac{\mathrm{RR}_{T_\mathrm{p}} \cdot X}{C_{T_\mathrm{p}}} \tag{5}$$

where X is the useful life.

4.2 Example for application of the proposed method

A case study of flood risk analysis is conducted targeting Point A (Fig. 6) based on the assumptions explained in the Sect. 4.1. Three plans for construction of flood walls were considered as a case study. Height of walls in three plans correspond to expected inundation depth in floods with return periods of 200 years (in Plan 1), 500 years (in Plan 2), and 1600 years (in Plan 3) in terms of 48 h rainfall, respectively. In this case study, we selected the rainfall scenario that resulted in the largest inundation area at the design scale (456 mm in 48 h rainfall) among those described in Sect. 3.1. The temporal distribution of rainfall was then adjusted so that the peak river discharge calculated from the selected rainfall scenario became identical to the peak river discharge used in the actual public flood hazard map for design rainfall for the Tama River basin. This rainfall pattern was also employed to generate rainfall scenarios for the other target return periods of 48 h rainfall. It has been confirmed that this rainfall scenario is one of the scenarios that provide the greatest inundation depth for the target site. We compared the maximum inundation depths at the target site when the levee was breached every 200 m for the design rainfall. We selected the levee breach point that resulted in the largest maximum inundation depth. The risk analysis was performed using the maximum inundation depths obtained from the flood inundation analysis considering the levee breach condition.

The results are summarized in Table 1. Note that Plan 0 In Table 1 is the case without construction of walls, and the flood protection level corresponds to 50 years in that case. According to Eq. (5), the optimal plan was Plan 1 which maximized the benefit-cost rate among the plans. This case study used a considerably simplified model for damage and

Table 1. Results of risk assessment for each protection plan.

Plan	Rainfall (mm per 48 h)	Return period (year)	EAD (JPY yr^{-1})	RR_{T_p} (JPY yr^{-1})	Maximum depth(m)	Construction cost (JPY)	Benefit-cost ratio ($\times \alpha X/\beta$)
0	385	50	$\alpha/50$	–	0	0	–
1	456	200	$\alpha/200$	0.015α	0.49	0.49β	0.031
2	500	500	$\alpha/500$	0.018α	0.79	0.79β	0.023
3	550	1600	$\alpha/1600$	0.019α	1.36	1.36β	0.014

construction costs. If the damage costs and flood countermeasures costs can be properly modeled according to the actual situation of the target business site, we can make more realistic examinations in the same procedure. Like this way, one can derive science-based information on benefit-cost relationships for each of considered flood protection measures, which can be considered useful for investment decision for increased flood protection capability of business entities.

5 Conclusion

In this work, a method to estimate comprehensive flood risk information for a business site was developed based on a series of flood inundation simulations for various rainfall scenarios. It was shown that the total inundation area is greatly different even for the same return period of rainfall when using rainfall scenarios with various spatiotemporal patterns. Characteristics of increase in maximum inundation depth with peak discharge varied greatly depending on the topography around the sites, which shows importance to analyze expected inundation processes in detail site by site in order to design robust flood protection measures. Combined with the risk analysis approach presented in Sect. 4, the proposed method can help decision-makers develop flood protection plans for individual business sites based on the hazard information derived from numerical simulations.

Several assumptions were made in this case study. By using a similar framework, it is possible to propose a reasonable level of countermeasures based on the simulations, taking into account the balance between the cost of implementing flood countermeasures and the risk reduction effect of the countermeasures. Depending on the actual situation of flood risk at the target location, it is possible to consider arbitrary scales of external forces that are insufficient from publicly available information. In this case study, we used rainfall patterns and levee breach points that gave the most severe flood inundation for the target site. However, in actual investment decisions, there are also other constraints such as budget, other rainfall patterns that give less flood inundation may also be considered in flood protection planning for private companies if they can accept risks caused by the relaxed assumption.

Although potential effectiveness was shown, there are still things to be improved in the proposed method, such as consideration of inundation from sewer systems or use of extreme rainfall scenarios derived from future climate experiments with consideration of climate change effects. Further studies are therefore needed to establish a comprehensive approach on investment decision support for improved BCPs against floods.

Data availability. The data that support the findings of this study are available from the corresponding author, upon reasonable request.

Author contributions. SN, KS, and NI designed the research framework. SN analyzed the data and prepared the draft. SN and DN improved the research design, academic objectives, literature review, and interpretation of simulation results. KS, NI, DN and YY contributed to improve the manuscript.

Competing interests. The contact author has declared that none of the authors has any competing interests.

Special issue statement. This article is part of the special issue "ICFM9 – River Basin Disaster Resilience and Sustainability by All". It is a result of The 9th International Conference on Flood Management, Tsukuba, Japan, 18–22 February 2023.

Review statement. This paper was edited by Mamoru Miyamoto and reviewed by two anonymous referees.

References

Fan, J. and Huang, G.: Evaluation of Flood Risk Management in Japan through a Recent Case, Sustainability, 12, 5357, https://doi.org/10.3390/su12135357, 2020.

JMA: Number of occurrences of hourly precipitation exceeding 50 mm in Japan (AMeDAS), https://www.data.jma.go.jp/cpdinfo/extreme/extreme_p.html, last access: 15 November 2022.

Kimura, T.: The Flood Runoff Analysis Method by the Storage Function Model, The Public Works Research Institute, Ministry of Construction, 1961 (in Japanese).

MLIT: Flood fighting act amendment, https://www.mlit.go.jp/river/suibou/suibouhou.html (last access: 17 October 2023), 2015a.

MLIT: Manual of determining probable maximum precipitation for making expected inundation (river and inland floods), https://www.mlit.go.jp/river/shishin_guideline/pdf/shinsuisoutei_honnbun_1507.pdf (last access: 17 October 2023), 2015b.

MLIT: Manual for making estimated flood inundation area map, 4th edn., https://www.mlit.go.jp/river/shishin_guideline/pdf/manual_kouzuishinsui_1710.pdf (last access: 17 October 2023), 2015c.

MLIT: Map viewer for estimated inundation at an arbitrary point, https://suiboumap.gsi.go.jp/ (last access: 17 October 2023), 2015d.

MLIT: Expected inundation area for Tama River system (design rainfall), https://www.ktr.mlit.go.jp/ktr_content/content/000647024.pdf (last access: 17 October 2023), 2016a.

MLIT: Expected inundation area for Tama River system (probable maximum rainfall), https://www.ktr.mlit.go.jp/ktr_content/content/000647812.pdf (last access: 17 October 2023), 2016b.

MLIT: Kuma River flood archive, Chapter 1. Weather overview, https://kumariver-r0207archive.jp/overview/detail/chapter1-1.html (last access: 17 October 2023), 2021.

MLIT: Measures based on the 2019 East Japan Typhoon (typhoon No. 19), https://www.ktr.mlit.go.jp/ktr_content/content/000840783.pdf (last access: 17 October 2023), 2022.

MLIT: Basic river management policy for Tama River system (Overview of basin and rivers for Tama River system), https://www.mlit.go.jp/river/basic_info/jigyo_keikaku/gaiyou/seibi/pdf/tama_3.pdf, last access: 17 October 2023a.

MLIT: Explanatory materials regarding changes in the basic river management policy for the Tama River system, https://www.mlit.go.jp/river/basic_info/jigyo_keikaku/gaiyou/seibi/pdf/tama_7.pdf, last access: 17 October 2023b.

Morita, M.: Flood risk analysis for determining optimal flood protection levels in urban river management, J. Flood Risk Manage., 1, 142–149, 2008.

Nagumo, N., Ohara, M., Kuribayashi, D., and Sawano, H.: The 2015 Flood Impact due to the Overflow and Dike Breach of Kinu River in Joso City, Japan, Journal of Disaster Research, 11, 1112–1127, https://doi.org/10.20965/jdr.2016.p1112, 2016.

Nohara, D., Takemon, Y., and Sumi, T.: Real-time flood management and preparedness: Lessons from floods across the western Japan in 2018, in: Advances in Hydroinformatics, edited by: Gourbesville, P. and Caignaert, G., Springer Water, Springer, Singapore, 287–304, https://doi.org/10.1007/978-981-15-5436-0_22, 2020.

Yamamoto, H., Kanemitsu, N., Koba, A., Tsujimoto, H., Shimai, K., Sakamoto, K., and Iwaya, K.: Damage Investigation of Flood Disaster in Rokkaku River, Saga Prefecture by Heavy Rainfall of Autumnal Rain Front of August, 2021, Journal of Japan Society for Natural Disaster Science, 41-2, 145–175, 2022.

Proc. IAHS, 386, 173–179, 2024
https://doi.org/10.5194/piahs-386-173-2024

Flood resilient landscapes: area-based solutions combine added value for society with flood risk management

Annemargreet de Leeuw[1], **Ellen Tromp**[1], **and Femke de Boer**[2]
[1]Deltares, Delft, 2600 MH, the Netherlands
[2]Reframing Studio, Amsterdam, 1053 VN, the Netherlands
Correspondence: Annemargreet de Leeuw (annemargreet.deleeuw@deltares.nl)

Received: 3 July 2023 – Revised: 31 October 2023 – Accepted: 24 February 2024 – Published: 19 April 2024

Abstract. Society faces challenges such as caring for sustainable agriculture, clean energy and restoring biodiversity, whilst developing housing and industries. Climate change meanwhile stresses the Dutch water management system, impacts flood risk management and fresh water supply. To ensure making the right decisions, which we will not regret in 100 years, we developed the concept of flood resilient landscapes. The concept of flood resilient landscapes confronts, with a perspective of long term development, desired socio-economic developments with carrying capacity and potential of underlying physical landscape conditions. The underlying principle is to create social added value while promoting or at least maintaining flood risk management, given (future) spatial and societal developments. The first results are so promising that the Dutch Flood Protection Programme aims to incorporate it. The flood resilient landscapes concept offers the prospect of keeping the Netherlands safe beyond 2100 at socially acceptable costs and with public support now and in the future and paves the way towards implementation throughout international deltas.

1 Introduction

The Netherlands is located at the delta of rivers Rhine, Meuse and Scheldt. Two-third of the country lies below sea-level, and if it were not for an extensive flood defence system, large part would be prone to flooding by hazard of high water at sea, river or lakes (Tromp et al., 2022). In the past, the Dutch responded reactively to floods. This changed after floods of 1953 by setting legal standards to flood defences and setting up a comprehensive flood defence system. In 2008, the second Delta Commission advised to take future uncertainties into account and therefore to anticipate even more rigorously (Delta Committee, 2008). Consequently, the Dutch adopted a risk-based approach (van der Most, 2014). Because of efficiency, managing flood risk in the Netherlands is focused on strengthening flood defences. Economic risk, local individual risk and societal risk aretranslated to statutory standards in terms of signal and lowest acceptable levels of flood (or failure) probability of levees. Every 12 years the regional water authorities assess whether the levees under their mandate are still up to standard. If not, the authorities can apply for funding for improvement at the Dutch Flood Protection Programme (DFPP) (Jorissen et al., 2016). Over 1500 km of levees and over 400 structures will have to be upgraded before 2050. The magnitude of the improvement challenge forces the DFPP to focus on effectiveness (increase of production rate) and on efficiency (reduction of cost per kilometre). However, the challenge that comes with reaching the set target is especially wicked due to high level of complexity, uncertainty, and conflicting interests (Tromp et al., 2022): flood risk issues are entwined with other local problems involving diverse stakeholders, whilst the sector is institutionally fragmented, and resources are distributed in a non-hierarchical way. The ever-increasing complexity is producing new challenges and demands (Tromp et al., 2019).

The current approach to managing flood risk in the Netherlands is layed down in the Water Act, which formalises governance (responsibilities of different government institutes in relation to each other), funds, applicable flood probability standards for flood defences and cyclic reporting to parlia-

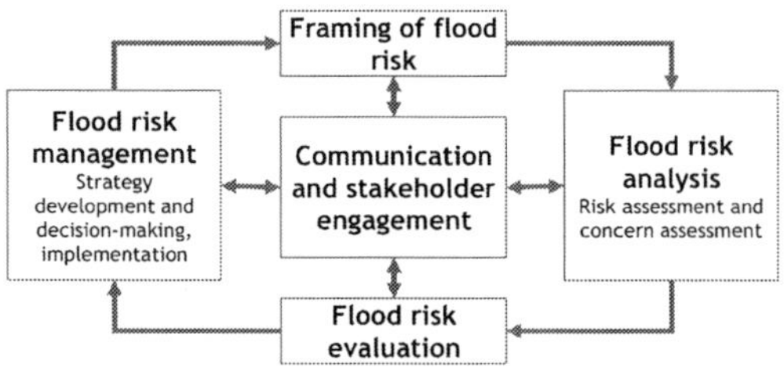

Figure 1. Framework for integrated flood management (APFM, 2017).

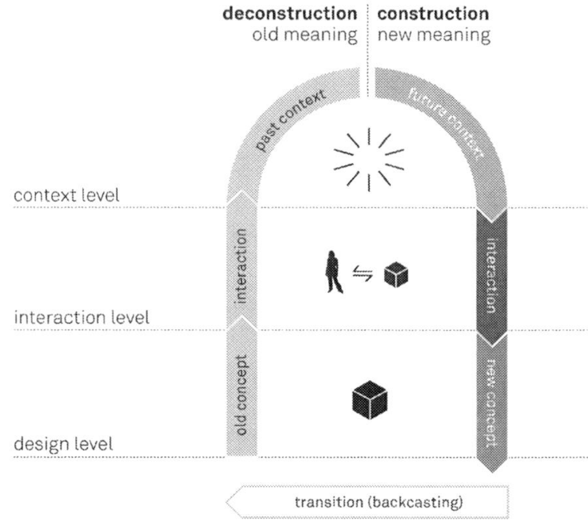

Figure 2. Following the vision in design approach of Hekkert and Van Dijk (2017), the left side of the model is the deconstruction phase, aimed at understanding the underlying values and world views behind current solutions, whereas the right side of the model shows the vision and design phase in which the aim is to develop a clear vision on the narrative(s) that can guide future goals, rules and solutions.

ment. A new Environment Act (expected to come in force 1 January 2024) will substitute this Water Act, by which flood defences will be treated as an integral component of rural and urban environment. This development entices a movement to scope flood risk management measures as part of a future oriented, sustainable spatial development in which value creation is key.

The complexity of planning sustainably for a future society, orientation on future, and the desire to strive for societal added value brought the following research question: *How can the use of design approaches in flood risk management give a perspective to a manageable approach to face these wicked problems?*

2 Frames and methods

2.1 Flood risk management framing

The way the Dutch have set up their flood risk management finds its origin in recent history and consecutive policy development. It corresponds with the European Framework Directive and its underlying Floods Directive 2007/60/EC that requires member states to report on preliminary flood risk assessments, areas of potential significant flood risk and flood hazard and risk maps, and to define flood risk management plans based on this information. Like other member states, the Netherlands have not made any notable changes to their administrative arrangements (EU, 2021).

The Associated Programme on Flood risk management of the World Meteorological Organisation (APFM, 2017) provides comprehensive guidance on how to plan for flood risk management in view of the Sendai Framework for disaster risk reduction (UNDP, 2016) and is equally useful for the implementation of the Floods Directive. The document states that planning for integrated flood management is an iterative, policy-cyclic process towards risk-based decision-making with a significant role for the involvement of stakeholders (Fig. 1).

In view of finding sustainable strategies, the flood risk analysis activity in this cycle comprises the definition of future scenarios and an analysis of the response of water and soil system to these scenarios. This analysis results in a definition of the current problem, with a view on future problems because of autonomous change. We will explain more about the Dutch future situation in Sect. 3.1. The challenge that we currently face, is how to perform framing for flood risk management in a more integrative, interactive, and future resilient way. The hypothesis is that design approaches could give (part of) the answer to the wicked problem faced, by anticipating to the future, instead of focusing on today's problems and challenges.

2.2 Vision in Design

Hekkert and Van Dijk (2017) published a guidance to designing, based on a vision. They described a standard approach to define this vision, based on an understanding of the changing patterns of concerns, attitudes, and behaviours in a complex future world. Their context-driven and human-centered approach is depicted in Fig. 2. Although the method was originally developed for industrial design, in the past decade the method has proved to help organizations, communities and citizens to understand their changing environment, envision alternative futures and to develop sustainable strategies to navigate from the present into a desired future. It was investigated how this design approach could also be applicable to the societal challenge of future proof flood risk management at landscape scale.

To this end, a desk study with additional interviews was performed to find clear and unclear, or unspoken elements that are part of our current solutions in Dutch flood risk management. In total eight experts were interviewed with expertise ranging from flood risk to delta biogeomorphology and from climate change to philosophy. A description of the current relation of Dutch society to flood risk management followed. This included a description of dominant narratives of that relation. At context level, a description of conditions and developments that led to the Dutch relation to flood risk management followed. The next level in the analysis is to find probable future narratives. A summary of the collected information can be found in Sect. 3.2.

3 Data and results

3.1 Qualitative description autonomous response of landscaped to climate change in the Netherlands

Following the work of the Dutch Delta program, the following typical Dutch main landscapes were defined, based on water and soil characteristics. Figure 3 gives an indication where these landscapes are in the Netherlands:

- Peatlands are located mainly in the west and north of the Netherlands. They suffer of subsidence due to drainage and oxidation of peat. The land use is mainly agriculture.
- (High) Sandy soils are mainly located in the higher east and south of the Netherlands. Thay are well drained for agricultural and urban use. The land use is mainly agriculture, nature and urban. These soils are vulnerable for drought.
- (Saltating) Coastal areas and deep polders are mainly located in the west and north of the Netherlands. They mostly consist of clay soil, which makes them very suitable for agriculture. These areas are vulnerable to floods and to salinization through surface- and groundwater.
- Main water system consist of main rivers, lakes and sea. The main water system is mainly designed for fast discharge of water, whereas the lakes are increasingly in use as strategic source of fresh water.
- Urban areas are mainly located in the west. Their drainage system is focussed on fast discharge of precipitation. High pavement rate in urban areas leads to vulnerability to flooding, drought and heat.

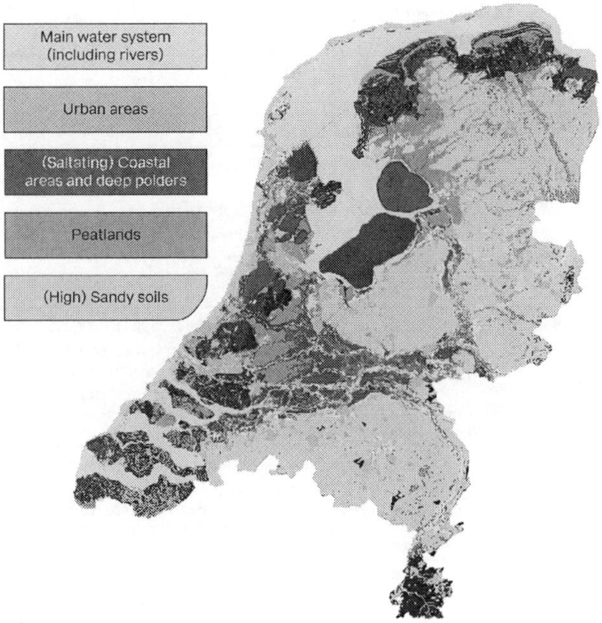

Figure 3. Indicative locations of landscapes in the Netherlands.

Climate change effects that drive responses of these landscapes are rising temperatures and dry periods, sea level rise, and intensifying precipitation. Experts on water- and soil management were challenged to specify in an expert elicitation session what emerging problems are to be foreseen for the Netherlands. The experts gave their answers under the condition of "business as usual" for land use functions.

Table 1 describes *major* responses, extracted from the full report of the session (Roosjen et al., 2022). To summarize: each type of landscape has their own challenge. Peat areas are susceptible to pluvial flooding and are vulnerable if too little freshwater is supplied. High sandy soils are susceptible to drought and heat stress and rely on supply of enough freshwater. Coastal areas and deep polders are vulnerable to flood risk and pluvial flooding. In the main water system, the challenge is to deliver enough discharge capacity in times of high water, and to deliver enough supply capacity in times of water shortage. Urban areas are vulnerable to both flood risk and pluvial flooding. These results correspond to other studies (Deltares et al., 2021; Schra, 2022). The question now is how much time there is before tipping points are reached and how much time is required to implement alternative pathways. Recent research showed (National Delta program, 2023) that the current flood risk strategy *technically* suffices to adapt to sea level rise. However, the societal willingness and acceptance of these technical measures is changing, also considering the current societal challenges the Dutch society (and worldwide on the Global North face). Recent research on decision making under deep uncertainty (Marchau et al., 2019) highlights the importance of back- and forecasting to develop strategies for e.g. freshwater supply seem to reach their tipping points earlier than flood risk management policy (National Delta program, 2023).

When focussing on flood risk, attention is drawn to peat lands, polder areas and the main water system. The situation will set boundary conditions to socio-economic development and gives direction to adaptation measures. If society wishes to maintain a current function under pressure, the function will need to adapt to new circumstances to remain feasible.

Table 1. Major effects of climate change in Dutch landscapes.

Dutch landscape	Expected response to climate change
Peatlands	Pluvial flooding
	High groundwater levels lead to low storage capacity for rainfall. Without intervention, more intense rainfall will directly lead to pluvial flood risk in peat areas. Subsidence in peat areas leads to high seepage pressure. To decrease the rate of subsidence in peat areas, raising groundwater levels is needed. This conflicts with other land use functions such as agriculture and urbanisation.
	Freshwater availability
	Freshwater demand in peat areas is high, to combat subsidence and to flush salty seepage. The availability of fresh water is under pressure because of climate change and increasing other freshwater demands. In some areas there will be an increased siltation of ground- and surface water. This will become problematic for the continuity of land use functions such as agriculture and nature.
(High) Sandy	Drought and heat stress
soils	The extensive drainage system in combination with sandy soils (which have little soil moisture capacity) lead to desiccation of high sandy areas. Climate change effects such as longer periods of drought and heat enhance this. More intensive use of land leads to an increase of drainage, which in turn leads to more dehydration.
	Fresh water availability
	Fresh water supply is now and will be in future under pressure because of the landscape design, which is focused on drainage. Climate change will lead to longer periods of droughts. Sea level rise and decrease of river discharges in dry season lead to a saltwater tongue entering the tidal river. The trend is inland moving. In the long run salt water can reach the inlet points of drinking water production sites. Deep phreatic layers now used for drinking water wells are becoming contaminated. Strategic groundwater storages are being depleted.
	Water quality effects.
	Drought and increasing temperatures lead to an increase of blue-green algae, botulism and high concentration of pollutants.
(Saltating)	Flood risk
coastal areas and deep	Sea level rise and subsidence lead to larger differences in water heads. This results in an increase of flood risk. High peak discharges of rivers, due to higher intensity rainfall, will lead to an increase of flood risk. The availability of sand for nourishment of the coast is uncertain in the long term.
polders	Pluvial flooding
	High groundwater levels decrease storage capacity of the soil. Without intervention, more intense rainfall will lead to higher pluvial flood risk.
Main water	Flood risk
system	More extreme precipitation leads to higher peak river discharge. Sea level rise leads to high water level conditions along coastal areas, and lead to decrease of the discharge capacity of rivers. This leads to higher water levels in tidal river areas and large lakes. In current policy, expected higher water levels lead to raising of flood protection and more room for the river projects. However, the main water system has limited space to cope with effects of climate change. To cope with higher sea levels, in current policy more sandy sediments are needed to replenish the coast fundament and dunes as a flood protection. A sustainable sediment balance in the rivers requires attention. Sea level rise might lead to drowning of the tidal flats of the Wadden Sea.
	Fresh water availability
	Buffering large volumes of water in the large reservoirs of the main water system (e.g. IJsselmeer) will more frequently be insufficient to provide enough water to different water use functions. Sea level rise and periods with low river discharges will lead to inward moving salt tongues. This will lead to diminished freshwater availability.
Urban areas	Flood risk
	Many urban areas are in risky places, because of attractiveness of nearby water. In the short-term urban areas in flood prone areas will face an increase of flood risk, mainly due to increase of peak river discharges. Sea level rise, increase of river discharges and subsidence lead to an increase of flood risk in urban areas in protected zones in the long term. Especially in urban areas space is scarce, including for flood risk management measures.
	Pluvial flooding
	Urban areas are more and more densely populated, which leads to an increase of paved area. Increasing rainfall intensities lead to large challenges in urban areas in managing flood risk. The target to build 900 000 homes leads to even further densification. Subsidence leads to high cost of management and maintenance of infrastructure and open space in urban areas. Social inequality adds to diversity in exposure and vulnerability to flood risk.

The alternative is to change the land use function. And if society wishes to construct built environment in peat lands or polder areas, special attention is needed to account for long term effects to create a sustainable design.

3.2 Societal context factors

Literature with a wide perspective but related to the attitude of Dutch society to flood risk management was reviewed, with a focus on finding current and future context factors and important themes.

This first step in the Vision in Design approach, the deconstruction phase, is aimed at understanding the Why behind the world of solutions that are currently used. Based on literature and experts interviews the dominant past and current narratives underlying flood risk management in The Netherlands were explored. An important insight was that there were four repetitive narratives in Dutch society (Jensen, 2020):

1. The technologic narrative tells how the Dutch created the Netherlands with hydraulic ingenuity, adaptive capacity and pioneer mind. It brings management perspective and hope, but the trust in technology also leads to being less prepared for disaster, of even to climate scepticism. This is the dominant narrative in the Netherlands today.

2. The apocalyptic narrative is a story of catastrophizing. The thought that the world will decease leads on one hand to a fatalistic attitude, but on the other hand fear motivates too: We will try hard to prevent a disaster like the floods of 1953 and base our policy on incidents.

3. The ecologic narrative tells us that nature is supreme and must be given the space it needs. We do not battle with water but live with water. To date stories of climate change and loss of biodiversity stimulate this narrative.

4. During past ages, the secular narrative declared floods as a collective punishment of God to the sin of Man. A current variant to this narrative is that man is increasingly aware of the influence of their own behaviour to climate change.

The currently dominant technologic narrative has led to trust in technology: the Dutch landscape, flood risk management and spatial planning are considered manufacturable. This leads to low awareness with the Dutch population on flood risk related to living below sea level: the Dutch are not prepared to respond to floods.

To understand the changing societal context, literature was reviewed to collect social, cultural, economic, technological, and other developments and trends that will impact people's behaviours and attitudes with regards to flood safety. Over 180 factors were found and enriched in eight interviews with leading experts. Verification was performed in a review team.

The research team structured these factors to describe potential future attitudes to flood safety. Inspired by Jacobs (2006) three main divides were found (de Boer et al., 2022):

1. Matterscape: the societal attitudes vary from belief in manufacturability to adapting to nature;

2. Powerscape: dealing with (climate) change can be organised by concentrating power or by organising decentralised power in smaller communities; and

3. Mindscape: in a fast-changing world people can find confidence being rooted in the past, or they trust in innovation and planning for the long term.

Realizing these three divides, eight logical narratives (Fig. 4) were constructed to represent future attitudes to flood risk management. One of these emerging narratives is the idea that water should flow rather than confining it between dikes. This *ecocentric* narrative is based on the idea of cooperation between "us" (humans) and "water" (sea, rivers). This might lead to more nature-based solutions and citizen participation. Likewise, seven other narratives were developed that move away from purely infrastructural flood protection. On a regional scale these narratives are used to design flood resilient landscapes.

4 Discussion and foresight

The investigation on system boundary conditions, arising from water and soil system and challenges that come with striving for sustainable, future-oriented developments covered not only flood risk and pluvial risk issues but also drought, heat and water quality issues. Integrated water management solutions and integrated spatial developments can leverage each other, e.g. creating space for water in or near urban area will benefit mitigation of urban heat stress. The generated overview can serve as a first start for deeper investigations of potential integrative solutions.

This research revealed four categories of factors, driving change in a landscape: (1) climate change, (2) societal challenges and transitions, (3) societal attitudes and (4) challenges related to maintain a sustainable water and soil system. Until now, our research covered only societal attitudes and challenges for a sustainable water and soil system.

The research brings new perspectives to (Dutch) flood risk management forward, also to understand how "wicked" the coupling of flood risk management and integrated spatial development is. So far in practice, we see too much emphasis on short-term solutions and incremental adaptation under the assumption that climate change will be smooth sailing (Tessler et al., 2015), and therefore on a technologic narrative. Consequently, disruptive events have at best provided the "evil" tipping points to transformative change (extreme rainfall in Europe, 2021). What we need is to identify potential moments of synergy and opportunities for radical reorientation (Lenton et al., 2019) and harness them to initiate

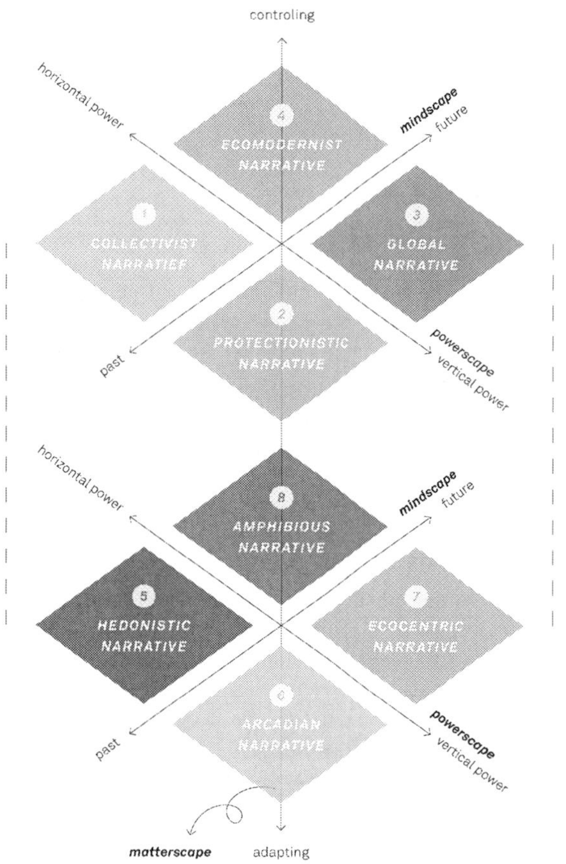

Figure 4. Narratives represent future societal attitudes to flood risk management in the Netherlands.

a transformative process for addressing climate change and other societal challenges. We believe that taking the values of future society into account, will ensure that the investments and decisions are no-regret. This is in line with the instruments of the decision-making under deep uncertainty (Marchau et al., 2019). From that perspective we believe that other narratives will coincide with the technologic one. Certainly, communication and stakeholder engagement will be shaped differently: instead from starting a discussion from a current flood risk management perspective the stakeholders will participate in a future value-driven approach. This will lead to a strategy that deals with all of the following: climate change, societal attitudes towards flood risk, emerging societal challenges, and sustainable ecosystem services of the natural system.

This first exploration generated insight and clarity as to what societal attitudes and perspectives towards flood risk management in the Netherlands could emerge. As such, bringing in a designer's approach already seems promising. In further experimental research the value of taking a designer's approach will be evaluated in real life pilot areas: Will it be possible to set up scoping when starting from a discussion on values in the area, on the effects of the current dominant perspective versus potentially desired narratives? Will we be able to use flood risk management as a leverage to transitions for large societal challenges vice versa, such as the transitions towards sustainable agriculture, the energy transition and housing? Which building blocks are needed to make a transition from the current situation towards a desired situation? Can we design a value-oriented, human-centered perspective and start working towards it? Are we able to define transformative pathways for future proof, integrated flood risk management in the Netherlands? In new research, we explore the application of the method further in real pilot areas. The stakeholders of these pilot areas experience the sense of urgency, and together with societal challenges this results in a window of opportunity. The flood resilient landscapes concept offers the prospect of keeping the Netherlands safe beyond 2100 at socially acceptable costs and with public support now and in the future, and paves the way towards implementation throughout international deltas, requiring strong stakeholder commitment and collaborative learning.

Data availability. No data sets were used in this article.

Author contributions. Conceptualization: AdL. Research on autonomous response of landscapes to climate change: AdL and ET. Research on potential societal context factors: FdB. All authors contributed to the final version of the manuscript and have read and agreed to the published version.

Competing interests. The contact author has declared that none of the authors has any competing interests.

Disclaimer. Publisher's note: Copernicus Publications remains neutral with regard to jurisdictional claims made in the text, published maps, institutional affiliations, or any other geographical representation in this paper. While Copernicus Publications makes every effort to include appropriate place names, the final responsibility lies with the authors.

Special issue statement. This article is part of the special issue "ICFM9 – River Basin Disaster Resilience and Sustainability by All". It is a result of The 9th International Conference on Flood Management, Tsukuba, Japan, 18–22 February 2023.

Acknowledgements. This research was made possible by the Strategic research funding to Deltares by the Ministry of Economic Affairs and Climate Policy and by the research funds of Knowledge for Flood Defences by the Ministry of Infrastructure and Water Management.

Financial support. This research has been supported by the Rijkswaterstaat (KPP Kennis voor Keringen) and the Ministerie van Economische Zaken en Klimaat, Topconsortium voor Kennis en Innovatie (Sito to Deltares).

Review statement. This paper was edited by Svenja Fischer and reviewed by two anonymous referees.

References

APFM (Associated Program on Flood Management): Selecting measures and designing strategies for integrated flood risk management, A Guidance Document, https://www.floodmanagement.info/publications/guidance%20-%20selecting%20measures%20and%20designing%20strategies_e_web.pdf (last access: 22 March 2024), 2017.

de Boer, F., Bizzarri, A., Gestel, M. van, Dijk, M. van, Dijk, S. van, Dijkstra, M., Oenema, A., Leeuw, A.M. de, Knoeff, H.: Future narratives in flood risk management, Part of framework flood resilient landscapes, Deltares, 50 pp., https://publications.deltares.nl/Deltares244.pdf (last access: 22 March 2024), 2022.

Delta Committee: Working together with water, a living land builds for its future, reportNational Delta Programme, 138 pp., https://english.deltaprogramma.nl/documents/publications/2008/09/03/working-together-with-water (last access: 22 March 2024), 2008.

Deltares, Bosch Slabbers and Sweco: Based on the watersystem: Boundaries to the constructability of our water- and soilsystem, report, Deltares, 50 pp., https://publications.deltares.nl/11206890.pdf (last access: 22 March 2024), 2021 (in Dutch).

EU: European Overview – 2nd Preliminary Flood Risk Assessments – Accompanying the document REPORT FROM THE COMMISSION TO THE EUROPEAN PARLIAMENT AND THE COUNCIL on the implementation of the Water Framework Directive (2000/60/EC), the Environmental Quality Standards Directive (2008/105/EC amended by Directive, 2013/39/EU) and the Floods Directive (2007/60/EC); Implementation of planned Programmes of Measures, New Priority Substances, Preliminary Flood Risk Assessments and Areas of Potential Significant Flood Risk; COM(2021) 970 final – SWD(2021) 970 final, European Commission, number COM(2021) 970 final, https://eur-lex.europa.eu/legal-content/EN/TXT/?uri=COM:2021:0970:FIN (last access: 22 March 2024), 2021.

Hekkert, P. and Van Dijk, M.: Vision in design, a guidebook for innovators, 2nd printing, BIS Publishers BV, ISBN 978-90-6369-371-8, 2017.

Jacobs, M. H.: The production of mindscapes: A comprehensive theory of landscape experience, thesis, Wageningen University, Wageningen University, ISBN 90-8504-520-7, 280 pp., https://edepot.wur.nl/40182 (last access: 22 March 2024), 2006.

Jensen, L.: The battle against water in language and picture: What stories do we tell? Narrative strategies related to water management and sea level rise, Radboud University, 48 pp., https://lottejensen.s3.amazonaws.com/uploads/spina/publication/file/980/Welke_verhalen_vertellen_we.pdf (last access: 22 March 2024), 2020 (in Dutch).

Jorissen, R., Kraaij, E., and Tromp, E.: Dutch flood protection policy and measures based on risk assessment, in: Proceedings of the 3rd European Conference on Flood Risk Management (FLOODrisk 2016), Lyon, France, 17–21 October 2016, E3S Web Conf., 7, 20016, https://doi.org/10.1051/e3sconf/20160720016, 2016.

Lenton, T. M., Rockström, J., Gaffney, O., Rahmstorf, S., Richardson, K., Steffen, W., and Schellnhuber, H. J.: Climate tipping points – too risky to bet against, Nature, 575, 592–595, https://doi.org/10.1038/d41586-019-03595-0, 2019.

Marchau, V. A. W. J., Walker, W. E., Bloemen, P. J. T. M., and Popper, S. W. (Eds.): Decision Making under Deep Uncertainty, from theory to practice, Springer International Publishing, https://doi.org/10.1007/978-3-030-05252-2, 2019.

National Delta Program: How can the Netherlands deal with sea level rise? Intermediate report of the knowledge program on sea level rise, National Delta program, 56 pp., https://www.deltaprogramma.nl/binaries/deltacommissaris/documenten/publicaties/2023/11/09/20231107-9525_tussenbalans-kennisprogramma-zeespiegelstijging_06-webversie/9525_Tussenbalans+kennisprogramma+Zeespiegelstijging_DV.pdf (last access: 22 March 2024), 2023 (in Dutch).

Roosjen, R., Niesten, M., and de Leeuw, A. M.: System Boundaries 2100, Part of framework flood resilient landscapes, Deltares, 55 pp., https://publications.deltares.nl/11208010_019_0001.pdf (last access: 22 March 2024), 2022 (in Dutch).

Schra, J., Kleinhans, M., Cohen, K., Haasnoot, M., and Middelkoop, H.: What does the Delta want?, Report 31171979, National Delta program, 43 pp., https://www.deltaprogramma.nl/binaries/deltacommissaris/documenten/publicaties/2022/04/01/wat-wil-de-delta-neogeografische-kaarten-van-het-nederlandse-laagland-in-een-toekomst-met-zeespiegelstijging/20220621+WatWilDeDelta_UU_incl+toelichtingkaarten_def.pdf (last access: 22 March 2024), 2022 (in Dutch).

Tessler, Z. D., Vörösmarty, C. J., Grossberg, M., Gladkova, I., Aizenman, H., Syvitski, J. P. M., and Foufoula-Georgiou, E.: Profiling risk and sustainability in coastal deltas of the world, Science, 349, 638–643, https://doi.org/10.1126/science.aab3574, 2015.

Tromp, E.: Enhancing Knowledge Transfer and Uptake in the Design Processes of Flood Defences, PhD thesis, Delft University of Technology, Delft, The Netherlands, https://doi.org/10.4233/uuid:df56c11b-5366-4720-9f58-015c9bf0316e, 2019.

Tromp, E., te Nijenhuis, A., and Knoeff, H.: The Dutch Flood Protection Programma: Taking innovations to the next level, Water, 14, 1460, https://doi.org/10.3390/w14091460, 2022.

UNDP: Sendai Framework for Disaster Risk Reduction (2015–2030), Preventionweb, 37 pp., https://www.preventionweb.net/files/43291_sendaiframeworkfordrren.pdf (last access: 22 March 2024), 2016.

van der Most, H., Tánczos, I., de Bruijn, K. M., and Wagenaar, D.: New Risk-based standards for flood protection in the Netherlands, in: Proceedings of the 6th International Conference on Flood Management, Sao Paulo, Brazil, 16–18 September 2014, ICFM conferences, http://eventos.abrh.org.br/icfm6/proceedings/ICFM6_Proceedings.zip (last access: 22 March 2024), 2014.

Proc. IAHS, 386, 181–187, 2024
https://doi.org/10.5194/piahs-386-181-2024

Flood damage model bias caused by aggregation

Seth Bryant[1,2], **Heidi Kreibich**[1], **and Bruno Merz**[1,2]

[1]GFZ German Research Centre for Geosciences, Section 4.4. Hydrology, Potsdam, Germany
[2]Institute of Environmental Science and Geography, University of Potsdam, Potsdam, Germany

Correspondence: Seth Bryant (bryant.seth@gmail.com)

Received: 19 July 2023 – Revised: 21 November 2023 – Accepted: 24 February 2024 – Published: 19 April 2024

Abstract. Flood risk models provide important information for disaster planning through estimating flood damage to exposed assets, such as houses. At large scales, computational constraints or data coarseness leads to the common practice of aggregating asset data using a single statistic (e.g., the mean) prior to applying non-linear damage functions. While this simplification has been shown to bias model results in other fields, the influence of aggregation on flood risk models has received little attention. This study provides a first order approximation of such errors in 344 damage functions using synthetically generated depths. We show that errors can be as high as 40 % of the total asset value under the most extreme example considered, but this is highly sensitive to the level of aggregation and the variance of the depth values. These findings identify a potentially significant source of error in large-scale flood risk assessments introduced, not by data quality or model transfers, but by modelling approach.

1 Introduction

With the increase in flood related disaster damages, the expansion of computation power, and the availability of global data, the development and application of meso- and macro-scale flood risk models has increased dramatically in the past decade (Ward et al., 2015). These flood risk models are often conceptualized as a chain of sub-models for the flood hazard, exposure of assets, and vulnerability to flooding; with each step bringing uncertainty (de Moel and Aerts, 2011). Vulnerability modelling, the last step in the chain, is generally found to be the most uncertain component in micro- and meso-scale models (de Moel and Aerts, 2011; Jongman et al., 2012). These findings are supported by work comparing modelled damages to those observed during flood events, where large discrepancies are regularly found between different models and against observations (Jongman et al., 2012; McGrath et al., 2015). Further challenges are introduced when such models are transferred to the macro-scale, where many exposed assets are aggregated through averaging into a single unit before vulnerability models are applied (Hall et al., 2005; Ward et al., 2015; Sairam et al., 2021). This process collapses heterogeneities within the aggregated unit (like variable flood depth), and poses poorly understood challenges to the accuracy of flood risk models.

Such scaling issues are not unique to flood risk models. Many fields find it convenient (or necessary) to simplify the system under study by averaging or aggregating some variable or computational unit (Denny, 2017). However, this assumption that system response is unaffected by averaging, is *false* for most real-world systems; a conundrum, widely called "Jensen's inequality" (Jensen, 1906) or "the fallacy of the average" which can be formalized as:

$$\overline{g(x)} \neq g(\overline{x}) \quad (1)$$

where g is a non-linear function and x is some response variable. Applied to flood vulnerability models, which are almost never linear (Gerl et al., 2016), Eq. (1) implies that aggregating or averaging assets (e.g., buildings) introduces new errors. It can be shown that the magnitude and direction of such errors are related to the variance of the response variable (σ_x^2) and the local shape of the function ($g''(\overline{x})$) (Denny, 2017).

Within a flood vulnerability model, flood damage functions (f) provide the mathematical relation between exposure and vulnerability variables (e.g., flood depth) and estimated flood damages (e.g., building repair costs) for a single asset. The most basic functions directly relate flood depth to damage – so-called depth-damage curves widely attributed to White (1945). Gerl et al. (2016) provides a comprehensive meta-analysis of 47 references containing flood damage functions found in the public literature. These functions cover a wide range of geography and sector including 7 continents and 11 sectors for example. The majority of functions identified were deterministic (96 %), multi-variable (88 %), and expressed loss relative to the total value of the asset (56 %). To provide a standardized library of these functions, each was harmonized to an intrinsic or common set of indicator variables leaving other unique indicators as default values (e.g., inundation duration). Following such a harmonization, Gerl et al. (2016) found significant heterogeneity in function shape and magnitude between authors for supposedly similar models. This partly explains the large discrepancies in risk model results reported by others (Jongman et al., 2012; McGrath et al., 2015).

Aggregation and scaling issues are rarely considered in the flood risk model literature. Jongman et al. (2012) compared results from six aggregated and two object-based asset models against observed damages from a 2002 flood in Germany and a 2005 flood in the U.K.. Asset data was developed from a 100 m land cover grid, which required adjusting the two object-based models by 61 %–88 % to reflect the average portion of building footprints within each grid cell. Model performance was mixed between the two studies, with the aggregated models over-predicting by a factor of two and object-based models under-predicting by 5 % on average for the German case. For the U.K. case, all models under-predicted.

In a recent large-scale study, Pollack et al. (2022) constructed a benchmark and aggregated analog models from roughly 800 000 single family dwellings and eight 30 m resolution flood depth grids with return periods ranging from 2- to 500-years. When only building attributes were aggregated, a small negative bias was observed (−10 %) while when hazard variables were also aggregated a large positive bias was found (+366 %) for annualized damage. Given the spatial correlation of building values and flood exposure found in their study area, they conclude that bias would be difficult to predict ex-ante.

In this paper we summarize work to improve our understanding of the effect of aggregation on a particular component of flood risk models: the flood damage function. We accomplish this by producing a first order approximation of the potential aggregation error for a general flood risk models, the first attempt of its kind we are aware of. To provide as broad an evaluation as possible, a library of 344 damage functions are evaluated against a single indicator variable: synthetically generated flood depth. These results are then analyzed to elucidate the potential significance and behaviour of aggregation on flood risk models.

Table 1. Summary of flood damage models from Gerl et al. (2016) showing the number of functions (f) per model.

model id	name	f count
3	FLEMO	242
4	IKSE	2
6	MERK	5
7	MURL	2
12	Neyshabouri (2012)	2
16	HAZUS-MH	35
17	MCM	17
23	Luino et al. (2009)	1
27	Toth et al. (2008)	6
37	Budiyono et al. (2015)	24
42	Dutta et al. (2003)	3
44	DSM	5

2 Methods

To evaluate the sensitivity of flood damage functions to the aggregation of input variables, a library of 344 damage functions are evaluated against synthetically generated water depths at various levels of aggregation. Statistics describing the difference between raw function outputs and the aggregated analogues are then computed on each function and each level of aggregation to describe the response of the function to aggregation.

2.1 Flood Damage Functions

For this study, we focus on direct tangible economic functions for estimating the relative loss to buildings from flood depth. From Gerl et al. (2016), 12 such models were identified, each composed of a collection of flood damage functions (f) with the primary independent variable as depth. To simplify the analysis, each of these is discretized by any secondary variables (i.e., variables other than flood depth) to produce a collection of 344 simple functions for relative damage from flood depth as summarized in Table 1. Each of these functions is monotonically increasing (i.e., $\mathrm{RL}_{i+1} \geq \mathrm{RL}_i$) and implemented in python as a lookup table (with linear interpolation applied between discrete depths (x) and their relative loss (RL) pairs). For out-of-range depth values (x) (i.e., those that are greater or less than the most extreme values provided in the lookup tables) either the maximum relative loss value ($f[x] = \mathrm{RL}_{\max}$) or a value of zero are returned ($f[x] = 0$) following our understanding of common practice. In other words, the function implementation caps the RL values to match those provided in the lookup table.

2.2 Synthetic Depths Generation

Each of the 344 selected discrete functions have a distinct shape, depth domain (from −1.22 to 10 m), and relative loss

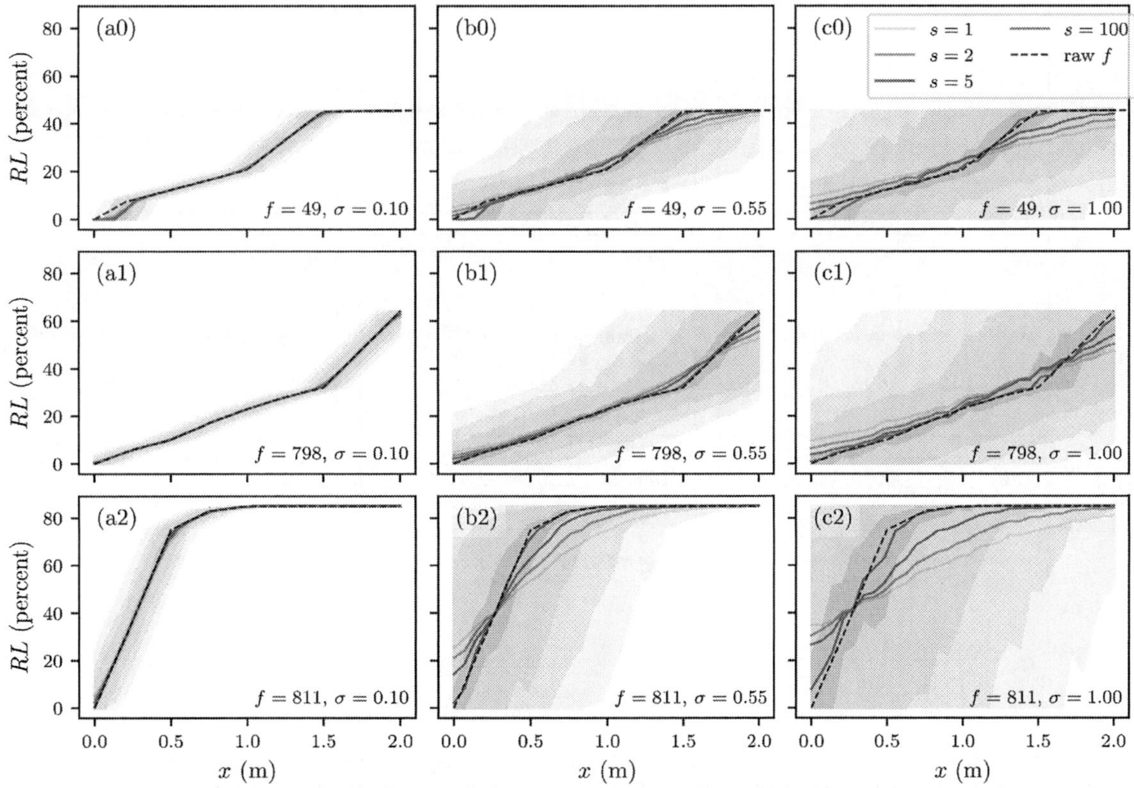

Figure 1. Relative loss vs. anchor depth (x_b) for three example functions and three synthetic depth generation standard deviations (σ) computed against the depth values and aggregation (s) described in the text. Shaded areas show the corresponding q_{95} and q_5 for the RL computed for each level of aggregation. Functions are from model 3, 3 and 37 from Table 1 for panel **(a)**, **(b)**, and **(c)** respectively.

domain (from 0.0 to 116, with a value greater than 100 indicating damages more severe than the estimated replacement value). In practice, each asset (e.g., house) would have its flood damage calculated with a specific function and the exposure variables (e.g., depth) sampled for each flood event. Under aggregation, multiple assets would be collapsed to a single node prior to function computation. The exposure (and vulnerability) indicators of this collapsed asset are generally computed through averaging of the original, singular assets. To provide a comprehensive analysis of the full range of exposure depths supported by each function, a range of depth values are generated using a normal distribution and then sampled and averaged according to the level of aggregation.

To generate the synthetic depth values, the independent variable domain (depths) is discretized from 0.0 to 2.0 to produce 30 anchor depth values (x_b). Each of these x_b values is used as the average or mean depth under evaluation, with lower values representing shallower floods and higher values (closer to 2.0) very deep floods. To represent the variance in depths within a group of assets for a given flood, raw synthetic depth values are then randomly generated around each x_b by sampling 2000 points from three normal distributions ($N[\sigma = (0.1, 0.55, 1.0), \mu = x_b]$). Higher variances ($\sigma^2$) represent steeper terrain and water surface gradients, while lower values represent flatter areas like a lowland floodplain. These raw values are then aggregated by randomly splitting the 2000 depth values into N_{12} groups of size $s_j = (1, 2, 5, 100)$ and computing the mean. In this way, 360 arrays of synthetic depth values covering a broad range of flood magnitudes, variance, and levels of aggregation are prepared.

2.3 Relative Loss Errors

Using each of the 360 synthetic depth arrays on each of the 344 functions, a set of relative loss arrays ($\mathrm{RL}_{x_b,s_j,\sigma,f}$) is computed, each associated with an anchor depth value (x_b), a level of aggregation (s), a standard deviation (σ) and a flood damage function (f). To compute the aggregation error potential of each function under various flood conditions, arrays are grouped by function (f), standard deviation (σ), and aggregation level (s). For each of these groups, the mean of the relative loss values is plotted against x_b. Second, for each of these four series, the area against the un-aggregated result is

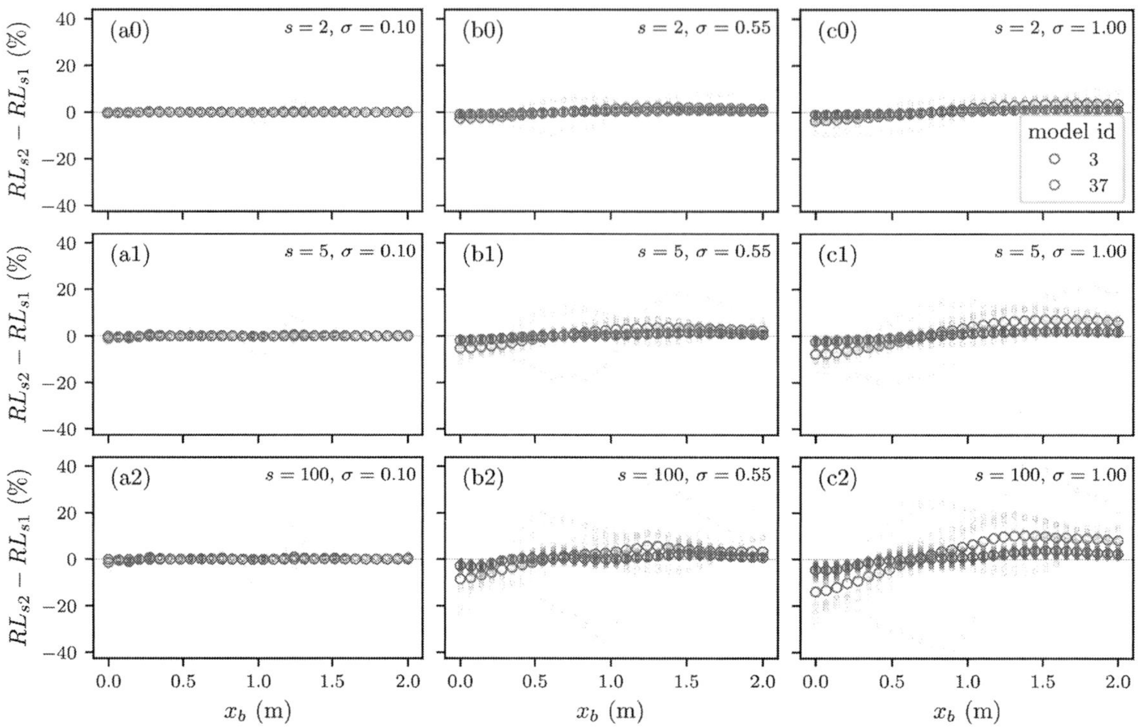

Figure 2. Relative loss error for three levels of aggregation (s) and three synthetic depth families (σ) for a selection of the models described in Table 1. Faint circles show each function (f) while the bold circles show the function ensemble mean for that model.

calculated as:

$$e_{s_j,\sigma,f} = \int_{x=0} \overline{\mathrm{RL}\left[x, s_1, \sigma, f\right]} - \overline{\mathrm{RL}\left[x, s_j, \sigma, f\right]} \quad (2)$$

where e is the aggregation error potential and the integrals are evaluated using the composite trapezoidal rule, a common method for estimating the area under a curve by discretizing into trapezoids (NumPy Developers, 2022).

3 Results and Discussion

Three example damage functions (f) are shown in Fig. 1 along with the mean and two quantiles (q_{95} and q_5) of the resulting loss values computed under four levels of aggregation against the generated depth values. These three damage functions were selected to demonstrate typical behaviour present in all functions. Looking from left-to-right, Fig. 1 shows how large variances (σ^2) employed in the synthetic depth generation act to spread the resulting relative loss values. Working in a similar but opposite manner, large aggregations (s) reduce the spread of relative loss values. This is intuitive if we consider that the averaging employed in the aggregation works like a filter to reduce variance.

A clock-wise rotation can also be seen in Fig. 1, where series with less aggregation (e.g., $s = 0$) have positive/negative deviation from the left/right tails of the raw functions. This can be explained by the treatment of out-of-range depth values (x). As explained above, when a synthetic depth value with negative exceedance is calculated (i.e., less than the minimum (x) provided in the lookup table), a relative loss of zero is returned. Away from these x range limits (e.g., $x_b = 1.0$), the normally distributed synthetic depths yield roughly similar positive and negative residuals from the monotonically increasing functions, which balance during the averaging. On the other hand, closer to the x range limits (e.g., $x_b = 0.0$) the residuals become unbalanced as those x values exceeding the range (e.g., $x_b < 0.0$) lose their relation with RL while those within range maintain the relation. At the low end of the range (e.g., $x_b = 0$), this phenomena produces a bias negatively related to aggregation and positively related at the high end (e.g., $x_b = 2.0$). This is intuitive if we consider the aggregated synthetic x values (i.e., $s > 1$) have had their variance reduced (through averaging), meaning fewer extreme x values and therefore the RL imbalance discussed above is less severe with aggregation. Because the depth values considered here are generated with a normal distribution, it is likely that some x values for low anchor depths (e.g., $x_b = 0.0$) would be below ground in a real flood risk model (some negative x values are realistic as buildings are typically elevated slightly above ground). In cases where such

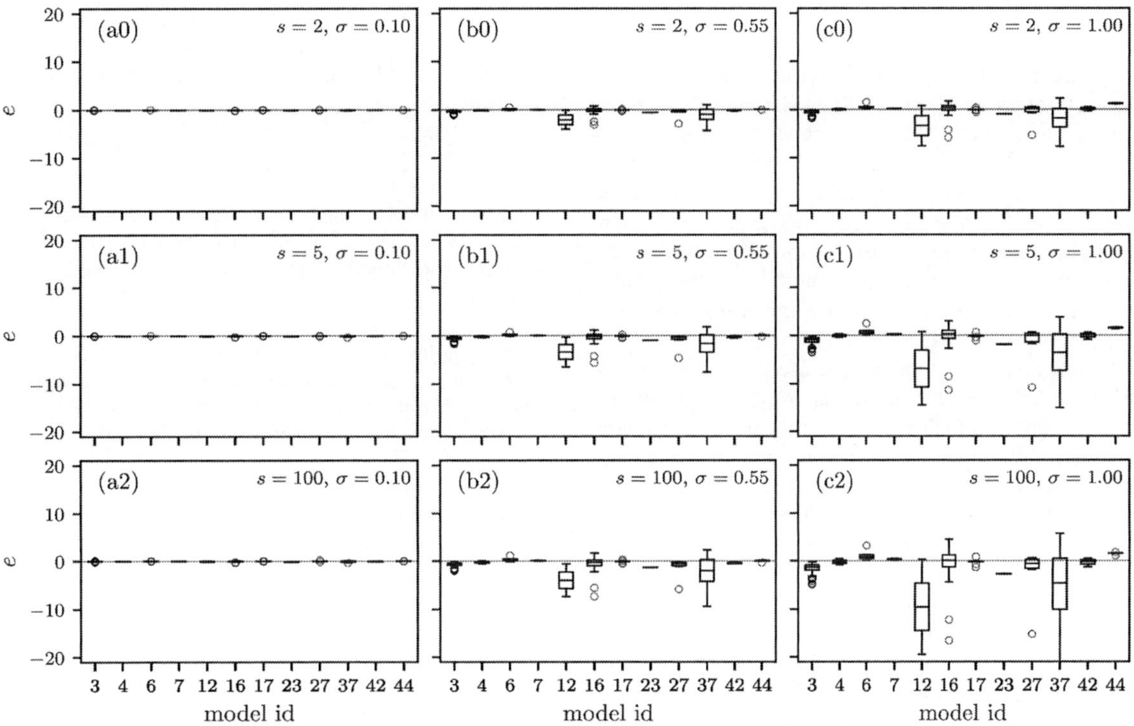

Figure 3. Aggregation error potentials calculated with Eq. (2) for each function (f) of the models described in Table 1.

ground water exposure is ignored or negligible (the case of most models), the bias around low anchor depths would be smaller than what we report here. A similar, but weaker argument could be made for large x_b values which traces back to the suitability of a normal distribution for representing exposure depths. This important issue was not investigated as part of our first order approximation.

Figure 2 shows the error in relative loss as a function of x_b for two example models from Table 1, each containing a set of damage functions. These models were selected to demonstrate behaviour present in all models. This shows that the sensitivity to aggregation errors varies by model or damage function (f). This is intuitive considering the diverse function shapes reported by Gerl et al. (2016). Looking at the general increase in absolute error magnitude from top left to bottom right of Fig. 2 suggests a positive relation with variance in depth values (σ^2) and the level of aggregation (s). This aligns with our understanding of Jensen's inequality if we consider that larger variances (σ^2) and aggregation both lead to a more severe compression of tail values about the mean. Figure 2 also shows that errors can be significant, approaching 40 % relative loss for the most extreme case considered here. In contrast, when variance in depth values (σ^2) is low, as in flat floodplains, the majority of functions have relative loss errors less than 5 % for the most extreme level of aggregation considered here ($s = 100$).

Examining the relation of error to x_b, Fig. 2 suggests generally negative errors for small depths ($x_b < 1.0$) and positive errors for larger depths ($x_b > 1.0$). This can be explained by the treatment of out-of-range depth values (x) and the clock-wise rotation shown in Fig. 1. The implications of this depth-dependent error could be significant for flood risk analysis aimed at informing flood protection investment decisions where a broad range of flood event magnitudes are assumed to have roughly similar errors (IWR and USACE, 2017). However, the significance of these findings is limited by our first order approach for generating synthetic depths from a normal distribution, which may not represent real exposure depths well.

To provide a simple metric for the 344 functions covered in Table 1, Eq. (2) was used to compute the total area of each error series (similar to what is shown in Eq. 2). The resulting aggregation error potentials are shown in Fig. 3. This shows that the sensitivity to aggregation errors varies widely between models and even between damage functions (f) within the same model. This is intuitive considering the diverse function shapes and the harmonization of variables necessary to directly compare the different models. Further, Fig. 3 shows the majority of models have mean error (e) values below zero, suggesting an overall negative bias across the full depth domain.

4 Conclusions

To better understand the potential and magnitude of errors introduced through averaging of flood risk models, a novel first order evaluation of 344 flood damage functions was performed using synthetically generated depth data. While the character and magnitude of aggregation errors will depend on the specifics of a given flood risk model, the general approach applied here demonstrates that low-depth floods tend to have negative errors while high-depth floods have positive errors. Further, we demonstrate that overall, error tends to be negative for the 344 damage functions considered.

The findings reported here provide useful information for flood risk modellers evaluating the appropriateness and extent of aggregation to include in their models. For example, in areas with high depth variance where models with large aggregation error potential (e on Fig. 3) are to be applied, aggregation should be minimized. While this can be difficult, minimizing aggregation can be achieved by constructing damage models where assets (e.g., buildings) are treated as individual elements within the model (rather than aggregated elements). Finally, this work demonstrates the potential severity of aggregation errors and how poorly these are understood, and therefore the need for further study.

Future work should evaluate the performance of the normal distribution applied here to synthetically generate depths. Also, a more generalizeable measure of curvature (e.g., local derivative) could be explored to more clearly classify and communicate the aggregation error potential of different functions. By extending the first order approximations developed here, the flood risk model domain could be segregated into areas with more or less sensitivity to aggregation errors. In this way, the accuracy of large-scale flood risk models could be improved without drastically increasing the computational requirements.

Code availability. Python scripts are provided here under the MIT license: https://doi.org/10.5281/zenodo.10810421 (Bryant, 2024).

Data availability. Function library is provided in Gerl et al. (2016).

Author contributions. SB prepared the manuscript, developed the concept, performed the analysis and computation. BM and HK reviewed the manuscript and supervised the work.

Competing interests. The contact author has declared that none of the authors has any competing interests.

Disclaimer. Publisher's note: Copernicus Publications remains neutral with regard to jurisdictional claims made in the text, published maps, institutional affiliations, or any other geographical representation in this paper. While Copernicus Publications makes every effort to include appropriate place names, the final responsibility lies with the authors.

Special issue statement. This article is part of the special issue "ICFM9 – River Basin Disaster Resilience and Sustainability by All". It is a result of The 9th International Conference on Flood Management, Tsukuba, Japan, 18–22 February 2023.

Acknowledgements. The authors thank Kai Schröter for participating in early discussions and helping with the damage function database. We would also like to thank the ICFM9 organizers for both the conference and funding this publication.

Financial support. The research presented in this article was conducted within the research training group "Natural Hazards and Risks in a Changing World" (NatRiskChange) funded by the Deutsche Forschungsgemeinschaft (DFG; grant no. GRK 2043/2).

Review statement. This paper was edited by Mohamed Rasmy and reviewed by two anonymous referees.

References

Bryant, S.: cefect/2210_AggFSyn: 2024-03-12: PIAHS publication, Zenodo [code], https://doi.org/10.5281/zenodo.10810421, 2024.

de Moel, H. and Aerts, J. C. J. H.: Effect of uncertainty in land use, damage models and inundation depth on flood damage estimates, Nat. Hazards, 58, 407–425, https://doi.org/10.1007/s11069-010-9675-6, 2011.

Denny, M.: The fallacy of the average: on the ubiquity, utility and continuing novelty of Jensen's inequality, J. Experiment. Biol., 220, 139–146, https://doi.org/10.1242/jeb.140368, 2017.

Gerl, T., Kreibich, H., Franco, G., Marechal, D., and Schröter, K.: A Review of Flood Loss Models as Basis for Harmonization and Benchmarking, PloS one, 11, e0159791, https://doi.org/10.1371/journal.pone.0159791, 2016.

Hall, J. W., Sayers, P. B., and Dawson, R. J.: National-scale Assessment of Current and Future Flood Risk in England and Wales, Nat. Hazards, 36, 147–164, https://doi.org/10.1007/s11069-004-4546-7, 2005.

IWR and USACE: Principles of Risk Analysis for Water Resources, Tech. rep., IWR, USACE, 298 pp., https://hdl.handle.net/11681/44744 (last access: 24 March 2024), 2017.

Jensen, J. L. W. V.: Sur les fonctions convexes et les inégalités entre les valeurs moyennes, Acta mathematica, 30, 175–193, publisher: Springer, 1906.

Jongman, B., Kreibich, H., Apel, H., Barredo, J. I., Bates, P. D., Feyen, L., Gericke, A., Neal, J., Aerts, J. C. J. H., and Ward, P. J.: Comparative flood damage model assessment: towards a European approach, Nat. Hazards Earth Syst. Sci., 12, 3733–3752, https://doi.org/10.5194/nhess-12-3733-2012, 2012.

McGrath, H., Stefanakis, E., and Nastev, M.: Sensitivity analysis of flood damage estimates: A case study in Fredericton, New Brunswick, Int. J. Disast. Risk Re., 14, 379–387, https://doi.org/10.1016/j.ijdrr.2015.09.003, 2015.

NumPy Developers: numpy.trapz – NumPy v1.26 Manual, https://numpy.org/doc/stable/reference/generated/numpy.trapz.html (last access: 12 March 2024), 2022.

Pollack, A. B., Sue Wing, I., and Nolte, C.: Aggregation bias and its drivers in large-scale flood loss estimation: A Massachusetts case study, J. Flood Risk Manage., 15, 4, https://doi.org/10.1111/jfr3.12851, 2022.

Sairam, N., Brill, F., Sieg, T., Farrag, M., Kellermann, P., Nguyen, V. D., Lüdtke, S., Merz, B., Schröter, K., Vorogushyn, S., and Kreibich, H.: Process-Based Flood Risk Assessment for Germany, Earth's Future, Wiley Online Library, 9, https://doi.org/10.1029/2021EF002259, 2021.

Ward, P. J., Jongman, B., Salamon, P., Simpson, A., Bates, P., De Groeve, T., Muis, S., de Perez, E. C., Rudari, R., Trigg, M. A., and Winsemius, H. C.: Usefulness and limitations of global flood risk models, Nat. Clim. Change, 5, 712–715, https://doi.org/10.1038/nclimate2742, 2015.

White, G. F.: Human Adjustment to Floods. A Geographical Approach to the Flood Problem in the United States, Ph.D. thesis, The University of Chicago, Chicago, 238 pp., 1945.

Proc. IAHS, 386, 189–195, 2024
https://doi.org/10.5194/piahs-386-189-2024

ICFM9 – River Basin Disaster Resilience and Sustainability by All

Designing and Developing Dynamic Decision Support Information for Disaster Response

Makoto Hanashima, Hiroaki Sano, and Yuichiro Usuda

Research Dept. for Disaster Information, National Research Institute for Earth Science and Disaster Resilience (NIED), Tsukuba-city, Ibaraki-pref., 305-0006, Japan

Correspondence: Makoto Hanashima (mhana@bosai.go.jp)

Received: 30 July 2023 – Revised: 28 November 2023 – Accepted: 24 February 2024 – Published: 19 April 2024

Abstract. Recently, Japan has been struck by extremely heavy rains and serious floods. To plan for disaster response, it is needed to consider various information such as precipitation, water level of rivers, weather forecasts, and so on. However, it requires high skill to integrate this information to utilize. Our study aims to provide the decision support information for both the national and the local governments by dynamic risk analysis to enforce disaster resilience. To demonstrate the effectiveness of disaster dynamics analysis, we have developed the "Dynamic Decision Support System for Disaster Response (DDS4D)". DDS4D synthesizes natural observation data, social observation data, and geospatial data to provide decision support information that fits decision maker's the situational awareness in real time. Verifying several information products generated by DDS4D in the actual flood in Japan, we confirmed that they could support decision making of government.

1 Introduction

The information sharing capability is the highest priority feature for the disaster information system. Even if the system has excellent functions, without information sharing capability, it may not perform properly in a serious disaster situation (UN-GGIM, 2017). To address this issue, we, National Research Institute for Earth Science and Disaster Resilience (NIED), have developed the "Shared Information Platform for Disaster Management (SIP4D)" to establish the inter-organizational common situational awareness under disaster situations (Hanashima et al., 2017). Previous studies have pointed out the importance of shared situational awareness in the coordination of the disaster response agencies (Bando and Machida, 2015; Kondo et al., 2019). Based on an extensive review of previous research, Sophronides et al. (2017) noted that shared situational awareness (SSA) is essential to prevent potential adverse effects and implement emergency response without delay in emergencies in which various organizations operate toward different goals. They also note that data acquisition from multiple sources and sharing information through network-centric systems contributes to the creation of a common operational picture which can support all the responding units to have the same understanding and awareness (shared situational awareness) of information and emergency status when conducting operations. The design concept of SP4D (Usuda et al., 2019) is to realize the above points made by Sophronides et al. as an actual system. The effectiveness of information sharing through SIP4D in establishing shared situational awareness in disaster response has been demonstrated in the actual disaster response (Ise et al., 2017).

Through our experience with SIP4D in several major disasters, we have found that shared situational awareness does not always activate to effective disaster response. In emergency situations, it is often difficult for the decision maker to process shared information, resulting in situations where information is shared but not fully utilized (Ise et al., 2022). This issue implies the need for a mechanism to assist decision makers, i.e., a decision support system, as well as shared situational awareness (Valachamy et al., 2022).

In a comprehensive review of decision support systems, Akter and Wamba (2019) state that in disaster response, a real-time data-driven analytics platform needs to be developed to prioritize urgent issues, prevent subsequent disasters, coordinate support organizations, and dispatch first responders. They also state that the major challenge is to develop a data integration protocol that ensures proper data collection, modelling, and notification systems.

Ka Hong et al. (2020) point out in their review that most current Disaster Management Decision Support Systems (DMDSS) focus on resource management as a method of managing tasks. They also state that, therefore, future DMDSS must include the study of task complexity to increase its reliability and accuracy in advising the task performer which task should be completed first. To address these issues, Zhang et al. (2019) have developed a system that automates decision support by weighting tasks using multi-criteria decision analysis models and GIS. This research is innovative in that it focuses on supporting group decision making and collaborative problem solving by combining various social vulnerability indices and social media data, rather than technical improvements in task management.

However, it is expected that even sophisticated decision support methods will not be fully effective during the imminent and initial phases of a disaster, because it is difficult to obtain sufficient information about the disaster situation. In addition, users often lack the skills to interpret the decision support information provided by the system and reflect it in their actions. This is not a problem of system functionality, but it is an issue that must be considered in the concept of decision support.

To address these issues, a technology-driven approach alone is not sufficient. This is because a technology-driven approach tends to focus on specific technologies and tools for handling information and may neglect to consider what is critical to information for decision-making (Neuvel et al., 2012; Usuda, 2021).

Through discussions with practitioners in charge of disaster management at the Cabinet Office (CAO), we have identified the following three simple requirements for information content for decision support.

a. Bringing information closer to the interests of decision makers.

b. Match the granularity of information to the decision maker's point of view.

c. Responding immediately to issues that arise in the field of disaster response.

We believed that by implementing a function in SIP4D to generate information that meets these requirements, the shared information should truly be useful for decision making in disaster response. To demonstrate this concept, we developed DDS4D as a proof-of-concept model and conducted a demonstration experiment using actual disaster data.

2 Fundamental design of DDS4D

Currently, information content that meets the previously mentioned requirements must be created manually. In the event of a catastrophe such as the Nankai Trough Earthquake, such a response will be impossible. Therefore, DDS4D must be able to generate the required information content without human intervention and on a regular basis.

DDS4D was designed by mapping the above three requirements to the following practical use cases.

To satisfy requirement "a", it is necessary to generate information that integrates disaster hazard data and the extent of expected damage.

For example, heavy rainfall as a hazard is provided by SIP4D in the form of grid data, but it is difficult to determine where and how much damage will occur from this data alone. In this case, the population of residents and number of buildings in the grid where heavy rainfall is occurring should be analysed together with rainfall data to provide information on estimated inundation damage.

Furthermore, if the estimated information is provided as grid data, The decision makers would have to replace it with the administrative areas. In this case, the grid data should be aggregated and provided by the administrative districts. This is the optimization of information granularity to satisfy requirement "b".

Finally, disaster response may require information that cannot be anticipated in advance. The 2019 Boso Peninsula typhoon caused massive power outages throughout Chiba Prefecture, requiring the identification and mapping of elderly welfare facilities that did not have their own power generation systems, but this type of information is not typically used in disaster response. To support decision making, such ad hoc information needs to be provided in a timely manner. This corresponds to requirement "c".

3 System overview of DDS4D

3.1 Functional requirements for decision making

In designing the DDS4D proof-of-concept model, we implemented the following three features to be able to demonstrate the previously mentioned requirements.

The first is "Foresight on potential municipalities to support". For the demonstration, DDS4D estimates damage using forecast data and estimate municipalities that should be supported. This analysis will be useful for quantitative decisions such as the number of dispatching supports. We show the example of this demonstration, supporting decisions based on Disaster Relief Act application predictions.

The second is "Nationwide, uniform, comprehensive and quantitative monitoring of abnormalities". For the demonstration, DDS4D continuously and automatically detects municipalities where evacuation instructions have not been issued even though there is a high risk of disaster. It will prevent omissions of evacuation instructions. We show the example of this demonstration, preventing delay of issuing the evacuation instructions.

The third is "Automatic detection of municipalities that will require support changes". For the demonstration, DDS4D detects that even if the estimated scale of damage is the same, there is a large difference in the response of basic local governments, such as the investigation of damage certification. This analysis will be utilized to detect the need to dispatch specialist staff, etc.

3.2 Architecture and functions of DDS4D

DDS4D is constructed from three sub systems, "DDS-DB", "DDS-SY" and "DDS-VI". The three subsystems operate autonomously and mutually collaborate with each other, triggered by data updates and user demand (see Fig. 1).

Disaster Dynamics Spatiotemporal Database, DDS-DB, is a large-size spatiotemporal database which records dynamic disaster data as time series geospatial data.

To be able to store all kinds of disaster dynamic data aggregated by SIP4D as spatiotemporal data with low latency at each update timing, DDS-DB has high-speed conversion capability for vector-type geospatial data between raster data. DDS-DB automates spatial processing between regular geospatial data and store the processing results in time series. It also has the ability to automatically generate the Representational State Transfer Application Programming Interface (REST API) for external systems to retrieve data from DDS-DB. Moreover, DDS-DB archives automatically the original data files over the medium to long term and easily restores the original data.

Disaster Dynamics Synthesizer, DDS-SY, is the agile data analysis infrastructure which operates the dynamic analysis based on the user-definable analysis scenario.

User can create and modify the scenarios according to the data processing flow. Once defined, the scenarios are automatically executed when the data are updated. The multiple spatial operations, the time-series operations, and the conditional judgments can be executed in parallel using the dynamic disaster data stored in DDS-DB as input. The processing results can be stored in DDS-DB as spatiotemporal datasets.

To provide an agile way to define a series of processes (analysis scenarios) arbitrarily by the user and automatically executed, it has simple Graphical User Interface (GUI) for creating, executing, and controlling analysis scenarios.

Disaster Dynamics Visualizer, DDS-VI, is the visualization sub-system for decision support that can draw maps and graphs in real time based on the information products created by DDS-SY. User can define the screen layout and presentation of information using GUI on their purpose.

To be able to display the DDS-VI, a web-based dashboard that lists dynamic disaster information for use by disaster response agencies, using dynamic disaster data according to an arbitrarily defined layout. To monitor updates to the data displayed in the DDS-VI and automatically update the contents of the display.

DDS-VI's GUI should be able to seamlessly display time-series data, and link lists, maps, and graphs to time-axis operations.

4 Demonstration of decision supporting functions

4.1 Rapid Evaluation of the Disaster Relief Act

Figure 2 shows the analytic framework for demonstration of the "Rapid evaluation of the Disaster Relief Act". This law is constituted to provide financial supports for affected municipalities and its support levels are defined based on the number of damaged households. DDS4D estimates the number of households that affected by flood and evaluates which municipalities are eligible to apply for the Disaster Relief Act. We defined the following process as the analysis scenario on DDS4D.

- Step 1: Analyze the number of buildings will be exposed heavy rainfall within the inundation hazardous areas.
- Step 2: Aggregate the number of buildings analyzed in "Step1" by municipalities.
- Step 3: Estimate the number of households will lose housings based on the result of "Step2" by applying the evaluation model.
- Step 4: Based on the result of "Step3", evaluate which clause of the Disaster Relief Act will be applied.

  If the rain fall forecast is used instead of the effective rainfall, these analytics could predict the status hours later.

  DDS4D performs this scenario every 30 min.

These information products can contribute to decision making at the national level, i.e., to make a nationwide estimate of the possible damage caused by the disaster and to support local governments; at the prefectural level, to support municipalities by making a quantitative estimate of the damage at an early stage of disaster response; and at the municipal level, to decide whether to apply for the Disaster Relief Act at an appropriate time.

Figure 3 shows actual screen images of DDS-View at the heavy rain in Miyagi prefecture, 16 July 2022.

Each three map shows actual, estimated, and predicted data for local governments which are applied the Disaster Relief Act, generated by the scenario shown in Fig. 2.

At 00:30 JST (Japan standard time) on 16 July, DDS4D evaluated that Osaki City might be suffered enough damage to be covered by the Disaster Relief Act and other municipalities will be suffered in next 6 h.

At 15:00 JST on 16 July, Miyagi prefecture applied the Disaster Relief Act to Osaki city and Matsushima town. The estimation time was 14.5 h earlier than the published time the act applied. It means that the municipalities could have more

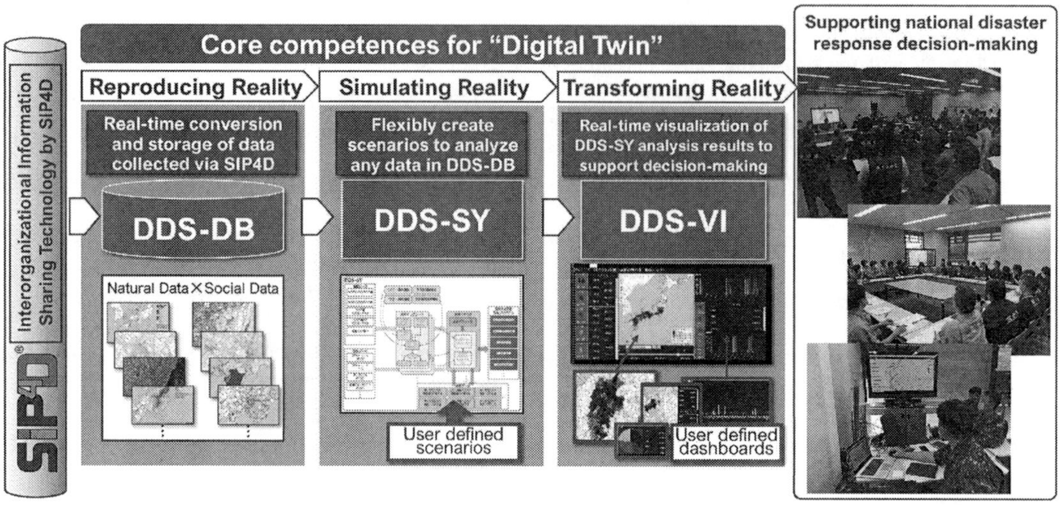

Figure 1. System design concept of "DDS4D". (All base maps are based on the GSI Maps published by Geospatial Information Authority of Japan).

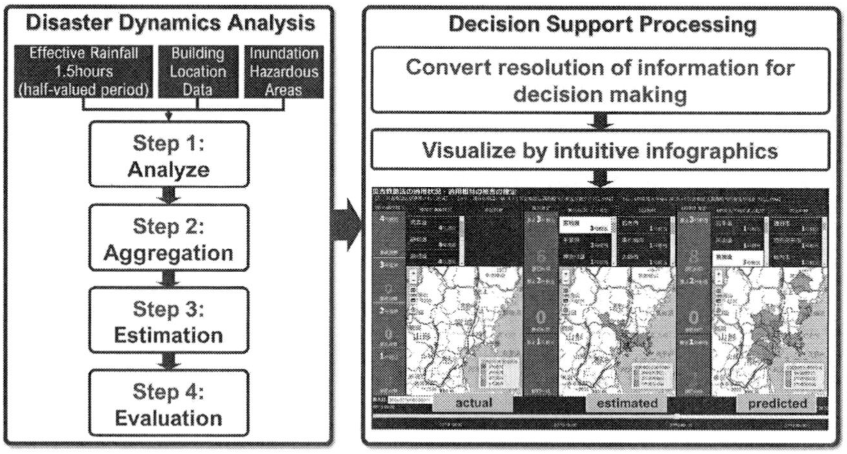

Figure 2. Analytic framework for "Rapid evaluation of the Disaster Relief Act". (All base maps are based on the GSI Maps published by Geospatial Information Authority of Japan).

time for preparation of disaster relief action when they refer the information products provided by DDS4D.

4.2 Demonstration of decision supporting functions

Figure 4 shows the analytic framework for demonstration of "Decision support for evacuation instruction".

According to CAO's evacuation information guidelines, it is recommended that municipalities issue the evacuation instruction when the sediment disaster warning information is announced. The government officers are supposed to keep track of municipalities that have been announced the sediment disaster warning information and have not yet issued the evacuation instruction and provide technical assistance as needed. However, it requires a great deal of effort to constantly check both the status of evacuation information issued by the more than 1700 municipalities nationwide and the status of sediment disaster warning information. Therefore, we have defined an analysis scenario that automatically performs this cross judgment and have DDS4D perform it every 10 min.

The decision to issue evacuation information is made by the mayor of the municipality and is not enforced by the government but detecting municipalities that are unable to decide due to a lack of information and confirming in advance whether assistance is required would be effective in preventing delays in issuing evacuation information.

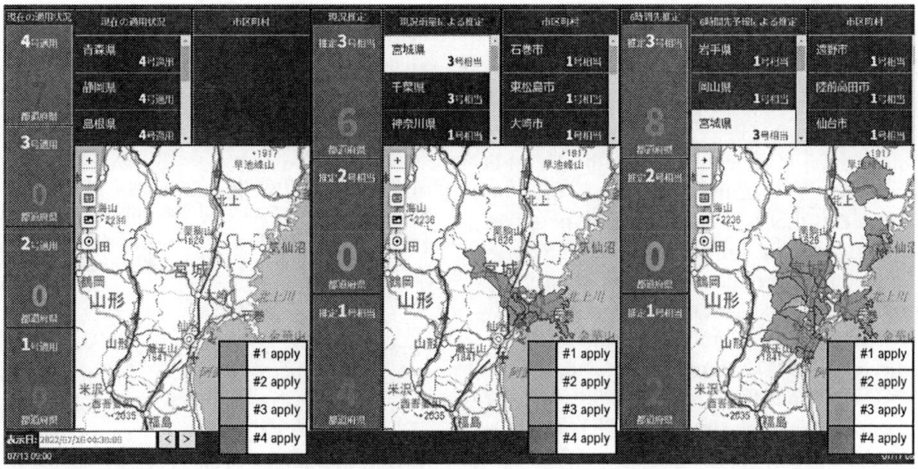

Figure 3. DDS4D shows the estimated results at 00:30 JST (Japan standard time), 16 July 2022. (The map shown on the centre) (All base maps are based on the GSI Maps published by Geospatial Information Authority of Japan).

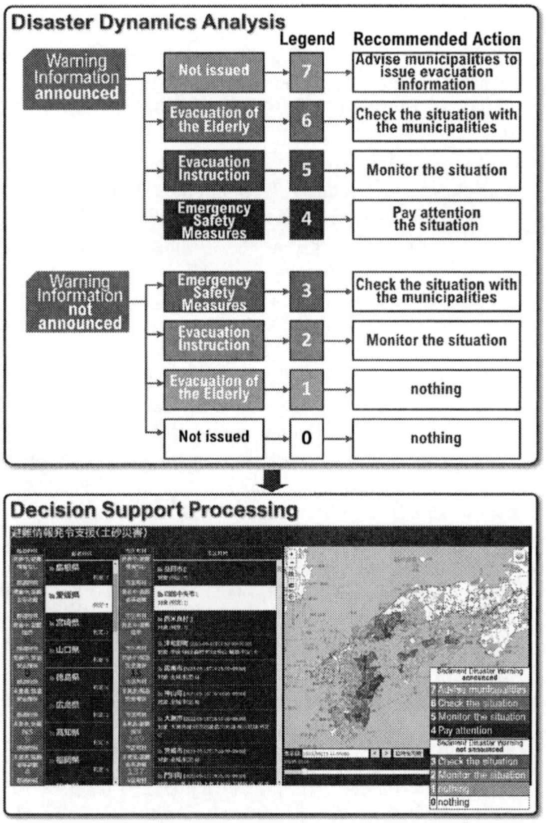

Figure 4. Analytic framework for demonstration of "Decision support for evacuation instruction". (The base map is based on the GSI Maps published by Geospatial Information Authority of Japan).

The decision to issue evacuation information is made by the mayor of the municipality and is not enforced by the government, but detecting municipalities that are unable to decide due to a lack of information and confirming in advance whether assistance is needed would be effective in preventing delays in issuing evacuation information.

When Typhoon No. 14 approached Japan in 2022, DDS4D detected delays in issuing evacuation notices in 41 municipalities. It can be inferred that some decision-making problems occurred in municipalities that remained unannounced for several hours.

By using DDS4D, government emergency management personnel can be relieved of the complicated task of comparing multiple sources of information to monitor delays in issuing evacuation orders. They can then focus on assisting communities that are hesitant to issue evacuation orders.

5 Conclusion

In this paper, we described the design concept and a practical demonstration case of DDS4D, a decision support system for decision makers in administrative disaster response. In the DDS4D decision support demonstration during the 2022 flood season, we developed about 20 kinds of dynamic analysis scenarios and verified their effectiveness by generating information content using actual disaster data. Although the cross-judgement to support the issuance of evacuation information is still being tested by government disaster management officials and specific examples are still being collected, this decision support information was developed through collaboration between the research team and government staff, an achievement that would not have been possible without the agile analysis scenario development capabilities of DDS4D.

The results of the DDS4D demonstration were presented to disaster management personnel in the three prefectures, and interviews were conducted. As a result, the three requirements established in the DDS4D design were evaluated as appropriate to support disaster response decision making. In addition, the information products generated by DDS4D were evaluated as unique and useful as decision support information. We believe that this fact demonstrates the effectiveness of the DDS4D design philosophy, i.e., the backcasting approach of defining the requirements of the information to be generated from the decision maker's point of view and reflecting them in the system architecture and demonstrates the originality of DDS4D. On the other hand, the development of DDS4D so far has focused on building an agile development platform for decision support and has not pursued the accuracy and validity of the analysis methods. Some disaster management personnel requested that the information be more persuasive and simpler to present to higher-level decision makers (e.g., prefectural governors). The future challenge is to improve the quality of analysis scenarios by introducing more accurate analysis methods and models with better explanatory power.

Code availability. Since the programs mentioned in this paper are under continuous development, there are no plans to release the source code at this time. The copyright of the program belongs to NIED.

Data availability. Because the data used in this paper include paid data, they cannot be made public due to data terms of use. For further information, please contact to the corresponding author.

Author contributions. MH was responsible for the conceptual design of DDS4D, directed the system development, and wrote the manuscript for this paper. HS was responsible for the development and demonstration of the disaster dynamics analysis methodology and contributed to the implementation of the analysis scenario presented in this paper. YU supervised the overall research and the conceptual framework of this paper.

Competing interests. The contact author has declared that none of the authors has any competing interests.

Special issue statement. This article is part of the special issue "ICFM9 – River Basin Disaster Resilience and Sustainability by All". It is a result of The 9th International Conference on Flood Management, Tsukuba, Japan, 18–22 February 2023.

Acknowledgements. The content of this paper was conducted as a part of "Enhancement of National Resilience against Natural Disasters" of the Cross-ministerial Strategic Innovation promotion Program (SIP) by the Council for Science, Technology, and Innovation.

Financial support. This research has been supported by the Council for Science, Technology and Innovation.

Review statement. This paper was edited by Mohamed Rasmy and reviewed by two anonymous referees.

References

Akter, S. and Wamba, S. F.: Big data and disaster management: a systematic review and agenda for future research, Ann. Oper. Res., 283, 939–959, https://doi.org/10.1007/s10479-017-2584-2, 2019.

Bando, M. and Machida, C.: For sharing disaster information in local government: Through inter-departmental cooperation of disaster prevention and medical affairs, J. Jpn. Soc. Civ. Eng. Ser F6 (Saf. Probl.), 71, I_205–I_212, https://doi.org/10.2208/jscejsp.71.i_205, 2015.

Hanashima, M., Sato, R., and Usuda, Y.: The Standardized Disaster-Information Products for Disaster Management: Concept and Formulation, J. Disaster Res., 12, 1015–1027, https://doi.org/10.20965/jdr.2017.p1015, 2017.

Ise, T., Takahashi, T., Sato, R., Sano, H., Isono, T., Hanashima, M., and Usuda, Y.: Consideration on Utilization of Information in Disaster Response Site – Based on Information Support for 2016 Kumamoto Earthquakes, J. Disaster Res., 12, 1028–1038, https://doi.org/10.20965/jdr.2017.p1028, 2017.

Ise, T., Hanashima, M., and Usuda, Y.: Current status and issues of information sharing in disaster response in Japan: Information linkage by "SIP4D," J. Disaster Res., 17, 976–984, https://doi.org/10.20965/jdr.2022.p0976, 2022.

Ka Hong, S. G., Ahmad, A., Raja Azman, R. F., and Faudzi, M. A.: Exploratory Review on Decision Support System for Disaster Management, in: 2020 8th International Conference on Information Technology and Multimedia (ICIMU), Selangor, Malaysia, 24–26 August 2020, 390–394, https://doi.org/10.1109/ICIMU49871.2020.9243314, 2020.

Kondo, Y., Ichikawa, M., Kondo, H., Koido, Y., and Otomo, Y.: Current disaster medicine in Japan and the change brought by information sharing, J. Disaster Res., 14, 292–302, https://doi.org/10.20965/jdr.2019.p0292, 2019.

Neuvel, J. M. M., Scholten, H. J., and van den Brink, A.: From spatial data to synchronised actions: The network-centric organisation of spatial decision support for risk and emergency management, Appl. Spat. Anal. Polic., 5, 51–72, 2012.

Sophronides, P., Papadopoulou, C.-A., Giaoutzi, M., and Scholten, H. J.: A Common Operational Picture in Support of Situational Awareness for Efficient Emergency Response Operations, Journal of Future Internet, 102, 10–35, https://doi.org/10.18488/journal.102.2017.21.10.35, 2017.

UN-GGIM: UN Strategic Framework on Geospatial Information and Services for Disasters, https://ggim.un.org/documents/UN-GGIM_Strategic_Framework_Disasters_final.pdf (last access: 14 November 2022), 2017.

Usuda, Y.: Forefront of disaster information sharing systems: Supporting disaster responses using SIP4D, IEICE Commun. Soc. Mag., 15, 192–199, https://doi.org/10.1587/bplus.15.192, 2021.

Usuda, Y., Matsui, T., Deguchi, H., Hori, T., and Suzuki, S.: The Shared Information Platform for Disaster Management – The Research and Development Regarding Technologies for Utilization of Disaster Information, J. Disaster Res., 14, 279–291, 2019.

Valachamy, M., Sahibuddin, S., Ahmad, N. A., and Bakar, N. A. A.: Critical success factors for geospatial data sharing in disaster management, IOP Conference Series: Earth and Environmental Science, 1064, 012038, https://doi.org/10.1088/1755-1315/1064/1/012038, 2022.

Zhang, Z., Hu, H., Yin, D., Kashem, S., Li, R., Cai, H., Perkins, D., and Wang, S.: A cyberGIS-enabled multi-criteria spatial decision support system: A case study on flood emergency management, Int. J. Digit. Earth, 12, 1364–1381, https://doi.org/10.1080/17538947.2018.1543363, 2019.

Proc. IAHS, 386, 197–202, 2024
https://doi.org/10.5194/piahs-386-197-2024

ICFM9 – River Basin Disaster Resilience and Sustainability by All

An approach to assess sediment disaster risk using soil moisture deficit: A case study about sediment disasters occurred during 2010–2011 in Niigata prefecture, Japan

Minjiao Lu[1], **Takahiro Ogura**[1], **Thandar Tun Zin**[1], **Shigeaki Oka**[2], **Shun-ichi Onitsuka**[2], **Naoaki Kawamura**[2], **Takahiro Yamamoto**[3], and **Koichi Komiyama**[3]

[1]Department of Civil and Environmental Systems, Nagaoka University of Technology, Nagaoka, 940-0041, Japan
[2]Tokyo Electric Power Company Holdings, Incorporated, Tokyo, 100-8560, Japan
[3]Department of Civil Engineering, National Institute of Technology, Nagaoka College, Nagaoka, 940-8532, Japan

Correspondence: Minjiao Lu (lu@vos.nagaokaut.ac.jp)

Received: 21 July 2023 – Revised: 14 November 2023 – Accepted: 24 February 2024 – Published: 19 April 2024

Abstract. Most sediment disasters are caused by surface failures, mostly induced by rainfall, and many of them cause extensive damages. Assessment of the risk of such disasters is crucial for the safety and security of society. This is becoming increasingly important in the era of global warming when torrential rains and localized heavy rains are on the increase. Instead of the index such as accumulated rainfall amount, we attempted to make a risk assessment using the soil moisture deficit, i.e., the remained capacity of the soil column to accept rainwater. First, a numerical model was developed to calculate soil moisture deficit in any or all tertiary meshes with existing routine data. The model was then applied to 446 tertiary meshes in Niigata Prefecture within which failures occurred during a period from 2010 to 2011. The model showed that the soil moisture deficit near the failure occurrence became smaller, i.e., the soil column was closer to complete saturation. This suggests that soil moisture deficit could be a good indicator of the risk. Furthermore, a probability distribution model was constructed from the minimum soil moisture deficit data near the failure occurrence. With this probability model, the threshold of soil moisture deficit to put an area on alert, corresponding to the disaster coverage ratio, can be easily calculated. The possibility of early warning is also demonstrated by combining the soil moisture deficit at a certain time with a short-time rainfall forecast for several hours in the future.

1 Introduction

Most sediment disasters are caused by surface failures, which are mostly rainfall-related phenomena except for large external forces such as earthquakes. Many of these phenomena can cause extensive damage. Assessing the risk of occurrence of such phenomena is extremely important for the safety and security of society.

The ability to predict when, where, how large, and under what conditions (in this case, rainfall, and soil) a sediment disaster will occur is the goal that many researchers have pursued. Many studies have been conducted based on the mechanism of land failure occurrence (e.g. Ono and Kazama, 2013; Danjo, 2017). On the other hand, statistical studies have been intensively conducted to investigate the relationship between factors related to sediment disasters, such as topographic factors, soil properties, and accumulated rainfall (e. g. Hayashi and Yamada, 2017).

Recently, there has been a growing trend to use the amount of water amount in the soil instead of accumulated rainfall (Okada et al., 2001). Okada et al. (2001) defined the amount of water amount in a tank model as the soil water index (SWI) and proposed a method to estimate the degree of danger by comparing it with the disaster history of the past 10 years. In this method, the entire country is divided into evenly spaced grid cells (tertiary meshes, about 1×1 km), and the three-

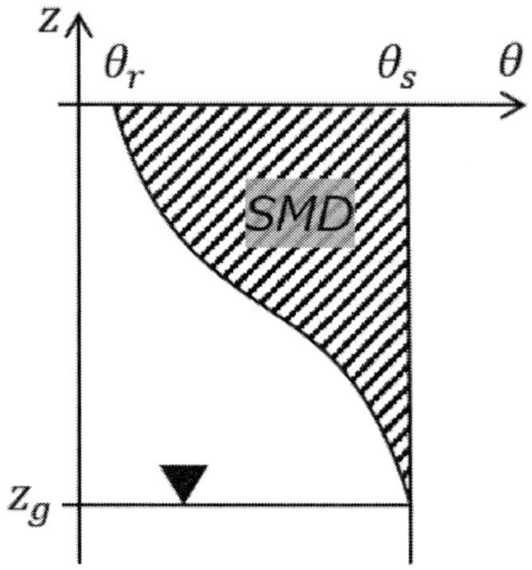

Figure 1. Definition of soil moisture deficit.

tank tank models (Ishihara and Kobatake, 1979) with fixed parameters are uniformly allocated for all grid cells. However, the creation of the pre-determined critical lines to judge the danger is technically quite difficult and requires advanced knowledge and skills in data collection and fine-tuning.

In this study, instead of accumulated rainfall or similar indexes, we attempt to evaluate the risk of sediment disaster using soil moisture deficit (hereafter referred to as SMD), i.e., the surplus capacity of the soil column to accept rainwater. First, we will develop a numerical model to calculate the SMD and make preparation for the model operation. Then, the model was applied to 446 sediment disasters that occurred in Niigata Prefecture over two years from 2010 to 2011 for each tertiary mesh within which the disaster occurred, and the dynamics of the SMD and its validity as an indicator of sediment disasters were confirmed. The model is then applied to the assessment and prediction of sediment disaster risk.

2 SMD and its modeling

As shown in Fig. 1, the soil moisture deficit for a semi-infinite vertical soil column, SMD, is the moisture deficit up to whole column saturation and is expressed as

$$\mathrm{SMD} = \int_{z_\mathrm{g}}^{0} (\theta_\mathrm{s} - \theta)\,\mathrm{d}z \tag{1}$$

where z is coordinate with the vertically upward direction as positive, z_g is the coordinate of the groundwater level, θ is the volumetric water content at z, and θ_s is the saturated volumetric water content, which is constant for homogeneous soils.

The water balance of the soil column can be expressed by

$$\frac{\mathrm{dSMD}}{\mathrm{d}t} = -P + E + R \tag{2}$$

where P, E and R are rainfall, evaporation, and runoff amount per unit of time, respectively, and t is time. Parametrization is required to estimate E and R. As can be seen from Eq. (2), this study does not consider snow-related processes. Attention should be paid to sediment disasters during the snow season in snowy regions.

Since forests, which cover about 70 % of Japan's land area, have very high permeability (Kuraji and Hoyano, 2004), only saturated surface runoff is considered in this study as follows:

$$R = \begin{cases} P - E & (\mathrm{SMD} = 0) \\ 0 & (\mathrm{SMD} > 0) \end{cases} \tag{3}$$

For evaporation, we use the evaporation efficiency, β

$$E = \beta E_\mathrm{p} \tag{4}$$

where E_p is the potential evaporation which is calculated by using the Makkink equation (Nagai, 1993) as described below. There are many studies on evaporation efficiency (e.g. Komatsu, 2003). Many of them use in some form the average saturation of a constant thickness of the surface layer (e.g. Merlin et al., 2016). Ogawa (2022) shows that in relatively quiescent environments, surface saturation can be used.

To obtain surface saturation, this study assumes that water infiltrated into the soil layer redistributes to deeper layers and quickly reaches a hydrostatic equilibrium where gravity and capillary forces balance each other. Hydrostatic equilibrium results in

$$\Psi + z = z_\mathrm{g} \tag{5}$$

where Ψ is the capillary potential. A moisture retention curve model will determine the soil water content profile and then the SMD in Eq. (1) for a z_g. In this study, the following Brooks-Corey model is used due to its simplicity

$$S_e = \frac{\theta - \theta_\mathrm{r}}{\theta_\mathrm{s} - \theta_\mathrm{r}} = \begin{cases} \left(\frac{\Psi_\mathrm{b}}{\Psi}\right)^{\lambda} & (\Psi < \Psi_\mathrm{b}) \\ 1 & (\Psi \geq \Psi_\mathrm{b}) \end{cases} \tag{6}$$

where S_e is the relative soil saturation, θ_r is the residual volumetric water content, Ψ_b is the bubbling pressure of the soil, and λ is a parameter. The relative soil saturation at the ground surface, $S_{e,0}$, can be derived from $\Psi = z_\mathrm{g}$ as

$$\beta = S_{e,0} = \begin{cases} \left(\frac{\Psi_\mathrm{b}}{z_\mathrm{g}}\right)^{\lambda} & (z_\mathrm{g} < \Psi_\mathrm{b}) \\ 1 & (z_\mathrm{g} \geq \Psi_\mathrm{b}) \end{cases} \tag{7}$$

Also, the state variable SMD can be calculated by

$$\mathrm{SMD}^* = \begin{cases} z_\mathrm{g}^* - 1 - \frac{1}{1-\lambda}\left[\left(z_\mathrm{g}^*\right)^{-\lambda+1} - 1\right] & \left(z_\mathrm{g}^* > 1\right) \\ 0 & \left(z_\mathrm{g}^* \leq 1\right) \end{cases} \tag{8}$$

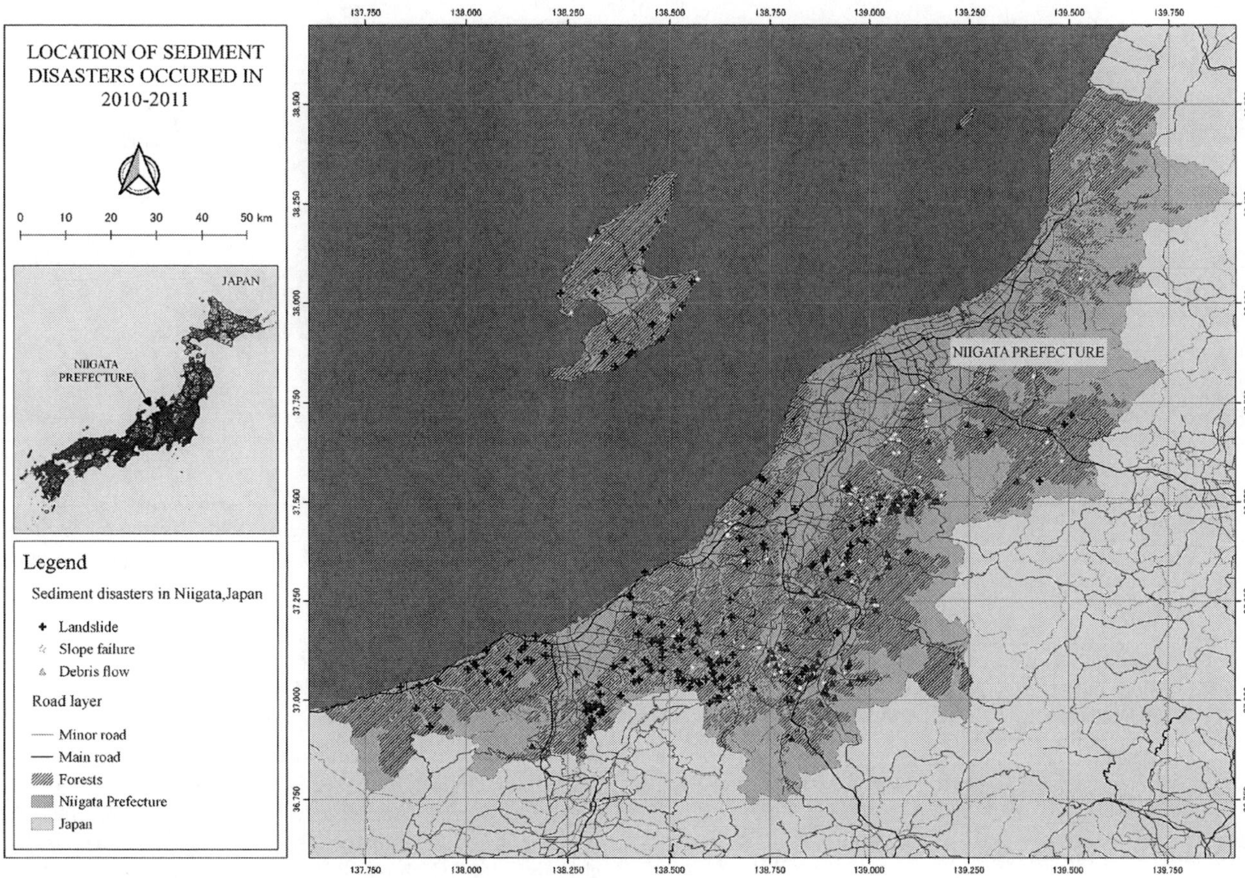

Figure 2. Sediment disasters occurring in Niigata prefecture in 2010–2011.

where

$$z_g^* = \frac{z_g}{\Psi_b}$$

$$\mathrm{SMD}^* = \frac{\mathrm{SMD}}{-\Psi_b(\theta_s - \theta_r)}$$

are the dimensionless groundwater level and the dimensionless SMD. As can be seen from Eq. (8), it is easy to obtain SMD from z_g. However, obtaining z_g from SMD requires the solution of nonlinear equations. A numerical solution is instead used to solve the following differential equation

$$\frac{\mathrm{d}z_g^*}{\mathrm{dSMD}^*} = \frac{1}{1 - \left(z_g^*\right)^{-\lambda}} \tag{9}$$

To avoid singularity at $z_g^* = 1$ in Eq. (9), the following approximation

$$z_g^* = 1 + \sqrt{\frac{2\mathrm{SMD}^*}{\lambda}} \tag{10}$$

is directly used when $\mathrm{SMD}^* \leq 0.0002$.

3 Research area and data

3.1 Research area

This study focuses on sediment disasters reported in Niigata Prefecture in the two years from 2010 to 2011, with 100 cases reported in 2010 and 346 cases in 2011. The spatial distribution of these data is shown in Fig. 2.

Sediment disasters are classified into three major categories: landslides, slope failures, and debris flows. In Niigata Prefecture, upon receiving a report, officials conduct a field survey and then classify the disaster and prepare a document. As shown in Table 1, a total of 446 sediment disasters (185 landslides, 127 slope failures, and 134 debris flows) were reported in the two years from 2010 to 2011. The exact time of occurrence cannot be determined in many cases. In such cases, we assume that the failures occurred during the 24 h period before the reporting time and check the analysis results during that period.

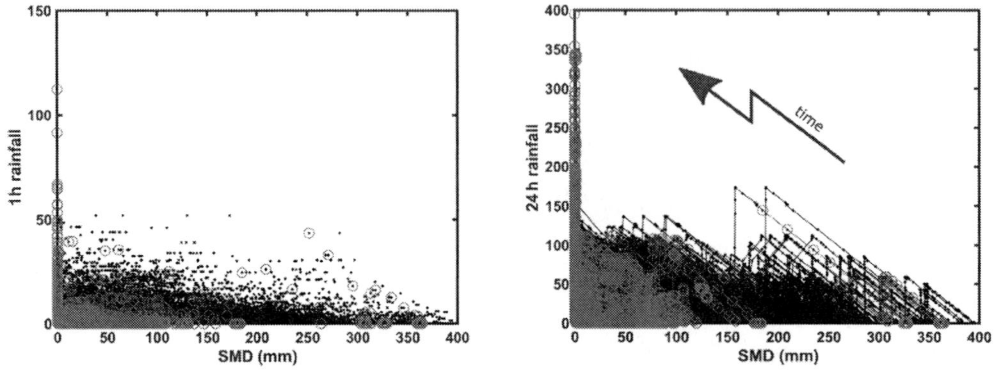

Figure 3. Relation between calculated SMD and 1 and 24 h cumulative rainfall.

Table 1. Number of sediment disasters in Niigata Prefecture during 2010–2011.

Sediment disaster	Season		Year
	Summer	Winter	
Landslide	108	77	185
Slope failure	110	17	127
Debris flow	127	7	134
Total	345	101	446

3.2 Soil data preparation

To run the above model, the physical properties of the soil, such as the parameters of the Brooks-Corey model, are required. In this study, the model is based on the soil code (hereinafter referred to as the KS code) in Land Classification Grid Data (LCGD) in Digital National Land Information (DNLI) and the Comprehensive Soil Classification Code (hereinafter referred to as the NK code) published by the National Agriculture and Food Research Organization (NARO) of Japan. From NK code and its associated volumetric water contents corresponding to pF0.0, pF1.5, pF2.7, and pF4.2, the Brooks-Corey model parameters (θ_s, θ_r, Ψ_b, and λ) are estimated by Nakayama's method (2020). A nationwide database of these parameters is made based on the relationship between NK and KS codes (Lu et al., 2023).

3.3 Meteorological data

We chose to use AMeDAS (Automated Meteorological Data Acquisition System) data from among the available rainfall data because its high observation accuracy, nationwide coverage, and long observation period contribute to stable model calculations and secure cases of sediment disasters.

From sunshine duration data, downward shortwave radiation can be calculated from time and location of AMeDAS stations (Shimazaki et al., 1995). Then the potential evaporation is calculated from the downward shortwave radiation by using Makkink's equation (Nagai, 1993). Its two parameters, a_m and b_m which are constants varying from region to region are set to $a_m = 0.95$ and $b_m = 0$ (Nagai, 1993).

4 Results and discussion

4.1 Calculation results

In this study, we implemented the SMD model on 446 grid cells of the DNLI tertiary meshes, within which the sediment disaster occurred, using the data described above. Since the initial value of soil moisture affects the results for a considerably long time (Rahman and Lu, 2015), we concatenate two years of data to its end to form four years of data and used the results of the last two years to reduce the influence of the initial value.

Figure 3 shows the relationship between the calculated SMD and the 1 and 24 h accumulated rainfall. As mentioned before, it is difficult to determine the exact time of occurrence of a sediment disaster. In this study, the 24 h before the time of reporting is referred to as the occurrence period, the rest the non-occurrence period, and the points in the occurrence period are indicated by red circles and those in the non-occurrence period by black dots. The data points of the same grid cell are connected by a line in time order to show the time development which is also shown schematically by blue lines and arrows in the figure. From this figure, it can be seen that SMD decreases with the increase of the accumulated rainfall. Many red points are concentrated at SMD = 0 at which sediment disasters are more likely to occur.

Figure 4 shows a histogram of the minimum SMD values during the occurrence period. The histogram bins are 50 mm wide, and the median value of each bin is indicated on the horizontal axis. The left shows all data and the right the data excluding SMD = 0. Similar results are also confirmed for the summer sediment disasters from May to November. Therefore subsequent analyses and discussions in this study will be conducted using year-round data.

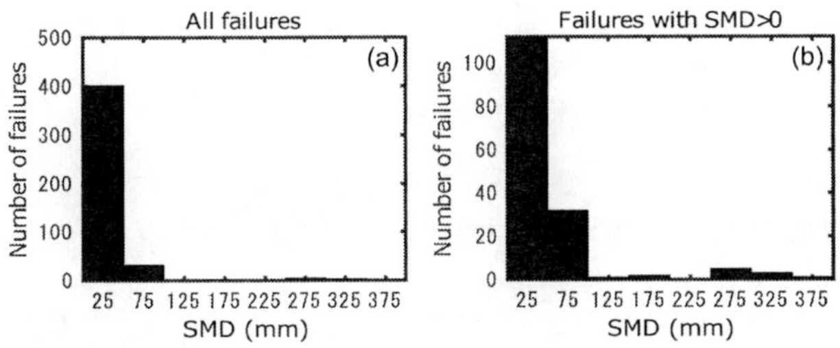

Figure 4. Histogram of SMD minimum values: **(a)** all data; **(b)** data excluding SMD = 0.

Figure 4 shows that about 90 % of sediment disasters occur at SMD ≤ 50 mm. To understand the distribution of the data other than SMD = 0, a histogram of the data excluding SMD = 0 is shown on the right panel. The occurrence of sediment disaster is mostly observed in wet conditions, namely small SMD. This shows that SMD is a good indicator of sediment disaster risk. The same trend is observed for all types of sediment disasters.

4.2 Probability distribution of SMD minima for disaster occurrence and its application

The density function of the probability distribution of the minimum SMD values within the occurrence period is proposed as follows:

$$f(\mathrm{SMD}) = p_0\delta(\mathrm{SMD}) + \frac{(1-p_0)}{\mathrm{SMD}_m}\exp\left(-\frac{\mathrm{SMD}}{\mathrm{SMD}_m}\right) \tag{11}$$

where p_0 is the probability of sediment disaster occurring at saturation, SMD_m is the average SMD of sediment disaster occurring before saturation is reached, and δ is the Dirac function. Therefore, the distribution function becomes

$$F(\mathrm{SMD}) = p_0 + (1-p_0)\left(1-\exp\left(-\frac{\mathrm{SMD}}{\mathrm{SMD}_m}\right)\right) \tag{12}$$

From this, the SMD threshold, SMD_c corresponding to the disaster capture ratio, p_c becomes

$$\mathrm{SMD}_c = \begin{cases} -\ln\left(\frac{1-p_c}{1-p_0}\right)\mathrm{SMD}_m & p_c > p_0 \\ 0 & p_c \le p_0 \end{cases} \tag{13}$$

when $p_c \le p_0$, $\mathrm{SMD}_c = 0$. If the alarm is issued when the SMD is smaller than SMD_c, the probability to capture the disaster occurrence is p_c.

Table 2 shows the distribution parameters obtained for each type. From this table, it can be seen that debris flows and landslides have similar characteristics of occurring when unsaturated, but debris flows tend to occur more frequently after being saturated. This difference may be due to the fact that debris flows need flow conditions in the river channel besides the failures on the slope. On the other hand, the occurrence probability of landslides after saturation is similar to that of slope failures, but the occurrence characteristics when unsaturated are quite different. They also occur in relatively large SMDs. All of the 10 cases on the right side of Figs. 3 and 4 (SMD > 200 mm) are slope failures. However, even if these 10 cases are excluded, SMD_m is still quite large. In the future, it may be necessary to consider different criteria for the alerting system. In the case of the disaster capture ratio, $p_c = 0.95$, SMD_c is 88.7 mm, and when SMD is below this value, we will be on alert.

Table 2. Parameters of probability distribution models for sediment disasters of all types.

	Landslide	Slope failure	Debris flow	Total
Total occurrence	185	127	134	446
Unstaturaed occurrence	86	49	23	158
p_0	0.535	0.614	0.828	0.646
SMD_m (mm)	26.8	85.9	27.7	45.3

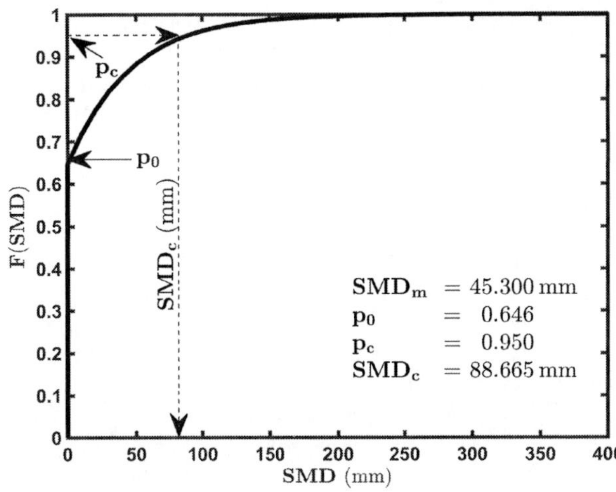

Figure 5. Derived statistical distribution function of minimum SMD and warning threshold SMD_c.

Figure 5 shows the derived statistical distribution function of minimum SMD Also the procedure to calculate warning threshold SMD_c using this distribution is illustrated.

As can be seen from Fig. 3, as accumulated rainfall increases, SMD decreases along the inverted 45° line. This means that the evaporation calculations in the model contribute to the formation of the initial SMD before rainfall, but are negligible during heavy rainfall events that induce sediment disasters. By utilizing these characteristics, it is possible to combine the current SMD with the P_{pred} of the short-time rainfall forecast for several hours in the future and use $\text{SMD} - P_{\text{pred}} < \text{SMD}_c$ to establish an early warning system.

5 Conclusions

In this study, we attempted to evaluate the risk of a sediment disaster using the SMD, i.e., the surplus capacity of the soil column to accept rainwater. First, a numerical model was developed to calculate the SMD. Datasets necessary for model operation are prepared. Then, the model was applied to 446 sediment disasters that occurred in Niigata Prefecture over two years from 2010 to 2011 for each tertiary mesh, and the SMD within the disaster occurrence period was confirmed to have decreased towards zero, namely saturation. It is suggested that the SMD can be a good indicator of the risk of sediment disaster.

A probability distribution model of the SMD was constructed based on the minimum value of SMD within the disaster occurrence period. This model can be used to easily calculate the threshold value of the SMD to be on alert, which corresponds to the disaster capture ratio. The possibility of combining the current SMD with short-time rainfall forecasts for several hours in the future to provide early warning is demonstrated.

Data availability. Data sets used in this manuscript are available upon request.

Author contributions. ML conceived the presented research. ML, TO and TTZ carried out coding, data manupilation and manuscript drafting. SOs, NK, TY and KK discussed research direction and results. All authors contributed to the final manuscript.

Competing interests. The contact author has declared that none of the authors has any competing interests.

Disclaimer. Publisher's note: Copernicus Publications remains neutral with regard to jurisdictional claims made in the text, published maps, institutional affiliations, or any other geographical representation in this paper. While Copernicus Publications makes every effort to include appropriate place names, the final responsibility lies with the authors.

Special issue statement. This article is part of the special issue "ICFM9 – River Basin Disaster Resilience and Sustainability by All". It is a result of The 9th International Conference on Flood Management, Tsukuba, Japan, 18–22 February 2023.

Acknowledgements. The authors would like to thank Department of Public Works, Niigata Prefecture, Japan for providing data and other assistance in conducting this study.

Review statement. This paper was edited by Kensuke Naito and reviewed by Dimitrios Fytanidis and one anonymous referee.

References

Danjo, T.: Slope failure mechanism and prediction method by rainfall, J. Japan Soc. Saf. Eng., 56, 463–469, https://doi.org/10.18943/safety.56.6_463, 2017.

Hayashi, S. and Yamada, T.: Setting method of the factor and the scale of heavy rainfall caused sediment-related disasters, J. Japan Soc. Nat. Disaster Sci., 36, 307–320, https://doi.org/10.24762/jndsj.36.3_307, 2017.

Ishihara, Y. and Kobatake, S.: Runoff model for flood forecasting, Bull. Disaster Prev. Res. Inst., 29, 27–43, 1979.

Komatsu, T. S.: Toward a robust phenomenological expression of evaporation efficiency for unsaturated soil surfaces, J. Appl. Meteorol., 42, 1330–1334, https://doi.org/10.1175/1520-0450(2003)042<1330:TARPEO>2.0.CO;2, 2003.

Kuraji, K. and Hoyano, H.: Green dams – forests, rivers, water cycle and disaster prevention, Tsukiji Shokan, Tokyo, 260 pp., ISBN: 978-4806713005, 2004.

Lu, M., Ogura, T., Zin, T. T., Oka, S., Onitsuka, S., Kawamura, N., Yamamoto, T., and Komiyama, K.: Sediment disaster risk assessment using soil moisture deficit, J. Japan Soc. Hydrol. Water Resour., 36, 118–128, https://doi.org/10.3178/jjshwr.36.1752, 2023.

Merlin, O., Stefan, V. G., Amazirh, A., Chanzy, A., Ceschia, E., Er-Raki, S., Gentine, P., Tallec, T., Ezzahar, J., Bircher, S., Beringer, J., and Khabba, S.: Modeling soil evaporation efficiency in a range of soil and atmospheric conditions using a meta-analysis approach, Water Resour. Res., 52, 3663–3684, https://doi.org/10.1002/2015WR018233, 2016.

Nagai, A.: Estimation of pan evaporation by Makkink equation, J. Japan Soc. Hydrol. Water Resour., 6, 238–243, https://doi.org/10.3178/jjshwr.6.3_238, 1993.

Nakayama, K.: Study on landslide risk assessment using soil moisture deficit, Nagaoka University of Technology, 136 pp., 2020.

Ogawa, K.: Relationship between evaporation efficiency and groundwater surface position in bare ground, Nagaoka University of Technology, 102 pp., 2022.

Okada, K., Makihara, Y., Shimpo, A., Nagata, K., Kunitsugu, M., and Saito, K.: Soil water index, Tenki, 47, 36–41, 2001.

Ono, K. and Kazama, S.: Assessment of rainfall-induced shallow landslides using a statistically-based and physically-based model, J. Japan Soc. Civ. Eng. Ser. B1, 69, I_925–I_930, https://doi.org/10.2208/jscejhe.69.I_925, 2013.

Rahman, M. M. and Lu, M.: Model spin-up behavior for wet and dry basins: A case study using the xinanjiang model, Water, 7, 4256–4273, https://doi.org/10.3390/w7084256, 2015.

Shimazaki, H., Lu, M., and Koike, T.: Investigation of a nation-wide real-time solar radiation calculation method based on AMeDAS sunshine duration data, in: Proceedings of the 50th Annual JSCE Conference, 200–201, 1995.

Proc. IAHS, 386, 203–208, 2024
https://doi.org/10.5194/piahs-386-203-2024

Satellite and UAV derived seasonal vegetative roughness estimation for flood analysis

Andre Araujo Fortes[1], **Masakazu Hashimoto**[2], **Keiko Udo**[1], **and Ken Ichikawa**[2,3]

[1]Department of Civil and Environmental Engineering, Tohoku University, Sendai, 980-8579, Japan
[2]International Research Institute of Disaster Science, Sendai, 980-8572, Japan
[3]Fukken Gijyutsu Consultants Co., Ltd., Sendai, 980-0012, Japan

Correspondence: Andre Araujo Fortes (andre.araujo.fortes.p6@dc.tohoku.ac.jp)

Received: 28 July 2023 – Revised: 11 September 2023 – Accepted: 24 February 2024 – Published: 19 April 2024

Abstract. One of the purposes of river management is the disaster protection of the nearby population. The effect of riparian vegetation on hydraulic resistance and conveyance capacity makes it a vital parameter for this purpose. With remote sensing techniques, vegetation information can be estimated. This paper's objective is to combine UAV and satellite imagery to obtain vegetation parameters with moderate resolution for hydraulic modeling, and to assess the seasonal effect of the vegetation on the Manning coefficient. Typhoon Hagibis was simulated with a 2D hydraulic model with a dynamic vegetative roughness estimation routine. Results demonstrate that this method achieved less error than the traditional static roughness value method of hydraulic modeling. The seasonal effect of the vegetation on the roughness was shown by a relationship between the percentage of vegetation cover and the average Manning in the stretch.

1 Background

On 12 October 2019, typhoon Hagibis hit Japan causing extreme rainfall and flooding several rivers across the territory (Kazama et al., 2021). River management is tasked to provide the safety of nearby populations from floods, thus accurate predictions of these events and mitigation measures must be made. Hydraulic models are used to provide information on flood risk and vulnerability, but many input data are needed, such as topography, flow data, and the Manning roughness coefficient, to provide accurate results (Bates, 2004).

Vegetation is the dominant factor in determining the roughness value (Ebrahimi et al., 2008), and their physical parameters are the primary determinants (Nikora et al., 2008). Studies show that plant density impacts the roughness while the vegetation is emergent (Aberle and Järvelä, 2013), and the ratio of the flow depth to the canopy height describes the roughness in the submerged state (Nikora et al., 2008). A good descriptor of plant density is the leaf area index (LAI) (Jalonen et al., 2013), which is defined as the total one-sided leaf area over ground unit area.

With vegetation mostly located in the floodplains, it is a common practice to adopt constant roughness values for channel and floodplains. This practice requires a time-consuming process of trial and error to obtain accurate results. According to Ebrahimi et al. (2008), the adoption of a dynamic Manning value varying with changes in flow and vegetation conditions can obtain better results than static Manning approaches. In addition, with seasonal changes in the vegetation parameters, variations in the hydraulic resistance also occur. Therefore, studies that combine the application of vegetation parameters on hydraulic models in real scenarios are very important.

With advancements in remote sensing techniques, obtaining vegetation physical parameters has become cheaper and less time-consuming (Fortes et al., 2022). The use of unmanned aerial vehicles (UAVs) equipped with cameras eased the identification and extraction of features such as vegetation height and their variation with the passing of seasons (van Iersel et al., 2018). Satellites have the advantage of more temporal frequency, covering a larger area, and with emission of more spectral bands, many indices can be obtained, although with less resolution than UAVs. Vegetation parameters can also be obtained from these sources (Gokool et al., 2022).

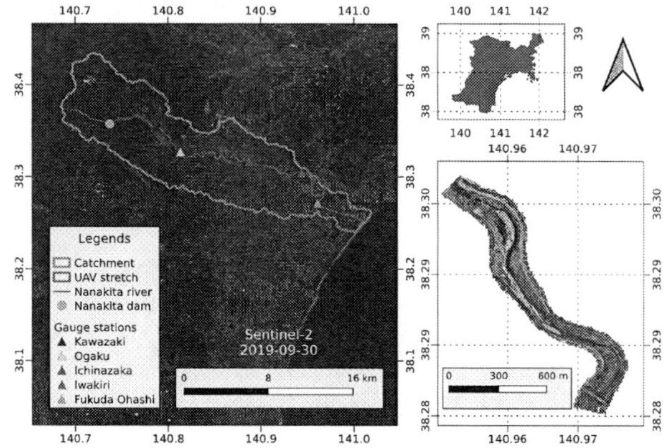

Figure 1. Study area, showing the location of the catchment and the location of the 2 km stretch, where the hydraulic model was applied.

Besides vegetation parameters, new technologies have allowed the acquisition of high-resolution spatially distributed water levels, such as the ICESat-2 satellite mission (Coppo Frias et al., 2023) and UAVs equipped with radar altimetry (Jiang et al., 2020), which can be useful in hydraulic modeling for calibration and validation.

The objective of this study is to combine UAV and satellite imagery to obtain LAI and vegetation height with moderate resolution for hydraulic modeling, with dynamic vegetative roughness calculation, and to assess the seasonal effect of the vegetation on the roughness parameter.

2 Study area

The hydraulic model was applied in a 2 km stretch of the Nanakita river, in the Miyagi prefecture of Japan. The river has 45 km of total length and, catchment area of 229 km^2. The mean annual discharge is 10 $m^3\,s^{-1}$ and the 100-year return period discharge is 1650 $m^3\,s^{-1}$ (Viet et al., 2006). According to Pilailar et al. (2003), the bed slope of Nanakita river is about 0.0016 in the upstream and 0.0003 in the downstream part of the river. Rivers in Japan are classified as class A or class B rivers, depending on their size. Class A rivers are managed by the national government and class B rivers, like Nanakita river, are managed by the local prefectures. UAV flights took place in the 2 km stretch in September 2019, and then monthly from April 2020 to March 2021. Figure 1 shows the catchment location and the location of the 2 km stretch of the river.

The vegetation in the stretch is mostly comprised of shrubs and tall grass species. The predominant population is *Pueraria montana var. Lobata* (Kudzu), *Miscanthus sinensis*, *Phragmites australis*, and *Solidago canadensis*.

During typhoon Hagibis, no overtopping of the flood control structures occurred in the 2 km stretch.

3 Methodology

To recreate the typhoon Hagibis flood event in the stretch, three 2D-hydraulic simulations were performed. One simulation using constant Manning values, from here referred to as Static Manning Simulation (SMS), and two simulations with dynamic Manning calculation. In the dynamic set, one simulation calculated the Manning value only in submerged vegetation scenario, with the method used in Fortes et al. (2022), from here referred to as Partial Dynamic Manning Simulation (PDMS), and the other calculated the Manning value for submerged and emergent vegetation scenarios, referred as Full Dynamic Manning Simulation (FDMS).

After the event recreation, the effect of seasonal variation of the vegetation on Manning was assessed by simulating the same event with vegetation conditions in each season. The UAV observation and sentinel-2 images from May 2020, November 2020, and January 2021 were used to represent spring, autumn, and winter, respectively. The recreation of the event used the observation from September 2019, representing the summer season. To perform the seasonal assessment, the FDMS model was used. The input data for the hydraulic simulation of each season was the same, differing only in the LAI and vegetation height values and distribution.

The upstream discharge used in the hydraulic models was obtained from Fortes et al. (2022), using a hydrologic model, covered in Sect. 3.1. Section 3.2 covers the hydraulic model used for the SMS and the topography used. The vegetative Manning algorithms are introduced in Sect. 3.3, and Sect. 3.4 explains the method to obtain the vegetation parameters.

3.1 Hydrologic simulation

The input discharge was obtained by running a hydrologic model, the Rainfall-Runoff Inundation (RRI) model, which is capable of simulating both rainfall-runoff and inundation phenomena (Sayama et al., 2012). It uses continuity and momentum equations as governing equations.

The input data was the catchment digital elevation model (DEM), obtained from MERIT Hydro (Yamazaki et al., 2019) with 90 m resolution, rainfall data from the radar/raingauge-analyzed (RA) precipitation (Ishizaki and Matsuyama, 2018) from 12–14 October 2019, land cover obtained from MLIT National Land Data Information (2014) with 100 m resolution and the Manning data, which was set 0.04 in river cells and in the slope cells, 0.5 for vegetation, 0.3 for urban areas and 0.04 in water bodies.

The outputs are the inundation map, water level, and discharge. The validation was obtained by comparing the simulated and observed water depth at the five water level gauge stations shown in Fig. 1.

3.2 Hydraulic model

A 2D hydraulic model was used for the event recreation in the stretch (Hashimoto et al., 2018). The model uses continuity and momentum equations as governing equations. The model's input data are the topography, the upstream discharge, and the Manning values. The outputs are the inundation and water level. The hydraulic model inputs as boundary conditions the outlet cells and their slope.

The computational domain for the hydraulic model was the 2 km UAV observed stretch of the river, shown in Fig. 1. The DEM used as topography input in all hydraulic simulations was constructed from 21 cross-sections provided by Miyagi prefecture. The final resolution of the DEM was 10 m. The Manning setting for the SMS was 0.022 for the channel and 0.038 for floodplains, as recommended by Miyagi prefecture. To distinguish channel and floodplain, a simulation using the average annual discharge of the river was performed. This method was chosen rather than making this classification from the topography because setting a height threshold to separate river and floodplain areas would not consider the slope progression of the river. Using the mean annual discharge was preferred because pre-calibrated roughness values for the river channel were available and could be used to observe where the water surface would accommodate in the stretch.

Outlet cells were selected as the boundary condition. The outlet discharge was calculated with Manning equation, and the slope was set as 0.0002, which is close to the average slope of the lower reaches of the river.

All models were validated by comparing simulated and observed water level profiles in 5 sections. The height was obtained by observation of photos taken during the event, the height of objects in the photographs was used to estimate the height difference between the top of the embankments and the watermarks.

3.3 Dynamic Manning routine

While the FDMS introduces the Manning calculation in the emergent state of the floodplain vegetation, the PDMS was performed using the algorithm developed by Fortes et al. (2022), which used only the submerged state to calculate the Manning values cellwise, blocking the water flow at the vegetated cells while in emergent state. The reason for considering using the PDMS is due to the coefficients needed by the FDMS, which are not always available.

3.3.1 Emergent vegetation state

The vegetation LAI was used for this state. The Darcy-Weisbach friction factor at each cell was calculated from Eq. (1), proposed by Järvelä (2004).

$$f = 4C_{D_x}\mathrm{LAI}\left(\frac{u}{u_x}\right)^x\left(\frac{h}{h_{\mathrm{veg}}}\right). \qquad (1)$$

where f is the friction factor, $C_{\mathrm{D}x}$ is the vegetation drag coefficient, "LAI" is the LAI value in the cell, x is a species-specific constant, u is the flow velocity, u_x is the minimum velocity used to find the x value, h is the water depth, and h_{veg} is the vegetation height. The $C_{\mathrm{D}x}$, x, and u_x were obtained from Aberle and Järvelä (2013).

The friction factor was then converted to the Manning coefficient using Eq. (2), from Box et al. (2021).

$$n = \sqrt{\left(\frac{f}{8gh^{-1/3}}\right)}. \qquad (2)$$

where g is the gravitational acceleration.

3.3.2 Submerged vegetation state

As explained in Fortes et al. (2022), for submerged state, the Manning coefficient was calculated by its relationship with the degree of submergence of the vegetation. The data was obtained from Japan Institute of country-ology and engineering (2002) and with regression analysis an exponential formula was obtained with a coefficient of determination (R^2) of 0.87, shown in Eq. (3).

$$n = 0.084 \cdot \left(h_{\mathrm{water}}\, h_{\mathrm{veg}}\right)^{-0.98} + 0.023. \qquad (3)$$

where h_{veg} is the height of the vegetation and h_{water} is the water depth.

3.4 Vegetation parameters

A k-means clustering algorithm was used in a sample of the UAV image based on the RGB values, then visual supervision was applied to obtain the vegetated cells and a multilayer Perceptron (MLP) algorithm was trained, and then used to predict the vegetated cells in the entire stretch.

The vegetation height was obtained by normalizing the digital surface model (DSM) values, which comprises both ground and above objects elevation, with the DEM values in the vegetated cells. A more detailed explanation of the method applied for the vegetation identification and height calculation is found in Fortes et al. (2022).

3.4.1 Leaf area index

The leaf area index was obtained by downscaling the MODIS LAI product from 500 to 10 m resolution using machine learning (ML), like in Gokool et al. (2022).

An MLP regressor model was trained using NDVI and EVI from MODIS as input variables and MODIS LAI as the target variable in the entire Tohoku region of Japan. To represent the summer seasons, MODIS images dating from 1 July 2020 to 30 September 2020 were averaged, then the model was trained. Afterward, Sentinel-2 NDVI and EVI with 10 m resolution from 30 September 2019 in the study area were used in the trained model to obtain the LAI values

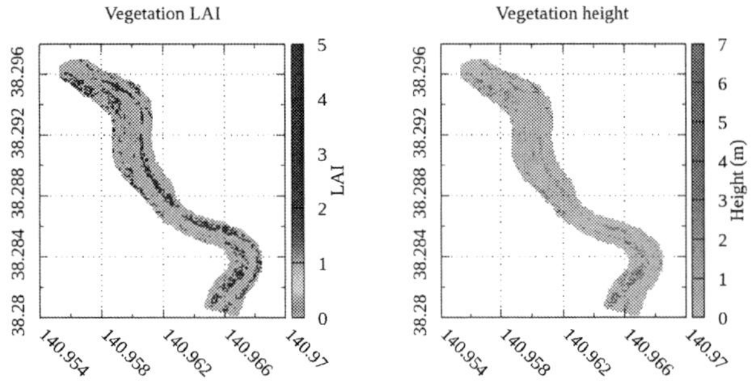

Figure 2. LAI and vegetation height in the stretch from the observation of September 2019 (summer season).

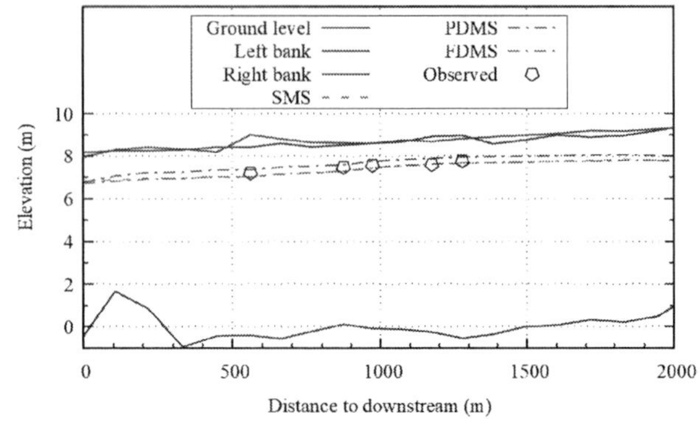

Figure 3. Longitudinal view of simulated water level and observed water level at five sections during peak inundation.

over the catchment. The LAI values were then confined to the vegetated cells in the UAV observation.

Different ML models were trained for each season with the same method, using the averaged MODIS values from 1 January to 31 March 2020 (winter), 1 April to 30 June 2020 (spring), and 1 October to 31 December 2020 (autumn). Sentinel-2 images for each season were used to predict their respective LAI values and distribution in the stretch.

4 Results and discussion

4.1 Vegetation conditions

The vegetation location was successfully obtained, with the MLP model achieving an accuracy of 0.99 for summer and spring and 0.96 for autumn and winter (Fortes et al., 2022).

The total vegetated area in September 2019 was 82 800 m^2, representing about 18 % of the total area of the stretch. The height of the vegetation was 1.7 m in average and the LAI was 2.43 in average. Figure 2 shows the estimated vegetation height and LAI in the stretch in the summer period, from the observations from September 2019.

4.2 Hydrologic simulation

As shown in Fortes et al. (2022), the RRI simulation could produce accurate results, with simulated water depth similar to the observed. Nash-Sutcliffe coefficient at each of the stations was −0.37 in Kawazaki, 0.77 in Ogaku, 0.88 in Ichinazaka, 0.73 in Iwakiri, and 0.62 in Fukuda Ohashi. Since most values ranged from 0.62 to 0.88, reaching values close to 1, the simulation was considered accurate, and the discharge obtained in the upstream section of the stretch was used in the hydraulic model. The peak discharge at the outlet was about 1250 $m^3 s^{-1}$, lower than the peak discharge of a 100-year return period flood event. In the stretch upstream, it was about 1050 $m^3 s^{-1}$.

4.3 Hydraulic simulations

The recreation of the typhoon event has shown a fair result when comparing simulated and observed water levels. The RMSE calculated for each simulation was 0.141 for SMS, 0.189 for PDMS, and 0.137 for FDMS. Figure 3 shows the peak water profile of the three simulations and the observed points.

The SMS and FDMS achieved the best results, closer to the observed water level. On the other hand, the PDMS achieved a higher water level, presenting an overestimation of the inundation. This happened due to the blockage of water flow in the emergent vegetation state. With Manning being dynamically calculated for both vegetation states, the FDMS presented a similar result when compared with the SMS, suggesting that the applied Manning algorithm is a valid substitute for calibrated static Manning values. The FDMS achieved the lowest RMSE, which agrees with the conclusions of Ebrahimi et al. (2008). This is important because the FDMS model does not require much calibration in terms of Manning value, with calibration being necessary only for the non-vegetated cells.

The seasonal changes of the vegetation conditions were as expected. Due to the lack of foliage, the vegetated area drastically reduced during the autumn and winter seasons. The peak vegetated area was in summer, followed by spring. The Manning values in the peak water level of each simulation were averaged for the entire terrain and compared with the ratio of the vegetated area over the entire stretch area.

Figure 4 shows the vegetated area in each season and the peak average Manning value. It clearly shows that the average Manning values strongly relate to the percentage of vegetated area in the stretch. Regarding the effect of the vegetation on the water level profile of each season, with the highest Manning, value, presented the highest water level profile of all seasons. The difference in water level between spring, autumn, and winter was not very perceptible. The highest difference in the water level profiles was between summer and

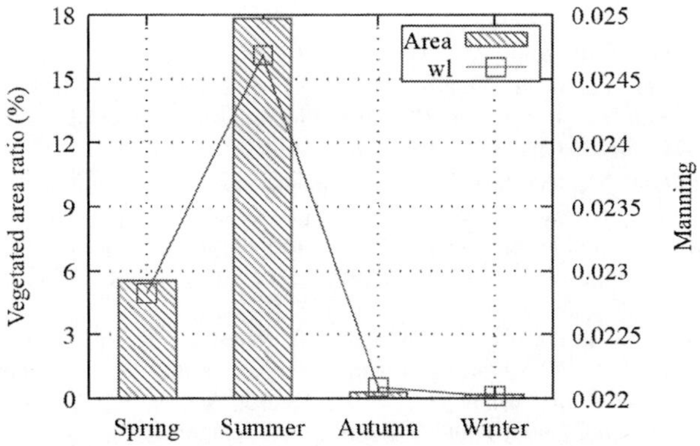

Figure 4. Vegetated area ratio and average peak Manning in all seasons.

winter, with 27.1 centimeters on average, and the lowest was between summer and spring, with 26.4 centimeters.

5 Conclusions

Machine learning proved to be an effective tool to acquire remote sensing vegetation data. The typhoon event could be accurately reconstructed in the study area. The FDMS presented the lowest RMSE value of 0.137, with a water level profile similar to the one produced by the SMS. This means that the Manning calculation algorithm worked well and was a good substitute for the static Manning of 0.038. The FDMS presented the advantage of requiring less calibration to achieve accurate results. Since the PDMS did not consider water flow in the emergent vegetation, it presented an overestimation of the inundation, being less applicable. The results show an advance in considering vegetation parameters to calculate Manning at a pixel level, but the models need to be tested in more rivers to fully validate the Manning algorithm.

The seasonal effect of the vegetation was demonstrated when the average Manning in the stretch in the peak was compared to the ratio of vegetated area, which demonstrated a strong relationship.

Code availability. Due to copyright restrictions, the code for the 2D flow model cannot be shared without the permission of the developers. The part for the vegetative roughness calculation can be shared upon request.

Data availability. The data can be available upon request.

Author contributions. AAF was responsible for Conceptualization, Investigation, Methodology, Software, Formal analysis, Visualization, and Writing – original draft preparation. The co-authors contributed as follows: MH participated in Methodology, Software, Data curation, Supervision, Validation, and Writing – review & editing, KU participated in Methodology, Validation, Supervision, Project administration, and Writing – review & editing, KI participated in Data curation, Resources, and Writing – review & editing.

Competing interests. The contact author has declared that none of the authors has any competing interests.

Special issue statement. This article is part of the special issue "ICFM9 – River Basin Disaster Resilience and Sustainability by All". It is a result of The 9th International Conference on Flood Management, Tsukuba, Japan, 18–22 February 2023.

Review statement. This paper was edited by Daisuke Harada and reviewed by two anonymous referees.

References

Aberle, J. and Järvelä, J.: Flow resistance of emergent rigid and flexible floodplain vegetation, J. Hydraul. Res., 51, 33–45, https://doi.org/10.1080/00221686.2012.754795, 2013.

Bates, P.: Remote sensing and flood inundation modelling, Hydrol. Process., 18, 2593–2597, https://doi.org/10.1002/hyp.5649, 2004.

Box, W., Järvelä, J., and Västilä, K.: Flow resistance of floodplain vegetation mixtures for modelling river flows, J. Hydrol., 601, 126593–126604, https://doi.org/10.1016/J.JHYDROL.2021.126593, 2021.

Coppo Frias, M., Liu, S., Mo, X., Nielsen, K., Ranndal, H., Jiang, L., Ma, J., and Bauer-Gottwein, P.: River hydraulic modeling with ICESat-2 land and water surface elevation, Hydrol. Earth Syst. Sci., 27, 1011–1032, https://doi.org/10.5194/hess-27-1011-2023, 2023.

Ebrahimi, N. G., Fathi-Moghadam, M., Kashefipour, S. M., Saneie, M., and Ebrahimi, K.: Effects of Flow and Vegetation States on River Roughness Coefficients, J. Appl. Sci., 8, 2118–2123, 2008.

Fortes, A., Hashimoto, M., Udo, K., Ichikawa, K., and Sato, S.: Dynamic Roughness Modeling of Seasonal Vegetation Effect: Case Study of the Nanakita River, Water, 14, 3649–3667, 2022.

Gokool, S., Kunz, R. P., and Toucher, M.: Deriving moderate spatial resolution leaf area index estimates from coarser spatial resolution satellite products, Remote Sens. Appl.-Soc. Environ., 26, 100746–100758, https://doi.org/10.1016/j.rsase.2022.100743, 2022.

Hashimoto, M., Yoneyama, N., Kawaike, K., Deguchi, T., Hossain, M. A., and Nakagawa, H.: Flood and Substance Transportation Analysis Using Satellite Elevation Data: A Case Study in Dhaka City, Bangladesh, J. Disaster Res., 13, 967–977, https://doi.org/10.20965/jdr.2018.p0967, 2018.

Ishizaki, H. and Matsuyama, H.: Distribution of the Annual Precipitation Ratio of Radar/Raingauge-Analyzed Precipitation to AMeDAS across Japan, Scientific Online Letters on the Atmosphere, 14, 192–196, https://doi.org/10.2151/sola.2018-034, 2018.

Jalonen, J., Järvelä, J., and Aberle, J.: Leaf Area Index as Vegetation Density Measure for Hydraulic Analyses, J. Hydraul. Eng., 139, 461–469, https://doi.org/10.1061/(ASCE)HY.1943-7900.0000700, 2013.

Japan Institute of country-ology and engineering: Manual of plans for river channel, Sankaidou, edited by: Yamamoto, K., ISBN-10 4381014928, ISBN-13 978-4381014924, 2002 (in Japanese).

Järvelä, J.: Determination of flow resistance caused by non-submerged woody vegetation, Int. J. River Basin Manage., 2, 61–70, https://doi.org/10.1080/15715124.2004.9635222, 2004.

Jiang, L., Bandini, F., Smith, O., Klint Jensen, I., and Bauer-Gottwein, P.: The Value of Distributed High-Resolution UAV-Borne Observations of Water Surface Elevation for River Management and Hydrodynamic Modeling, Remote Sens., 12, 1171, https://doi.org/10.3390/rs12071171, 2020.

Kazama, M., Yamakawa, Y., Yamaguchi, A., Yamada, S., Kamura, A., Hino, T., and Moriguchi, S.: Disaster report on geotechnical damage in Miyagi Prefecture, Japan caused by Typhoon Hagibis in 2019, Soils Found., 61, 549–565, https://doi.org/10.1016/J.SANDF.2020.12.001, 2021.

National Land Data Information: National Land Data Information [data set], https://nlftp.mlit.go.jp/ksj/gml/datalist/KsjTmplt-L03-b_r.html (last access: 12 August 2021), 2014.

Nikora, V., Larned, S., Nikora, N., Debnath, K., Cooper, G., and Reid, M.: Hydraulic Resistance due to Aquatic Vegetation in Small Streams: Field Study, J. Hydraul. Eng., 134, 1326–1332, https://doi.org/10.1061/(ASCE)0733-9429(2008)134:9(1326), 2008.

Pilailar, S., Sakamaki, T., Izumi, N., Tanaka, H., and Nishimura, O.: The characteristic change of fine particulate organic matter due to a flood in the Nanakita River, Proceedings of hydraulic engineering, 47, 1033–1038, https://doi.org/10.2208/prohe.47.1033, (2003).

Sayama, T., Ozawa, G., Kawakami, T., Nabesaka, S., and Fukami, K.: Rainfall–runoff–inundation analysis of the 2010 Pakistan flood in the Kabul River basin, Hydrol. Sci. J., 57, 298–312, https://doi.org/10.1080/02626667.2011.644245, 2012.

van Iersel, W., Straatsma, M., Addink, E., and Middelkoop, H.: Monitoring height and greenness of non-woody floodplain vegetation with UAV time series, ISPRS J. Photogramm. Remote, 141, 112–123, https://doi.org/10.1016/J.ISPRSJPRS.2018.04.011, 2018.

Viet, N., Tanaka, H., Nakayama, D., and Yamaji, H.: Effect of Morphological Changes and Waves on Salinity Intrusion in the Nanakita River Mouth, Proceedings of Hydraulic Engineering, 50, 139–144, https://doi.org/10.2208/prohe.50.139, 2006.

Yamazaki, D., Ikeshima, D., Sosa, J., Bates, P. D., Allen, G. H., and Pavelsky, T. M.: MERIT Hydro: A High-Resolution Global Hydrography Map Based on Latest Topography Dataset, Water Resour. Res., 55, 5053–5073, https://doi.org/10.1029/2019WR024873, 2019.

Proc. IAHS, 386, 209–216, 2024
https://doi.org/10.5194/piahs-386-209-2024

Open Access PIAHS

Attributing weather patterns to Davao River extreme rainfall from Reanalysis and GCM

Ralph Allen Acierto, Tomoki Ushiyama, and Toshio Koike

International Center for Water Hazard and Risk Management, Public Works Research Institute, Tsukuba, Japan

Correspondence: Ralph Allen Acierto (acierto@icharm.org)

Received: 21 July 2023 – Revised: 20 September 2023 – Accepted: 25 February 2024 – Published: 19 April 2024

Abstract. Extreme rainfall or heavy rainfall events (HREs) causes significant socio-economic damages annually affecting local development especially in developing countries. Thus, assessing changes in frequency and magnitude of HREs under climate change using global climate model (GCM) projections became ubiquitous to hydrological impact studies. Here, we present a framework for evaluating GCM's ability in reproducing the seasonal frequency of HREs in Davao River basin and the associated weather patterns that led to HREs. Our results show that HREs in MRI-AGCM 3.2S occurred 81 % in DJF and 4 % in JJA, which showed over(under) estimation bias during DJF(JJA) season compared to ERA5 HREs that show occurrence of 50 % in DJF and 18 % in JJA. Furthermore, we examined the weather pattern and anomalies that led to anomalous conditions of the season specific HREs in Davao River basin, which showed MRI-AGCM3.2S was able to reproduce the general structure of anomalous conditions fairly well on both seasons in comparison with ERA5. However, the slight over(under)estimation of the surface anomalous conditions in DJF(JJA) are directly proportional to the over(under)estimation in rainfall magnitude in the basin.

1 Introduction

Climate information is crucial to water resource and disaster risk reduction management due to the high sensitivity of water resource systems to climate variabilities worsened by global warming (Clark et al., 2016). So, investigating future climate and changes in flood hazard for basin-scale assessment using projections from general circulation models (GCM) has been widely adopted (e.g. Wood et al., 1997; Fowler et al., 2007; Teutschbein and Seibert, 2010; Breuer et al., 2017). One of the key factors in determining future flood hazard are the occurrence of extreme events such as heavy rainfall events, which received particular focus due to potential increase of future disaster damages and likely increase shown by GCM projections (Pfahl et al., 2017; Neelin et al., 2022). For hydrological impact assessments, rainfall time-series including the heavy rainfall events come from dynamically downscaled GCM projections to provide physically consistent variables and subsequently bias-corrected. However, naive application of dynamical downscaling of GCM projections and subsequent application of bias-correction can potentially lead to unintended consequences of over-correction of climate model simulations and overconfidence in the climate change signal from global models (Clark et al., 2016). Recent reviews (Clark et al., 2016; Meresa et al., 2022) on providing hydrological extremes advocate in characterizing and reducing uncertainties within each of the steps of the impact assessment modeling chain. Furthermore, naive application of bias-correction techniques disregards the fact that its application fundamentally assumes a skillful GCM (or RCM) as input (Maraun, 2016).

Essentially, credible climate information relies on the understanding of climate model biases (Maraun et al., 2017) to appropriately reduce the uncertainty coming from utility of downscaling methods and bias-correction in providing for hydrological applications. However, commonly used GCM assessments typically focus on either purely statistical or purely process-oriented indices. Furthermore, common bias-correction methods were highly developed on the statistical or mathematical but it remains underdeveloped in considering process-related information (Maraun et al., 2017).

In this study, we present a framework combining the utility of statistics and process-related understanding in evaluating a GCM's ability to reproduce heavy rainfall events in a study river basin coincident with significant large-scale conditions. We present a framework for evaluating high-resolution GCMs for representing large-scale conditions that are coincident with the extreme rainfall or heavy rainfall event (HRE) in a river basin. We used ERA5 as reference dataset, which has a comparable model resolution of around 27 km, for evaluating the 20 km resolution model MRI-AGCM 3.2 (MRI-AGCM 3.2S). We also illustrate a framework for linking significant hydrological events or heavy rainfall event (HREs) to the large-scale or synoptic features in the region. There are two main components of this framework: (1) assessment of GCM's ability to reproduce the seasonal frequency of the annual HREs; and (2) identification of large-scale/synoptic conditions in ERA5 for seasonal HREs in the river basin and compare it with large-scale/synoptic conditions from MRI-AGCM 3.2S for respective seasons. The framework establishes the baseline for identifying significant weather patterns related to HREs in Davao river basin using ERA5. Then, it utilizes the identified weather patterns of seasonal HREs as reference for assessing the capability of a high-resolution GCM called MRI-AGCM 3.2S to reproduce similar weather patterns of seasonal HREs.

This type of assessment is necessary for defining the limitation of high-resolution GCMs in reproducing seasonal HREs and associated weather patterns, which can be used for in event based scenario modeling under a storyline approach (Shepherd, 2019) for climate adaptation studies. This assessment also informs critical information to the GCM's utility for dynamical downscaling and appropriate selection and/or necessity of bias-correction method.

2 Data and methodology

2.1 Data

In this study, the following datasets were used:

1. Daily rainfall data from synoptic stations of Philippine Atmospheric, Geophysical and Astronomical Services Administration (PAGASA) from 1980 to 2016 for Davao, Malaybalay, General Santos, Cotabato, Surigao, Lumbia, Butuan, Hinatuan, Zamboanga, and Dipolog. Rainfall data were interpolated using Cressman-barnes method to make a gridded rainfall dataset that was masked by Davao River shapefile to get the basin rainfall.

2. Daily surface variables (rainfall, mean-sea-level-pressure or MSLP, surface wind components u10 and v10) and pressure-level variables (zonal and meridional wind, temperature, relative humidity, geopotential height from fifth generation of ECMWF atmospheric reanalyses (ERA5, Hersbach et al., 2020) from 1979 to 2018.

3. Same variables selected in ERA5 from Japan's Meteorological Research Institute (MRI) Atmospheric General Circulation Model (AGCM) 3.2S (Mizuta et al., 2012) with a period of 1979–2003. This GCM is a part of the on-going efforts in increasing the accuracy of GCM projections such as the on-going efforts on the World Climate Research Programs' Comprehensive Model Inter-comparison Project (CMIP). The MRI-AGCM outputs have been used in various programs such as CORDEX (Gutowski et al., 2016), HighResMIP(Haarsma et al., 2016), and Japan's domestic assessment studies (Mori et al., 2021) creating regional climate information for impact assessment and adaptation studies (Ito et al., 2020). MRI-AGCM 3.2 has showed good reproducibility on the regional precipitation globally (Ito et al., 2020) and improvements to representation of tropical cyclones (Murakami et al., 2011, 2012).

2.2 Methodology

We selected HREs in the river basin using basin-averaged rainfall from raingauge observation using annual block maxima (ABM) and peak-over-threshold (POT) method. For the POT method, we used a 99th percentile threshold. To provide comparison with basin-averaged rainfall from raingauge, we calculated the area-averaged (area is between 7 and 8° latitude and 125 to 126° longitude) rainfall for ERA5 and MRI-AGCM 3.2S.

First, we selected HREs using ABM and POT method were analysed to characterize the seasonal frequency in the river basin from raingauge, ERA5, and MRI-AGCM 3.2S. We focused on the dominant monsoon seasons in the months of June–July–August (JJA) for southwest monsoon and December–January–February (DJF) for northeast monsoon.

Second, we used the POT method for selecting seasonal HREs for ERA5 and MRI-AGCM 3.2S for respective periods shown in Table 1. We calculated the composites of surface variable for the HREs of JJA and DJF season. We assessed the ability of MRI-AGCM 3.2S in reproducing HREs through comparison with ERA5. Composite weather patterns were calculated as the mean of corresponding variables during the selected HREs. In addition, anomaly composites were also calculated and were based on 1979–2003 seasonal climatology. We focused on anomalous wind and pressure characteristics of the HRE composites to understand its difference with seasonal conditions over Davao river basin.

2.3 Study Area and basin rainfall characteristics

The study area (shown in Fig. 1) is the Davao river basin with an area of 1700 km^2 is located in Mindano island of

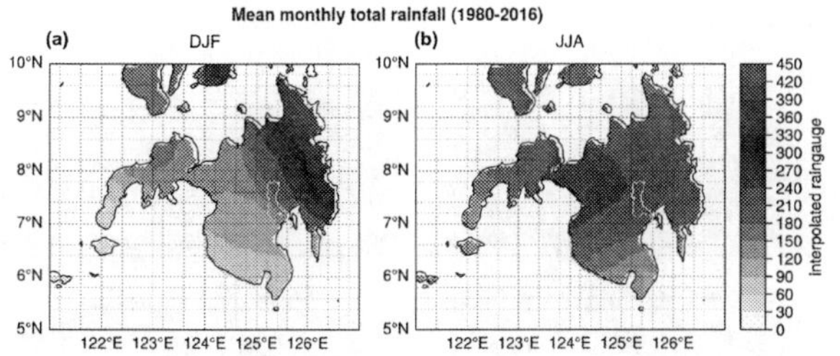

Figure 1. Mean monthly total rainfall for DJF and JJA seasons in Mindanao Island using interpolated rainfall data from PAGASA stations.

the Philippines. Figure 1 shows the summary of rainfall over the Davao River basin, where the basin-averaged climatology and annual maximum rainfall was extracted within the river basin. Wet period occurs during months of May to October while the dry period occurs in the months of November to April of the next year. As shown in Fig. 1, DJF season shows rainfall gradient consistent with a northeast monsoon flow with a maximum rainfall occurring on the eastern coasts, which is near the northeastern boundary of the Davao River basin. For JJA season, the island receives higher volume of rainfall with smaller variations spatially but with some maximum located on the west-northwestern coasts coinciding with south-southwest monsoon flows, which is on western boundary of the river basin. The study basin was selected due to the occurrence of HREs in DJF and JJA season that has very distinct large-scale conditions.

3 Results and discussion

This section is organized into two main sections: Davao River HRE statistics (Sect. 3.1), and Climatology of seasonal HREs from ERA5 and MRIAGCM3.2S (Sect. 3.2). Section 3.1 discusses on two main points: (1) reliability of ERA5 as a reference dataset by comparing it with the raingauge dataset; and (2) assessment of the GCM as compared with the ERA5 dataset. On the other hand, Sect. 3.2 focuses on the description of synoptic conditions coincident with the occurrence of HREs in the study basin. We primarily use composite analysis to compare MRI-AGCM 3.2S with ERA5.

3.1 Davao River HRE statistics

The extreme rainfall or heavy rain events (HREs) for Davao River basin are selected using ABM method, which is summarized in Tables 1 and 2. As shown in Table 1, raingauge observations show that the highest frequency of HRE events occur 38 % in December–January–February (DJF) season and 24 % in June–July–August (JJA) season.

To establish ERA5 as a reference, ERA5 was compared with the raingauge observations using the same period from

Table 1. Percent Frequency (Events) of Selected HRE using ABM.

Season	Raingauge (RG)	ERA5	MRI-AGCM 3.2S
DJF	38 % (14)	50 % (20)	80.8 % (21)
JJA	24 %(9)	12.5 % (5)	3.85 % (1)
SON	16 %(6)	17.5 %(7)	3.85 % (1)
MAM	22 %(8)	20 %(8)	11.5 % (3)

Note: Data (period) – raingauge (1980–2016), ERA5 (1979–2018), and MRI-AGCM3.2S (1979–2003)

Table 2. 1980–2016 Percent Frequency (Events) of ABM HREs.

Season	Raingauge (RG)	ERA5
DJF	37.84 % (14)	48.65 % (18)
JJA	24.32 % (9)	10.81 % (4)
MAM	21.62 % (8)	21.62 % (8)
SON	16.22 % (6)	18.92 % (7)

1980 to 2016, which is summarized in Fig. 2a and shown in Table 2. As shown by the boxplot comparison showing the mean-normalized HRE rainfall amount, ERA5 shows a comparable range but with slightly higher variance than raingauge observations. ERA5 shows good correspondence with seasonal frequency of the selected HREs but with over(under)estimation for DJF(JJA) seasons. The frequency bias is partly due to the coarse resolution of ERA5. Overall, ERA5 shows a good skill in representing the seasonal frequency and mean-normalized HRE rainfall amount.

In a similar manner, HREs were calculated for MRI-AGCM 3.2S and was compared against selected HREs from ERA5 using the period 1979–2003. Figure 2b shows that the mean-normalized HRE rainfall of MRI-AGCM 3.2S has comparable interquartile range but shows higher outliers indicated by the wider min-max range of the boxplots. Looking at seasonal frequency of the HREs from MRI-AGCM 3.2S, highest frequency of HREs occur in DJF season with almost twice as much as HREs in ERA5. For JJA season, the seasonal frequency of MRI-AGCM 3.2S is underestimated by 4 times as much as the ERA5 HREs. The higher frequency of DJF seasons of the selected HREs from MRI-AGCM 3.2S indicate that the rainfall magnitude during this season is significantly larger than compared to other seasons. Here, we illustrate that a season specific HREs has significant overestimation bias for DJF season.

3.2 Climatology of HREs in ERA5 and MRI-AGCM 3.2S

We focused on the two seasons (DJF, JJA) with dominant seasonal monsoon and have the highest frequency of HREs based on raingauge observations. Additionally, the characteristics of HREs and associated weather patterns of each season can be segregated and frequency bias from AMB/POT selection using the entire period can be avoided. Fig-

Table 3. 1979–2003 Percent Frequency (Events) of ABM HREs.

Season	ERA5	MRI-AGCM 3.2S
DJF	42.31 % (11)	76.92 % (20)
JJA	15.38 % (4)	3.85 % (1)
MAM	19.23 % (5)	11.54 % (3)
SON	19.23 % (5)	3.85 % (1)

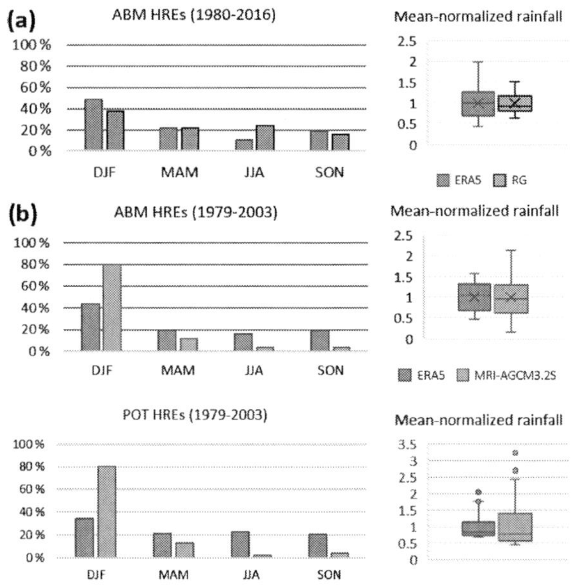

Figure 2. Seasonal Frequency and Mean-normalized rainfall of HRE events: **(a)** comparison of raingauge with ERA5; **(b)** comparison of ERA5 with MRI-AGCM 3.2S.

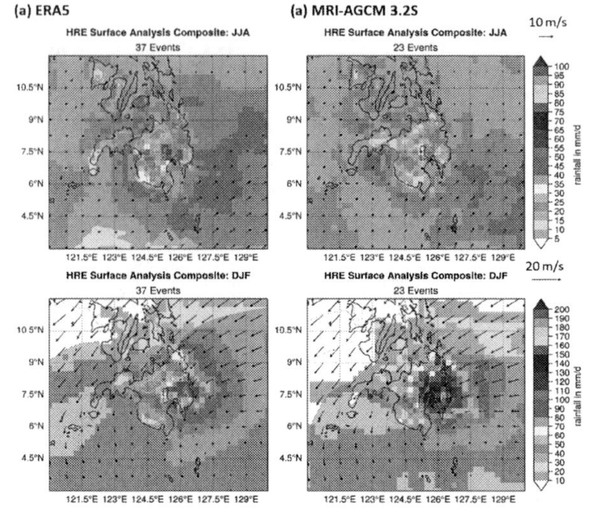

Figure 3. HRE composites of surface variables (rainfall, wind in vectors, MSLP in contours) over Mindanao island for **(a)** ERA5 and **(b)** MRI-AGCM 3.2S in JJA and DJF seasons.

ure 3 shows the rainfall over Mindanao island of the HRE composites for each season, which shows MRI-AGCM 3.2S under(over)-estimates rainfall magnitude during JJA(DJF) season in comparison with ERA5.

In this section, we describe the climatology of weather patterns associated with HREs during DJF and JJA seasons. A wider analysis area, which is referred here as monsoon region covering 90 to 150° longitude and 10° S to 45° N, is used to show the relevant synoptic conditions and anomalies as shown in Fig. 4. We also discuss how the associated conditions can impact the magnitude of rainfall in the river basin.

3.2.1 Climatology of DJF season HREs

DJF season composites show good correspondence between MRI-AGCM 3.2S (Fig. 4d) and ERA5 (Fig. 4c). The extent of the SH-MCL pressure system and northeast monsoon flow are very similar with ERA5. This is also supported by similar positive pressure anomaly on the north of Luzon Island and negative pressure anomaly located on the Mindanao island as shown in the anomaly composites (Fig. 4g and h).

Despite the similarities, minor differences can be seen on the location and extent of the positive pressure anomalies by MRI-AGCM 3.2S and ERA5 in Fig. 4g and h. ERA5 exhibits a distinct positive pressure anomaly indicated by the +2 Pa contour centered just below Honshu island of Japan, while MRI-AGCM 3.2S shows a more positive pressure anomaly indicated by +3 Pa contours located more landward in the Korean peninsula that protrudes beyond Kyushu Island as shown by the +2 Pa countours.

On both datasets, the positive-negative pressure anomaly system over Philippine archipelago causes an anomalous northeasterly flow that ends on the negative pressure low located in Mindanao island. However, due to the slight difference in the positive pressure anomaly configuration, we can observe a slightly different characteristic of the anomalous northeasterly flow from ERA5 and MRI-AGCM 3.2S. ERA5 exhibits a northeasterly flow induced by the additional positive-negative gradient between north of Luzon and Mindanao island. In comparison, the anomalous northeasterly flow shown by MRI-AGCM 3.2S is slightly stronger due to the higher pressure gradient induced by the positive pressure anomaly located more landward. Comparing the anomalous positive-negative systems, ERA5 shows a 3Pascal difference that is lower than shown by MRI-AGCM 3.2S with 5Pascal difference. Higher-pressure gradient shown by MRI-AGCM 3.2S than ERA5 resulted in higher anomalous rainfall caused by a stronger anomalous cyclonic circulation near Mindanao island.

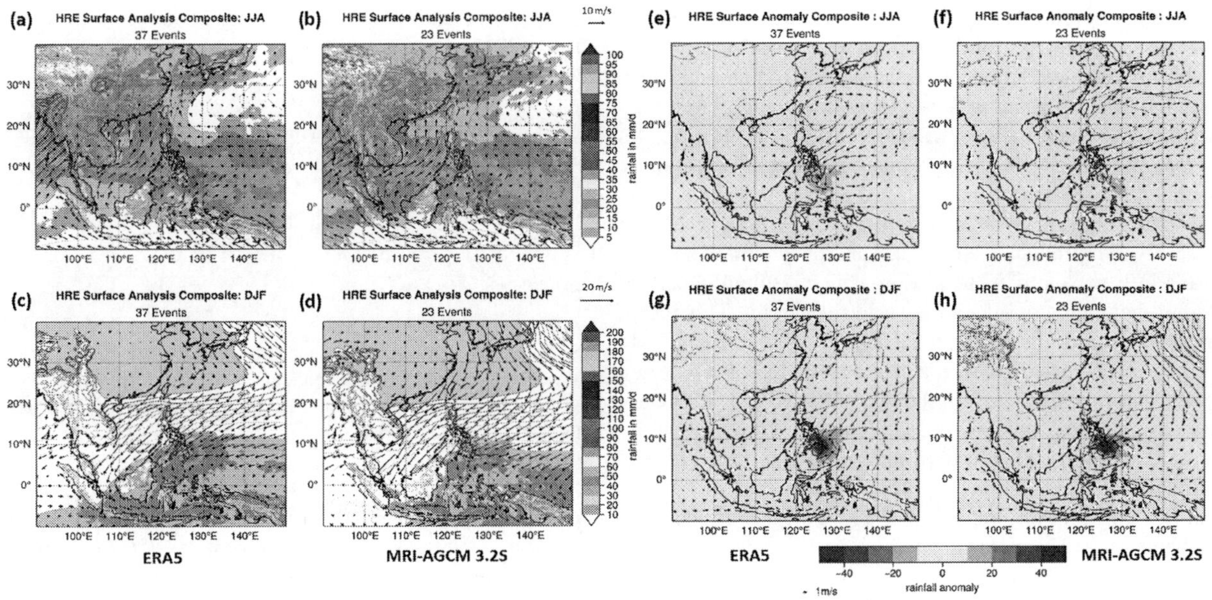

Figure 4. Comparison showing composites and anomalies of selected HREs for respective seasons.

3.2.2 Climatology of JJA season HREs

JJA season composites (ERA5 in Fig. 4a and MRI-AGCM 3.2S in Fig. 4b) show comparable general pressure system but with distinguishable differences. ERA5 shows an Asiatic Low with some relative high-pressure area over Tibetan Plateau. In contrast, MRI-AGCM 3.2S shows a more extended area of relative high pressure located on the north of Tibetan Plateau. The southwest monsoon flow is evident in both ERA5 and MRI-AGCM 3.2S.

However, ERA5 shows two main rainfall distribution following the southwest monsoon flows, which are situated over Myanmar coasts extending landward and from Gulf of Thailand towards western coast of Philippines. In contrast, MRI-AGCM 3.2S shows a dominant rainfall distribution following the southwest monsoon flow from Bay of Bengal traversing to Gulf of Thailand up to the western coasts in the Philippines, with rainfall maxima located on the southernmost tip of Myanmar. Furthermore, Western North Pacific subtropical high (WNPSH) that usually coincides with relative higher mean-sea-level-pressure located along the 20–30° N latitude band has a more coherent structure in ERA5 than in MRI-AGCM 3.2S composites.

Although both ERA5 and MRI-AGCM 3.2S show a positive pressure anomaly shown by the +2 Pa contours (Fig. 4e and f) acting like a blocking mechanism located on the north of Luzon island, they show slightly different configuration in terms of pressure gradient that translates to slight difference in anomalous circulation leading to a difference in rainfall magnitude as a consequence. In ERA5, the positive pressure anomaly has almost circular configuration that acts like an anticyclone block inducing an anomalous northeasterly flow that eventually feeds into an anomalous cyclonic circulation on the eastern coasts of Mindanao island. The resulting anomalous circulation causes the anomalous rainfall over the Davao River basin. In contrast, MRI-AGCM 3.2S shows a more flattened anomalous anticyclone block located in the North of Luzon that induces a more easterly anomalous flow contributing to a weaker cyclonic circulation near Mindanao island. The weaker anomalous cyclonic circulation causes lower anomalous rainfall in Davao river basin that eventually translates to an underestimation of rainfall amount for JJA HREs as compared with ERA5.

4 Conclusions

The ability of GCMs to reproduce the statistics of the present climate is considered as a measure of credibility to project potential future scenarios. Based on our analysis, seasonal frequency occurrence of annual HREs in MRI-AGCM 3.2S are overestimated (underestimated) HRE occurrence in DJF (JJA) season compared with ERA5. The relative magnitude of rainfall of HREs determines the seasonal frequency estimated from ABM or POT selected HREs that can impact the basis for extreme events selection for hydrological assessments.

By focusing on dominant monsoon climatology and using POT method for each season, we were able to understand the weather patterns or large-scale/synoptic conditions during the seasonal HREs in the river basin using ERA5 as reference and evaluated the MRI-AGCM3.2S against it. Using composite analysis, we found that MRI-AGCM 3.2S has good ability of reproducing relevant large scale and/or synoptic features on both seasons: monsoon trough or ITCZ, Western North Pacific subtropical High (WNPSH), and converging wind flows of south-westerly to westerly winds from the Bay of Bengal and easterly wind from the Coral Sea in JJA season; and Siberian High-Maritime Continent Low pressure system, ITCZ, and northeasterly to easterly winds associated with cold-surge in DJF season. MRI-AGCM 3.2S has better ability in reproducing large-scale and/or synoptic features during DJF season than JJA season, which resulted in better rainfall magnitude of HREs in Davao river basin. This indicates that projections related to HREs during DJF season has a higher reliability than during JJA season. In addition, the anomaly composite analysis showed MRI-AGCM 3.2S can also reproduce the anomalous conditions in comparison with ERA5. However, it also showed that small differences in the configuration and location of these anomalous conditions can lead to significant differences on the rainfall magnitude and frequency of HRE in Davao river basin. It should be noted that Davao river basin is relatively a small river basin with an area under 2000 km^2 so small difference in configuration of the synoptic conditions that lead to anomalous conditions can significantly impact rainfall in the basin. For future work, there is a need to classify the HREs for each season of dominant synpotic features like monsoon, typhoon, and cold surge events. Utility of indices to represent the synoptic conditions or anomalous patterns should also be identified and used to quantify the difference of the GCM with ERA5. In addition, the periods used for the composite analysis for ERA5 and MRI-AGCM3.2S are different which may be sensitive but had not been addressed here.

This study and future work on mechanisms and climatology of HREs in river basin scale are critical information in understanding biases from GCM and its potential impact in characterizing and predicting basin hydro-climatic hazards used in disaster management and climate change assessment studies for adaptation investment planning and decision making. In addition, the framework combines statistics and process oriented evaluation tool for understanding heavy rainfall events in GCMs, which are focused on the GCM's ability to replicate its the seasonal frequency and associated synoptic conditions. It also provides basis for constructing storyline approach (Shepherd et al., 2018; Shepherd, 2019) on both research and science communication aspects of climate change studies.

Code availability. Data analysis and visualization in the study used the following python packages: xarray (https://doi.org/10.5281/zenodo.598201, Hoyer et al., 2024), Pandas (https://doi.org/10.5281/zenodo.3509134, The pandas development team, 2024), MetPy (https://doi.org/10.5281/zenodo.160750, May et al., 2016), cartopy (https://doi.org/10.5281/zenodo.1182735, Elson et al., 2023), and ProPlot (https://doi.org/10.5281/zenodo.3873878, Davis, 2021).

Data availability. Data used in the study are ERA5 (Hersbach et al., 2020) and MRI-AGCM 3.2S (Mizuta et al., 2012). Rainfall station data from The Philippine Atmospheric, Geophysical and Astronomical Services Administration (PAGASA) used in this study can be requested directly from their office.

Author contributions. RAA: Conceptualization, Methodology, Software, Validation, Formal analysis, Investigation, Data Curation, Writing – Original Draft, Visualization. TU: Conceptualization, Resources, Writing – Review and Editing, Supervision. TK: Conceptualization, Supervision.

Competing interests. At least one of the (co-)authors is a guest member of the editorial board of *Proceedings of IAHS* for the special issue "ICFM9 – River Basin Disaster Resilience and Sustainability by All". The peer-review process was guided by an independent editor, and the authors also have no other competing interests to declare.

Disclaimer. Publisher's note: Copernicus Publications remains neutral with regard to jurisdictional claims made in the text, published maps, institutional affiliations, or any other geographical representation in this paper. While Copernicus Publications makes every effort to include appropriate place names, the final responsibility lies with the authors.

Special issue statement. This article is part of the special issue "ICFM9 – River Basin Disaster Resilience and Sustainability by All". It is a result of The 9th International Conference on Flood Management, Tsukuba, Japan, 18–22 February 2023.

Acknowledgements. The authors would like to acknowledge support from PAGASA for providing the daily rainfall data used and Meteorological Research Institute of Japan for providing MRI-AGCM 3.2S dataset used in this study.

Financial support. This research has been partly supported by the TOUGOU and SENTAN Program of MEXT, Japan (grant no. JPMXD0717935561).

Review statement. This paper was edited by Daisuke Harada and reviewed by Tomohito J. Yamada and one anonymous referee.

References

Breuer, L., Gosling, S. N., Yang, T., Hoffmann, P., Hattermann, F. F., Krysnaova, V., Wada, Y., Su, B., Masaki, Y., Müller, C., Daggupati, P., Stacke, T., Fekete, B., Motovilov, Y., Vetter, T., Flörke, F., Liersch, S., Donnelly, C., and Samaniego, L.: Sources of uncertainty in hydrological climate impact assessment: a cross-scale study, Environ. Res. Lett., 13, 015006, https://doi.org/10.1088/1748-9326/aa9938, 2017.

Clark, M. P., Wilby, R. L., Gutmann, E. D., Vano, J. A., Gangopadhyay, S., Wood, A. W., Fowler, H. J., Prudhomme, C., Arnold, J. R., and Brekke, L. D.: Characterizing Uncertainty of the Hydrologic Impacts of Climate Change, Current Climate Change Reports, 2, 55–64, https://doi.org/10.1007/s40641-016-0034-x, 2016.

Davis, L. L. B.: ProPlot (v0.9.5), Zenodo [code], https://doi.org/10.5281/zenodo.3873878, 2021.

Elson, P., Sales de Andrade, E., Lucas, G., May, R., Hattersley, R., Campbell, E., Dawson, A., Little, B., Raynaud, S., scmc72, Snow, A. D., Comer, R., Donkers, K., Blay, B., Killick, P., Wilson, N., Peglar, P., lgolston, lbdreyer, Szymaniak, A. J., Berchet, A., Bosley C., Davis, L., Krasting, F. J., Bradbury, M., Kirkham, D., stephenworsley, and Havlin, C.: SciTools/cartopy: v0.22.0 (v0.22.0), Zenodo [code], https://doi.org/10.5281/zenodo.1182735, 2023.

Fowler, H. J., Blenkinsop, S., and Tebaldi, C.: Linking climate change modelling to impacts studies: Recent advances in downscaling techniques for hydrological modelling, Int. J. Climatol., 27, 1547–1578, https://doi.org/10.1002/joc.1556, 2007.

Gutowski Jr., W. J., Giorgi, F., Timbal, B., Frigon, A., Jacob, D., Kang, H.-S., Raghavan, K., Lee, B., Lennard, C., Nikulin, G., O'Rourke, E., Rixen, M., Solman, S., Stephenson, T., and Tangang, F.: WCRP COordinated Regional Downscaling EXperiment (CORDEX): a diagnostic MIP for CMIP6, Geosci. Model Dev., 9, 4087–4095, https://doi.org/10.5194/gmd-9-4087-2016, 2016.

Haarsma, R. J., Roberts, M. J., Vidale, P. L., Senior, C. A., Bellucci, A., Bao, Q., Chang, P., Corti, S., Fučkar, N. S., Guemas, V., von Hardenberg, J., Hazeleger, W., Kodama, C., Koenigk, T., Leung, L. R., Lu, J., Luo, J.-J., Mao, J., Mizielinski, M. S., Mizuta, R., Nobre, P., Satoh, M., Scoccimarro, E., Semmler, T., Small, J., and von Storch, J.-S.: High Resolution Model Intercomparison Project (HighResMIP v1.0) for CMIP6, Geosci. Model Dev., 9, 4185–4208, https://doi.org/10.5194/gmd-9-4185-2016, 2016.

Hersbach, H., Bell, B., Berrisford, P., Hirahara, S., Horányi, A., Muñoz-Sabater, J., Nicolas, J., Peubey, C., Radu, R., Schepers, D., Simmons, A., Soci, C., Abdalla, S., Abellan, X., Balsamo, G., Bechtold, P., Biavati, G., Bidlot, J., Bonavita, M., De Chiara, G., Dahlgren, P., Dee, D., Diamantakis, M., Dragani, R., Flemming, J., Forbes, R., Fuentes, M., Geer, A., Haimberger, L., Healy, S., Hogan, R. J., Hólm, E., Janisková, M., Keeley, S., Laloyaux, P., Lopez, P., Lupu, C., Radnoti, G., de Rosnay, P., Rozum, I., Vamborg, F., Villaume, S., and Thépaut, J.-N.: The ERA5 global reanalysis, Q. J. Roy. Meteor. Soc., 146, 1999–2049, https://doi.org/10.1002/qj.3803, 2020.

Hoyer, S., Roos, M., Joseph, H., Magin, J., Cherian, D., Fitzgerald, C., Hauser, M., Fujii, K., Maussion, F., Imperiale, G., Clark, S., Kleeman, A., Nicholas, T., Kluyver, T., Westling, J., Munroe, J., Amici, A., Barghini, A., Banihirwe, A., Bell, R., Hatfield-Dodds, Z., Abernathey, R., Bovy, B., Omotani, J., Mühlbauer, K., Roszko, M. K., and Wolfram, P. J.: xarray (v2024.02.0), Zenodo [code], https://doi.org/10.5281/zenodo.598201, 2024.

Ito, R., Nakaegawa, T., and Takayabu, I.: Comparison of regional characteristics of land precipitation climatology projected by an MRI-AGCM multi-cumulus scheme and multi-SST ensemble with CMIP5 multi-model ensemble projections, Prog. Earth Planet. Sc., 7, 77, https://doi.org/10.1186/s40645-020-00394-4, 2020.

Maraun, D.: Bias Correcting Climate Change Simulations – a Critical Review, Current Climate Change Reports, 2, 211–220, https://doi.org/10.1007/s40641-016-0050-x, 2016.

Maraun, D., Shepherd, T. G., Widmann, M., Zappa, G., Walton, D., Gutiérrez, J. M., Hagemann, S., Richter, I., Soares, P. M., Hall, A., and Mearns, L. O.: Towards process-informed bias correction of climate change simulations, Nat. Clim. Change, 7, 764–773, https://doi.org/10.1038/nclimate3418, 2017.

May, R., Arms, S., kpozsonyi, Haberlie, A., Marsh, P., deeply-cloudy, mmorello1, Goebbert, K., Leeman, J., and Helmus, J. J.: metpy/MetPy: 0.4 (v0.4.0), Zenodo [code], https://doi.org/10.5281/zenodo.160750, 2016.

Meresa, H., Tischbein, B., and Mekonnen, T.: Climate change impact on extreme precipitation and peak flood magnitude and frequency: observations from CMIP6 and hydrological models, Natural Hazards, 111, 2649–2679, https://doi.org/10.1007/s11069-021-05152-3, 2022.

Mizuta, R., Yoshimura, H., Murakami, H., Matsueda, M., Endo, H., Ose, T., Kamiguchi, K., Hosaka, M., Sugi, M., Yukimoto, S., Kusunoki, S., and Kitoh, A.: Climate simulations using MRI-AGCM3.2 with 20-km grid, J. Meteorol. Soc. Jpn., 90, 233–258, https://doi.org/10.2151/jmsj.2012-A12, 2012.

Mori, N., Takemi, T., Tachikawa, Y., Tatano, H., Shimura, T., Tanaka, T., Fujimi, T., Osakada, Y., Webb, A., and Nakakita, E.: Recent nationwide climate change impact assessments of natural hazards in Japan and East Asia, Weather and Climate Extremes, 32, 100309, https://doi.org/10.1016/j.wace.2021.100309, 2021.

Murakami, H., Wang, B., and Kitoh, A.: Future change of western North Pacific typhoons: Projections by a 20-km-mesh global atmospheric model, J. Climate, 24, 1154–1169, https://doi.org/10.1175/2010JCLI3723.1, 2011.

Murakami, H., Wang, Y., Yoshimura, H., Mizuta, R., Sugi, M., Shindo, E., Adachi, Y., Yukimoto, S., Hosaka, M., Kusunoki, S., Ose, T., and Kitoh, A.: Future Changes in Tropical Cyclone Activity Projected by the New High-Resolution MRI-AGCM*, J. Climate, 25, 3237–3260, https://doi.org/10.1175/JCLI-D-11-00415.1, 2012.

Neelin, J. D., Martinez-Villalobos, C., Stechmann, S. N., Ahmed, F., Chen, G., Norris, J. M., Kuo, Y. H., and Lenderink, G.: Precipitation Extremes and Water Vapor: Relationships in Current Climate and Implications for Climate Change, Current Climate Change Reports, 8, 17–33, https://doi.org/10.1007/s40641-021-00177-z, 2022.

Pfahl, S., O'Gorman, P. A., and Fischer, E. M.: Understanding the regional pattern of projected future changes in extreme precipitation, Nat. Clim. Change, 7, 423–427, https://doi.org/10.1038/nclimate3287, 2017.

Shepherd, T. G.: Storyline approach to the construction of regional climate change information, P. Roy. Soc. A, 475, 834–860, https://doi.org/10.1098/rspa.2019.0013, 2019.

Shepherd, T. G., Boyd, E., Calel, R. A., Chapman, S. C., Dessai, S., Dima-West, I. M., Fowler, H. J., James, R., Maraun, D., Martius, O., Senior, C. A., Sobel, A. H., Stainforth, D. A., Tett, S. F., Trenberth, K. E., van den Hurk, B. J., Watkins, N. W., Wilby, R. L., and Zenghelis, D. A.: Storylines: an alternative approach to representing uncertainty in physical aspects of climate change, Clim. Change, 151, 555–571, https://doi.org/10.1007/s10584-018-2317-9, 2018.

Teutschbein, C. and Seibert, J.: Regional climate models for hydrological impact studies at the catchment scale: A review of recent modeling strategies, Geography Compass, 4, 834–860, https://doi.org/10.1111/j.1749-8198.2010.00357.x, 2010.

The pandas development team: pandas-dev/pandas: Pandas (v2.2.1), Zenodo [code], https://doi.org/10.5281/zenodo.3509134, 2024.

Wood, E. F., Lettenmaier, D., Liang, X., Nijssen, B., and Wetzel, S. W.: Hydrological Modeling of Continental-Scale Basins, Annu. Rev. Earth Planet. Sc., 25, 279–300, https://doi.org/10.1146/annurev.earth.25.1.279, 1997.

Proc. IAHS, 386, 217–222, 2024
https://doi.org/10.5194/piahs-386-217-2024

Long-range streamflow prediction using a distributed hydrological model in a snowfed watershed

Abdul Moiz and Akiyuki Kawasaki

Institute for Future Initiatives, The University of Tokyo, Tokyo, 113-8654, Japan

Correspondence: Abdul Moiz (amoiz93@hotmail.com)

Received: 1 August 2023 – Revised: 20 November 2023 – Accepted: 24 February 2024 – Published: 19 April 2024

Abstract. Inadequate planning of spring snowmelt discharge can lead to wastage of water resources for various purposes such as hydropower and lead to reduced capacity of the dams to control floods during rainy season. In this research we analyze how much can the predictive skill of long-range forecasts be improved by using a distributed hydrological model. We used the Water and Energy Budget-based Distributed Hydrological Model with improved snow physics (WEB-DHM-S) for generating long-range forecasts with a lead time of up to 3 months for the case of Kurobe River Basin in Japan. The predictive skills of two sets of simulations were compared (i) climatology and (ii) ensemble stream flow prediction (ESP). In the case of ESP, the initial conditions of WEB-DHM-S are updated using real-time datasets from Radar-AMeDAS, AMeDAS and JRA55. We found that the model initial conditions are particularly important during the spring snowmelt season and can improve the forecast predictive skills quite significantly compared with the climatology.

1 Introduction

Streamflow forecasts play an important role in adequate management of water resources, such as through the operation of dam reservoirs for flood control or hydropower generation. The predictive skill of hydrometeorological forecasts depends upon the model structure/parameterization, meteorological forecasts and the model initial moisture (snow, soil) conditions (Koster et al., 2010).

While the accuracy of meteorological forecasts at shorter lead times is improving, their accuracy at longer lead times is still unreliable. With the advancement in the development of physically based distributed hydrometeorological models it is becoming possible to estimate initial moisture (snow, soil) conditions quite accurately. These initial conditions can play a very important role in improving the predictive skills of long-range meteorological forecasts. Hydropower dam reservoir operators often require long-range forecasts for planning reservoir operations. The simplest way to obtain these forecasts is to use climatology, which is quite commonly used by hydropower reservoir operators. However, hydropower dam reservoirs are often located in mountainous regions where snowmelt runoff plays an important role in hydropower production. Here, the initial conditions of snow (in addition to soil) can play an important role in improving the predictive skills of the long-range forecasts.

The objective of this study is to evaluate the importance of initial moisture (snow, soil) conditions in improving the hydrometeorological forecasts compared with the use of climatology. At first, we develop a long-range (3-month lead time) forecasting system using historical meteorological observation (available in real-time) and the Water and Energy Budget-based Distributed Hydrological Model with improved snow physics (WEB-DHM-S). Then we compared the predictive skills of this forecasting system with climatology for hindcasts initialized in different seasons to evaluate the contribution of the initial conditions in the predictive skills of streamflow forecasts.

2 Study Area

For the purpose of this study, we used the case of Kurobe River basin in Japan, a snowfed watershed which receives heavy snowfall in the winter season. The snowmelt runoff provides water to a cascade of hydropower dam reservoirs operated by Kansai Electric Power Company (KEPCO).

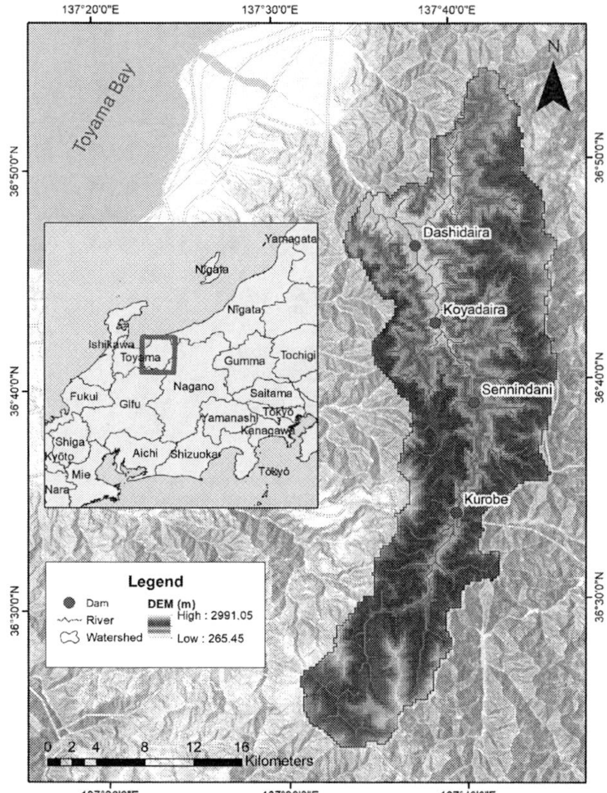

Figure 1. Kurobe River basin and the location of hydropower dams in Kurobe River basin.

There are a total of four hydropower dams in this river basin (see Fig. 1). Kurobe Dam is the largest dam out of the four with the highest regulating capability. Since the inflows to the downstream dams are heavily influenced by the operation of Kurobe Dam, in this study we have only considered the area upstream of Kurobe Dam reservoir.

3 Methodology and Data

3.1 Description of WEB-DHM-S

For hydrological modelling of the watershed and for hydrometeorological forecasting we used the latest iteration of WEB-DHM-S in this study called the 4-component WEB-DHM-S model. The original WEB-DHM-S model is capable of representing turbulent fluxes (water and energy) and can simulate snow processes and variables such as snow depth, snow density and spatial snow cover in addition to river discharge (Shrestha et al., 2010).

The latest iteration of WEB-DHM-S has been reframed into four main components (i) the land surface component, (ii) the vertical soil moisture solver, (iii) hillslope runoff processes, and (iv) kinematic wave routing. In addition, the latest iteration of WEB-DHM-S also allows saving a "snapshot" of the model state variables (including the snow variables) and use them as initial conditions for forecasting. Furthermore, the new version of the model is also more computationally efficient due to the use of parallel programming. These features make this new version of the model very suitable for real-time hydrometeorological forecasting.

3.2 Model setup and datasets

Datasets can be divided into two categories, (i) static datasets and (ii) dynamic datasets. Static datasets include the Digital Elevation Model (DEM) from Japan Geospatial Institute, the soil type from Food and Agriculture Organization and the land use type from United States Geological Survey. Dynamic datasets include the Leaf Area Index and the Fraction of Photosynthetically Active Radiation, both of which were obtain from MODIS sensors. Further dynamic datasets include the meteorological forcings, which are very important for forecasting. In this study the precipitation data was obtained from Radar-AMeDAS (Automated Meteorological Data Acquisition System), which has a resolution of 1 km and a temporal resolution of 30 min. Air temperature is based on the station observations from AMeDAS and is corrected for elevation by using the vertical temperature profile from JRA55 (Japanese 55-year reanalysis). The remaining meteorological variables such as wind speed, relative humidity, downward longwave and shortwave radiation etc. were also obtained from JRA55. The WEB-DHM-S model was set at a resolution of 250 m and all of the datasets were regridded to this resolution. More details about the datasets and data pre-processing are explained by Moiz et al. (2022).

For model validation and evaluation of the skills of forecasts, we used the observed inflow to Kurobe dam, point snow depth at Kurobe dam and the satellite observed snow cover. Kurobe dam inflow and point snow depth data were obtained from KEPCO and the snow cover data was obtained from MODIS sensor.

3.3 Experiments

We conducted four different kinds of experiments to evaluate the importance of the model initial conditions in the predictive skills of hydrometeorological forecasts. 13 years of data was used from 2008–2020. In all of these experiments, the model is already fully calibrated and validated against river discharge, point snow depth and satellite snow cover data (see Moiz et al., 2022 for details). These experiments are visually explained in Fig. 2a–d.

In the first forecast experiment, the model is run continuously for 13 years from 2008–2020. The output of the model in this case is called Perfect Forecast and acts as the upper bound for the skill of forecast, as the skill of any forecast cannot be better than this forecast. During this process, the states (initial conditions) of the model are also saved for

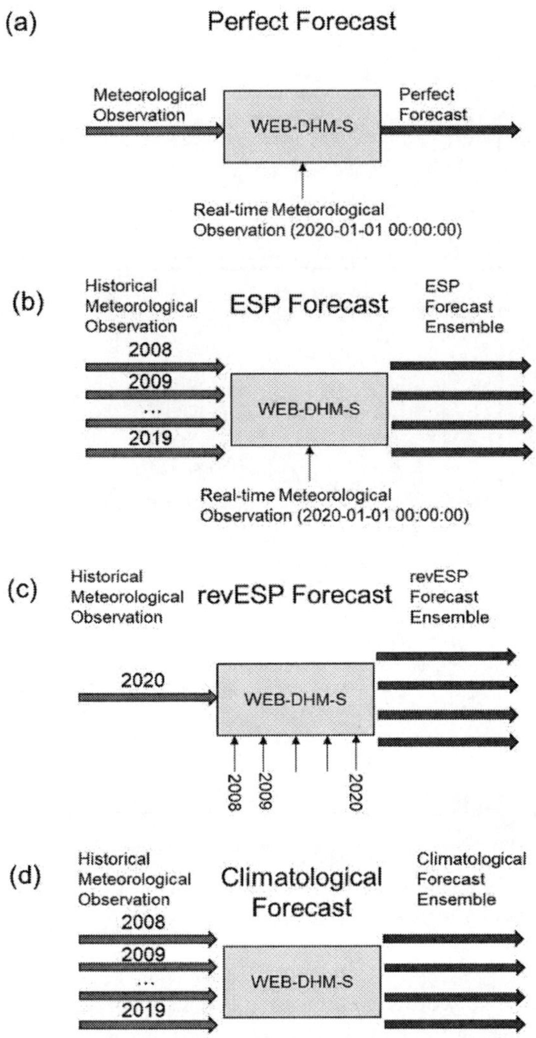

Figure 2. Setting of all the different experiments used to generate forecasts **(a)** Perfect Forecast **(b)** ESP Forecast **(c)** revESP Forecast and **(d)** Climatological Forecast. Here, the red arrows represent meteorological inputs, the blue arrows show hydrometeorological outputs and the black arrows show the meteorological data used for model initialization.

every month. The second forecast experiment is called ESP (Ensemble Streamflow Prediction) Forecast. In this case, the model is initialized using previously saved states for each time when the forecast is issued (every month). The ensemble meteorological forecast in this case is obtained by resampling the historical data using a leave-one-year-out approach.

For example, for the forecast issued on January 2017, meteorological data for all the years except 2017 is used leading to 12 ensemble members. In the case of revESP (reverse ESP) experiment, the reverse is true. An ensemble of initial conditions is used to generate hydrometeorological forecasts using the leave-one-year-out approach, but the perfect meteorological forecast is used. For example, for the forecast issued on January 2017, the initial conditions for January from all the years except 2017 are used while the meteorological forecast is based on the year 2017. In the case of the ESP Forecast experiment, the entire skill of the forecast is contributed by the initial conditions whereas in the case of revESP Forecast experiment, the entire skill is contributed by meteorological forecasts. In a final experiment called the Climatological Forecast, both the initial condition and the meteorological forecast are obtained using the leave-one-year-out approach. This forecast acts as the lower benchmark as this is entirely based on climatological information.

3.4 Forecast evaluation metrics

The ESP forecast is evaluated against the perfect forecast and climatological forecast using the continuous ranked probability score (CRPS) and the continuous ranked probability skill score (CRPSS).

$$\mathrm{CRPS} = \mathrm{CRPS}(P, x_a) = \int_{-\infty}^{\infty} [P(x) - P_a(x)]^2 \mathrm{d}x \qquad (1)$$

$$\mathrm{CRPSS} = 1 - \frac{\mathrm{CRPS}_{\mathrm{Forecast}}}{\mathrm{CRPS}_{\mathrm{Reference}}} \qquad (2)$$

Here, $P(x)$ is cumulative density function of the ensemble forecast, whereas the $P_a(x)$ is the cumulative density function of the observation (in this case, the perfect forecast). The value of CRPS ranges from 0 (perfect) to $+\infty$ (no skill) and the values of CRPSS ranges from $-\infty$ (no skill) to 1 (perfect). In case of CRPSS, climatology is used as the reference. A value of 1 means perfect skills, value of 0 means no skills compared to climatology and a negative value indicates that the forecast is has less skills than the climatology.

We then further also compared the ESP and revESP by comparing their relative mean squared errors (MSE) with each other following Wood and Lettenmaier (2008).

$$\mathrm{RelativeMSE}[\mathrm{ESP}] = \frac{\mathrm{MSE}[\mathrm{ESP}]}{\mathrm{MSE}[\mathrm{Clim}]} \qquad (3)$$

A relative MSE close to 0 indicates a higher predictive skill compared to climatology whereas a relative MSE close to 1 indicates a poor predictive skill similar to that of climatology.

4 Results and Discussion

4.1 Model calibration and validation

The WEB-DHM-S model was calibrated for the year 2011 and then validated for 2012–2014. The calibration process involved tuning the soil parameters, the snow process parameters as well as the radar snowfall correction parameters. In

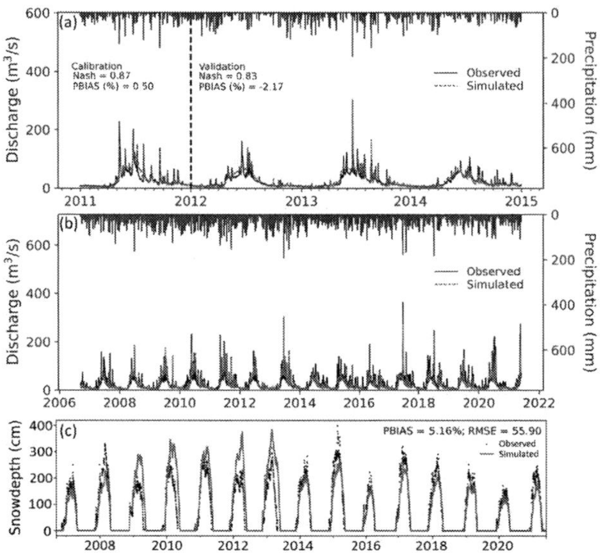

Figure 3. Comparison of the WEB-DHM-S river discharge simulation and observation. **(a)** Calibration (2011) and Validation (2012–2014). **(b)** Entire simulation period (2008–2020).

general, the model showed a high accuracy as discussed in Moiz et al. (2022). We further also validated the entire time series of river discharge simulation using data from 2008–2020. The model exhibited reasonable accuracy with a Nash coefficient of 0.79 and a percentage BIAS of −12.02 %.

The model was also validated against point snow depth observation (see Fig. 3c) and satellite snow cover (not shown here), both of which were also reasonably simulated. Snow depth at Kurobe can reach a maximum of around 4 m depth and starts melting in April where it contributes to Kurobe dam inflow. Most of this snow melts until the end of May. Therefore, in these two months initial condition of snow is very important for predicting snow melt runoff.

4.2 Long-range hydrometeorological forecasts

In all the experiments, forecasts are issued at the beginning of every month with a lead-time of 3-months. In the case of ESP forecast, for example, the initial condition is obtained using actual meteorological forcings and running the model up to the real-time (in this case, the time when the forecast is to be issued). An example of the forecast issued on 1 May 2010 is shown in Fig. 4.

From Fig. 4a it is quite evident that the use of initial condition leads to a much more reasonable forecast of point snow depth as compared with climatology. In Fig. 4b while the discharges at the beginning of the forecast are quite similar, they quickly become very different over time. Here it can be seen the ESP performs significantly better whose skill is mainly attributed to the initial condition. This difference can

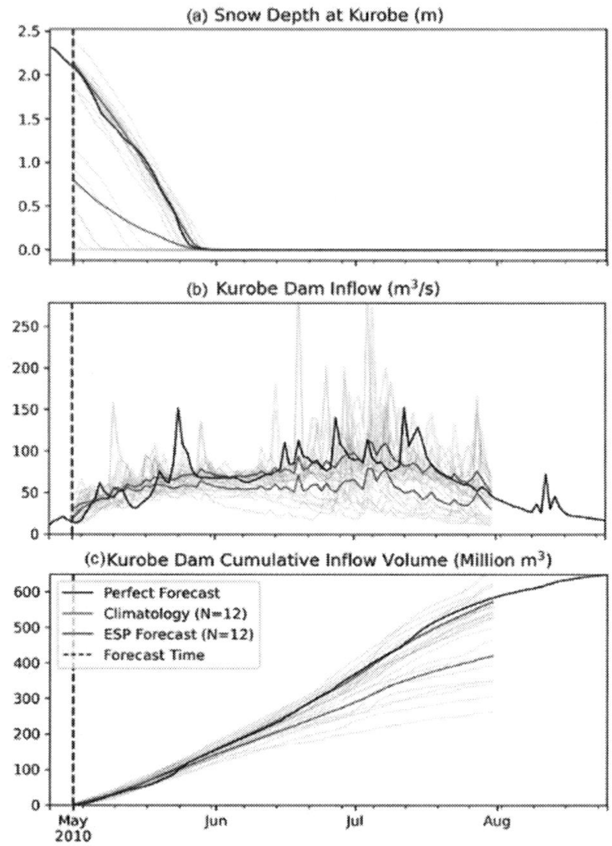

Figure 4. Comparison of ESP forecast with climatology and perfect forecast for forecast issued on 1 May 2010.

lead to very different forecasted cumulative reservoir inflow volumes at the end of the forecast horizon, which can have consequence for reservoir operations. The month of May is dominated by snowmelt runoff, so the initial condition of the snowpack can play a very important role in contributing to the skill of the forecast.

4.3 Comparison of ESP and climatology metrics

Figure 5 shows the CRPSS metric for the ESP forecasts average over 13 years and groups into four seasons (MAM = March–April–May; JJA = June–July–August; SON = September–October–November; DJF = December–January–February) depending upon the time the forecast was issued.

In general, the ESP is more skilful than the climatology for 8, 75, 11 and 10 d of lead time for MAM, JJA, SON and DJF respectively. In the case of MAM, the skill quickly drops after the first 8 d. In the case of JJA, the forecast remains more skilful for a very long lead time. We expect that this is largely due to the contribution of the forecast issued in May and June, where the initial condition of snow play an im-

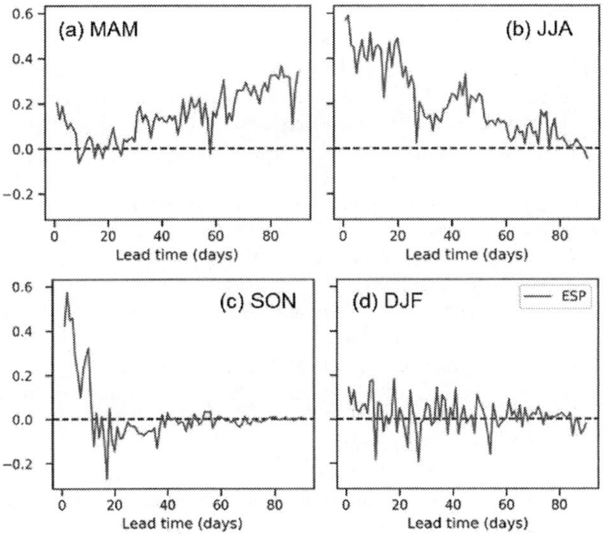

Figure 5. CRPSS for ESP forecast for different seasons.

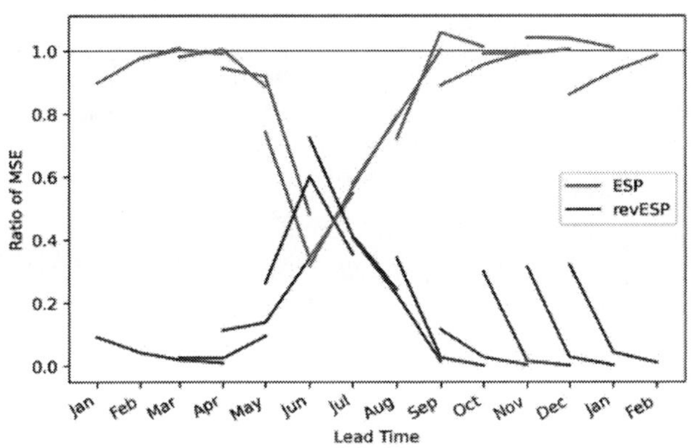

Figure 6. Comparison of relative MSE for ESP and revESP.

portant role. Furthermore, here during the rainy season, the soil moisture initial condition must also play a very important role, to predict floods. The performance during SON and DJF is very poor and we think that this is likely due to the reason that the river discharge in this season is very low, since most of the precipitation received is in the form of snow.

4.4 Comparison of ESP and revESP

Figure 6 shows the comparison of the relative MSE for ESP and revESP. The climatology and perfect forecasts experiments are conducted as benchmarks. Climatology is used to calculate Ratio of MSE. Climatology has a Ratio of MSE equal to 1, whereas perfect forecast has a ratio of MSE equal to 0. ESP and revESP generally lie between these two benchmarks. The revESP has a lower relative MSE compared with ESP which is very close to 1.

However, the situation is quite different for the forecasts issued in April, May, June, and July. Here, the relative MSE for ESP is sometimes higher and sometimes comparable to that of revESP. This means that for forecasts issued in these months, the model initial conditions play an important role. The decreasing (increasing) values of relative MSE for ESP (revESP) in April and May are due to the reason that most of the snow at Kurobe melts during these months, therefore the snow initial condition is important in these months. During JJA some snow exists at higher elevations. Furthermore, these months receive a lot of rainfall. Both of these factors contribute to soil moisture. Therefore, in JJA the initial condition of soil moisture in addition to snow is an important contributor to the skill of ESP forecast.

5 Conclusions

In this paper, we explained the development of a long-range (3-month) hydrometeorological prediction system for a snowfed watershed in Japan, using the latest iteration of WEB-DHM-S. We used 13-years of hindcast simulations to determine the accuracy of these forecast and more specifically answer, whether initial condition play an important role in contributing towards the skill of the forecasts. We found that the use of initial conditions can lead to the forecasts which are more skilful than climatology by 8–75 d of lead-time depending upon the season. Highest increase in skills was during the time when most of the snow melts. We believe that this research can lead to the development of long-range forecasting systems which can guide reservoir operators in snowfed watersheds for effective water resources management.

Data availability. JRA55 dataset is from Kobayashi et al. (2015) and downloaded using the Data Integration and Analysis System (DIAS) (https://diasjp.net/en, last access: 10 June 2021). Radar-AMeDAS dataset was also downloaded from DIAS (https://diasjp.net/en, last access: 9 August 2021). Radar-AMeDAS can also be downloaded from a publicly open database (http://database.rish.kyoto-u.ac.jp/arch/jmadata/data/jma-radar/synthetic/original/, JMA, 2021). AMeDAS data can be downloaded from JMA's website (https://www.data.jma.go.jp/gmd/risk/obsdl/index.php, JMA, 2020).

Author contributions. AM contributed to conceptualization, methodology, formal analysis and writing. AK contributed to conceptualization, supervision, and funding acquisition.

Competing interests. The contact author has declared that neither of the authors has any competing interests.

Special issue statement. This article is part of the special issue "ICFM9 – River Basin Disaster Resilience and Sustainability by All". It is a result of The 9th International Conference on Flood Management, Tsukuba, Japan, 18–22 February 2023.

Financial support. This research was support by KEPCO, the DIAS project and the Ministry of Education, Culture, Sports, Science and Technology (MEXT).

Review statement. This paper was edited by Mohamed Rasmy and reviewed by two anonymous referees.

References

JMA – Japan Meteorological Agency: AMeDAS station air temperature dataset, https://www.data.jma.go.jp/gmd/risk/obsdl/index.php, last access: 5 November 2020.

JMA – Japan Meteorological Agency: Radar-AMeDAS precipitation dataset, http://database.rish.kyoto-u.ac.jp/arch/jmadata/data/jma-radar/synthetic/original/, last access: 9 August 2021.

Kobayashi, S., Ota, Y., Harada, Y., Ebita, A., Moriya, M., Onoda, H., Onogi, K., Kamahori, H., Kobayashi, C., Endo, H., and Miyaoka, K.: The JRA-55 reanalysis: General specifications and basic characteristics, J. Meteorol. Soc. Jpn. Ser. II, 93, 5–48, https://doi.org/10.2151/jmsj.2015-001, 2015.

Koster, R. D., Mahanama, S. P. P., Livneh, B., Lettenmaier, D. P., and Reichle, R. H.: Skill in streamflow forecasts derived from large-scale estimates of soil moisture and snow, Nat. Geosci., 3, 613–616, https://doi.org/10.1038/ngeo944, 2010.

Moiz, A., Zhongwang, W., Naseer, A., Kawasaki, A., Acierto, R. A., and Koike, T.: Improving snow-process modeling by evaluating reanalysis vertical temperature profiles using a distributed hydrological model, J. Geophys. Res.-Atmos., 127, e2021JD036174, https://doi.org/10.1029/2021JD036174, 2022.

Shrestha, M., Wang, L., Koike, T., Xue, Y., and Hirabayashi, Y.: Improving the snow physics of WEB-DHM and its point evaluation at the SnowMIP sites, Hydrol. Earth Syst. Sci., 14, 2577–2594, https://doi.org/10.5194/hess-14-2577-2010, 2010.

Wood, A. W. and Lettenmaier, D. P.: An ensemble approach for attribution of hydrologic prediction uncertainty, Geophys. Res. Lett., 35, 1–5, https://doi.org/10.1029/2008GL034648, 2008.

Proc. IAHS, 386, 223–228, 2024
https://doi.org/10.5194/piahs-386-223-2024

Evaluation of satellite precipitation products for real-time extreme river flow modeling in data scarce regions

Mihretab G. Tedla[1,2], **Mohamed Rasmy**[1,2], **Toshio Koike**[1,2], **and Li Zhou**[3]

[1]International Centre for Water Hazard and Risk Management under the auspices of UNESCO (ICHARM) & Public Works Research Institute (PWRI), Tsukuba 305-8516, Japan
[2]National Graduate Institute for Policy Studies (GRIPS), Tokyo 106-8677, Japan
[3]Institute for Disaster Management and Reconstruction, Sichuan University, Chengdu, China

Correspondence: Mihretab G. Tedla (mihretab1@gmail.com)

Received: 24 July 2023 – Revised: 19 October 2023 – Accepted: 25 February 2024 – Published: 19 April 2024

Abstract. Inadequacy of spatio-temporal hydro-climatic data limits the efficacy of hazard monitoring and disaster risk reduction activities in disaster-prone areas. Various satellite missions are recently providing climate data, but prior evaluation and enhancement of these data are necessary for a reliable application. In this study, we conducted performance evaluation and enhancement of three real-time satellite precipitation products (SPPs) (GSMaP, GPM-IMERG, and PERSIANN) for flood modeling in the Blue Nile basin. The bias correction improved the original SPPs, with the largest improvement being for factors generated from 10 d mean data. Flood event hydrograph indicated satisfactory results of error metrics on the devastating flood event of 2012. Employing reliable physical–based distributed hydrologic models provided longer lead time and high-accuracy flood simulation. Furthermore, the results indicate that integrating available initial observed precipitation data improved the efficiency of SPPs simulation, and hence are applicable in operational flood monitoring.

1 Introduction

Floods are major water-related disasters that affect millions of people around the world. As non-structural measures, flood forecasting and real-time monitoring require gathering essential quantitative information for effective water resource management and disaster risk reduction. In recent times, active remote sensing instruments have greatly improved, providing detailed information with a high spatio-temporal resolution. Consequently, the number of available satellite precipitation products (SPPs) has increased, along with the quality of the data they generate (Illingworth et al., 2015). However, to reliably employ SPPs in hydrological modelling, it is necessary to validate and enhance their performance beforehand. This evaluation is essential because the accuracy of SPPs is influenced by various factors such as sampling uncertainty, indirect observation errors, and type of inversion algorithm (Lo Conti et al., 2014).

Moreover, the demand for real-time datasets in disaster risk monitoring is on the rise, especially among policymakers and disaster relief officials, as it enables evidence-based decision-making. In addition, real-time data are invaluable for researchers and scientists studying disaster patterns and trends. Numerous studies have assessed the effectiveness of satellite-borne precipitation datasets in rivers flow modelling (Gebregiorgis and Hossain, 2013; Kubota et al., 2007; Ushio et al., 2009). Moreover, Zhou et al. (2021, 2022) developed SPPs bias correction methods to improve the applicability of the SPPs to an evaluation of the adequacy of observation stations for hydrologic modeling. Real-time bias correction requires the determination of an effective correction factor based on past simulations for usage in various applications. These bias correction techniques can be utilized for correcting real-time SPPs as well as climate forecasts. In this study, we applied spatiotemporal windows to perform bias correction and evaluate both the original and bias-corrected SPPs.

Impact assessment of floods and drought requires reliable physical–based Distributed Hydrologic Models (DHMs) for a comprehensive understanding of the various hydrological

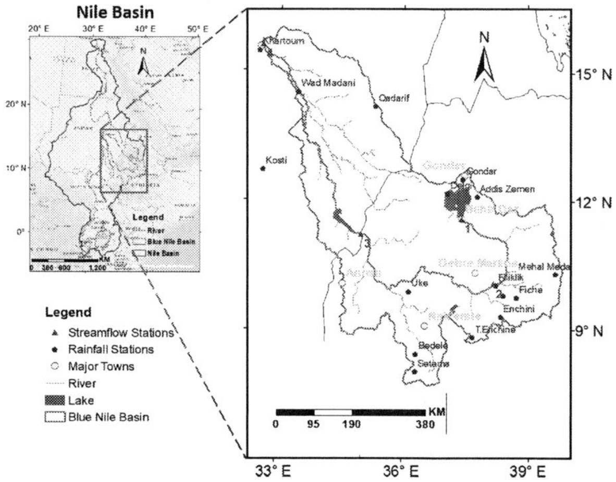

Figure 1. The Blue Nile basin and observation stations. Base map source: ESRI.

components. We employed the Water Energy Budget Rainfall Runoff Inundation (WEB-RRI) model, which incorporates water and energy budget processes to model flood onset timing, peak flood discharge, and inundation characteristics (Rasmy et al., 2019). Hydrological extreme projections in the Blue Nile basin indicated the need for actions on climate change adaptation and disaster risk reduction (Tedla et al., 2022). Moreover, the Blue Nile Basin is not immune to flash floods, which can cause severe damage and loss of life. Hence the entire approach is expected to assist climate change adaptation efforts by providing hydrological extremes forecasting of a catchment. The objective is to enhance the accuracy and lead-time of near real-time SPP on extreme flood modeling with a high degree of confidence. The combined application of real time data, bias corrections and simulation models is expected to enhance flood early warning and disaster risk reduction activities in transboundary river basin.

2 Study area

The Blue Nile, the largest tributary of the Nile Basin (Fig. 1), which covers a catchment area of about 312 000 km^2 and provides 62 % of the flow at the Aswan Dam (World Bank, 2006). The basin is characterized by rugged topography and a considerable variation in altitude, ranging from about 350 m above sea level (m a.s.l.) at Khartoum to over 4250 m a.s.l. in the Ethiopian highlands (Easton et al., 2012).

3 Materials and methods

3.1 Gauged data

Observed daily precipitation data from 24 observation stations upstream of the Blue Nile river basin have been collected for the period 2003–2012 (Fig. 1). Similarly, observed river discharge data have been collected from the Abbay near Kessie flow observation station. The flow at this station is used for model setup by calibration (2003–2005) and validation (2006–2012) periods.

3.2 Satellite precipitation products (SPPs)

Three open-source SPPs with high spatial and temporal resolution have been used for flood modelling evaluation. The SPPs are selected for their availability at a real time. As shown in (Fig. 2), the grid based SPPs provide good spatial distribution compared to the point based observed datasets (Fig. 2a).

3.2.1 GSMaP-NRT

The Global Satellite Mapping of Precipitation-Near-Real-Time (GSMaP-NRT) version 6 data, provided by Earth Observation Research Center, Japan Aerospace Exploration Agency (JAXA), was obtained at a daily time step with $0.1° \times 0.1°$ resolution and 4 h latency, Fig. 2b. The GSMaP is produced based on a Kalman filter-based CMORPH model that refines the precipitation rate propagated based on the atmospheric moving vector derived from MW sensors (Ushio et al., 2009). The data can be accessed from https://sharaku.eorc.jaxa.jp/GSMaP/ (last access: 14 April 2022).

3.2.2 IMERG-Early

The Integrated Multi-satellitE Retrievals for Global Precipitation Measures (IMERG) algorithm combines information from the Global Precipitation Measurement (GPM) satellite constellation to estimate precipitation over the majority of the Earth's surface. For this study, we used the IMERG Early Precipitation Level 3 ($0.1° \times 0.1°$ resolution) V06 at daily intervals derived from the half-hourly, Fig. 2c (Tan et al., 2019). The latest product for half hourly in near-real time at 4 h latency was downloaded from: https://disc.gsfc.nasa.gov/ (last access: 26 May 2022).

3.2.3 PERSIANN-CCS

Precipitation Estimation from Remotely Sensed Information using Artificial Neural Networks-Clouds Classification System (PERSIANN-CCS), a real-time global precipitation product was obtained with high-resolution ($0.04° \times 0.04°$), Fig. 2d. It enables the categorization of cloud-patch features based on cloud height, areal extent, and variability of texture estimated from satellite imagery (Nguyen et al., 2019). The data can be accessed from: https://chrsdata.eng.uci.edu/ (last access: 26 August 2022).

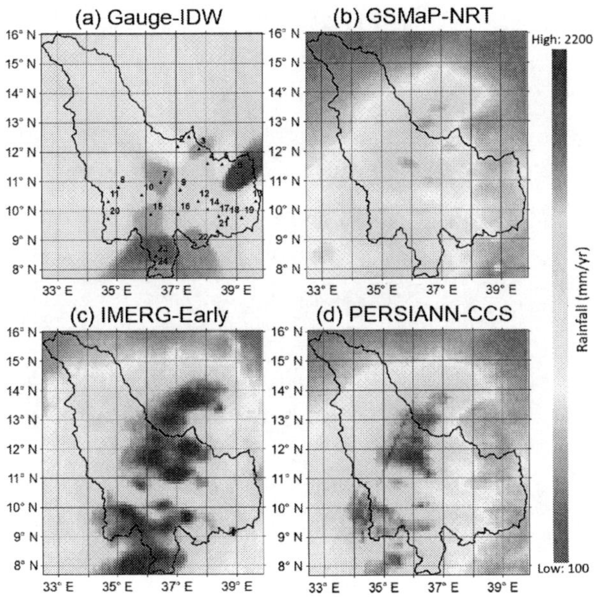

Figure 2. Annual average precipitation in Blue Nile basin between (2003–2012).

3.3 SPPs bias-correction method

The rainfall bias correction method for SPPs correction was selected here in light of the availability and distribution of observation data. For the application of bias correction factors, we used the Ratio bias correction at each gauge stations. The factors are computed at a 10 d and monthly average time step (rather than the original daily data window). This approach aids in eliminating errors in observation and outliers in daily observation. The technique involves utilizing each observation stations to convert three closest SPP grids into point data. Accordingly, the correction equations are modified as necessary to accommodate this modification.

$$F_k = \frac{P_{g,k}}{P_{s,k}} \quad (1)$$

where F_k is the correction factor; $P_{g,k}$ is the gauge, and $P_{s,k}$ is the SPP estimates for k stations average.

After that, the bias-corrected SPPs are calculated as

$$P_{scor(i,j)} = F_k \cdot P_{s(i,j)} \quad (2)$$

where $P_{scor(i,j)}$ represents the corrected estimates of the original SPPs ($P_{s(i,j)}$) at grid cell (i, j).

3.4 Hydrologic model forcing data

Model inputs such as air temperature, wind speed, shortwave downward radiation, longwave radiation, specific humidity, and surface pressure, were obtained from the Japanese 55 year Reanalysis (JRA-55) data prepared by the Japan Meteorological Agency (JMA) (Kobayashi et al., 2015). Global products for terrestrial evapotranspiration (ET) were prepared using MODIS version 08. The MOD16A2 terrestrial ET products were estimated using an improved algorithm by Mu et al. (2011). These data were interpolated to model grid resolutions of 1 arcmin.

3.5 Discharge simulation and evaluation

For extreme flow simulation, the SPPs were simulated using the Water Energy Budget-based Rainfall-Runoff-Inundation (WEB-RRI) model, a physically based distributed model. WEB-RRI is structured into four major modules to evaluate water and energy fluxes, low flow estimation, flood onset timing, peak flood discharge, and inundation characteristics (Rasmy et al., 2019). The model estimates overland flow by separating surface and groundwater with the 2-D diffusive wave module derived from the continuity, momentum and Manning's equations. The river routing of river channel within each slope grid cell is estimated using a 1-D diffusive wave equations.

For performance evaluation, categorical statistical indices, such as the threat score (TS), false-alarm ratio (FAR), probability of detection (POD), and strike ratio (SR) which evaluates how well the SPPs match to the gauged data, were employed. Non-categorical statistical error metrics measures, Correlation Coefficient (CC), Nash–Sutcliff equation (NSE), Mean Average Error (MAE), Mean Bias Error (MBE) and root-mean-squared error (RMSE) were used for model performance (i.e., the strengths and weaknesses of the model). The peak flood events hydrograph was evaluated for extreme flood event from observed discharge with discharge from SPPs simulation using the statistical error metrics.

4 Results and discussion

4.1 SPPs performance with gauged data

The grid-based satellite precipitation products (SPPs) were transformed into point data using the inverse distance weighting method for evaluating with the observation data. When comparing the original SPPs estimates against the observation data, it was observed that the IMERG-Early original data overestimate both the mean and peak rainfall. In contrast, the GSMaP-NRT and PERSIANN-CCS estimates were observed to be underestimates (as depicted in Fig. 3).

Further, categorical and non-categorical indices analyses results of the original SPPs comparison indicates that all the SPPs (GSMaP-NRT, IMERG-Early and PERSIANN-CCS) exhibited poor correlation with the observation data. These findings indicate the necessity for improvement of the SPPs.

Consequently, a bias correction was carried out to enhance the accuracy of the SPPs and assesses the performance difference between the two windows datasets (i.e. 10 d mean and monthly). To calculate the bias correction factor for each SPP, regression analysis was employed with the observation

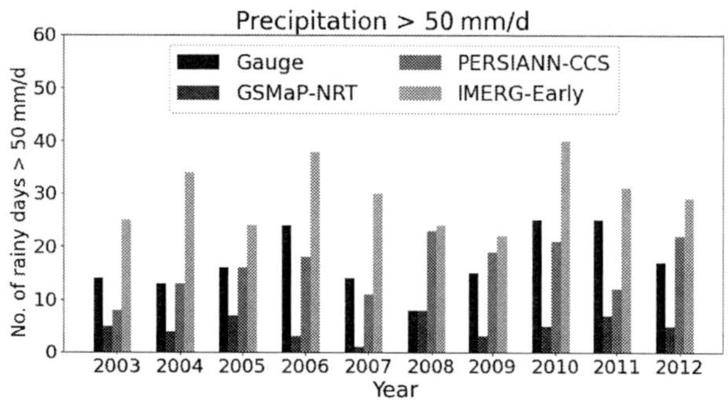

Figure 3. SPPs point scale comparison on extreme events.

Table 1. Evaluation indices metrics for basin average SPPs.

Evaluation type		GSMaP-NRT	IMERG-Early	PERSIANN-CCS
Non-Categorical	CC	0.38	0.41	0.33
	RMSE	7.69	8.95	8.56
	BIAS	−0.99	0.59	−0.7
	MAE	3.53	4.2	3.94
Categorical	Threat Score	0.57	0.56	0.52
	FAR	0.36	0.41	0.25
	POD	0.83	0.9	0.63
	Strike ratio	7.42	7.13	7.62

data. The three stations, near each observation location, are used to convert the gridded SPPs into point data for generating correction factors through regression. The scatterplots of SPPs for the two correction windows are presented in Fig. 4.

4.2 Flow evaluations

The WEB-RRI model was calibrated and validated, and the result indicates satisfactory range of error indices, as presented in Table 2.

Afterwards, discharge simulation of original and bias corrected SPPs was conducted to evaluate the performance of bias correction in reproducing the hydrograph of the flood event period. The results demonstrated that the bias correction significantly enhanced the simulation skills for each SPPs with the extreme flow hydrograph and error metrics (Fig. 5).

The results of each SPP, original and bias corrected simulations (with 10 d and monthly correction factors) performance were assessed with peak hydrograph and error metrics. These results indicate that the 10 d mean window performed better in improving the accuracy and reliability of the SPPs for discharge simulation.

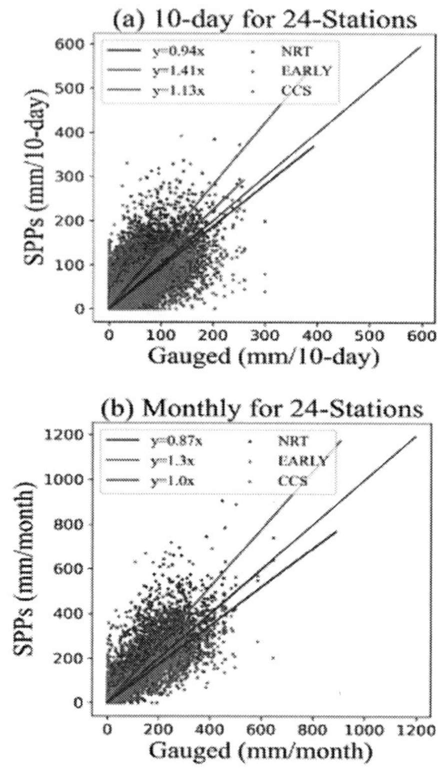

Figure 4. Regression graph of observed and SPPs rainfall.

Table 2. Model calibration and validation.

Simulation for	NSE	MBE ($m^3 s^{-1}$)	RMSE ($m^3 s^{-1}$)
Calibration	0.53	91.51	602.57
Validation	0.67	1.63	509.76

4.3 Flood event simulation

To enhance the applicability of real time flood estimation with the catchment characteristics, we conducted an investigation on the lead time and accuracy of SPPs in WEB-RRI model. This investigation considered an observed extreme event that occurred in July 2012. As shown in Fig. 6, real time dataset simulations provides lead time of SPPs up to 78 h with higher accuracy at the downstream of the Blue Nile in Sudan. Moreover, when the model was driven by real-time SPPs integrated with initial observed conditions, the accuracy of the model significantly improved, resulting to an acceptable error metrics index with longer lead time.

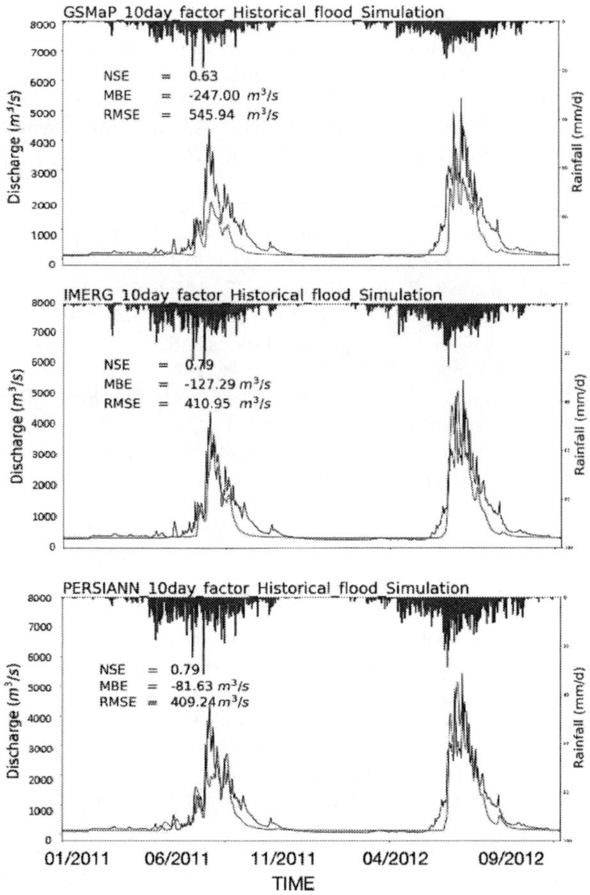

Figure 5. Flow hydrograph for observation and bias corrected SPPs with the best performing correction factors.

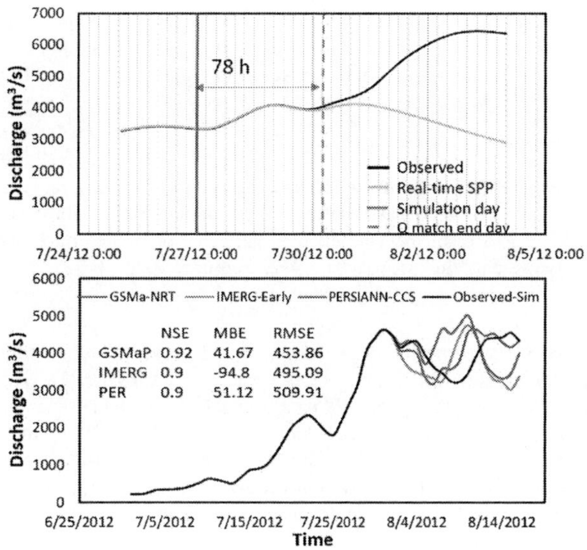

Figure 6. Performance comparison of (a) real-time SPPs lead time and (b) integrated simulation of observed and SPPs.

5 Conclusion

From the results obtained in this study, the following conclusions can be drawn regarding the applicability of SPPs to flood monitoring:

- The applicability of bias correction methods significantly improved the accuracy of the SPPs in simulating river flow and floods modelling.
- The study utilized three independent real-time SPPs (GSMaP, GPM–IMERG, and PERSIANN) for the simulation of river flow and floods. The results revealed that all are applicable with variations in the performance of these products.
- By evaluating the lead time and accuracy of SPPs, it was found that the models driven by real-time SPPs can predict floods with acceptable errors ranges for timely and reliable flood forecasting and disaster preparedness.
- The findings of this study have practical implications for informed decision making in flood preparedness, mitigation, and response in transboundary river basins.

Code and data availability. The hydrologic model (WEB-RRI) codes and descriptions are detailed in the subsequent publication: Rasmy et al. (2019). The SPPs data were obtained from open-source datasets, with access links provided in the data description section. Observed data requests should be directed to the respective institutions. Some of the datasets and codes employed for data processing are available at Zenodo [codes] via the following links: https://doi.org/10.5281/zenodo.10835234 (Tedla, 2024).

Author contributions. Conceptualization, investigation, methodology, software and visualization, MGT and MR; Data curation, formal analysis, MGT and LZ; Funding acquisition, project administration, resources, supervision and validation, MR and TK; writing – original draft, MGT; writing – review and editing, MGT, MR and TK.

Competing interests. At least one of the (co-)authors is a guest member of the editorial board of *Proceedings of IAHS* for the special issue "ICFM9 – River Basin Disaster Resilience and Sustainability by All". The peer-review process was guided by an independent editor, and the authors also have no other competing interests to declare.

lished maps, institutional affiliations, or any other geographical representation in this paper. While Copernicus Publications makes every effort to include appropriate place names, the final responsibility lies with the authors.

Special issue statement. This article is part of the special issue "ICFM9 – River Basin Disaster Resilience and Sustainability by All". It is a result of The 9th International Conference on Flood Management, Tsukuba, Japan, 18–22 February 2023.

Acknowledgements. The authors would like to thank GSMaP (JAXA), GPM-IMERGE (NASA) and the PERSIANN (UCI-CHRS) for providing the SPP data for analysis, model simulations and evaluation. We thank the Ethiopian National Meteorology Institute and the Ministry of Water and Energy for the observed datasets. We also acknowledge anonymous reviewers for their comments and suggestions that improved this manuscript.

Review statement. This paper was edited by Mamoru Miyamoto and reviewed by two anonymous referees.

References

Easton, Z. M., Awulachew, S. B. Steenhuis, T. S., Habte, S. A., Zemadim, B., Seleshi, Y., and Bashar, K. E.: Hydrological process in the Blue Nile, IWMI Books, Reports H045313, International Water Management Institute, 2012.

Gebregiorgis, A. S. and Hossain, F.: Understanding the Dependence of Satellite Rainfall Uncertainty on Topography and Climate for Hydrologic Model Simulation, IEEE T. Geosci. Remote, 51, 704–718, https://doi.org/10.1109/TGRS.2012.2196282, 2013.

Illingworth, A. J., Barker, H. W., Beljaars, A., Ceccaldi, M., Chepfer, H., Clerbaux, N., Cole, J., Delanoë, J., Domenech, C., Donovan, D. P., Fukuda, S., Hirakata, M., Hogan, R. J., Huenerbein, A., Kollias, P., Kubota, T., Nakajima, T., Nakajima, T. Y., Nishizawa, T., Ohno, Y., Okamoto, H., Oki, R., Sato, K., Satoh, M., Shephard, M. W., Velázquez-Blázquez, A., Wandinger, U., Wehr, T., and van Zadelhoff, G.-J.: The EarthCARE satellite: the next step forward in global measurements of clouds, aerosols, precipitation, and radiation, B. Am. Meteorol. Soc., 96, 1311–1332, https://doi.org/10.1175/BAMS-D-12-00227.1, 2015.

Kobayashi, S., Ota, Y., Harada, Y., Ebita, A., Moriya, M., Onoda, H., Onogi, K., Kamahori, H., Kobayashi, C., and Endo, H.: The JRA-55 reanalysis: General specifications and basic characteristics, J. Meteorol. Soc. Jpn. Ser. II, 93, 5–48, 2015.

Kubota, T., Shige, S., Hashizume, H., Aonashi, K., Takahashi, N., Seto, S., Hirose, M., Takayabu, Y. N., Ushio, T., Nakagawa, K., Iwanami, K., Kachi, M., and Okamoto, K.: Global precipitation map using satellite-borne microwave radiometers by the GSMaP project: Production and validation, IEEE T. Geosci. Remote, 45, 2259–2275, https://doi.org/10.1109/TGRS.2007.895337, 2007.

Lo Conti, F., Hsu, K.-L., Noto, L. V., and Sorooshian, S.: Evaluation and comparison of satellite precipitation estimates with reference to a local area in the Mediterranean Sea, Atmos. Res., 138, 189–204, https://doi.org/10.1016/j.atmosres.2013.11.011, 2014.

Mu, Q., Zhao, M., and Running, S. W.: Improvements to a MODIS global terrestrial evapotranspiration algorithm, Remote Sens. Environ., 115, 1781–1800, https://doi.org/10.1016/j.rse.2011.02.019, 2011.

Nguyen, P., Shearer, E. J., Tran, H., Ombadi, M., Hayatbini, N., Palacios, T., Huynh, P., Braithwaite, D., Updegraff, G., Hsu, K., Kuligowski, B., Logan, W. S., and Sorooshian, S.: The CHRS Data Portal, an easily accessible public repository for PERSIANN global satellite precipitation data, Sci. Data, 6, 180296, https://doi.org/10.1038/sdata.2018.296, 2019.

Rasmy, M., Sayama, T., and Koike, T.: Development of water and energy Budget-based Rainfall-Runoff-Inundation model (WEB-RRI) and its verification in the Kalu and Mundeni River Basins, Sri Lanka, J. Hydrol., 579, 124163, https://doi.org/10.1016/j.jhydrol.2019.124163, 2019.

Tan, J., Huffman, G. J., Bolvin, D. T., and Nelkin, E. J.: IMERG V06: Changes to the Morphing Algorithm, J. Atmos. Ocean. Tech., 36, 2471–2482, https://doi.org/10.1175/JTECH-D-19-0114.1, 2019.

Tedla, M. G.: Satellite Precipitation Products data performance evaluation with observed datasets, Zenodo [code and data set], https://doi.org/10.5281/zenodo.10835234, 2024.

Tedla, M. G., Rasmy, M., Tamakawa, K., Selvarajah, H., and Koike, T.: Assessment of Climate Change Impacts for Balancing Transboundary Water Resources Development in the Blue Nile Basin, Sustainability, 14, 15438, https://doi.org/10.3390/su142215438, 2022.

Ushio, T., Sasashige, K., Kubota, T., Shige, S., Okamoto, K., Aonashi, K., Inoue, T., Takahashi, N., Iguchi, T., Kachi, M., Oki, R., Morimoto, T., and Kawasaki, Z.-I.: A Kalman Filter Approach to the Global Satellite Mapping of Precipitation (GSMaP) from Combined Passive Microwave and Infrared Radiometric Data, J. Meteorol. Soc. Jpn. Ser. II, 87A, 137–151, https://doi.org/10.2151/jmsj.87A.137, 2009.

World Bank: Africa Development Indicators 2006: From the World Bank Africa Database, The World Bank, 2006.

Zhou, L., Rasmy, M., Takeuchi, K., Koike, T., Selvarajah, H., and Ao, T.: Adequacy of Near Real-Time Satellite Precipitation Products in Driving Flood Discharge Simulation in the Fuji River Basin, Japan, Appl. Sci., 11, 3, https://doi.org/10.3390/app11031087, 2021.

Zhou, L., Koike, T., Takeuchi, K., Rasmy, M., Onuma, K., Ito, H., Selvarajah, H., Liu, L., Li, X., and Ao, T.: A study on availability of ground observations and its impacts on bias correction of satellite precipitation products and hydrologic simulation efficiency, J. Hydrol., 610, 127595, https://doi.org/10.1016/j.jhydrol.2022.127595, 2022.

Proc. IAHS, 386, 229–235, 2024
https://doi.org/10.5194/piahs-386-229-2024

Application of Entrainment Velocity Concept for Evaluation of River Morphology: A Numerical Model Study

Robin K. Biswas[1], **Shinji Egashira**[2], **and Daisuke Harada**[2]
[1]Bangladesh Water Development Board, Dhaka, 1205, Bangladesh
[2]International Centre for Water Hazard and Risk Management, Tsukuba-shi, Ibaraki-ken 305-8516, Japan
Correspondence: Robin K. Biswas (biswasrobin@gmail.com)

Received: 29 July 2023 – Revised: 7 November 2023 – Accepted: 24 February 2024 – Published: 19 April 2024

Abstract. Conventionally, fine sediment transport is evaluated in terms of advection-diffusion equation of suspended sediment in which the erosion rate of sediment from the bedload layer plays a significant role. The present study aims to evaluate the riverbed evolution and associated morphological changes in fine sediment dominated Bhairab River of Bangladesh by a depth integrated two-dimensional numerical model. The erosion rate of bed sediment has been evaluated by two approaches. The first approach applies the fall velocity of the sediment particle and equilibrium reference sediment concentration. The other employs the entrainment velocity concept. The results obtained from numerical simulations are verified by the field investigations. The results suggest that the riverbed evolution and morphological changes predicted with the entrainment velocity concept show better agreement with the data obtained from the field investigations.

1 Introduction

Most of the riverbeds in Bangladesh are composed of fine-grained sediment that is easily entrained, slowly settled, remained in suspension, transported by flow over a long distance before settling and actively involved in the riverbed evolution and morphological changes. Many researchers such as Celik and Rodi (1988) have studied the mechanics of fine sediment transport and associated morphological changes taking the support of numerical model simulations. Depth-averaged equations comprised of governing equations for water flow, mass conservation for bedload and suspended sediment have been employed in these studies. The mass conservation of the suspended sediment is evaluated by advection–diffusion equation that contains erosion rate or sediment entrainment rate term and deposition rate term. The advection-diffusion equation for suspended sediment in non-equilibrium condition is as follows:

$$\frac{\partial ch}{\partial t}+\frac{\partial cuh}{\partial x}+\frac{\partial cvh}{\partial y}=\frac{\partial}{\partial x}\left(\varepsilon_x h\frac{\partial c}{\partial x}\right)+\frac{\partial}{\partial \mathrm{x}}\left(\varepsilon_y h\frac{\partial c}{\partial y}\right)+E-D. \quad (1)$$

where c is the depth-averaged concentration of suspended sediment in the water column, h is the thickness of upper flow layer where sediment remain in suspension, u and v are the depth-averaged velocities in the x and $y$directions, and ε_x and ε_y are the mixing coefficients in the x and y directions, E is the entrainment rate or erosion rate of the sediment from the bedload layer (hereinafter erosion rate term), D is the deposition rate of sediment from water column above the bedload layer.

Several formulas are available in literature for the erosion rate term emanating from the difference in the treatment of the sediment kinematics and the equilibrium reference sediment concentration (c_{ae}). Many researchers such as, Itakura and Kishi (1980), Van Rijn (1984, 1986), Garcia and Parker (1991) have proposed formulae for determining c_{ae} and computed erosion rate as $E=c_{\mathrm{ae}}w_0$, where

w_0 is the fall velocity of the sediment particle. Garcia and Parker (1991) compared seven relations proposed by Einstein (1950), Engelund and Fredsoe (1976), Smith and McLean (1977), Itakura and Kishi (1980), Van Rijn (1984), Celik and Rodi (1984), Akiyama and Fukushima (1986) for entrainment rate of non-cohesive sediment into suspension and concluded that the relations proposed by van Rijn and Smith and McLean perform show better agreement against a common set of experimental data. Findings from investigations on the mechanics of suspended sediment-laden flow as well as data from the experimental and field observations have revealed that the conventional methods overestimate c_{ae} (Itakura and Kishi, 1980; Garcia and Parker, 1991). Harada et al. (2022) compared the equilibrium reference suspended sediment concentration obtained using the methods proposed by them and the formulas proposed by Van Rijn (1986), Itakura and Kishi (1980) and Garcia and Parker (1991). They found that the formula proposed by Van Rijn (1986), Itakura and Kishi (1980) and Garcia and Parker (1991) exhibit higher sediment concentration for sediment particles having w_0/u_* in the range of 0.01 to 0.1.

Noticeable deviations are noted among studies when c_{ae} from field observations and experimental investigations are plotted against w_0/u_* where, u_* is the shear velocity and defined as $u_* = \sqrt{ghi_e}$ and i_e is the energy slope. Moreover, riverbed having sediment diameter several tens of micrometres and w_0/u_* falls out lower range, difficulties arise to apply c_{ae} chart and predicted erosion rates by the conventional formulae are relatively high (Harada and Egashira, 2020; Harada et al., 2022). Density stratification during the high flow conditions significantly affects the suspended sediment transport (Wright and Parker, 2004). The density stratification effects tend to be greater in large, low-slope rivers. To solve such difficulties, Harada and Egashira (2020) have introduced the entrainment velocity concept that rooted on the density stratified flows. They also have developed a method to evaluate erosion rate for very fine sediment and validated it by data from flume tests.

Bedload transport is crucial to maintain stable channel in sand bed rivers. Lateral diffusive transport of suspended sediment from the central region of the river increases the suspended sediment concentration besides deposition rate near the bank region. The deposited sediment returns to the central region as later bedload under the influence of gravity. A stable channel cross section is achieved when the lateral diffusive transport of suspended sediment is balanced by the lateral bedload (Parker, 1978; Ikeda and Izumi, 1991). Estimation of the bedload transport depend on the functional form of bedload transport rate formulae. Most of the usually applied formulae such as, formulae proposed by Meyer-Peter and Müller (1948), Ashida and Michiue (1972) can be expressed in terms of shear stress excess or its equivalent and have a functional form of $\tau_*^{1.5}$ for high bed shear stress, here τ_* is the non-dimensional bed shear stress. Bedload rates using such formulae for high bed shear stress is independent of sediment size. There are bedload rate formulae with its functional form of $\tau_*^{2.5}$ (e.g., Egashira et al., 1997) in which the bedload rates are contrariwise proportional to sediment particle size. Harada and Egashira (2020) has proposed the visco-coulomb model to evaluate bedload rates for very fine sediment which has been derived from the velocity profile in bedload layer and is expressed as follows:

$$q_b = \frac{c_s}{6}\frac{u_* h_s}{\upsilon} u_* h_s. \tag{2}$$

where q_b is the bedload rate, c_s is the depth-averaged concentration of sediment in the bedload layer, h_s is the thickness of bedload layer, ν is the kinematic viscosity. The bedload layer thickness is expressed as follows:

$$\frac{h_s}{h} = \frac{\tan\theta}{(\sigma/\rho - 1)\, c_s \tan\phi}. \tag{3}$$

where $\tan\theta$ is the bed slope that, under non-uniform flows slowly varying in space, can be replaced by energy slope, σ and ρ is the mass density of sediment and water respectively, ϕ is the angle of friction.

The present study aims to evaluate the riverbed evolution and associated morphological changes in a fine sediment-dominated Bhairab River of Bangladesh using a depth-averaged two-dimensional model. To do so, we have estimated the erosion rate by applying the entrainment velocity concept as well as by the conventional method. The bedload rates are evaluated by bedload formulae having functional form $\tau_*^{1.5}$ and $\tau_*^{2.5}$ as well as by the visco-coulomb model. Moreover, the applicability of newly proposed formula of erosion rate by Harada and Egashira (2020) is discussed comparing with existing formulae.

2 Characteristic of study area

Bhairab River an alluvial meandering river is one of the right bank distributaries of the Ganges River that originates at Murshidabad district of West Bengal, India and enters into Bangladesh near Meherpur district. The total length within Bangladesh and average width is about 90 km and 60 m, respectively. The study reach is about 25 km long. Based on Landsat images from 1978 to 2016, bank lines of the Bhairab River have minor shifting tendency over the past thirty-eight years. Currently, Bhairab is disconnected from the Ganges River. Thus, the river is no more fed by the water and sediment from the Ganges River. The sediment load originates mainly from the riverbed which gives an opportunity to understand the entrainment rate of the fine sediment from the riverbed. Figure 1 shows that the downstream part of the river more meandering in nature comparing to the upstream part. The longitudinal bed slope of is 3.5 cm km^{-1}. The riverbed is composed of sediment having $D_{50} = 46$ µm and average suspended sediment concentration varies from 50 to 70 mg L^{-1}. Figure 2shows that there are no significant changes in grain

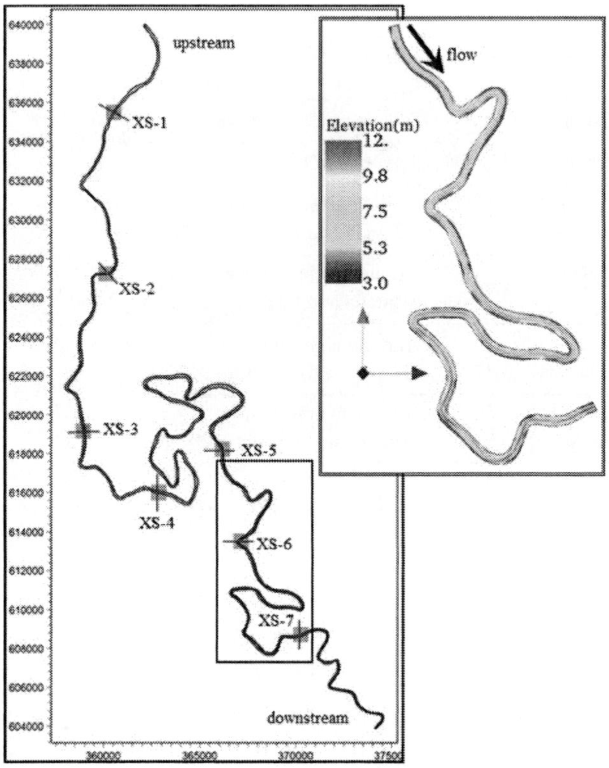

Figure 1. Planform of the part of Bhairab River within Bangladesh and elevation profile of study reach together with locations of the measured cross-sections.

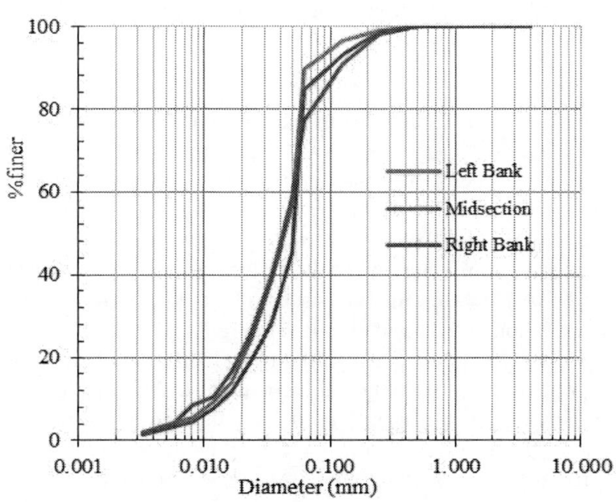

Figure 2. Grain size distribution of bed and bank materials.

size distributions of riverbed and bank materials and thus between the floodplain and riverbed. In 2000, the river experienced maximum discharge of $140\,\mathrm{m^3\,s^{-1}}$. However, minimum flow was found to be less than $1\,\mathrm{m^3\,s^{-1}}$.

3 Governing equations for numerical model

To assess riverbed evolution and morphological changes, we have employed mass conservation equations for upper flow layer with suspended sediment, and for lower flow layer in which bedload dominates, besides the governing equations for the water flow. The governing equations are as follows:

$$\frac{\partial h}{\partial t}+\frac{\partial (uh)}{\partial x}+\frac{\partial (vh)}{\partial y}=0. \tag{4}$$

$$\frac{\partial (hu)}{\partial t}+\frac{\partial (huu)}{\partial x}+\frac{\partial (huv)}{\partial y}+gh\frac{\partial (h+z_\mathrm{b})}{\partial x}+\frac{\tau_x}{\rho}-\frac{\partial (h\sigma_{xx})}{\partial x}-\frac{\partial (h\tau_{yx})}{\partial y}=0. \tag{5}$$

$$\frac{\partial (hv)}{\partial t}+\frac{\partial (huv)}{\partial x}+\frac{\partial (hvv)}{\partial y}+gh\frac{\partial (h+z_\mathrm{b})}{\partial y}+\frac{\tau_y}{\rho}-\frac{\partial \left(h\tau_{xy}\right)}{\partial x}-\frac{\partial (h\sigma_{yy})}{\partial y}=0. \tag{6}$$

$$\frac{\partial z_\mathrm{b}}{\partial t}+\frac{1}{1-\lambda}\left(\frac{\partial q_{\mathrm{b}x}}{\partial x}+\frac{\partial q_{\mathrm{b}y}}{\partial y}+E-D\right)=0. \tag{7}$$

where where, t is the time, x and y are the coordinate in flow direction and transverse to flow direction, g is the acceleration due to gravity, σ_{xx}, σ_{yy}, τ_{xy} and τ_{yx} are the depth-averaged Reynolds stresses, τ_x and τ_y are the x and y components of bed shear stress (τ_b), z_b is the river bed elevation, $q_{\mathrm{b}x}$ and $q_{\mathrm{b}y}$ are the x and y components of the bedload rate (q_b), λ is the porosity. The bed shear stress and depth averaged Reynolds-stresses are evaluated by conventional method.

We have applied three methods: (a) conventional formula (Ashida and Michiue, 1972) (b) visco-coulomb formula (Harada and Egashira, 2020) and (c) Egashira et al. (1997) formula to evaluate the bedload rates. The erosion rate is evaluated by conventional method (Itakura and Kishi, 1980) as well as by entrainment velocity concept (Harada and Egashira, 2020). Equilibrium sediment concentration (c_e) and entrainment velocity (W_e) for the latter approach that rooted on density stratified flow and are expressed as follows:

$$c_\mathrm{e}=\frac{W_\mathrm{e}}{w_0}c_\mathrm{s},\ \frac{W_\mathrm{e}}{u}(=e)=\frac{K}{R_{i*}},\ R_{i*}=\left(\frac{\sigma}{\rho}-1\right)c_\mathrm{s}\frac{gh}{u^2}. \tag{8}$$

$$c_\mathrm{e}=\frac{K}{w_0/u_*}\sqrt{\frac{8}{f}}\frac{1}{(\sigma/\rho-1)}\frac{u^2}{gh}=\frac{K}{w_0/u_*}\left(\frac{8}{f}\right)^{3/2}\frac{i_\mathrm{e}}{(\sigma/\rho-1)}. \tag{9}$$

where R_{i*} is the Richardson number, e is the entrainment coefficient, f is the Darcy–Weisbach friction factor that defined as $u/u_*=\sqrt{8/f}$, $K=1.5\times10^{-3}$ (Egashira and Ashida, 1980). Nays2dh solver (Asahi et al., 2013) with appropriate modifications is applied for conducting the numerical simulations with the help of International River Interface Cooperative (iRIC) graphical user interface. Finite difference

Table 1. Numerical simulations conditions adopted in the present study.

Case	Flow	Bedload treatment	Erosion rate
1-1	Constant ($160\,m^3\,s^{-1}$)	Ashida and Michiue (1972)	Itakura and Kishi (1980)
1-2	Hydrograph	Ashida and Michiue (1972)	Itakura and Kishi (1980)
2-1	Constant ($160\,m^3\,s^{-1}$)	Egashira et al. (1997)	Turbulent entrainment (Harada and Egashira, 2020)
2-2	Hydrograph	Egashira et al. (1997)	Turbulent entrainment (Harada and Egashira, 2020)
3-1	Constant ($160\,m^3\,s^{-1}$)	Density current (Harada and Egashira, 2020)	Turbulent entrainment (Harada and Egashira, 2020)
3-2	Hydrograph	Density current (Harada and Egashira, 2020)	Turbulent entrainment (Harada and Egashira, 2020)

scheme with general coordinate systems is employed to conduct the numerical computations.

4 Results and discussions

4.1 Calculation conditions

A total of 6 simulations have been conducted considering uniform sediment bed having D_{50} of 46 µm. Two types of flow (a) constant discharge of $160\,m^3\,s^{-1}$ and (b) flow hydrograph as shown in Fig. 3 are supplied from the upstream boundary. The flow conditions are defined in order to understand the erosion-deposition process during the peak discharge as well as under the non-uniform flow condition. The computational domain is long enough. There are no time series of measured water levels at the downstream boundary. Therefore, it is assumed that the water surface at downstream boundary is parallel to the river bed surface and the flow condition in the downstream boundary is defined as uniform flow. The initial bed elevation is defined based on the measured bathymetric data. There are no measured sediment data at the upstream boundary. Therefore, the sediment boundary condition is defined by assuming an exponential distribution of suspended sediment in vertical direction and equilibrium sediment concentration profile for the conventional method. The computational domain is divided into 11 420 grids having $\Delta x = 35$ m and $\Delta y = 12$ m. The time step for numerical simulation is 0.5 s. The details of numerical simulations conditions are shown below in Table 1.

4.2 Riverbed evolution and morphological changes

Figure 4 is the results showing the riverbed evolution in terms of riverbed aggradation and degradation obtained by numerical simulations using bedload formulae having functional form proportional to $\tau_*^{1.5}$ (Ashida and Michiue, 1972). The erosion rate of bed sediment is evaluated by conventional method following the formula of Itakura and Kishi (1980).

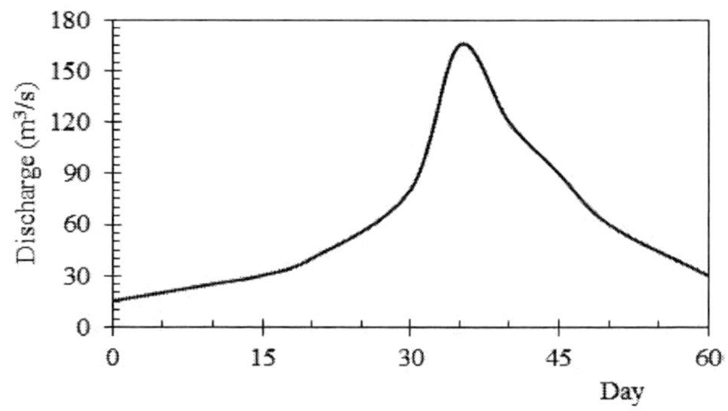

Figure 3. Flow hydrograph supplied at upstream boundary during numerical simulations.

Figure 4a is the resultant riverbed level changes after passing a constant discharge ($160\,m^3\,s^{-1}$) for 20 d, whereas Fig. 4b shows the results after passing the hydrograph as shown in Fig. 3. Figure 4a shows that riverbed degradation takes place actively along the channel centre with riverbed aggradation in bank region. Figure 4b shows riverbed degradation around the bend. However, in both cases, more than 7 m of river bed aggradation and degradation is observed.

Figure 5a shows the results of riverbed aggradation and degradation that are obtained from numerical simulations by employing visco-coulomb model for the bedload rate. Figure 5b is the results showing riverbed evolution by bedload a formula having functional form proportional to $\tau_*^{2.5}$. In both cases, erosion rate is evaluated by applying the entrainment velocity concept. The results indicate active bed lowering around the bend. The bed degradation in conventional method is about 3 times higher comparing to the same in the entrainment velocity concept.

The difference instigates due to the treatment of sediment entrainment. The results suggest that the equilibrium reference sediment concentration determined by the conventional

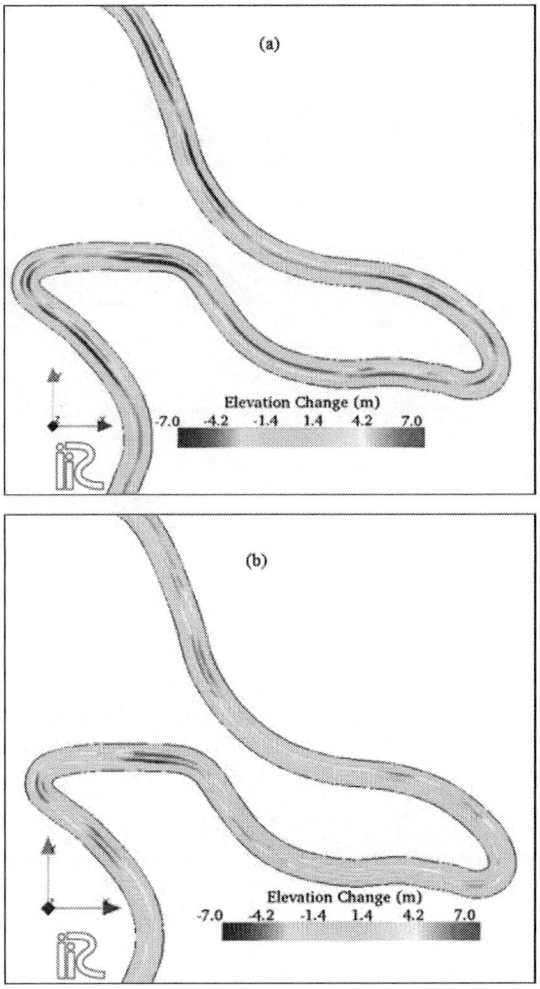

Figure 4. Riverbed evolution based on Ashida and Michiue (1972) and Itakura and Kishi (1980) formulae for **(a)** constant discharge, **(b)** for hydrograph.

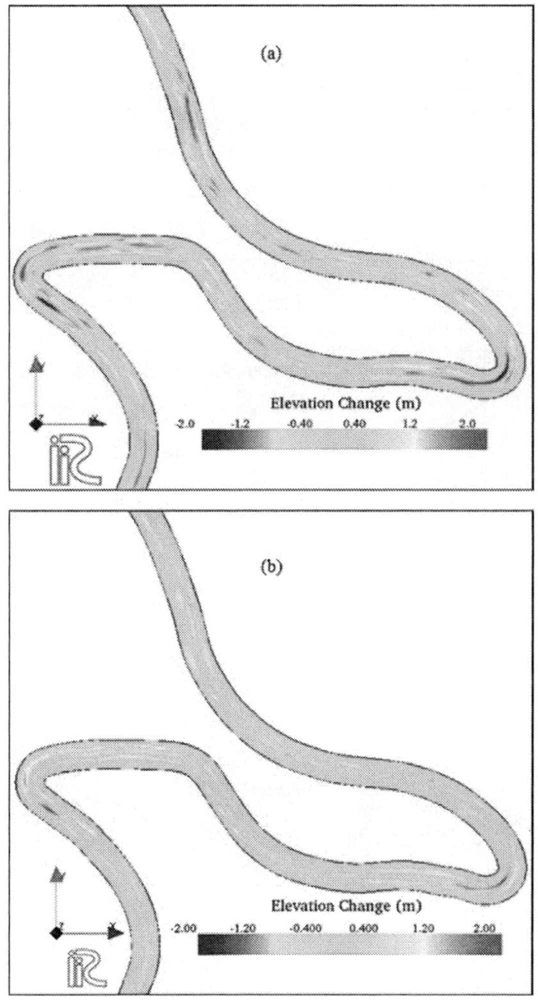

Figure 5. Riverbed evolution for constant discharge **(a)** visco-coulomb and **(b)** Egashira et al. (1997) formula with entrainment velocity concept.

method is comparatively high. Figure 6 is the actual changes in cross-section in the field at XS-5, XS-6 and XS-7 between 2016 to 2020. XS-5 is located at the upstream and XS-7 is located at downstream. The observed data shows very small riverbed degradation at each of the cross-sections. This indicates the erosion and deposition process are almost balancing each other. Figure 7 shows simulated cross-section at XS-7 after 20 d for a constant discharge of $160\,m^3\,s^{-1}$. The results suggest active bed degradation at the channel centre for the computation with conventional method. According to Harada et al. (2022), equilibrium reference suspended sediment concentration predicted with the orthodox approaches exhibits higher sediment concentration for fine sediment particles. The erosion rate which is the product of equilibrium reference suspended sediment concentration and the particle fall velocity results comparatively higher value. The simulated results also show similar tendency which differ from the data obtained from the field observations The results obtained with the entrainment velocity concept are found to capture the bed degradation process and show better agreement with the data from field investigations.

5 Summary

Suspended sediment transportation has a remarkable non-equilibrium characteristic. The present study discusses riverbed evolution and associated morphological changes based on the results of numerical simulations and field observation in a reach of Bhairab River in Bangladesh where the riverbed is composed of very fine sediment. Results of

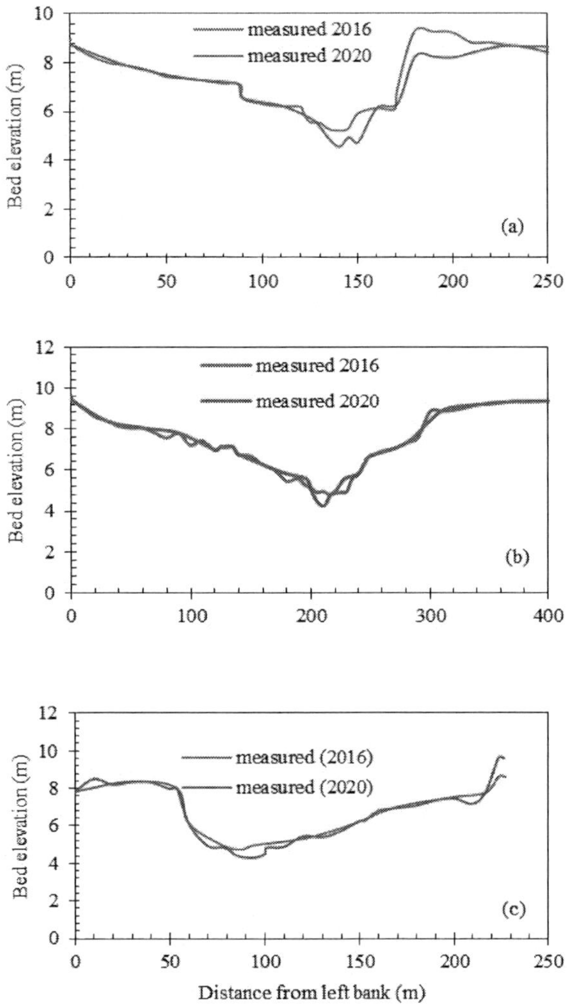

Figure 6. Changes in riverbed between 2016 and 2020 at **(a)** XS-5 **(b)** XS-6 and **(c)** XS-7 based on measured data.

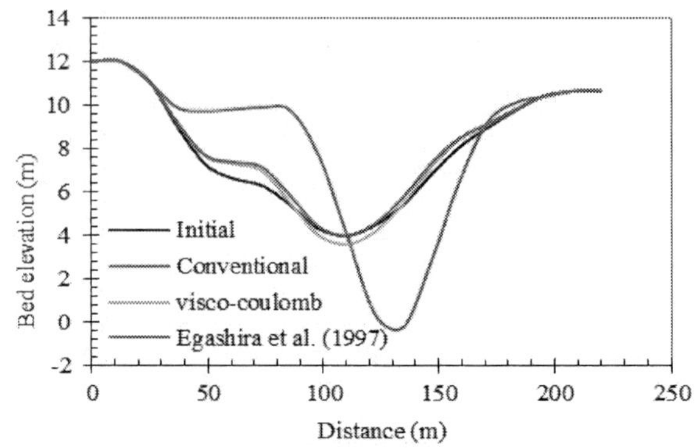

Figure 7. Comparison between river cross-sections at XS-7 based on simulated data using conventional method, visco-coulomb and Egashira et al. (1997) using a constant discharge of $160\,\mathrm{m^3\,s^{-1}}$.

the present study primarily support the following main conclusions:

1. Erosion rates evaluated with entrainment velocity concept and equilibrium sediment concentration produce noticeable dissimilarities in the channel planform from identical initial channel geometry, riverbed slope, sediment size distribution and roughness conditions after experiencing similar flow condition.

2. The predicted bed degradation with the conventional method is about 3 times higher comparing to the same in the entrainment velocity concept.

3. The cross-section profiles obtained from numerical simulations with entrainment velocity concept are consistent with the data from field investigations.

4. It is extremely difficult to examine the applicability of the bed-load equation to a suspended sediment dominated river. The present simulations have been conducted based on measured bathymetric and grain size distribution data only. Considering the issue of limited measured data and uncertainty of sediment transport, the results will be further verified based on the rigorous field observations.

Code availability. The numerical simulations for flow and channel changes have been conducted using the International River Interface Cooperative (iRIC) software platform. Nays2DH module has been applied in the study. The codes are not publicly available. However, details of the software such as the Solver Manual (Shimizu and Takebayashi, 2011) and User's 30 Manual (iRIC, 2022) can be accessed from the relevant webpage of iRIC (https://i-ric.org/en/, last access: 18 March 2024).

Data availability. Data and meta-data which are utilized in the present study are available upon request to the corresponding author.

Author contributions. RKB, SE and DH conceived of the presented idea and contributed to the development of the theory. RKB collected the data from the field level, performed the numerical computations. All the authors verified the analytical methods. SE encouraged RKB to investigate the applicability of the entrainment velocity concept and supervised the findings of this study. All authors discussed the results and contributed to the final manuscript.

Competing interests. At least one of the (co-)authors is a guest member of the editorial board of *Proceedings of IAHS* for the special issue "ICFM9 – River Basin Disaster Resilience and Sustainability by All". The peer-review process was guided by an independent editor, and the authors also have no other competing interests to declare.

Disclaimer. Publisher's note: Copernicus Publications remains neutral with regard to jurisdictional claims made in the text, published maps, institutional affiliations, or any other geographical representation in this paper. While Copernicus Publications makes every effort to include appropriate place names, the final responsibility lies with the authors.

Special issue statement. This article is part of the special issue "ICFM9 – River Basin Disaster Resilience and Sustainability by All". It is a result of The 9th International Conference on Flood Management, Tsukuba, Japan, 18–22 February 2023.

Acknowledgements. The authors would like to express sincere gratitude to Bangladesh Water Development Board (BWDB), Institute of Water Modelling (IWM), Bangladesh and International Center for Water Hazards and Risk Management (ICHARM), Japan for their support throughout the study. The data and information provided by IWM have played an important role in this study. The authors also express thankfulness to the anonymous reviewers for their constructive comments.

Review statement. This paper was edited by Kensuke Naito and reviewed by two anonymous referees.

References

Akiyama, J. and Fukushima, Y.: Entrainment of noncohesive sediment into suspension, 3rd Int. Symp. on River Sedimentation, edited by: Wang, S. Y., Shen, H. W., and Ding, L. Z., Univ. of Mississippi, 804–813, ISBN 0937099007, 1986.

Asahi, K., Shimizu, Y., Nelson, J., and Parker, G.: Numerical simulation of river meandering with self-evolving banks, J. Geophys. Res.-Earth, 118, 2208–2229, https://doi.org/10.1002/jgrf.20150, 2013.

Ashida, K. and Michiue, M.: Study on hydraulic resistance and bedload transport rate in alluvial streams, Proc. JSCE, 206, 59–69, 1972.

Celik, I. and Rodi, W.: A deposition entrainment model for suspended sediment transport, Report SFB 210/T/6, Universitat Karlsruhe, Karlsruhe, West Germany, 1984.

Celik, I. and Rodi, W.: Modeling suspended sediment transport in non-equilibrium situations, J. Hydraul. Eng., 114, 1157–1197, https://doi.org/10.1061/(ASCE)0733-9429(1988)114:10(1157), 1988.

Einstein, H. A.: The bed load function for sediment transportation in open channels. Technical Bulletin 1026, U.S. Dept. of Agric, Soil Conservation Service, Washington DC, 1950.

Egashira, S. and Ashida, K.: Studies on the structures of density stratified flows, Bulletin of DPRI Kyoto Univ., 29, 165–198, 1980.

Egashira, S., Miyamoto, K. and Itoh, T.: Constitutive equations of debris flow and their applicability, Int. Conf. on Debris-Flow Hazards Mitigation: Mechanics, Prediction, and Assessment, Proceedings, 340–349, https://doi.org/10.13140/2.1.4623.7122, 1997.

Engelund, F. and Fredsoe, J.: A sediment transport model for straight alluvial channels, Nordic Hydrol., 7, 293–306, https://doi.org/10.2166/nh.1976.0019, 1976.

Garcia, M. and Parker, G.: Entrainment of bed sediment into suspension, J. Hydraul. Eng., 117, 414–435, https://doi.org/10.1061/(ASCE)0733-9429(1991)117:4(414), 1991.

Harada, D. and Egashira, S.: Erosion rate of bed sediment in terms of entrainment concept, Proc. of the 22nd IAHR-APD Congress, Sapporo, Japan, 14–16 September 2020, ID 1-2-3, 2020.

Harada, D., Egashira, S., Ahmed, T. S., and Ito, H.: Entrainment of bed sediment composed of very fine material, Earth Surf. Proc. Land., 47, 3051–3061, https://doi.org/10.1002/esp.5442, 2022.

Ikeda, S. and Izumi, N.: Stable channel cross section of straight sand rivers, Water Resour. Res., 27, 2429–2438, https://doi.org/10.1029/91WR01220, 1991.

iRIC: iRIC Developer's manual, Release 3.0.0, https://iric-solver-dev-manual.readthedocs.io/_/downloads/en/v3_en/pdf/ (last access: 18 March 2024), 2022.

Itakura, T. and Kishi, T.: Open channel flow with suspended sediments, J. Hydraul. Div., 106, 1325–1343, https://doi.org/10.1061/JYCEAJ.0005483, 1980.

Meyer-Peter, E. and Müller, R.: Formulas for bed-load transport, IAHSR 2nd meeting, Stockholm, Sweden, 7–9, VI. 1948, appendix 2 ,39–64, 1948.

Parker, G.: Self-formed straight rivers with equilibrium banks and mobile bed, Part 1. The sand-silt river, J. Fluid Mech., 89, 109–125, 1978.

Shimizu, Y. and Takebayashi, H.: Nays2DH Solver Manual, https://i-ric.org/webadmin/wp-content/uploads/2018/12/Nays2DH_SolverManual_English_v4.pdf (last access: 18 March 2024), 2011.

Smith, J. D. and McLean, S. R.: Spatially averaged flow over a wavy surface, J. Geophys. Res., 82, 1735–1746, https://doi.org/10.1029/JC082i012p01735, 1977.

Van Rijn, L. C.: Sediment transport, Part II: Suspended load transport, J. Hydraul. Eng., 110, 1613–1641, https://doi.org/10.1061/(ASCE)0733-9429(1984)110:11(1613), 1984.

Van Rijn, L. C.: Mathematical modeling of suspended sediment in non-uniform flows, J. Hydraul. Eng., 112, 433–455, https://doi.org/10.1061/(ASCE)0733-9429(1986)112:6(433), 1986.

Wright, S. and Parker, G.: Density stratification effects in sand-bed rivers, J. Hydraul. Eng., 130, 783–795, https://doi.org/10.1061/(ASCE)0733-9429(2004)130:8(783), 2004.

Proc. IAHS, 386, 237–249, 2024
https://doi.org/10.5194/piahs-386-237-2024

Open Access PIAHS

A tale of two floods: Hawkesbury-Nepean valley floods of February 2020 and March 2021

Wendy Sharples[1], Katayoon Bahramian[1], Kesav Unnithan[2,3,4], Christoph Rüdiger[1], Jiawei Hou[1], Christopher Pickett-Heaps[1], and Elisabetta Carrara[1]

[1]Research, Science and Innovation Group, Australian Bureau of Meteorology, Melbourne, 3008, Australia
[2]ITB-Monash Research Academy, Mumbai 400076, India
[3]Department of Civil Engineering, Indian Institute of Technology Bombay, Mumbai 400076, India
[4]Department of Civil Engineering, Monash University, Clayton, VIC 3168, Australia

Correspondence: Wendy Sharples (wendy.sharples@bom.gov.au)

Received: 18 July 2023 – Revised: 18 October 2023 – Accepted: 24 February 2024 – Published: 19 April 2024

Abstract. The Hawkesbury-Nepean valley is one of the largest coastal basins in NSW. It supports the local agriculture industry and is an important environmental asset. Due to its narrow sandstone gorges, which create natural choke points, floodwaters from its major tributaries can rapidly back up, rise and spill out onto the flood plain. Thus, the valley is flood-prone, with a history of disastrous events, aggravated by a constrained road network for evacuation.

Two flood events occurred in the Hawkesbury-Nepean valley in 2020 and 2021, however, the impact of each of those events was different in terms of lives lost (2 fatalities compared to none) and economic losses (more than AUD 2 billion compared to less than AUD 1 billion). In this study, reasons for the variation in impacts are explored by determining an inundation likelihood map, derived using a combination of the height above nearest drainage method and streamflow forecasts, and considering antecedent hydrological and climate conditions.

1 Introduction

Floods along the eastern coast of Australia have always featured in Australia's natural hazard profile since records began, due to the presence of intra-annual or seasonal as well as interannual climate drivers which drive flood weather conditions in that area. These climate drivers are projected to be exacerbated by climate change, where heavy rainfall events are projected to increase in the East Coast region (National Hydrological Projections, East Coast Assessment Report 2022; Matic et al., 2022), changing the severity and frequency of floods, which will negatively impact socio-economic and ecosystem resilience.

The intra-annual or seasonal drivers, shown in Fig. 1, include:

1. East coast lows (ECLs), which are intense low-pressure systems that occur on average several times each year off the eastern coast of Australia. ECLs will often rapidly intensify overnight making them one of the more dangerous weather systems, where individual east coast lows usually last for a few days. ECLs are most common during autumn and winter with a maximum frequency in June (Kiem et al., 2016).

2. Tropical cyclone seasons (TCS). The TCS officially runs November–April, but often TC activity is variable throughout the season. Storm surges accompanying TCs can exacerbate riverine flooding along the coast, where storm surges often accompany TCs (Needham et al., 2015).

3. Monsoon seasons (MS). MS is associated with cloudy conditions, lengthy periods of heavy rain, occasional thunderstorms and strong winds, often causing flooding. In addition, the Madden-Julian Oscillation can influence the timing and intensity of "active" monsoon periods in northern Australia, leading to enhanced rainfall – both in intensity and duration (Wheeler et al., 2009).

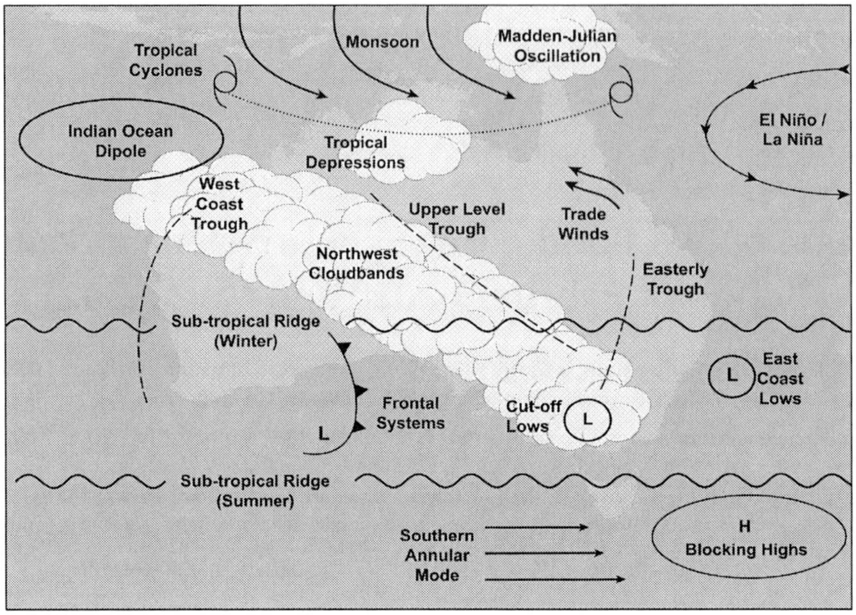

Figure 1. Major climate drivers for flood weather (Source: © Bureau of Meteorology, 2010).

The interannual drivers, shown in Fig. 1, include:

1. La Niña conditions. There is an established relationship between La Niña strength and rainfall, where the greater the sea surface temperature and Southern Oscillation Index (SOI) difference from normal, the larger the rainfall response (Power and Callaghan, 2016). For example, the six wettest winter–spring periods on record for eastern Australia occurred during La Niña years, and there have been more TCs in the Australia region, with twice as many making landfall (Alexander and Hayman, 2008). La Niña conditions result in an increased likelihood of flooding.

2. A negative Indian Ocean Dipole (IOD). A negative IOD often results in more rainfall than average over southeastern Australia. In conjunction with La Niña conditions, rainfall is above average over large parts of Australia. For example, during 2010, under La Niña conditions, a negative IOD developed, and combined to produce heavy rainfall and widespread flooding across eastern Australia (Lim and Hendon, 2017).

3. Landscape antecedent conditions. Wet antecedent conditions or saturated soil and full water stores can increase the flood maxima generated from an event such as heavy rainfall, TC or storm surge (Wasko et al., 2020).

The Hawkesbury-Nepean valley located along the southeast coast of New South Wales (NSW), is one of the largest coastal basins in NSW (Fig. 2a and b). It supports the local agriculture industry and hence is an important environmental asset. Flood risk in this valley has been described as the highest in NSW, if not Australia (Infrastructure NSW, 2021). This risk arises from several exposure and vulnerability factors, including the large and growing population, a constrained road network which poses challenges for evacuation, and low levels of flood awareness (Infrastructure NSW, 2021). In addition, the Hawkesbury-Nepean valley's unique topography and river network are environmental factors exacerbating flood risk, due to the contribution of 5 major tributaries or "taps" flowing into 1 river system – the "drain" – which is constrained by narrow downstream gorges (Fig. 2c). Due to the narrow sandstone gorges, which create natural choke points, floodwaters from its major tributaries can rapidly back up, rise and spill out onto the flood plain. This causes floodwaters to back up across floodplains to considerable depths –known as the "bathtub" effect (Fig. 2c). Thus, the valley is flood-prone, with a history of disastrous events, with over 100 floods recorded from 1799 up until 2021 (Infrastructure NSW, 2021).

Two flood events occurred in the Hawkesbury-Nepean valley in 2020 and 2021, with near identical climate conditions, however, the impact of each of those events was different in terms of lives lost (2 fatalities compared to none) and economic losses (over two billion AUD compared to less than one billion AUD) (source: Infrastructure NSW, 2021; Insurance Council of Australia, 2021). In this study, the reasons for the variation in impacts are explored by determining:

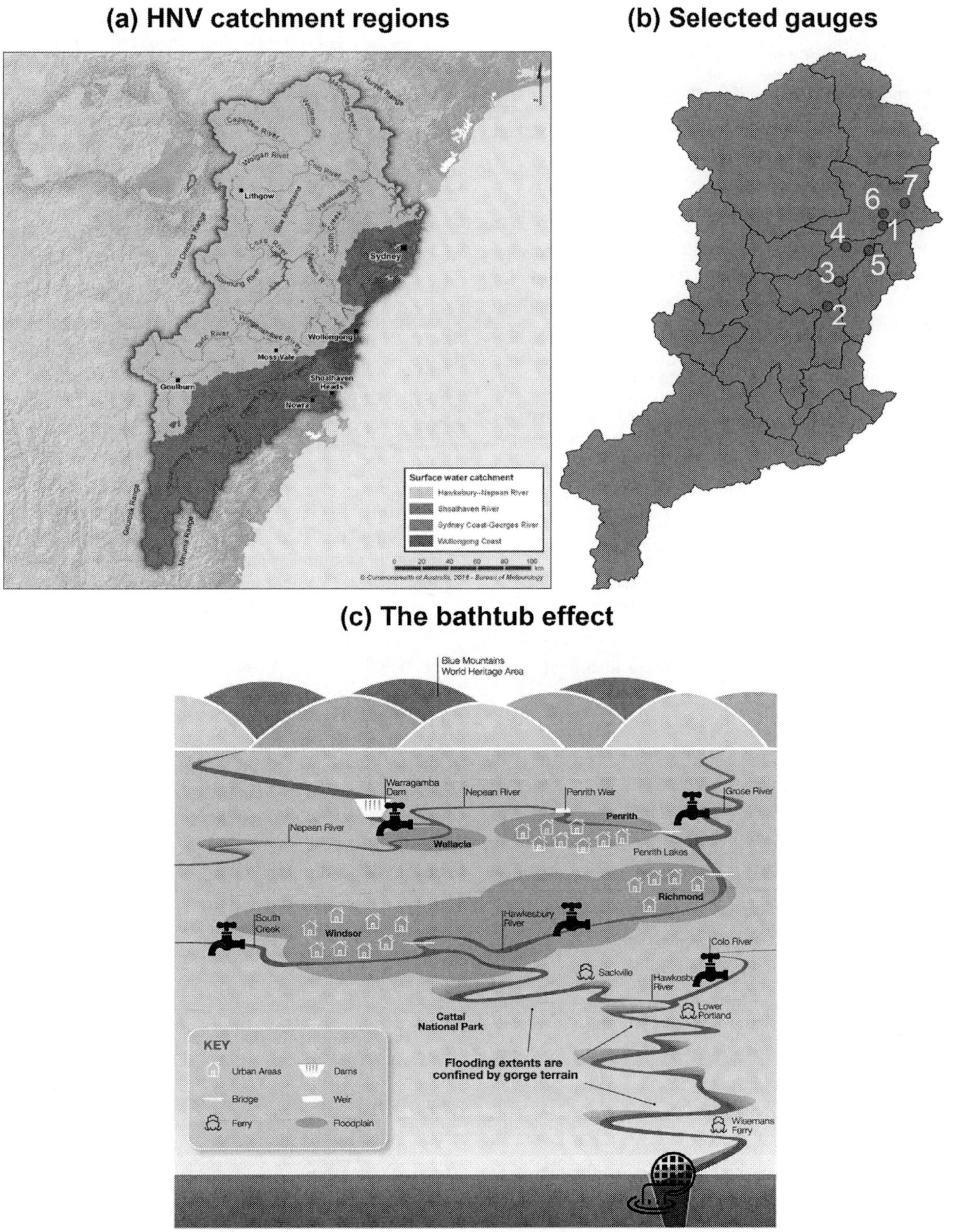

Figure 2. (a) NSW catchment boundaries and rivers along the eastern coast, including the Hawkesbury-Nepean catchment in yellow. **(b)** Sub-catchment boundaries within the Hawkesbury-Nepean catchment and selected gauging stations: 1 – Sackville, 2 – Wallacia Weir, 3 – Penrith, 4 – North Richmond, 5 – Windsor PWD, 6 – Colo Junction and 7 – Webbs Creek. **(c)** Illustration of the "bathtub effect" with taps and drain, unique to the Hawkesbury-Nepean valley (adapted from Infrastructure NSW, 2021).

i. the antecedent landscape conditions,

ii. the climatic conditions, and

iii. the inundation extent and depth,

for each event in order to gauge and compare the resultant hazard footprint.

2 Methods

This study employs the height above nearest drainage (HAND) methodology along with analysis of observational datasets (both grid and point based) to compare the two flood events.

2.1 Probabilistic inundation mapping methodology

The height above nearest drainage (HAND) approach to flood inundation mapping is a simple approach based on the height of the flood depth (Fig. 3a). The HAND methodological workflow can be described in the following steps:

1. The contributing catchment to a defined outlet is rasterized to the resolution of the supporting DEM.
2. Flow direction and accumulation in each grid cell is derived using a D8 algorithm (O'Callaghan and Mark, 1984).
3. In-stream and out-of-stream cells are identified using the flow line vector (river defined by the DEM).
4. The relief between all out-of-stream cells from their nearest in-stream cell is calculated to define a HAND raster.
5. From a given flow rate, a rating curve can be used to convert flow to water-level.
6. Subtracting the HAND raster from the water-level raster yields a water-level above-surface raster. All values greater than zero show the estimated flood height at each cell.

The height of the riverbank versus the height of the flood depth where the difference between each is taken as the volume of water spilling out into the flood plain (Fig. 3a).

An illustration of this workflow is given in Fig. 3 top panel (b) (Lababidi, 2021). The probabilistic approach takes the HAND depth uncertainty into account and generates an ensemble based on the increments between the lower and upper 95th interval bounds, producing an ensemble of water-level above-surface rasters which are then turned into probabilities for each cell based on the number of times that cell is above zero. The HAND approach used in this study uses:

- a varying Mannings roughness coefficient to estimate streamflow, as a function of hydraulic radius and reach slope (Marcus et al., 1992),
- a HAND threshold position of 0.25 m (in order to calculate a HAND raster; Unnithan et al., 2020),
- uncertainty in bankfull geometry to calculate an ensemble of HAND rasters,
- a 1D vector flow routing scheme (Yamazaki et al., 2011), and,
- captures water management and backwater effects (explained in Sect. 2.2; Mizukami et al., 2016).

2.2 Capturing water management and backwater effects

Due to the unique "bathtub effect" in the Hawkesbury-Nepean catchment, it is important that backwater effects or upstream flooding caused by downstream conditions such as channel restriction and/or high flow in a downstream confluence stream are considered. In addition, given the large dam in this catchment, the dam outflows also need to be considered.

Backwater effects are incorporated into the HAND approach via the 1D flow routing scheme (Yamazaki et al., 2011; Mizukami et al., 2016). During a flood event, the dam storage is monitored and if the storage has not yet reached capacity, the upstream discharge contributions of the reach where the dam is located are set to be zero.

2.3 Data

The HAND methodology calls for a hydrologically conditioned digital elevation model (DEM) and gridded runoff or streamflow data. The DEM used was the Bureau of Meteorology's Geofabric (Fig. 3c and d) which is a national dataset based on mapped hydrological features and a 1 s digital elevation model (DEM). Geofabric v3 (Bureau of Meteorology Water Information Services, 2015) contains the following products used in this study:

- Geofabric Surface Network: Network representation of hydrological features, including over 2.5 million stream segments and 6000 gauging stations.
- Geofabric Surface Catchments: Catchment boundaries derived from a 1 s Digital Elevation Model, with approximately 7 million base level catchments.

The gridded runoff data used is hydrological simulation data from the Bureau of Meteorology's operational landscape model, AWRA-L v7 (https://www.awo.bom.gov.au, last access: 18 October 2023, Frost et al., 2021). It simulates the flow of water through the landscape with precipitation entering the grid cell through vegetation and soil moisture stores and then out of the grid cell as evapotranspiration, runoff or deep drainage to the groundwater. Thus, runoff generated via AWRA-L considers antecedent conditions.

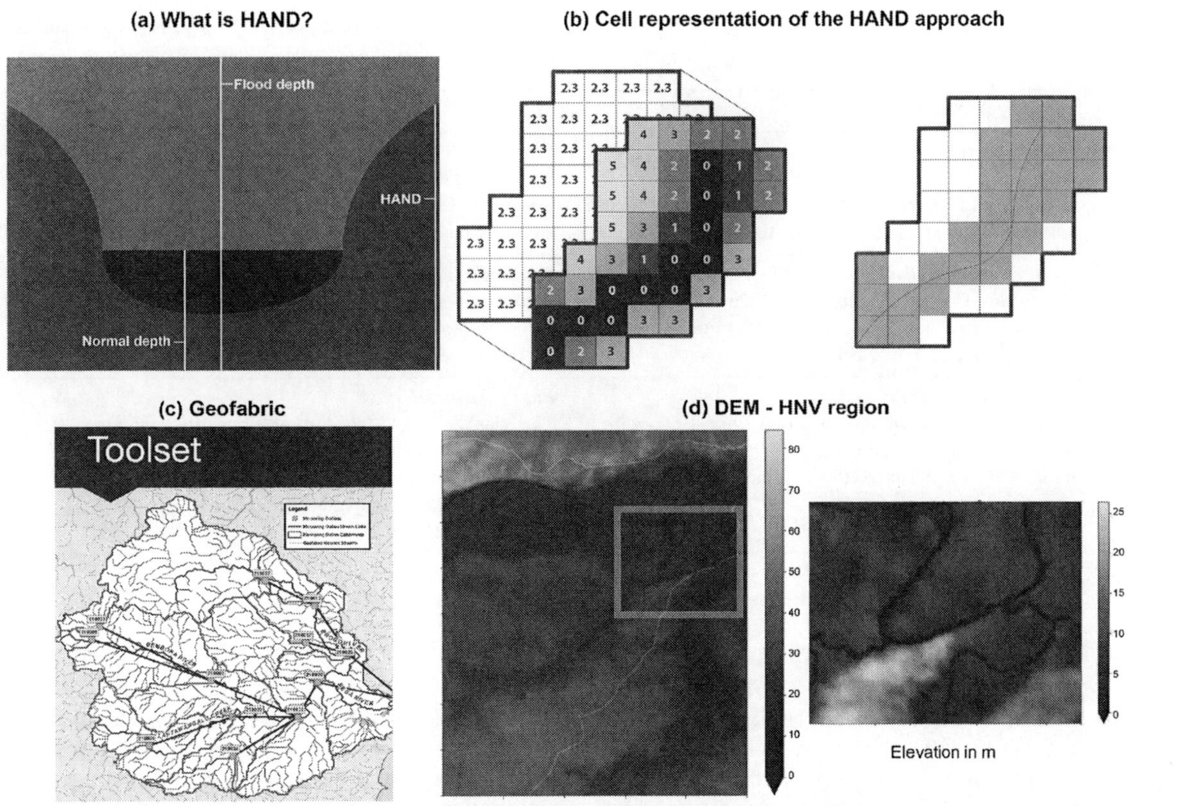

Figure 3. **(a)** HAND illustration for a given streambed. **(b)** The cell representation of the HAND workflow (adapted from Lababidi (2021)). **(c)** Illustration of the Geofabric toolset. **(d)** DEM of the Hawkesbury-Nepean valley Sackville (station 1) region in with sub-catchment boundaries in white (left panel) and zoomed in from orange box (right panel).

All datasets used in the analysis are:

1. the Australian Water Availability Project rainfall data (Bureau of Meteorology AWAP, 2022),
2. the AWRA-L v7 root zone soil moisture data (Frost et al., 2021),
3. the Bureau of Meteorology streamflow gauging station data for a selection of "high impact" locations (Bureau of Meteorology Flood Warning Services, 2022),
4. the Warragamba Dam storage data (Water NSW, 2022),
5. the global bankfull width and depth database (Andreadis et al., 2013).

3 Results

In this section we compare flood characteristics, catchment antecedent conditions, climatic conditions, and hazard footprints between the two flood events.

3.1 Comparison of flood characteristics

A comparison between the two flood events in the Hawkesbury-Nepean valley; detailing rainfall, climate drivers, and flood characteristics are described in Table 1. Flood characteristics are described by a classification system used by Australia's emergency services as well as a chance of occurrence which is expressed as a likelihood of a flood of a certain classification occurring in a given year. Floods are classified via a three-tiered scheme: "Minor", "Moderate" and "Major" where the classification levels at each station are decided by the jurisdictions and are based on water levels which cause certain impacts.

3.2 Comparison of antecedent conditions

The February 2020 floods came off the back of the notorious "Black Summer", a period of which, due to its intensity, size, duration, and uncontrollability, was considered a megafire (Geary et al., 2022). Black Summer occurred due to an extended period of El Niño events, causing a prolonged period of drought and very dry landscape conditions (Squire et al., 2021). This can be seen in Fig. 4 where the root zone soil

Table 1. Flood event characteristics for the February 2020 and March 2021 floods in the Hawkesbury-Nepean valley (Infrastructure NSW, 2021).

February 2020 flood event	March 2021 flood event
Climate drivers: A coastal trough with embedded low-pressure circulations hovered off the NSW coast – not intense enough to be classified as an East Coast Low.	*Climate drivers*: Persistent presence of a coastal trough, generating rain for over a week – not intense enough to be classified as an East Coast Low.
Rainfall characteristics: Heavy rain from 6–13 February where some locations recorded more than 500 mm. The most intense rainfall occurred over 12 h on 9 February.	*Rainfall characteristics*: Heavy rain from 18–23 March where some locations recorded more than 500 mm. The most intense rainfall occurred over 24 h to 09:00 LT on 21 March.
Flood characteristics: This was the first "Moderate" flood since 1992, with a likelihood of occurrence of 20 % chance per year. The flood peaked at 9.3 m (AHD) at Windsor on the 9th.	*Flood characteristics*: This was the first "Major" flood since 1990, with a likelihood of occurrence of 5 %–10 % chance per year. The flood had two distinct peaks, one on 21 March and one on 23 March but overall peaked at 12.9 m (AHD) at Windsor on 21 March.
Warragamba Dam releases: No releases were made over the course of the flood.	*Warragamba Dam releases*: Over 450 GL (gigalitres) were released from the dam over the course of the flood.

moisture (from AWRA-L v7) in the Hawkesbury-Nepean catchment is mostly between – and 0 and 0.3 fraction fullness (Fig. 4a, rootzone soil moisture plot) and the Warragamba Dam (which services the Hawkesbury-Nepean catchment) storage is at ~ 40 % or ~ 1000 GL (Fig. 4c and d, bar charts, panels, Water NSW). In comparison, the March 2021 floods occurred in a La Niña year, where soil moisture conditions were wet, hovering between 0.5 and 0.8 fraction fullness for most of the catchment (Fig. 4b, rootzone soil moisture plot) and the Warragamba Dam storage is at ~ 90 % or ~ 2500 GL (Fig. 4, bar chart; Water NSW, 2022).

3.3 Comparison of rainfall

The rainfall totals and average daily rainfall over both events, calculated using the climate data from AWAP, are similar, particularly when comparing the average daily rainfall across both events (Fig. 5c and d). However, the February 2020 flood event rainfall totals and average daily rainfalls are slightly higher over the Hawkesbury-Nepean catchment (Fig. 5).

3.4 Comparison of water levels

The resulting streamflow generated from both events can be extrapolated from water level plots or hydrographs shown in Fig. 6. These gauging stations are: 1 – Sackville, 2 – Wallacia Weir, 3 – Penrith, 4 – North Richmond, 5 – Windsor PWD, 6 – Colo Junction and 7 – Webbs Creek, considered to be "high impact" locations, where their locations are shown in Fig. 2b. The hydrographs from the 2021 event show that the excess streamflow generated is larger than for the 2020 event, where water level peaks are higher (except for Wallacia Weir) and the duration of the event is longer across the seven gauging stations. Even though the actual peak between events is very similar between events for station 2 (Wallacia Weir), due to the high concentration of rainfall in that region for the 2020 event (see Fig. 5a), however, given the longer duration of the March 2021 event, more water is flowing onto the flood plane comparatively. For individual stations, it is interesting to note that the peak of the hydrograph of station 6 (Colo Junction) in the February 2020 event arrives significantly earlier than the peaks for other stations, possibly due to the after effects of the bushfires on the land cover, as well as the timing of rainfall flowing into the Colo River region.

3.5 Comparison of flood extent and depth

Probabilistic inundation extents and depths (where depth is calculated relative to the bankfull elevation) are produced via the aforementioned methodology (Sect. 2) for the Sackville region of the Hawkesbury River (validation shown in Appendix A). Both the depth and probability of inundation for the 2021 event are higher (darker colors) than for the 2020 event and the inundation extent covers a larger area for the 2021 event, as expected from the flood classifications ("Moderate" in 2020 vs. "Major" in 2021).

4 Discussion

The description of climate conditions and rainfall totals were remarkably similar between the two flood events; Febru-

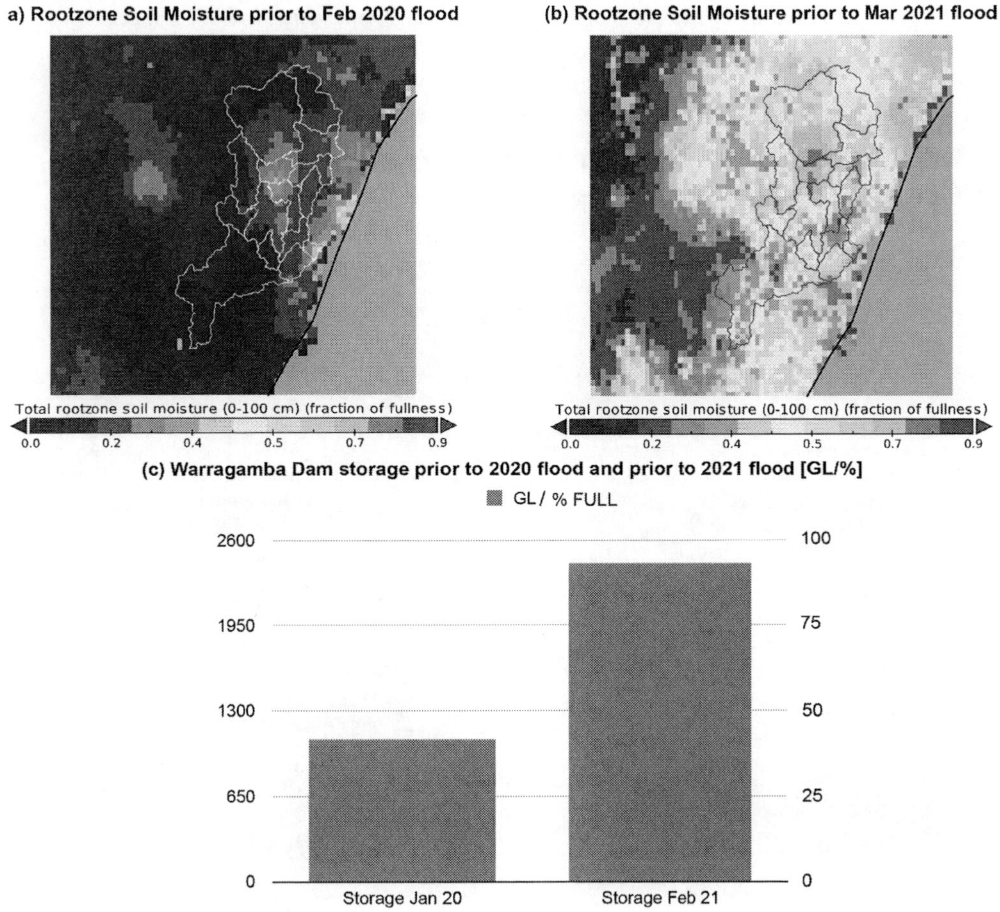

Figure 4. Antecedent conditions. **(a)** Soil moisture prior to February 2020 **(c)** Water storage in GL and percentage fullness of Warragamba dam prior to February 2020 and March 2021 event.

ary 2020 and March 2021 (Table 1). In fact, the rainfall totals were slightly higher in certain regions during the February 2020 event (Fig. 5). Post event reviews have classified the February 2020 event as a 1 in 5 event, but the March 2021 event as a 1 in 20 event (Infrastructure NSW, 2021). This is also consistent with the inundation maps generated via the HAND technique (Fig. 7) where the depth and probability of inundation extent are larger for the 2021 flood event.

The impacts between the two events are outlined in Table 2. Table 2 shows that in comparison, the impacts of the March 2021 Flood Event are more severe in terms of affected areas, damages and fatalities. This tallies with the event classifications (Infrastructure NSW, 2021) as well as the HAND probabilistic inundation and depth maps (Fig. 7).

Upon examination of the antecedent soil moisture conditions, the dam storages and rainfall totals (Figs. 4 and 5), the major differences between the events are the antecedent soil moisture and dam storages. This is an important finding to note when considering that even though heavy rainfall events are projected to increase under climate change scenarios, the annual average rainfall is projected to decrease in NSW (Matic et al., 2022). This means that soil moisture conditions may be drier and affect flood maxima in much the same way as for the February 2020 flood event. Several studies examining the relationship between flood maxima and antecedent catchment conditions also agree with this study's findings – dry catchment conditions decrease flood maxima and conversely wet catchment conditions increase flood maxima (Sharma et al., 2018; Wasko et al., 2020). Thus, future work examining projected flooding in the Hawkesbury-Nepean catchment would need to also consider dam outflows and catchment antecedent conditions – a whole water cycle approach, rather than solely projecting climate fields.

The limitations of this study include the assumptions and shortcomings of the implementation of the HAND methodology employed, where according to our validation (see Appendix A), the number of cells inundated is overestimated. Further tuning would be enhanced by using additional events

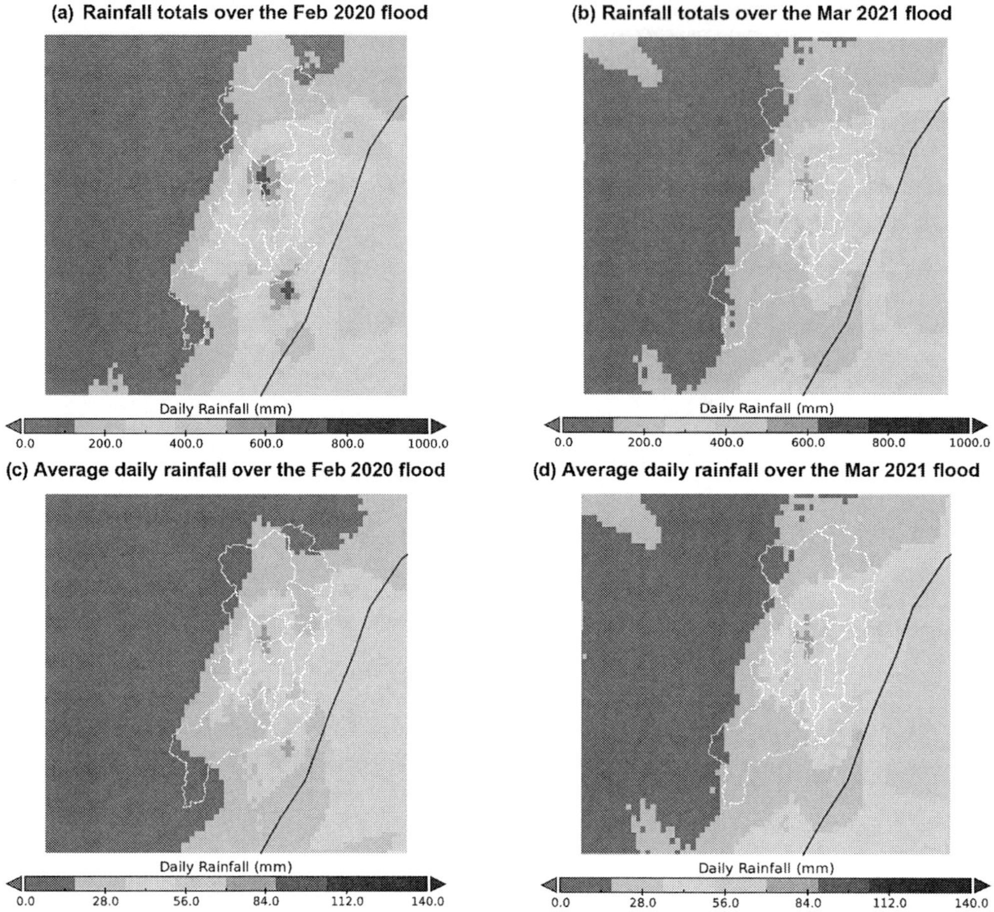

Figure 5. Rainfall totals; summed over the flood events and daily averages. **(a)** Totals over the February 2020 flood, **(b)** Totals over the March 2021 flood. **(c)** Averaged daily totals for the February 2020 flood. **(d)** Averaged daily totals for the March 2021 flood.

Table 2. Flood impacts for the February 2020 and March 2021 floods in the Hawkesbury-Nepean valley (Infrastructure NSW, 2021; Insurance Council of Australia, 2021).

February 2020 flood event	March 2021 flood event
Affected areas and communities: Most impacts were constrained to low lying areas. A few bridges closed, around 65 properties were affected, less than 200 people were ordered to evacuate.	*Affected areas and communities*: All rural and urban fringe areas. About 600 dwellings and 300 commercial/industrial buildings (most on rural land) were affected. Key roads and bridges flooded. Over 2000 people were subject to evacuation orders.
Damages estimates: Estimated economic losses less than AUD 1 billion across the East Coast. Approximately 10 000 insurance claims across NSW.	*Damages estimates*: Estimated economic losses over AUD 2 billion across the East Coast. Approximately 110 000 insurance claims across NSW.
Fatalities: Zero lives lost.	*Fatalities*: Two lives were lost during the flood event.

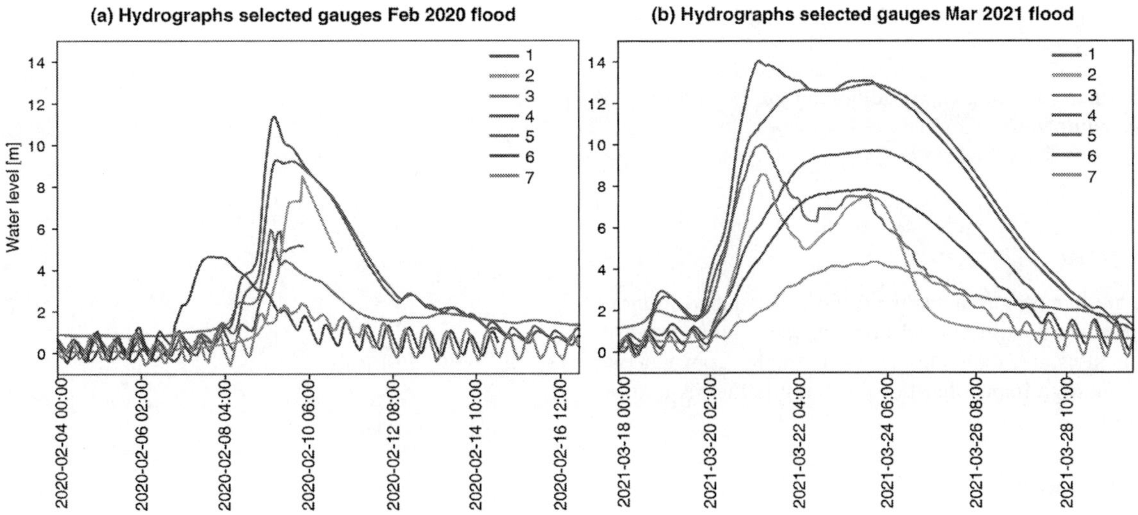

Figure 6. Hydrographs for the 7 selected gauging stations for the February 2020 **(a)** and March 2021 **(b)** floods in Hawkesbury-Nepean valley. Gaps in the data are due to the cessation of monitoring when flood levels decrease to back below minor.

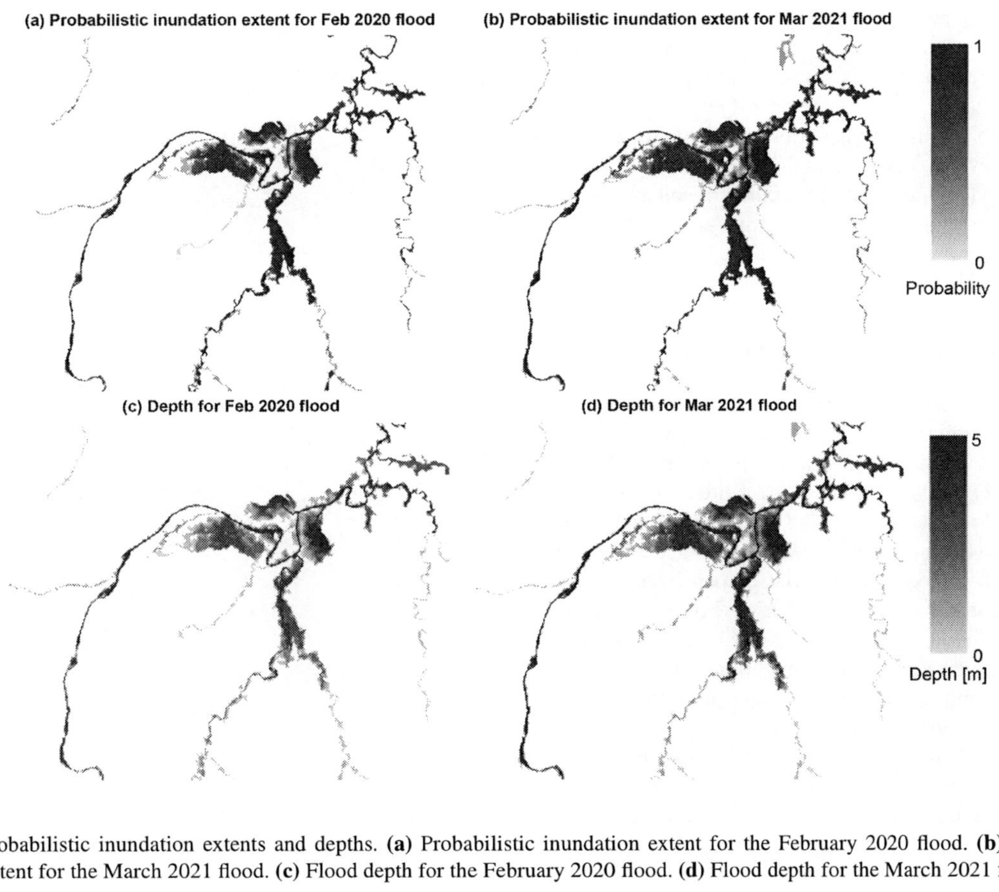

Figure 7. Probabilistic inundation extents and depths. **(a)** Probabilistic inundation extent for the February 2020 flood. **(b)** Probabilistic inundation extent for the March 2021 flood. **(c)** Flood depth for the February 2020 flood. **(d)** Flood depth for the March 2021 flood.

in the same region. In addition, the HAND methodology could be improved by:

- Further hydrological conditioning of the DEM plus blending with LiDAR.
- Routing the gridded runoff before ingesting.
- Using a more sophisticated routing and backwater scheme.

There are many more fields we could have studied in terms of antecedent conditions and flood characteristics, however the soil moisture, dam storage, hydrographs from gauging stations at high impact locations, and rainfall totals told the full story.

5 Conclusions

Two flood events occurred in the Hawkesbury-Nepean valley in 2020 and 2021, with near identical climate conditions and thus rainfall patterns, however, the impacts from these events were classed as 1 in 5 and 1 in 20 events respectively; where the 2021 event resulted in 2 more lives lost, above AUD 1 billion more in damages and approximately 10 times more insurance claims. This study found that the antecedent catchment conditions prior to the flood events played a significant role in limiting the hazard footprint depth and extent of the February 2020 event, and thus its impact. Where drier conditions led to no dam outflows, filling of soil moisture and ground water storage deficits, which in turn led to less excess streamflow generated, and therefore less area inundated across the catchment.

Appendix A: Validation and verification

The validity of the HAND results over both events is examined by comparison with Sentinel 2 data across the Sackville area of the catchment, on 10 February 2020 and 23 March 2021. The metrics used for validation are the: Critical Success Index (CSI), Youden's Index (YI), and Cohens Kappa (CK). The three metrics were used to get a more holistic measure of performance where CSI penalizes false positives in comparison to the other metrics (Fig. A1 top panels), YI is the likelihood of an inundated cell in the model in cells that are inundated versus those that are not inundated in the observations, and CK is a statistic that is used to measure inter-rater reliability (and intra-rater reliability) for qualitative (categorical) items (Schaefer, 1990; Fluss et al., 2005; Ben-David, 2008). The stringent validation metrics show the HAND methodology is performant, with most metrics greater than 0.5 in value across both events. It does however show that the HAND methodology employed in this study tends to overestimate the number of inundated cells.

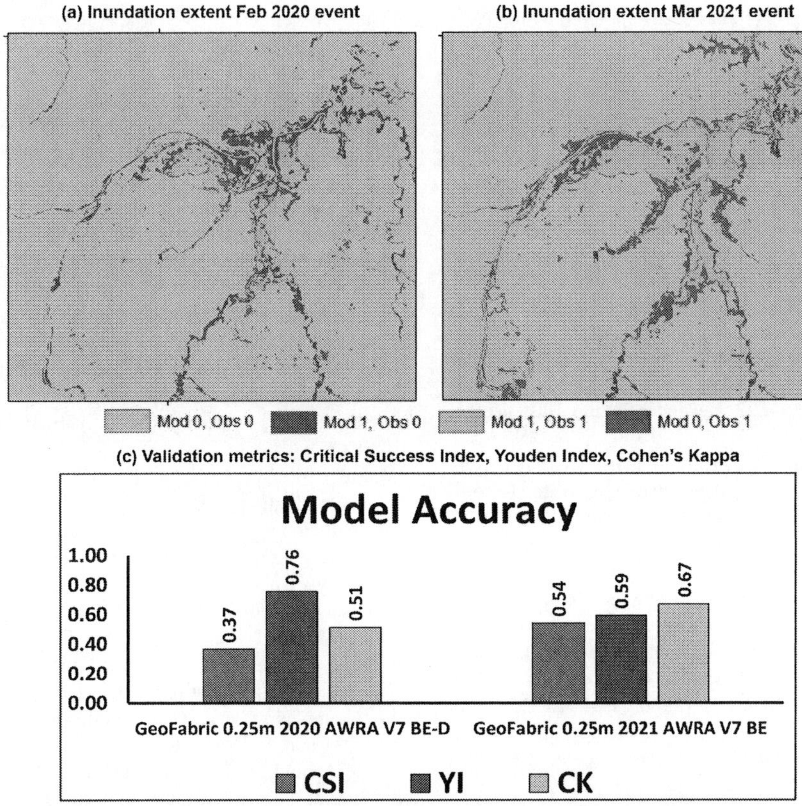

Figure A1. Flood event validation for the February 2020 and March 2021 floods in Hawkesbury-Nepean valley. **(a)** Spatial plot of CSI validation for the February 2020 flood. **(b)** Spatial plot of CSI validation for the March 2021 flood. **(c)** All validation metrics calculated over the validation area presented in a bar chart.

Code and data availability. Data and software available on request depending on licensing agreements.

Author contributions. WS: Conceptualization, Methodology, Formal analysis, Writing – original draft, Visualization. KB: Conceptualization, Methodology, Writing – review & editing. KU: Conceptualization, Investigation, Software, Methodology. CR: Conceptualization, Investigation, Methodology. JH: Methodology, Writing – review & editing. CPH: Writing – review & editing. EC: Conceptualization, Funding acquisition.

Competing interests. The contact author has declared that none of the authors has any competing interests.

Disclaimer. Publisher's note: Copernicus Publications remains neutral with regard to jurisdictional claims made in the text, published maps, institutional affiliations, or any other geographical representation in this paper. While Copernicus Publications makes every effort to include appropriate place names, the final responsibility lies with the authors.

Special issue statement. This article is part of the special issue "ICFM9 – River Basin Disaster Resilience and Sustainability by All". It is a result of The 9th International Conference on Flood Management, Tsukuba, Japan, 18–22 February 2023.

Acknowledgements. This work was funded by the Australian Climate Service. The Australian Climate Service provides improved information, data, intelligence and expert advice on climate risks and impacts to support and inform decision-making. The Australian Climate Service is a partnership between the Bureau of Meteorology, CSIRO, the Australian Bureau of Statistics, Geoscience Australia. The authors would like to acknowledge the reviewers who greatly improved this paper. The authors would also like to acknowledge the support given by the Australian Bureau of Meteorology, in particular, the Hydrology Science section.

Financial support. This research has been supported by the Australian Climate Service.

Review statement. This paper was edited by Mamoru Miyamoto and reviewed by two anonymous referees.

References

Alexander, B. and Hayman, P.: The Regional Institute – Can we use forecasts of El Nino and La Nina for frost management in the Eastern and Southern grains belt?, ASA 2008 14th AAC, http://www.regional.org.au/au/asa/2008/concurrent/farming-uncertain-climate/5623_alexanderbm.htm (last access: 18 October 2023), 2008.

Andreadis, K. M., Schumann, G. J.-P., and Pavelsky, T.: A simple global river bankfull width and depth database, Water Resour. Res., 49, 7164–7168, https://doi.org/10.1002/wrcr.20440, 2013.

Ben-David, A.: About the relationship between ROC curves and Cohen's kappa, Eng. Appl. Artif. Intel., 21, 874–882, https://doi.org/10.1016/j.engappai.2007.09.009, 2008.

Bureau of Meteorology: Australia's Climate Drivers, http://www.bom.gov.au/climate/about/ (last access: 1 January 2010), 2010.

Bureau of Meteorology AWAP: Australian Gridded Climate Data (AGCD)/AWAP, v1.0.0 Snapshot (1900-01-01 to 2022-12-31), Bureau Data Catalogue, http://www.bom.gov.au/metadata/catalogue/19115/ANZCW0503900567 (last access: 18 October 2023), 2022.

Bureau of Meteorology Flood Warning Services: Flood Warning Services, National flood forecasting and warning service, Water Information, Bureau of Meteorology, http://www.bom.gov.au/water/floods/floodWarningServices.shtml (last access: 18 October 2023), 2022.

Bureau of Meteorology Water Information Services: Australian Hydrological Geospatial Fabric|Bureau Data Catalogue, http://www.bom.gov.au/water/geofabric/ (last access: 18 October 2023), 2015.

Fluss, R., Faraggi, D., and Reiser, B.: Estimation of the Youden Index and its associated cutoff point, Biometr. J., 47, 458–472, https://doi.org/10.1002/bimj.200410135, 2005.

Frost, A. J., Ramchurn, A., and Smith, A.: The Australian Landscape Water Balance model (AWRA-L v7) Technical Description of the Australian Water Resources Assessment Landscape model version 7, https://awo.bom.gov.au/assets/notes/publications/ (last access: 18 October 2023), 2021.

Geary, W. L., Buchan, A., Allen, T., Attard, D., Bruce, M. J., Collins, L., Ecker, T. E., Fairman, T. A., Hollings, T., Loeffler, E., Muscatello, A., Parkes, D., Thomson, J., White, M., and Kelly, E.: Responding to the biodiversity impacts of a megafire: A case study from south-eastern Australia's Black Summer, Divers. Distrib., 28, 463–478, https://doi.org/10.1111/ddi.13292, 2022.

Infrastructure NSW: Hawkesbury-Nepean River March 2021 Flood Review – Final Report, © Infrastructure NSW, https://www.infrastructure.nsw.gov.au/media/3315/hnr-march-2021-flood-review.pdf (last access: 18 October 2023), 2021.

Insurance Council of Australia: Annual report/Insurance Council of Australia Limited, https://insurancecouncil.com.au/ (last access: 18 October 2023), 2021.

Kiem, A. S., Twomey, C., Lockart, N., Willgoose, G., Kuczera, G., Chowdhury, A. F., Manage, N. P., and Zhang, L.: Links between East Coast Lows and the spatial and temporal variability of rainfall along the eastern seaboard of Australia, J. South. Hemisph. Earth Syst. Sci., 66, 162–176, https://doi.org/10.22499/3.6602.006, 2016.

Lababidi, S.: An evaluation of the Height Above Nearest Drainage (HAND) flood mapping methodology with the implementation of multivariant linear regression algorithm, Northeaster University, https://doi.org/10.17760/D20409241, 2021.

Lim, E. P. and Hendon, H. H.: Causes and Predictability of the Negative Indian Ocean Dipole and Its Impact on la Niña during 2016, Sci. Rep., 7, 1–11, https://doi.org/10.1038/s41598-017-12674-z, 2017.

Marcus, W. A., Roberts, K., Harvey, L., and Tackman, G.: An evaluation of methods for estimating Manning's n in small mountain streams, Mount. Res. Dev., 12, 227–239, https://doi.org/10.2307/3673667, 1992.

Matic, V., Bende-Michl, U., Hope, P., Srikanthan, S., Oke, A., Khan, Z., Thomas, S., Sharples, W., Kociuba, G., Peter, J., Vogel, E., Wilson, L., and Turner, M.: East Coast – National Hydrological Projections Assessment report, https://awo.bom.gov.au/assets/notes/publications/ (last access: 18 October 2023), 2022.

Mizukami, N., Clark, M. P., Sampson, K., Nijssen, B., Mao, Y., McMillan, H., Viger, R. J., Markstrom, S. L., Hay, L. E., Woods, R., Arnold, J. R., and Brekke, L. D.: MizuRoute version 1: A river network routing tool for a continental domain water resources applications, Geosci. Model Dev., 9, 2223—2228, https://doi.org/10.5194/gmd-9-2223-2016, 2016.

Needham, H. F., Keim, B. D., and Sathiaraj, D.: A review of tropical cyclone-generated storm surges: Global data sources, observations, and impacts, Rev. Geophys., 53, 545–591, https://doi.org/10.1002/2014RG000477, 2015.

O'Callaghan, J. F. and Mark, D. M.: The extraction of drainage networks from digital elevation data, Comput. Vis. Graph. Image Process., 28, 323–344, https://doi.org/10.1016/S0734-189X(84)80011-0, 1984.

Power, S. B. and Callaghan, J.: The frequency of major flooding in coastal southeast Australia has significantly increased since the late 19th century, J. South. Hemisp. Earth Syst. Sci., 66, 2–11, https://doi.org/10.1071/es16002, 2016.

Schaefer, J. T.: The Critical Success Index as an Indicator of Warning Skill, Weather Forecast., 5, 570–575, https://doi.org/10.1175/1520-0434(1990)005<0570:tcsiaa>2.0.co;2, 1990.

Sharma, A., Wasko, C., and Lettenmaier, D. P.: If Precipitation Extremes Are Increasing, Why Aren't Floods?, Water Resour. Res., 54, 8545–8551, https://doi.org/10.1029/2018WR023749, 2018.

Squire, D. T., Richardson, D., Risbey, J. S., Black, A. S., Kitsios, V., Matear, R. J., Monselesan, D., Moore, T. S., and Tozer, C. R.: Likelihood of unprecedented drought and fire weather during Australia's 2019 megafires, npj Clim. Atmos. Sci., 4, 64, https://doi.org/10.1038/s41612-021-00220-8, 2021.

Unnithan, S. L., Biswal, B., and Rüdiger, C.: Flood inundation mapping by combining GNSS-R signals with topographical information, Remote Sens., 12, 3026, https://doi.org/10.3390/RS12183026, 2020.

Wasko, C., Nathan, R., and Peel, M. C.: Changes in Antecedent Soil Moisture Modulate Flood Seasonality in a Changing Climate, Water Resour. Res., 56, e2019WR026300, https://doi.org/10.1029/2019WR026300, 2020.

Water NSW: Greater Sydney Dam Levels – WaterNSW, https://www.waternsw.com.au/nsw-dams/nsw-storage-levels/greater-sydney-dam-levels (last access: 18 October 2023), 2022.

Wheeler, M. C., Hendon, H. H., Cleland, S., Meinke, H., and Donald, A.: Impacts of the Madden-Julian oscillation on australian rainfall and circulation, J. Climate, 22, 1482–1498, https://doi.org/10.1175/2008JCLI2595.1, 2009.

Yamazaki, D., Kanae, S., Kim, H., and Oki, T.: A physically based description of floodplain inundation dynamics in a global river routing model, Water Resour. Res., 47, W04501, https://doi.org/10.1029/2010WR009726, 2011.

Proc. IAHS, 386, 251–258, 2024
https://doi.org/10.5194/piahs-386-251-2024

Open Access PIAHS

Countermeasures against flood damage of buildings and their applicability

Nozomu Kiuchi[1] **and Taku Nakano**[2]

[1]National Institute for Land and Infrastructure Management, 1 Tachihara, Tsukuba, 305-0802, Japan
[2]Building Research Institute, 1 Tachihara, Tsukuba, 305-0802, Japan

Correspondence: Nozomu Kiuchi (kiuchi-n92ta@mlit.go.jp)

Received: 1 August 2023 – Revised: 21 September 2023 – Accepted: 24 February 2024 – Published: 19 April 2024

Abstract. In Japan, devising ways to live in areas with a possibility of flooding, to regulate and guide buildings and land use should be considered to respond to future climate change. This article investigates the cost-effectiveness of flood damage countermeasures at the building and site level based on the flood risk of the location. In the research, 3 cases were assumed; (1) when constructing a new wooden detached house, (2) when renovating an existing condominium, and (3) when a business office moves into the first floor of a RC building and a model examination of flood countermeasures was carried out and demonstrated. By calculation and comparison of "average expected payback period", it was found reasonable to first prepare for flooding at a minimum of 30 to 50 cm above ground level, and then to consider countermeasures in stages for greater flooding depths.

1 Significance of building inundation countermeasures in flood risk reduction in Japan

In Japan, typhoons and torrential rains are causing unprecedented flood damage also in urban areas. When houses encounter a flood, depending on the degree of flooding, residents may face a life-threatening danger, and most of all, they will have to deal with the heavy burden of cleaning, tidying up, and restoring buildings and household goods soiled or damaged, under sometimes outages of lifelines such as electricity, water, and sewage.

The impact of climate change on a global scale is behind the frequent and severe flood damage, and it is important for us as a society to adapt and prepare. Within the concept of "River Basin Disaster Resilience and Sustainability by All" responding to future climate change, proposed in Japan, regulations and guidance of buildings and land use within in areas of possible flooding are mentioned, and it is therefore necessary to clarify the applicability of countermeasures against flood damage of buildings.

2 Significance of cost-effectiveness evaluation of anti-flood measures for buildings

With respect to countermeasures to prevent flood damage of buildings, efforts in Europe and the U.S. are leading the way, including the publication of a guidebook (FEMA, 2014) in the U.S. and a book published by the Royal Institute of British Architects (Barsley, 2020). Various efforts have also been initiated recently in Japan.

However, it is not clear what level of countermeasures should be targeted for each location in these efforts, other than using hazard maps and maps of anticipated flooding zones as given conditions. Flood maps based on the Flood Prevention Law in some areas show events of rare occurrence, compared to average building life span of Japan (roughly 30 years), with large inundation areas and depths in urban areas, making it difficult to make realistic assumptions.

For instance, although 26 % of households residing in the assumed inundation zone (by the Housing Bureau, Ministry of Land, Infrastructure, Transport and Tourism), based on rainfall as a benchmark for river maintenance (Level 1), the maximum assumed depth is below 1st floor level (GL + 0.5 m), more than half of the houses actually inundated during 2001–2020 year were inundated below 1st floor level (according to the Statistics of Flood Damage of the above

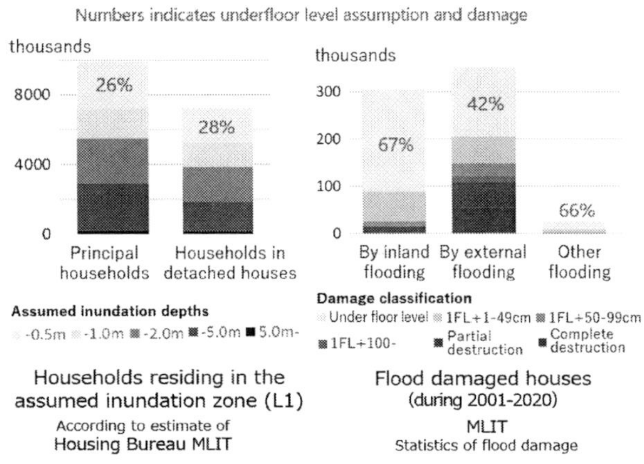

Figure 1. Assumed damage and statistical damage.

Ministry), as shown in Fig. 1. Moreover, while 7 292 014 households live in areas where the maximum expected inundation depth exceeds the above-mentioned 1st floor level, 218 863 households (about 1/33) actually experienced inundation damage (river related flooding) during the above 20-year period, which roughly translates to the above-floor inundation probability of 1/660. Obviously, this rarity becomes even more pronounced, if a larger assumed inundation area and depth based on assumed maximum rainfall (Level 2) is taken in account.

In Japan, as in most other countries, anti-flood measures at the building and site level are often viewed from the perspective of mitigating property damage, etc., rather than ensuring occupants' lives, and we therefore considered it important to evaluate the cost-effectiveness of such measures based on more specific inundation risk information of the building's location.

3 Purpose and method of the study

There are various types of buildings and construction activities, with different hardware and software issues expected to be involved in flood countermeasures. So, we targeted detached houses, condominiums (self-owned multi-family houses), and business premises as building types. For all the above types, we examined the cost-effectiveness of flooding countermeasures based on the risk of the location, etc.

The study of the cost-effectiveness of flooding countermeasures was based on a desk study (model study) based on a model plan, with the cooperation of architectural design firms with extensive experience in the design of the relevant buildings. We followed the undermentioned procedure for all the building types.

i. Interviews and on-site surveys were conducted on the actual status of inundation damage and inundation countermeasures for the relevant buildings, and the preconditions for the study, such as the direction of model plan preparation and outline of inundation countermeasures, etc. were organized. The targets were selected from the applicable building types that experienced major water damage (7 cases in Kanto, Chubu, Chugoku, and Kyushu region) from 2015–2020 with various levels of flooding, and interviews were conducted during 2019–2020 to more than 20 of residents, condominium management associations, business firms, building engineers, renovation companies, volunteer supporters, etc. who experienced restoration and reconstruction of damaged buildings.

ii. Model "standard plans" without flooding countermeasures or a typical plan for existing buildings (in the case of renovation) were prepared, and a trial design was conducted to provide detailed enough information to estimate the cost of realization.

iii. Several alternative flooding "countermeasure plans" or flood retrofit plans were considered as options to the above "standard plans". Trial designs were also conducted for each of these alternatives so as to enable estimation of the realization cost. Regarding the inundation countermeasures, realistic and generally applicable measures were applied.

iv. Based on the trial designs of the "standard plans" and "countermeasure plans", the cost to realize (building, renovation, etc.) the plans were determined by requesting estimates from builders or by estimating with reference to similar cases. The increase in the cost of the "countermeasure plan" from the cost of the "countermeasure plan" was defined as the "cost" required to provide anti-flood measures.

$$\begin{aligned}\text{Additional cost} &= \text{Cost to realize countermeasure plan}\\ &\quad - \text{Cost to realize standard plan} \qquad (1)\end{aligned}$$

v. For the standard plan and the countermeasure plans, the cost of repairing the building and the amount of damage to movable property such as furniture, fixtures, and merchandise (restoration cost) that would be required in case of inundation were estimated for each flood depth, and the difference between the standard plan and the countermeasure plans were defined as the "effect" associated with the flood countermeasures.

$$\begin{aligned}\begin{array}{c}\text{Reduction in}\\\text{restoration cost}\end{array} &= \text{Restoration cost of standard plan}\\ &\quad - \text{Restoration cost of countermeasure plan} \qquad (2)\end{aligned}$$

vi. In order to take into account, the frequency of flooding, the expected annual average amount of damage reduction based on the probability of occurrence was calculated according to the calculation method of the "Flood

Control Economic Study Manual" by MLIT (2020).

$$\text{Expected reduction in restoration cost} = \sum^{\text{Inundation depth}} (\text{Reduction in restoration cost} \times \text{Occurence probability}) \quad (3)$$

vii. The "average expected payback period" for the cost of inundation countermeasures was calculated by dividing the cost of inundation countermeasures by the expected annual average damage reduction amount. By comparing this index, the cost-effectiveness of anti-flood measures (corresponding to the risk of flooding) can be compared for individual sites for the building type in question.

$$\text{Expected payback period} = \frac{\text{Additional cost}}{\text{Expected reduction in restoration cost}} \quad (4)$$

The "average expected payback period" is intended to be used as a guide for understanding the relative necessity of anti-flood measures, and therefore, is not intended to be calculated strictly. In addition, such as the mental distress of residents, loss of business premises, social costs, etc. are not taken into account, and should be considered in accordance with the purpose of use.

For the frequency (probability) of flooding by inundation depth, data equivalent to the assumed inundation depths (50 m grid) for internal and external flooding based on different rainfall probabilities which are published in the Shiga Prefecture "Neighborhood Safety Level Map" were used. At the time the study was initiated, this was the such data available in Japan. Details of this data is described by the provider (Taki et al., 2019). Within the three scenarios included in this data, inundation depths for each location, the data from "overtopping breaches", which are said to account for 70 % of all inundation events, was used. Grid meshes with low to high risk of inundation were selected based on the conditions of a 2-year probability of a maximum inundation depth of less than 50 cm and a 200-year probability of 50 cm or more, and then meshes with a household density below a certain level (1 per grid for Sect. 4.1, 5 per grid for Sect. 4.2 and 4.3) were excluded and included in the study, depending on the target building type. The inundation characteristics of the grid used in Sect. 4.1 are shown in Fig. 2.

4 Study of each building type

4.1 Wooden detached houses

First, a study was conducted on detached houses, which are the dominant type of residence in rural areas, small and medium-sized cities, and suburban areas of large cities, where the percentage of inundation damage is high. Since many houses are made of wood in Japan and it is considered extremely difficult to take anti-flood measures this type of building through renovation, we assumed constructing of

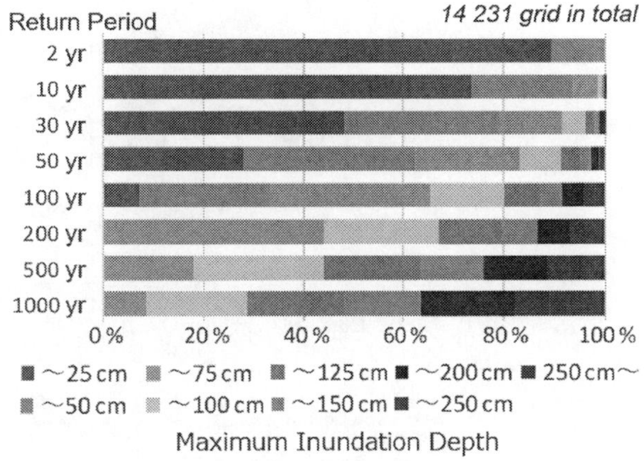

Figure 2. Percentage of maximum inundation depth by return period (probability) of the subject grid for wooden detached house study (Sect. 4.1).

a new wooden detached house (2 stories and 99 m^2 floor space).

Based on interviews and foreign sources mentioned in Sect. 2, the following 3 flooding countermeasure plans in Fig. 3 were considered, to evaluate the cost-effectiveness: facilitate-restoration plan, waterproof building, and elevated foundation plans.

"Facilitate-restoration plan" focused on post-disaster smooth and inexpensive restoration, according to the "wet floodproofing" concept. Underfloor level (Level 1) inundation prevention and improvement of under floor workability, and reducing embroiled work in the components such as treatment of walls when removing floorboards, etc., were considered.

"Waterproof building plan" intends to prevent immersion from flood, according to "dry floodproofing" concept, of water up to the bottom of the hip window height (Level 2), to reduce damage to the building and goods. Planned to shut off water from the hip wall (by RC wall) and opening (by watertight board, etc.) and prevent the backflow of drainage (by check valve).

"Elevated foundation plan" is another type of "dry floodproofing" which elevates the height of the first floor to the top of hip wall height of the standard plan using high foundation, to prevent immersion and reduce damage to building and goods. External stairs steps shall be increased for entry and exit from the outside.

Figure 3 shows the summary of cost study following the former part of the procedure (i–v) explained in Sect. 3. Cross-sectional views outline each plan, and building cost for each plan as estimated are shown. The repair cost for maximum of 3 inundation levels as estimated are shown; Level 1 (ground level to 1st floor level), Level 2 (1st floor level to bottom-of-hip-window level), and Level 3 (bottom-of-hip-window level to top-of-hip window level). This is equivalent to building

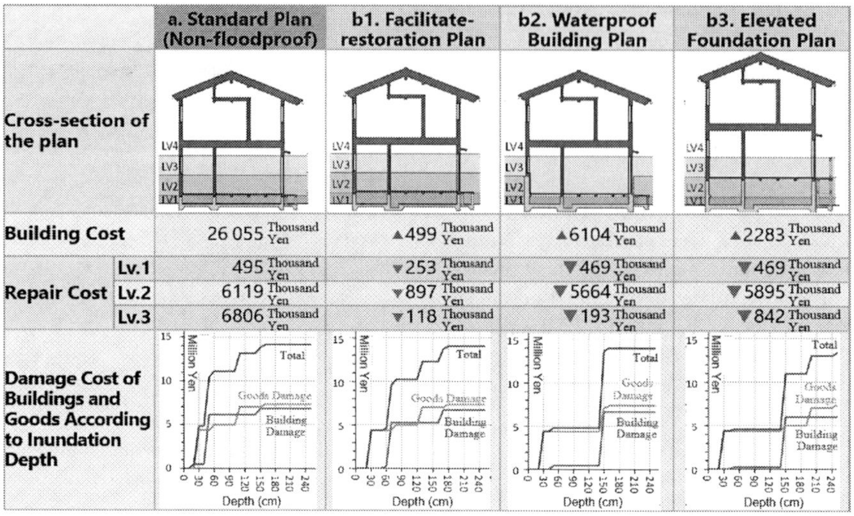

Figure 3. Summary of additional building and recovery costs for wooden detached houses.

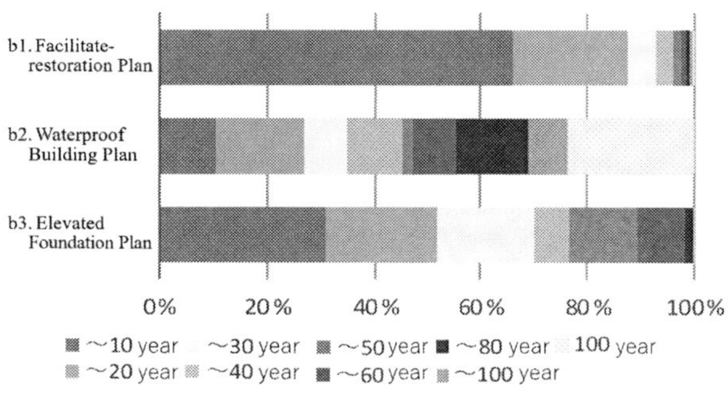

Figure 4. Average expected payback period (detached house).

damage cost, and damage costs for goods (such as furniture and electronic goods) according to inundation depth are also estimated and added to compose total damage cost.

Using these results, the expected payback period was calculated for each of the targeted grids using the procedure described in the latter part (vi–vii) of Sect. 3, and the percentages of grids by expected payback period for 3 flooding countermeasure plans are shown in Fig. 4.

From the studies, depending on the inundation characteristics of the site as shown in Fig. 2, the payback period for "Facilitate-restoration plan" were comparatively short, the period for "Elevated foundation plan" comparatively longer, and the period for "Waterproof building plan" comparatively longest. However, "Waterproof building plan" has advantages those other plans do not have, such as having fewer steps from the ground level, and within a grid of about 1/4, the expected payback period is less than 20 years.

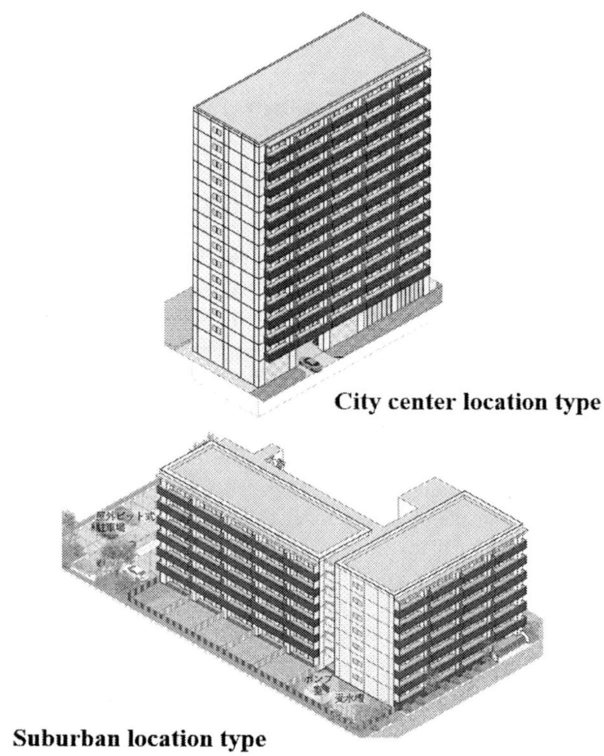

Figure 5. 2 model types of condominiums.

4.2 Condominiums

Next, target building type is condominiums, which occupy a large proportion of residences in urban areas of large cities. The study focused on the flood-proof renovation of existing condominiums, which has been of growing interest since the damages of high rise condominiums caused by 2019 Typhoon Hagibis of functional shutdown due to flooding in the basement electrical room. The study was based on the following two types of condominiums of different locations but with almost the same number of housing units (just over 60 units, Fig. 5).

"City center location type" is assumed to be located in city center or near stations, and is a mixed-use high-rise (14 stories) with a small commercial facility on the first floor, and facilities such as electricity and water supply are located on the basement level, with a mechanical parking lot (with underground pit) inside the building.

"Suburban location type" is assumed to be located in suburban residential are, and is a middle-rise (7 stories) condominium without a basement floor, and water supply facilities are located on the first floor of an outdoor annex and electrical facilities on the first floor of an indoor building. In addition, a mechanical parking lot (with an underground pit) was assumed to be installed outdoors to provide the same number of parking spaces as the number of dwelling units.

Countermeasures were considered, using mainly sandbags, watertight plates, watertight doors, and watertight valves for 3 levels of inundation assumption; Low for Level 1 (up to 30 cm from ground level), Medium for Level 2 (30–50 cm from ground level), and High for Level 3 (50–150 cm from ground level). Flood path and inundation areas were speculated for combinations of 2 location types, 3 level of countermeasures, and 4 level of inundations, with additional Level 4 (150–200 cm from ground level). Also, repair costs, and countermeasure installation position, for such as watertight plates, and costs were studied following Sect. 3 procedures as for a detached house.

Figure 6 shows the summary of countermeasure and repair (damage) costs for different level of flooding and countermeasures, and the percentage of average expected payback period for the countermeasure plans are shown. For "City center location type", the cost-effectiveness of low-level countermeasures, using sandbags, is comparatively quite high, and is influenced by the fact that entrance halls, underground floors, and indoor pit-type parking lots are not inundated by 30–50 cm flood depths. For "Suburban location type", the cost-effectiveness of low-level measures is still comparatively high, but the difference is limited compared to the "City center location type". This is due to the fact that outdoor pit-type parking lots, which are difficult to prevent from being flooded by sandbags, have a larger damage amount of JPY 44 million at a flooding depth of 60 cm.

As a result of the study, it was found that there is a high need to prioritize measures to protect vulnerable facilities such as entrance halls, and pit-type parking lots.

4.3 Business premises

Finally, from the perspective of recovery and reconstruction of flood-damaged areas, we considered it important not only to protect homes and livelihoods, but also the business premises that support the local economy and liveliness, so we conducted a study for this building type. We assumed the case of a small business premises occupying the first floor of an RC building (City center location type condominium) in occasion of interior construction (infill).

Considering the diversity of room and equipment configurations and fixtures by business type, 4 business types and models were assumed: offices, retail stores, restaurants, and small clinics. Countermeasures were considered, and mainly, watertight plates, reinforced concrete exterior walls were used for low level countermeasure plans of inundation up to 50 cm above ground level. A medium level countermeasure plans uses such measures for inundation up to 100 cm above ground level, and in addition, secondary watertight compartment was planned (except for retail shops, which was found difficult due to plan configuration) to be installed in the room, in case the primary watertight compartment does not function. Figure 7 shows the image of the medium level countermeasure plans.

Repair costs, and countermeasure installation position, for such as watertight plates, and costs were studied following Sect. 3 procedures as for a detached house and condominiums. In addition, to estimate the cost-effectiveness of the secondary watertight compartment, it is necessary to assume the case where the primary watertight compartment does not function, and here the estimation was made under the assumption that the probability of functioning is half (described as "Partial work" in Fig. 8).

Figure 8 shows the summary of countermeasure and repair (damage) costs for different level of flooding and countermeasures, and the percentage of average expected payback period for the countermeasure plans are shown.

The construction cost ratio of the standard plan is assumed to be; Real estate office (1.00) $<$ Retail shop (1.04) $<$ Restaurants (1.40) $<$ Dentist (1.85), and the proportion of the additional cost of the measures decreases as the construction cost of the proposed standard increases. For business premises such as dental clinics, which have many small rooms and built-in furniture, high construction costs, and a concentration of extremely expensive medical equipment, the results indicate that even the flooding countermeasure proposal 2, which has a large additional cost, can recover the cost of the countermeasure in a relatively short time if a certain amount of flooding is expected. On the other hand, for industries that do not have expensive equipment, such as real estate brokerage offices, advanced flood countermeasures may not be a realistic option because the cost of the countermeasures may not be recovered in the expected number of years of operation, and therefore, light countermeasures with little additional cost are generally considered appropriate.

		City center location				Surburban location			
Countermeasure		None	Low	Medium	High	None	Low	Medium	High
Above cost (mil. yen)		–	390	17 750	30 650	–	4050	21 700	32 150
Repair cost according to inundation level (mil. yen)	Level 4	61 000	△390	0	0	95 000	△50	0	▼46 800
	Level 3	61 000	△390	0	▼56 500	95 000	△50	0	▼46 800
	Level 2	28 000	▼26 110	▼28 000	▼28 000	48 500	△50	▼45 800	▼46 000
	Level 1	19 000	▼17 110	▼19 000	▼19 000	25 000	▼23 750	▼24 000	▼24 000

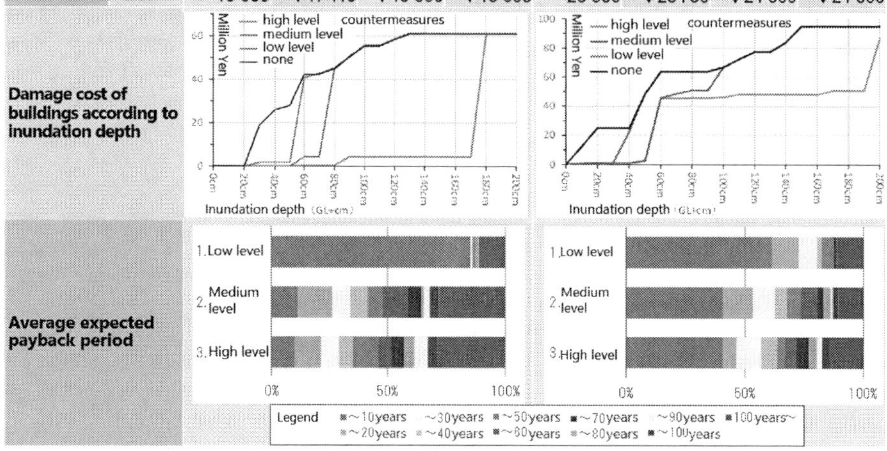

Figure 6. Summary of additional building and recovery costs, etc. for condominiums.

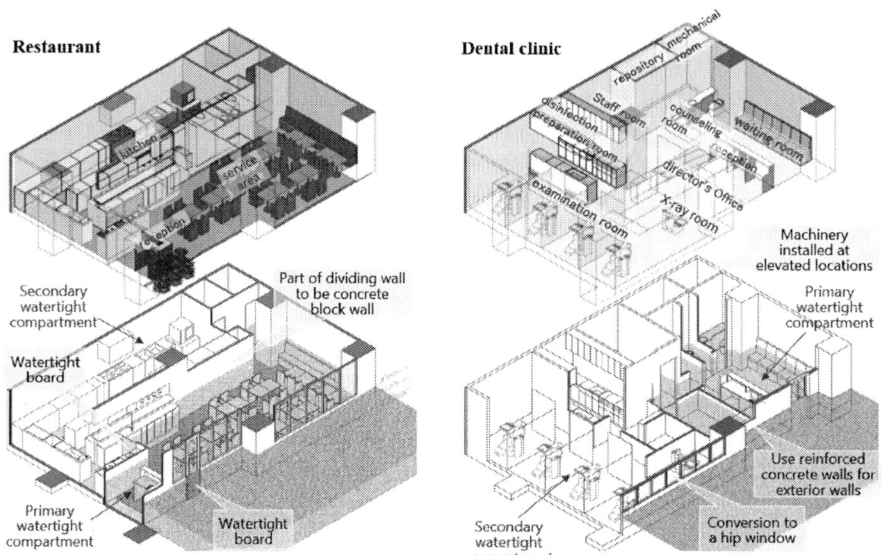

Figure 7. Isometric image of floodproofing of business premises as of restaurant and dental clinic.

As a result, the applicability of flooding countermeasures depends on the following factors; the flood risk of the location, the amount of expensive fixtures and equipment that can cause damage at shallow flood levels, whether such fixtures and equipment are located in an area capable to form a secondary watertight compartment, whether the employees can install watertight panels, and whether cost recovery can be considered over a long period of time. As for the 4-business premises considered, small clinics, that meet the latter four conditions, the high applicability of medium level anti-flood measures are confirmed, and in other business premises low level measure seems more applicable. (Note that "medium level" countermeasure plan does not include secondary watertight compartment, and is not in that sense medium level.)

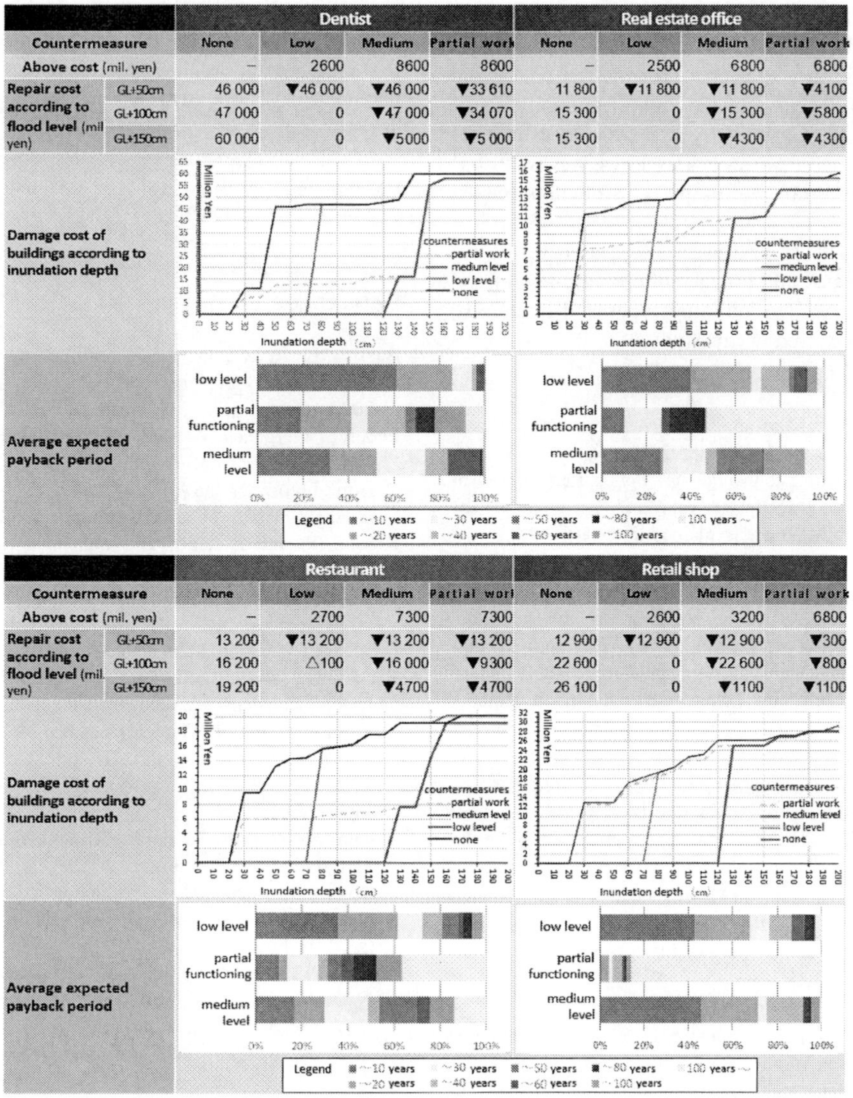

Figure 8. Summary of additional building and recovery costs etc. for business premises.

5 Conclusion

In this article, it was demonstrated through the presentation of major trial design drawings, etc., that flooding countermeasures with a certain degree of applicability can be realized in the three building types and situations considered. In addition, we demonstrated that it is possible to study the cost-effectiveness of inundation countermeasures by using the information of maximum inundation depth for each stage of the return period, such as of "Neighborhood Safety Level Map" by Shiga Prefecture, using the analysis methods and procedures described in this report.

In relatively many cases where flooding risk is recognized, it was found reasonable to first prepare for flooding at a minimum of 30 to 50 cm above ground level, which can be addressed with sandbags (water bags) or simple watertight boards, and then to consider countermeasures in stages for greater flooding depths, depending on the type of flooding risk. For example, low-level flooding countermeasure plan for the business premises studied in Sect. 4 is designed to reduce the risk of flooding for inundation level of up to 50 cm above ground level, and it was found feasible for nearly 40 % of the targeted area, and it can be seen from Fig. 2 that in many cases this is not the inundation depth based on the assumed maximum rainfall of 1000 year return period. It means that even if larger inundation depth estimates are shown in the hazard maps of expected maximum rainfall or planned scale rainfall, there is possibility to reduce the expected annual damage by implementing measures by preparing for more frequent inundation.

The most effective way to prepare for the frequency and depth of inundation depends on the type of inundation risk of the subject site, the form and cost of the measures required for the building conditions, and the vertical distribution of equipment and movable property inside and outside the building.

It is expected that these findings will be used in the future to advance the study and practice of building-level flood countermeasures for various building types and situations, including those not covered in this study.

Data availability. Most of the underlying research data are presented in detail in Kiuchi and Nakano (2023), in Japanese, and the authors are currently working to provide its English translation. Another data of "Neighborhood Safety Level Map" is provided by the author of Kentaro Taki (2019) and explained in the article.

Author contributions. NK: Conceptualization, Investigation, Formal analysis, Methodology, Writing, Supervision and Writing-review and editing. TN: Data Curation and Formal analysis. All authors discussed the results and commented on the manuscript.

Competing interests. The contact author has declared that neither of the authors has any competing interests.

Note. This paper is an abstract for English readers of a report (Kiuchi and Nakano, 2023), which developed and extended papers (Kiuchi et al., 2021, 2022, 2023), all written in Japanese.

Special issue statement. This article is part of the special issue "ICFM9 – River Basin Disaster Resilience and Sustainability by All". It is a result of The 9th International Conference on Flood Management, Tsukuba, Japan, 18–22 February 2023.

Acknowledgements. The authors express gratitude to co-authors of Kiuchi et al. (2021), (2022) and (2023) for contributing to the partial studies of this article, and Kentaro Taki for providing "Neighborhood Safety Level Map" equivalent data.

Review statement. This paper was edited by Daisuke Harada and reviewed by two anonymous referees.

References

Barsley, E.: Retrofitting for Flood Resilience, A Guide to Building & Community Design, RIBA Publishing, ISBN 9781859467343, 2020.

Federal Emergency Management Agency (FEMA): Homeowner's Guide to Retrofitting; Six Ways to Protect Your Home from Flooding, USA, https://www.fema.gov/sites/default/files/2020-08/FEMA_P-312.pdf (last access: 20 March 2024), 2014.

Kiuchi, N. and Nakano, T.: Studies on Floodproofing Plans of Buildings and Their Cost-effectiveness, Report of the Building Research Institute, 153, ISSN 0451-6486, https://www.kenken.go.jp/japanese/contents/publications/report/153/index.html (last access: 20 March 2024), 2023 (in Japanese).

Kiuchi, N., Tsuchimoto, T., Nakano, T., Imai, N., Ishiyama, Y., Inoue, T., Meno, F., and Watanabe, S.: Study on the Floodproofing Plans of Wooden Detached Housings and Their Cost-effectiveness Evaluation, AIJ Journal of Technology and Design, 27, 499–504, https://doi.org/10.3130/aijt.27.499, 2021 (in Japanese).

Kiuchi, N., Nakano T., Fujiki, R., and Yamaki, S.: Model Study of Existing Condominiums on the Floodproof Retrofitting and Their Cost-effectiveness Evaluation, AIJ Journal of Technology and Design, 28, 442–447, https://doi.org/10.3130/aijt.28.442, 2022 (in Japanese).

Kiuchi, N., Imai, N., Yamasaki, Y., Okamoto, Y., Inoue, T., and Nakamura, R.: Model Study of Floodproofing plans and Their Cost-effectiveness Evaluation of Premises Occupying 1st floor of an RC Building, AIJ Journal of Technology and Design, 29, 451–458, https://doi.org/10.3130/aijt.29.453, 2023 (in Japanese).

Ministry of Land, Infrastructure, Transport and Tourism (MLIT): The Flood Control Economy Investigation Manual, 60, 2020 (in Japanese).

Taki, K., Yamashita, K., Hirayama, N., and Takanishi, S.: Strategy Planning to Enhance the Community's Resilience for the Flood Disaster on the Site Risk Assessment, Adv. River Eng., 25, 79–84, https://doi.org/10.11532/river.25.0_79, 2019 (in Japanese).

Proc. IAHS, 386, 259–264, 2024
https://doi.org/10.5194/piahs-386-259-2024

Open Access PIAHS

A Study on Improving Disaster Mitigation Awareness by Simulated Flood Experience Using VR Videos

Tsuyoshi Koyabu[1,2] **and Masatoshi Denda**[3,4]

[1]Department of Engineering Mechanics and Energy, University of Tsukuba, Ibaraki, 305-8573, Japan
[2]Information Systems Division, IDEA Consultants, Inc., Tokyo, 154-8585, Japan
[3]Department of Engineering Mechanics and Energy, University of Tsukuba, Ibaraki, 305-8573, Japan
[4]Water-related hazard group National research agency Public Works research Institute, Tsukuba city, 305-8516 Japan

Correspondence: Tsuyoshi Koyabu (koyabu@ideacon.co.jp)

Received: 21 July 2023 – Revised: 3 November 2023 – Accepted: 24 February 2024 – Published: 22 April 2024

Abstract. In recent years, the frequency of floods has been increasing every year, causing considerable damage, in particular, to human life due to delays in evacuation. In this research, a web page containing a questionnaire was created to investigate the changes in human action during floods to contribute to efficient disaster prevention and mitigation activities by residents and municipality officers in charge of disaster management. The subjects were made to watch videos created using Virtual Reality (VR) technology. The questionnaire on the web page was used to verify whether watching the VR videos would change people's attitudes toward disaster mitigation in the event of a flood. The results showed that the subjects who watched the VR videos were more likely to choose the correct action and the decision time to choose the action was earlier. These results indicate that VR videos can be used to raise awareness of efficient disaster mitigation actions during floods.

1 Introduction

In Japan, water-related disasters occur every year due to climate change. The torrential rain in western Japan in 2018 resulted in more than 200 deaths. One of the factors contributing to the high death toll was delayed evacuation (Ushiyama et al., 2019). To implement early evacuation actions, it is important to effectively raise awareness regarding water-related disasters and provide disaster education to a wider population. There are various methods of disaster education for water-related disasters. It has been demonstrated that experiencing floods through virtual reality (VR) videos has proved to be effective (Kuribayashi et al., 2020). Researchers have studied the effects of using VR technology to enhance the evacuation awareness. They allowed people to imagine flood scenarios and experience evacuation during a flood using VR (Nagumo et al., 2022). Furthermore, research is being conducted to visualize the effects of river level forecasts in an easy-to-understand manner using VR to promote evacuations during floods (Morooka et al., 2022). In addition, flood disaster virtual reality applications have been developed in Japan (Fujimi and Fujimura, 2020), China (Pan et al., 2020), and Indonesia (Purnomo et al., 2022), demonstrating the effectiveness of using VR applications as flood education. There are operability issues to experience the application, especially for the elderly, who need to become accustomed to using the head-mounted display and controllers. On the other hand, viewing videos can be easily implemented by anyone. Disaster prevention education through VR videos is an easy mechanism that can be used worldwide to personalize the lessons learned videos to be passed on. VR videos have changed the way people experience floods, allowing "first-hand" experience by providing a virtual experience of flooding. In this study, an online questionnaire was created to investigate potential changes in the behaviors during a flood by viewing VR videos, which has been rapidly advancing in recent years. We aimed to prevent and mitigate disasters and disaster responses by residents and responsible officials effectively and efficiently. The VR videos were created by the Public Works Research Institute based on the lessons learned

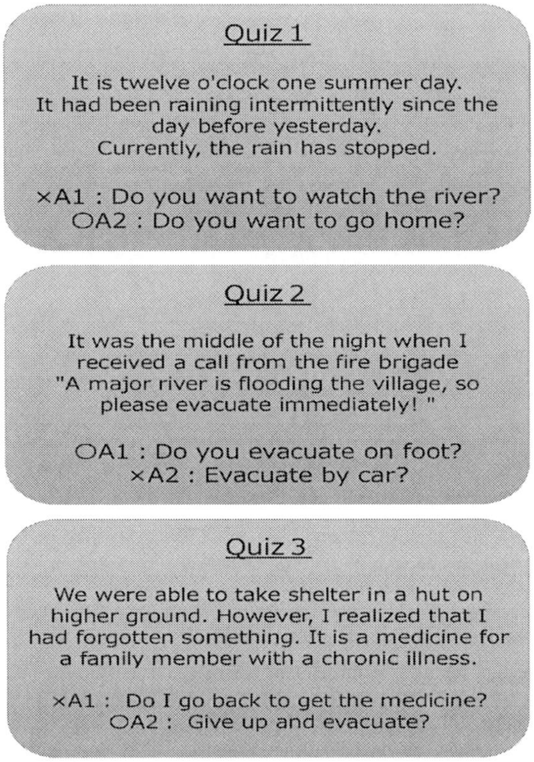

Figure 1. Content of the quiz on evacuation actions.

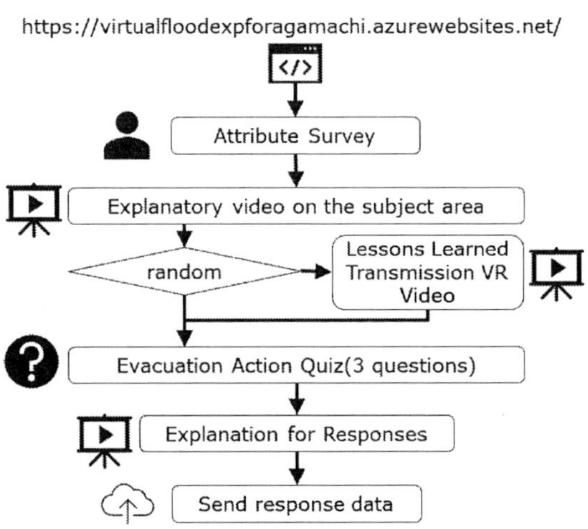

Figure 2. Structure of the webpage for survey response.

from actual flooding experiences. The information obtained from the questionnaire webpage was used to verify whether viewing VR movies containing lessons learned from previous disasters changed people's outlook toward disaster mitigation and floods.

Figure 3. A scene from the VR video of quiz 1.

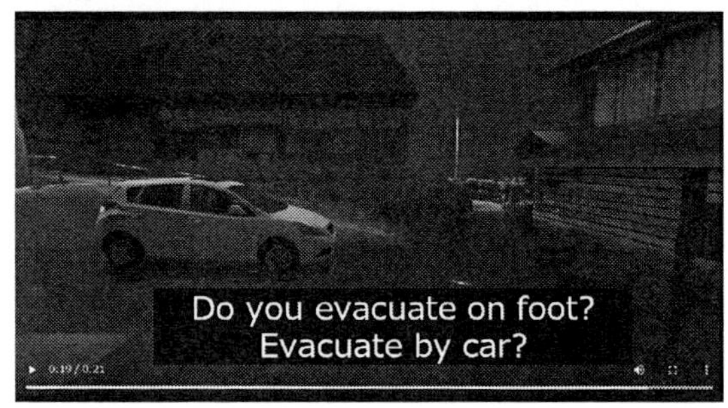

Figure 4. A scene from the VR video of quiz 2.

2 Experimental Methods

2.1 Web Questionnaires

A questionnaire was filled by subjects who had and had not viewed the video of the lessons using VR. The quiz asked participants to select one of two options for three different situations: during a river flood, just before flooding, and after flooding (Fig. 1).

Note that because the quiz was designed for a certain model district, all subjects were asked to watch a video explaining the target district before answering the questions. The survey was conducted via the internet by creating a webpage for the survey. It consisted of a demographic survey of the subjects, a video explaining the target area, and answers to three quizzes (Fig. 2).

The quiz on the evacuation action was a video. It was answered after watching the video (Figs. 3–5). After answering each quiz, the participants were asked to judge whether they had taken the correct action or not. They were also asked to watch a video explaining the lessons learned from previous floods, so that they understood the correct course of action.

2.2 Characteristics of Subjects

The total number of respondents was 29. Viewing of the lessons-learned VR videos was randomly assigned on the webpage. Thirteen subjects viewed the lessons-learned VR

Table 1. Percentage of correct answers to quiz on evacuation actions.

Quiz 1	average percentage of correct answers
VR video viewing available	77 %
No VR video viewing	75 %
Quiz 2	Percentage of Correct Answers
VR video viewing available	100 %
No VR video viewing	100 %
Quiz 3	Percentage of Correct Answers
VR video viewing available	100 %
No VR video viewing	94 %
ALL	Percentage of Correct Answers
VR video viewing available	92 %
No VR video viewing	90 %

Figure 5. A scene from the VR video of quiz 3.

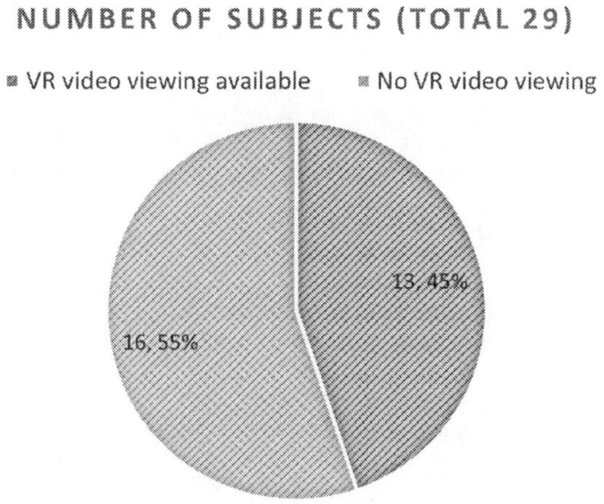

Figure 6. Number of subjects (29 in total) in the experiment to verify the effectiveness of the lessons learned via VR.

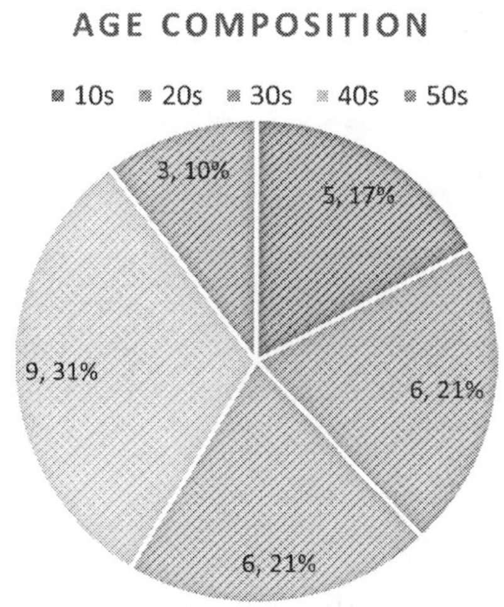

Figure 7. Age composition of the subjects.

videos, and 16 subjects did not view the videos (Fig. 6). The age of the subjects ranged from teens to 50s (Fig. 7).

3 Results and Analysis

3.1 Percentage of Correct Answers

The average percentage of correct answers on the quiz on subjects' evacuation actions was 92 % and 90 %, respectively, for subjects who did and did not watch the lessons-learned VR videos (Table 1 and Fig. 8). In addition, the percentage of correct answers for quizzes 1 and 3 was higher after viewing the lessons-learned VR videos. Quiz 2 was answered correctly by all participants.

A two-sample t-test was conducted to determine whether the difference in the percentages of correct responses with and without viewing the lesson-learned VR video was significant, assuming that the variances were different. The resulting p-value was 0.635, which was not less than the significance level of 0.05, and therefore was not considered to be a significant difference.

The percentage of correct answers did not change depending on whether the respondents in the age groups 11–20 and 51–60 had watched the VR video of the lessons learned. Conversely, the percentages of correct answers increased for participant ages 11–20 after watching the VR video of the lessons learned. However, the percentage decreased for those in the 31–50 age group. The trend remained identical for each quiz. There was no clear trend (Figs. 9 and 10).

3.2 Response Time

Next, the time required by participants to answer the questions was compared. For the answers on the webpage, we

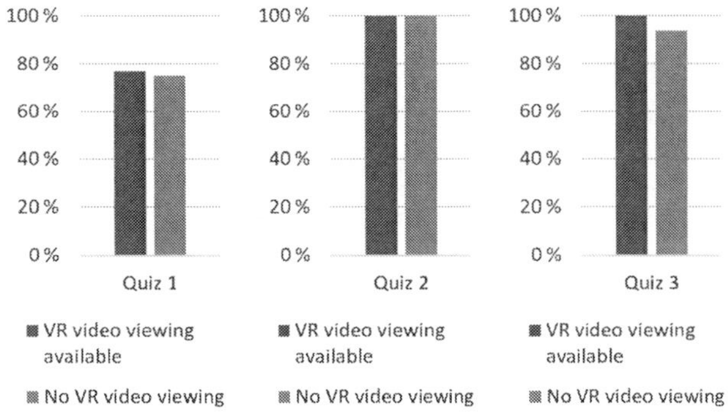

Figure 8. Percentage of correct answers to quiz on evacuation actions.

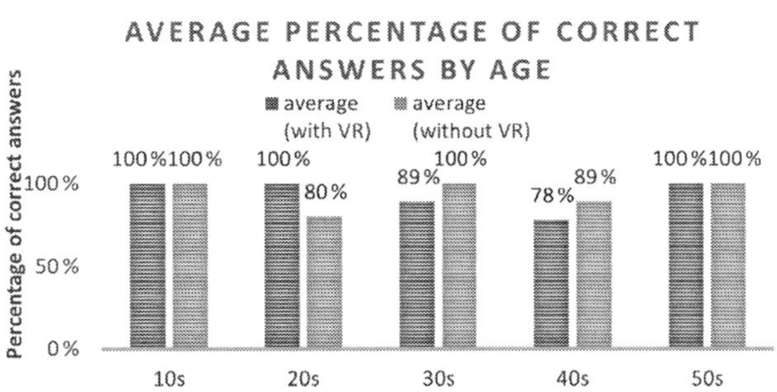

Figure 9. Average percentage of correct answers by age.

measured how long it took the respondents to answer the questions from the time they started watching the video. The duration of the question videos was approximately 2000 ms. In all quizzes, subjects who watched the lessons-learned VR videos responded more quickly (Table 2 and Fig. 11).

A two-sample t-test was also conducted for the response time, assuming that the variances were different. The results showed a significant difference for quiz 2, with a p-value of 0.043, which is below the 0.05 level of significance. Quiz 2 asked the participants to choose the means of evacuation. The correct answer is to evacuate on foot rather than by car. All subjects chose the correct means of evacuation; however, the decision was made more quickly by subjects who had viewed the lessons-learned VR video.

Younger teens quickly chose appropriate actions after viewing the lessons-learned VR videos (Fig. 12).

They selected appropriate actions quickly in quizzes 1 and 2. In quiz 3, 50s quickly selected the appropriate action by viewing the lessons-learned VR video (Fig. 13).

4 Discussion and Future Research

The results of the survey using a questionnaire webpage based on the video viewing showed that viewing the lessons-learned transmission VR video changed the awareness regarding the course of conduct during floods. Participants chose appropriate actions more quickly. This may be attributed to the fact that the subjects were given more information by the more realistic VR videos explaining the situation before and after the flood, in addition to text and voice explanations of the situation. The lessons-learned VR videos included the following information about the flood.

- It is not raining; however, the major rivers have rising water levels.

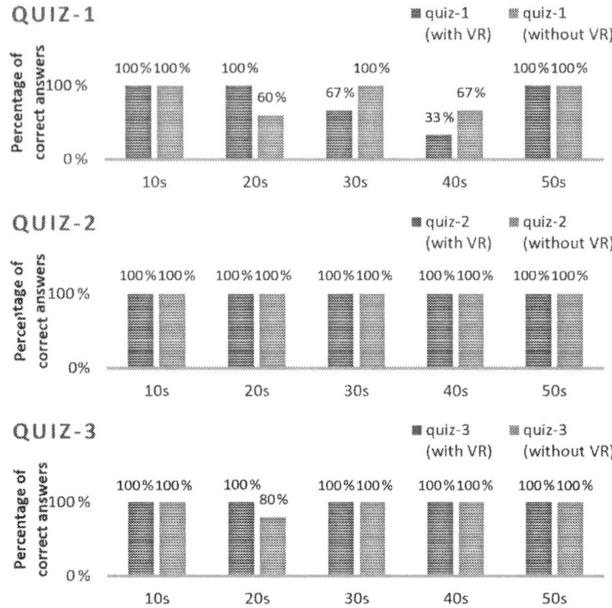

Figure 10. Percentage of correct answers for each quiz by age.

Table 2. Response time (ms) for quiz on evacuation actions.

Quiz 1	Average response time (ms)
VR video viewing available	6701
No VR video viewing	9645
Quiz 2	Average response time (ms)
VR video viewing available	3423
No VR video viewing	5323
Quiz 3	Average response time (ms)
VR video viewing available	3735
No VR video viewing	4044
ALL	Average response time (ms)
VR video viewing available	4620
No VR video viewing	6338

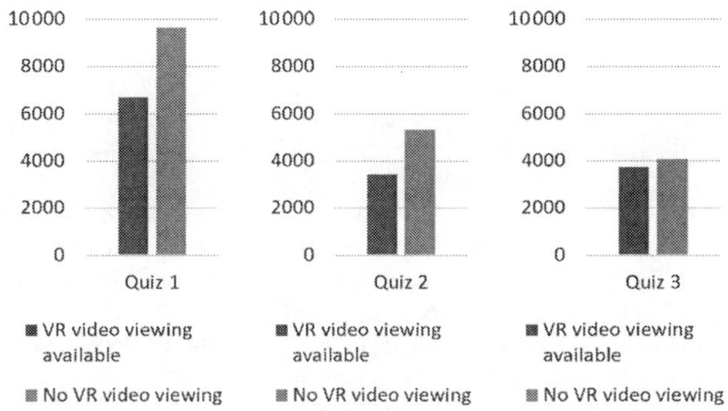

Figure 11. Response time (ms) for the quiz on evacuation actions.

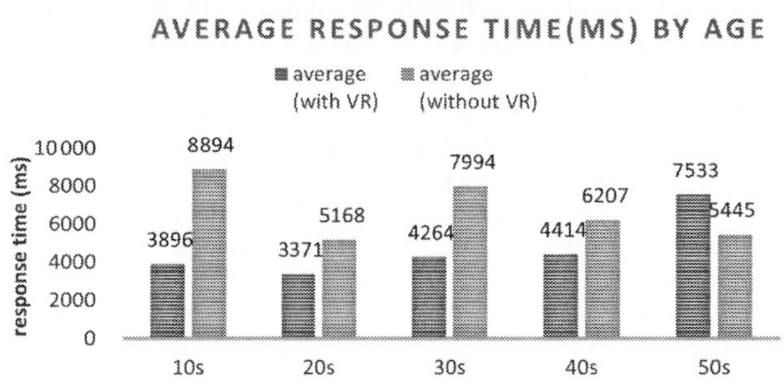

Figure 12. Average response time by age.

- Cars in parking lots are not flooded; however, the surrounding streets are flooded.
- Surrounding roads and low houses are flooded at night.

The situational descriptions influenced the way the recipients of the information perceive the situation during a water disaster, depending on their awareness of disaster preparedness. Some subjects will respond to at least some important information (information with greater weighting) with the least amount of information and will choose to evacuate (Fig. 14).

The important information that elicits a reaction is strongly visual. The realistic visual situation created by VR deepens the understanding of the choice of evacuation action. It provides a better evaluation of the information and helps to decide to evacuate. Although the evaluation of the information is subjective, prior disaster prevention education and experience with actual disasters can provide more responsive information and better evaluation information. In the future, we would like to conduct experiments in a virtual space using VR technology to investigate what type of information leads to evacuation actions and factors influencing individual judgments. Although this is an initial experiment and the number of subjects is small, we plan to increase the number of subjects with different attributes in the future.

Since VR video viewing via the Internet can be effectively used in areas around the world where flood damage can occur, we would like to consider creating VR videos based on flood-damage cases and lessons learned not only in Japan but also internationally, and expanding the use of such videos to other districts in the future.

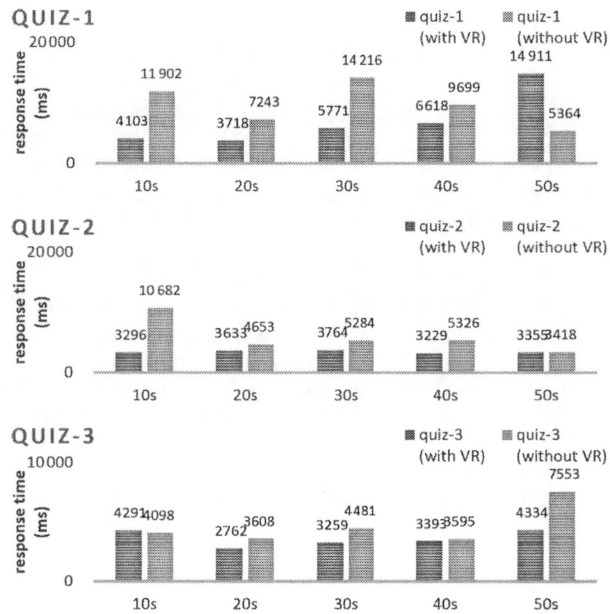

Figure 13. Average response time for each quiz by age.

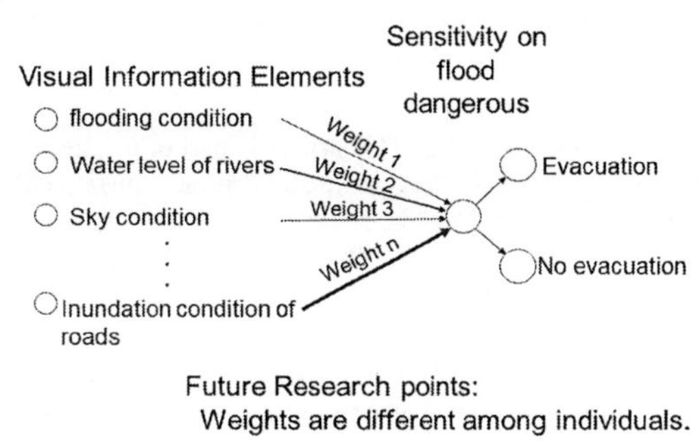

Figure 14. Schematic diagram of the decision path leading to evacuation actions.

5 Summary

The results of the study of changes in actionable attitudes pre- and post-flooding by viewing the lessons-learned VR video revealed the following.

- There is no clear effect of viewing the lessons-learned transmission VR video on the ability to choose appropriate actions.

- Viewing the VR video of the lessons-learned can help people to quickly determine the appropriate evacuation actions.

Data availability. The data is not publicly accessible. These data are about the psychological processes of individual subjects and must be treated with extra care in terms of research ethics.

Author contributions. TK and MD: Conceptualization; TK and MD: methodology; TK: software; TK: formal analysis; TK and MD: investigation; TK: writing – original draft preparation; MD: writing – review and editing; MD: supervision.

Competing interests. The contact author has declared that neither of the authors has any competing interests.

Disclaimer. Publisher's note: Copernicus Publications remains neutral with regard to jurisdictional claims made in the text, published maps, institutional affiliations, or any other geographical representation in this paper. While Copernicus Publications makes every effort to include appropriate place names, the final responsibility lies with the authors.

Special issue statement. This article is part of the special issue "ICFM9 – River Basin Disaster Resilience and Sustainability by All". It is a result of The 9th International Conference on Flood Management, Tsukuba, Japan, 18–22 February 2023.

Acknowledgements. We would like to thank the local government officials and residents of Aga Town for their tremendous contribution and cooperation during the field survey and interviews. We would also like to thank the students of the University of Tsukuba and the staff of IDEA Consultants, Inc. for their cooperation in the survey.

Review statement. This paper was edited by Kensuke Naito and reviewed by Jagriti Mishra and two anonymous referees.

References

Pan, Z., Cheok, A. D., Müller, W., and Zhang, M.: A virtual reality training system for flood security, Trans. Edutainment, XVI, 126–134, Springer Berlin/Heidelberg, https://doi.org/10.1007/978-3-662-61510-2_12, 2020.

Fujimi, T. and Fujimura, K.: Testing public interventions for flash flood evacuation through environmental and social cues: The merit of virtual reality experiments, Int. J. Disast. Risk Re., 50, 101690, doi.org/10.1016/j.ijdrr.2020.101690, 2020.

Morooka, Y., Tsuchiya, S., Qian, C., and Takeshita, T.: Developing a river water level predicting system with 3D viewer using virtual reality technology, Advances in River Engineering, 28, 37–42, https://doi.org/10.11532/river.28.0_37, 2022.

Nagumo, N., Denda, M., Harada, D., Koike, T., Egashira, S., and Shinya, T.: An experiment of offering virtual evacuation experience program for flood damage mitigation, 51, https://doi.org/10.14866/ajg.2022a.0_51, 2022.

Purnomo, F. A., Yoeseph, N. M., Koesuma, S., Hartono, R., Ristania, S., and Riasti, B. K.: Virtual reality flood disaster with variations of water levels in the early warning system, The Institute of Electrical and Electronics Engineers, Inc. (IEEE) Conference Proceedings, https://doi.org/10.1109/APICS56469.2022.9918727, 23–24 August 2022.

Kuribayashi, D., Ohara, M., Koyabu, T., and Sawano, H.: A study on improving flood awareness and motivation for disaster mitigation by flood simulated experience using image video, Journal of Disaster Information Studies, 18, 35–45, https://doi.org/10.24709/jasdis.18.1_35, 2020.

Ushiyama, M., Honma, M., Yokomaku, S., and Sugimura, K.: Characteristics of victims caused by heavy rainfall disaster in July, 2018, Journal of Japan Society for Natural Disaster Science, 38, 29–54, https://doi.org/10.24762/jndsj.38.1_29, 2019.

Proc. IAHS, 386, 265–270, 2024
https://doi.org/10.5194/piahs-386-265-2024

Developing a flood monitoring system by utilizing real-time satellite rainfall estimates and water energy budget-based rainfall-runoff inundation model in West Africa

Abdul Wahid Mohamed Rasmy[1], Maksym Gusyev[2], Katsunori Tamakawa[1], Miho Ohara[1], and Toshio Koike[1]

[1]International Center for Water Hazards and Risk Management (ICHARM), PWRI, Tsukuba, Ibaraki, Japan
[2]Institute of Environmental Radioactivity (IER), Fukushima University, Fukushima, Japan

Correspondence: Abdul Wahid Mohamed Rasmy (awmrasmy@gmail.com)

Received: 20 July 2023 – Accepted: 25 February 2024 – Published: 7 May 2024

Abstract. The African regions have been affected by frequent flood-related disasters. To support decision-makers in flood disaster management, a real-time flood monitoring system was established in the Niger and Volta basins under the Water disaster platform to enhance climate resilience in Africa (WADiRE-Africa). To bridge a gap between data unavailability and inaccessibility, the near-real-time satellite rainfall data (GSMaP-NRT) was evaluated and found to have a very low correlation (< 0.4) with gauge data at daily scales and a higher correlation (> 0.7) at monthly scales. The statistically bias-corrected GSMaP-NRT data was used in the Water and Energy Budget-based Rainfall-Runoff-Inundation (WEB-RRI) model to generate hydrological responses of the basins. As the simulated river discharges were overestimated, an additional correction method was employed for the GSMaP-NRT data to match well with observed discharges. The prototype system provides real-time flood information in both basins.

1 Introduction

Flood disasters have magnified over the last decade due to global changes (e.g. global warming, population increase, and land-use changes). Though Flood Early Warning System (FEWS) services are proven life-savers and crucial for flood risk management, those services are woefully insufficient for those who need them most, and half the world is not prepared for disasters (https://wmo.int/media/news/un-warns-half-world-not-prepared-disasters, last access: 19 April 2024).

Especially, the West African regions have been affected by frequent flood-related disasters (Floodlist News in Africa, 2020) hindering the region's sustainable development and requiring the capacity to predict, monitor, and mitigate flood disasters in the countries of the Niger and Volta basins. Despite various efforts by international donor organizations to bridge gaps in in-situ data, modeling technology, and capacity building in the West Africa region, these gaps remain the main challenge preventing a setup of FEWS to produce reliable outputs for policymakers in the West Africa region. In addition, the Niger and Volta basin countries lack flood-disaster management practices to prepare for disasters in advance. Given this situation, and to increase flood disaster preparedness and resilience, UNESCO IHP with help from the Government of Japan initiated the regional project titled "Water Disaster Platform to Enhance Climate Resilience in Africa" (WADiRE-Africa). The project was implemented in the Niger and Volta River basins by International Centre for Water Hazard and Risk Management (ICHARM), UNESCO Intergovernmental Hydrological Programme (IHP), and AGRHYMET Regional Center in collaboration with the Niger Basin Authority (NBA) and the Volta Basin Authority (VBA).

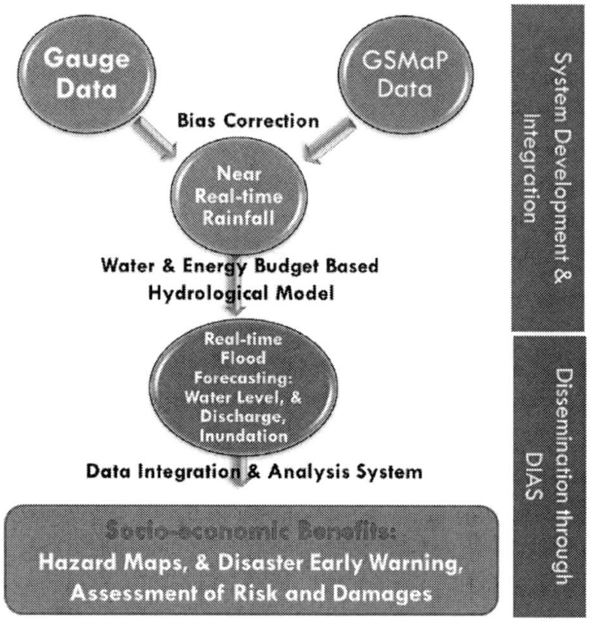

Figure 1. FEWS components.

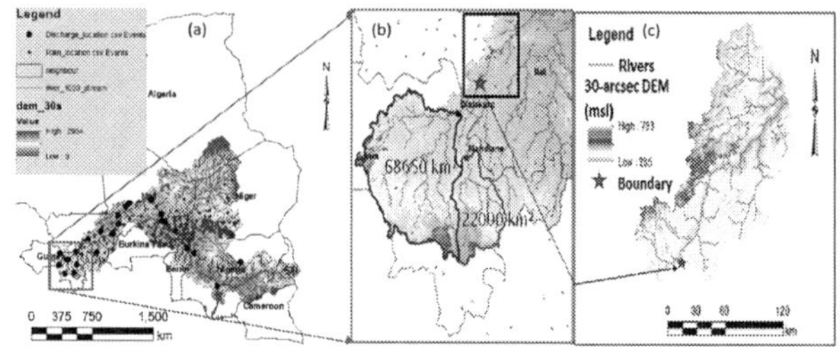

Figure 2. The Niger River basin **(a)**, the upstream area **(b)**, and the hot spot model for Bamako city **(c)**.

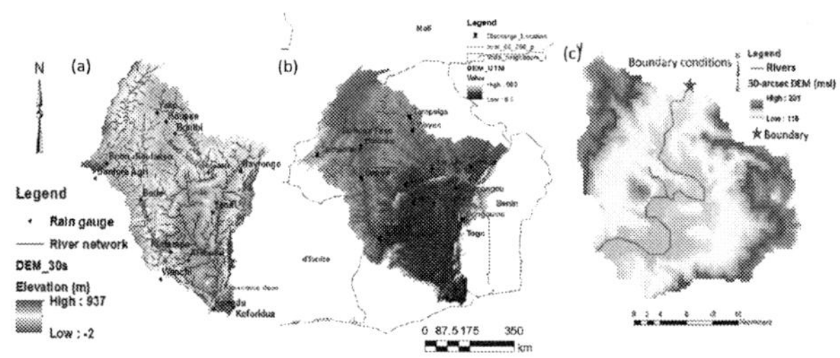

Figure 3. The Volta River basin and rain gauges **(a)** discharge gauges **(b)** and hot-spot model for the Mango city **(c)**.

2 Methodology and data

Preferably, a FEWS System (IFIS) should have the following essential components: (a) adequate real-time rainfall observations, (b) quantitative rainfall prediction (not incorporated in this project), (c) a runoff and inundation simulator, and (d) a data analysis and dissemination system (Rasmy et al., 2023). Figure 1 is the schematic diagram of major FEWS components and the data integration.

Real-time rainfall data with adequate spatial and temporal resolutions were unavailable in these basins. Global satellite precipitation products (SPPs) can contribute to water-related disaster management by improving their quality by merging with available ground gauges for hydrological applications in poorly gauged basins. This study used near-real-time GSMaP products, which are available at 0.1° spatial and 1 h temporal resolutions (Kubota et al., 2020). In addition, available ground rainfall and discharge data were obtained and used for satellite rainfall data calibration and model verification.

Moreover, physically-based Distributed Hydrological Models (DHMs) with the capability of simulating rainfall-runoff-inundation processes are very effective tools for obtaining crucial flood-related information (i.e., peak flow, inundation depths, and extents). this study employed the Water and Energy Budget-based Rainfall-Runoff-Inundation (WEB-RRI) model as a flood simulator, which has several merits, including (a) physical formulations for evapotranspiration fluxes and soil moisture dynamics to improve the accuracy of soil-water storages and reduce model-related uncertainties; (b) the compatibility with numerical weather models to estimate basin responses to model forcing data directly; and (c) restarting capability for warm-up simulations (Rasmy et al., 2019).

The WEB-RRI models were developed in the river basin scale (the Niger model at 4 km grid and the Volta model at 2 km grid) and flood hot spot scale (at 0.45 km grid) (Figs. 2 and 3). Flood hot spot scale models have nested within the regional scale WEB-RRI models by using upstream boundary condition river discharge, which was simulated by the regional scale WEB-RRI models.

Data Integration and Analysis System (DIAS) started in 2006 to collect, store, and analyze global and local data in combination with socio-economic data, and convert data into information useful for disaster/crisis management and to make this information available within Japan and overseas (https://diasjp.net/en/about/, last access: 19 April 2024). This study used the DIAS resources for data archiving, integration, model simulation, and information dissemination for the Volta and the Niger River basins.

3 Results and discussion

3.1 Satellite rainfall analysis

Figure 4 is the scatter plot for the comparison between the observed data and GSMaP-NRT data at all the rain gauge locations at daily and monthly scales during 2002–2006 in the Volta River basin. As shown in the figures, the daily correlation between observed data and GSMaP-NRT data has a very low correlation (< 0.3), however, the monthly correla-

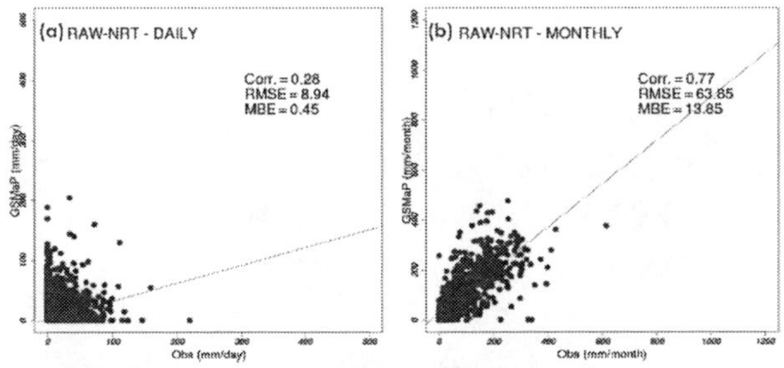

Figure 4. Comparison of in-situ and GSMaP-NRT data at daily and monthly scale during 2002–2006 in the Volta River basin.

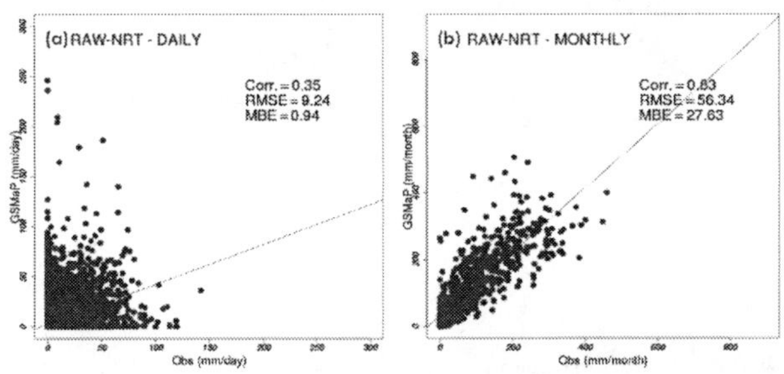

Figure 5. Comparison of in-situ and GSMaP-NRT data at daily and monthly scale during 2003–2007 in the Niger River basin.

tion values are very much higher (> 0.7). Similarly, the daily correlation between observed data and GSMaP-NRT data in the Niger basin has a very low correlation (< 0.4), however, the monthly correlation values are very much higher (> 0.8) as shown in Fig. 5. These results indicate that the applicability of GSMaP-NRT monthly data is reliable and has a high potential to be used for monthly based assessments and water resources management, however, the GSMaP-NRT data has to be bias-corrected to utilize for flood monitoring and mapping applications in these poorly gauged basins reliably.

Due to the unavailability of real-time ground data, we have estimated statistical bias correction factors based on 10 d averaged products and implemented these factors in real-time flood monitoring and mapping applications. The bias correction method used in this study is the simple gradient fitting to unity. Figure 6 is the plots that compare the 10 d products of GSMaP-NRT and ground gauges before and after the bias correction for the selected rain gauges in the Volta River basin. The statistics are shown in the same plot. In this method, the correlation between GSMaP-NRT and ground observation will not be changed, however, the slope or gradient of the points will be adjusted to unity. The correction factors were calculated at each gauging point and distributed to basin scales by the Thiessen polygon method as shown in Fig. 7.

3.2 Hydrological simulations

3.2.1 Model calibration

The rainfall and Discharge data from 1965 to 1970 at the limited number of stations were obtained from the Volta River Basin Authority and used for model calibration. The discharge data at Yarugu and Waisi stations were used as a boundary condition to simulate discharge data at the Nawuni station due to the unavailable of water release data from several dams in this basin (Fig. 3). Similarly, the discharge data at the Boromo station was used as a boundary condition to derive discharge data at the Dapaola station.

As shown in Fig. 7a, the simulated discharge at the Nuwuni gauging point is well comparable with the observed discharges during the 1965 to 1970 period. Particularly, the low-flows and peak-flows are well-matched with observed low-flows and peak-flows. The estimated performance statistics (Nash = 0.84, RMSE = 139 $m^3\,s^{-1}$) are very high in this station during the simulation period. On the other hand, the simulated discharges for the same model parameters at the Dapola station were higher and the estimated performance statistics (Nash = 0.36, RMSE = 113 $m^3\,s^{-1}$) were relatively poor in this station during the simulation period. The poor Nash value at this location was due to the unavailability of the dam data from the upstream regions. Unfortunately, there was no ground rainfall data was available from the upstream part of the Niger basin for the model calibration and verification. As a result, the same calibrated parameters from the Volta Basin were used for the Niger Basin simulation.

3.2.2 Model validation for GSMaP data

Figure 9 compares the observed and simulated river discharges derived from bias-corrected GSMaP-NRT data during 2004–2006 at Nawuni and Dapola stations. As shown in Fig. 9a1 and b1, the bias-corrected GSMaP-NRT data overestimated the peak flow, particularly during June, July, and September. To improve the hydrological responses of the basin, the rainfall data was further modified without changing the model assuming that the model is well calibrated using the ground data. The modification factors were identified by trial and error method for better matching of discharges, and the factors were January–May = 0.25, June = 0.5, and July = 1. As shown in Fig. 9a2 and b2, after the additional modification of the GSMaP-NRT data, the simulated discharges were well-matched with observed discharges at both locations.

Similarly, the bias-corrected GSMaP-NRT data, which was derived from the 10 d correction factors and additional factors (January–May = 0.25, June = 0.5, July = 1) used for simulating hydrological responses of the Niger basin. The model simulated discharge outputs were compared at the Mandiana and Dialakuro stations (Fig. 2b). As shown in Fig. 10, the simulated discharges were well-matched with observed discharges at both locations during 2016–2018.

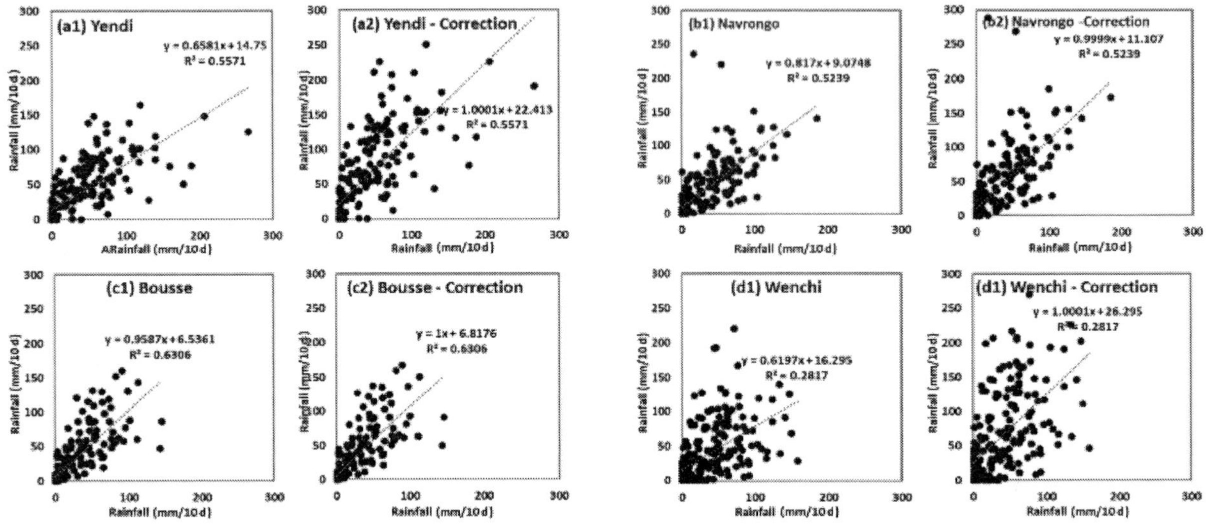

Figure 6. Comparison of in-situ and GSMaP-NRT data at 10-daily and bias-correction using gradient fitting method in the Volta basin.

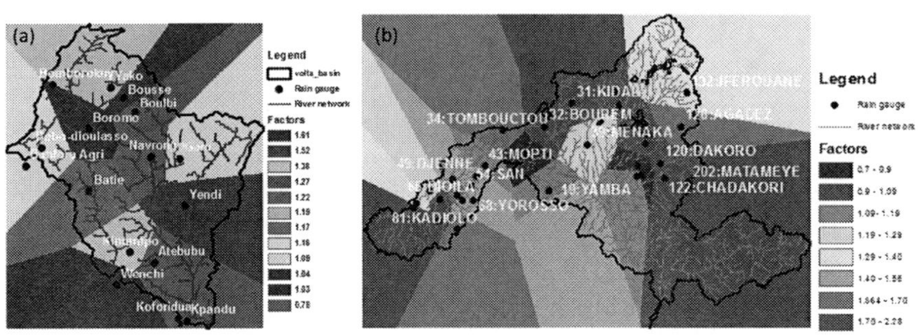

Figure 7. The distributed statistical bias-correction factors for **(a)** the Volta River basin and **(b)** the Niger River basin.

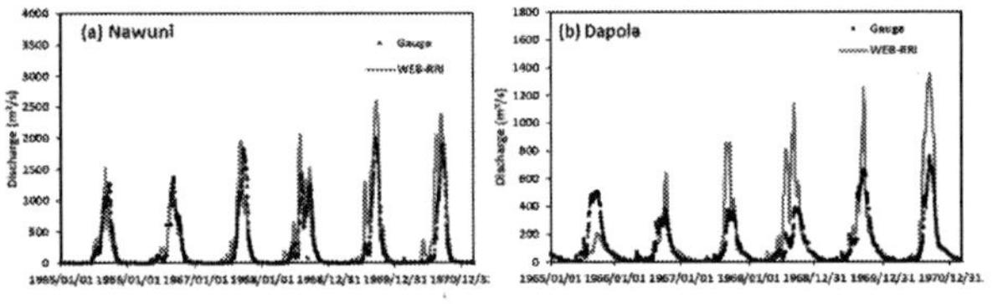

Figure 8. Comparison of observed and simulated river discharges derived from observed rain data from 1965 to 1970 at **(a)** Nawuni and **(b)** Dapola Stations.

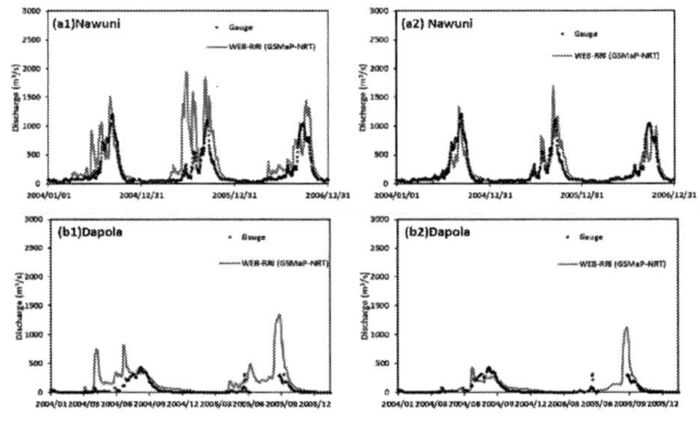

Figure 9. Comparison of observed and simulated river discharges derived from bias-corrected GSMaP-NRT data at **(a1)** Nawuni for 10 d factors, **(a2)** Nawuni for 10 d factors and additional factors, **(b1)** Dapola for 10 d factors, and **(b2)** Dapola station for 10 d factors and additional factors.

Therefore, it is confirmed that the selected bias correction method for GSMaP-NRT data is applicable for generating reliable flow information in both basins.

3.3 Implementation of FEWS in DIAS

FEWS in the Volta and the Niger River basin were implemented on Data Integration and Analysis System (DIAS) as shown in Fig. 11. As shown in the figures, the near real-

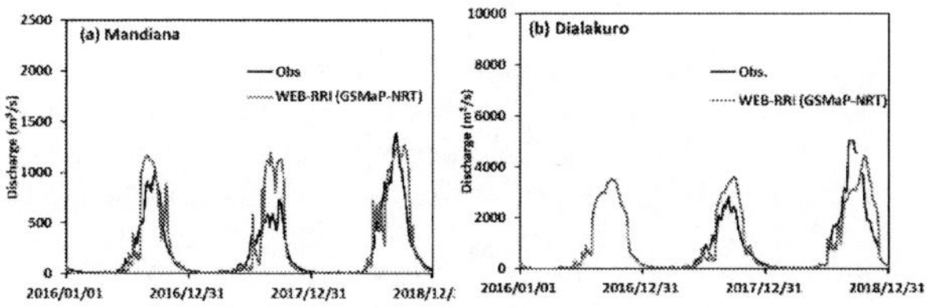

Figure 10. Comparison of observed and simulated river discharges derived from bias-corrected GSMaP-NRT data during 2016–2018 using 10 d statistical factors and additional factors as similar to the Volta River basin at **(a1)** Madiana station and **(a2)** Dialakuro station.

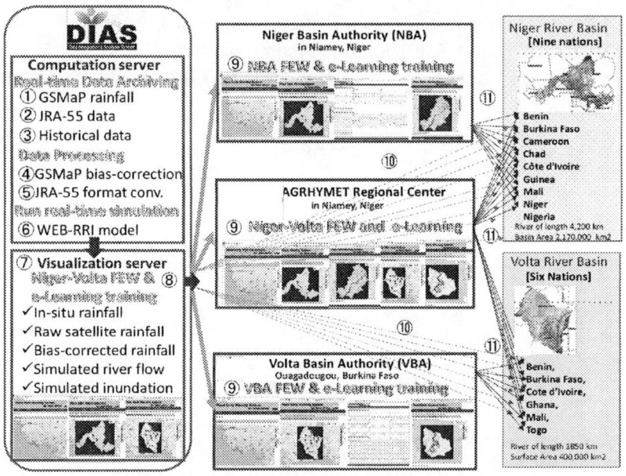

Figure 11. Schematic diagram of the regional FEWS in DIAS for the Niger and Volta River basins (DIAS logo was from https://diasjp.net/, last access: 19 April 2019).

time GSMaP rainfall and other data are archived on DIAS (Steps 1–2). In Step 3, historical rainfall data of Niger-Volta basins are manually uploaded on DIAS, which also allows the automatic uploading of the real-time in-situ data of each gauge. In Steps 4–5, DIAS converts archived data into the unified data format and conducts a bias correction of the GSMaP rainfall. The WEB-RRI models of Niger-Volta basins are run and simulated outputs of flood discharge and inundation depth are visualized on DIAS (Step 7). In Step 8, model outputs and in-situ, raw, and bias-corrected GSMaP rainfall are processed for downloading.

4 Summary and conclusion

The FEWS services are crucial for flood risk management and the sustainable development of the nations. The Niger and the Volta River basins in West Africa do not have adequate data to implement FEWS for the effective management of water-related disasters. This study developed an FEWS in both river basins by utilizing available ground data, near-real-time satellite rainfall products, and a water and energy budget-based hydrological model.

The daily correlation between observed data and GSMaP-NRT data has a very low correlation (< 0.4), however, the monthly correlation values are much higher (> 0.7) in both river basins. To utilize the GSMaP-NRT data for flood monitoring and mapping applications in these poorly gauged basins reliably, the statistical bias correction factors based on 10 d averaged products were estimated and implemented these factors in real-time flood monitoring and mapping applications. The WEB-RRI model was developed at both basins to simulate the hydrological responses of the basins. It is found that discharge estimated for the bias-corrected GSMaP-NRT data has overestimated the discharges, particularly during June, July, and September. To improve the hydrological responses of the basin, the rainfall data was further modified by trial and error method by matching the simulated discharges with observed discharges. It is found that the distributed 10 d correction factors together with additional rainfall factors (January–May = 0.25, June = 0.5, July = 1) can simulate discharges well as compared to observed discharges at Madiana and Dialakoro of Niger basin and Nawuni and Dapola of Volta basin.

The FEWSs in the Volta and the Niger River basin were implemented in DIAS by utilizing the statistical bias-correction method and the calibrated WEB-RRI model in both Niger and Volta River basins. The FEWS system executes the near-real-time GSMaP satellite rainfall bias-correction processes and feeds to the WEB-RRI model with natural flow conditions (i.e. without considering upstream dam impacts due to the absence of dam inflow and outflow data). The future activity of this research will focus on improving the system further by archiving more data from relevant authorities in the West African region.

Code and data availability. Data and code used in the study are available and can be obtained from the corresponding author.

Author contributions. All the authors contributed to the conceptualization, methodology, analysis, and writing.

Competing interests. At least one of the (co-)authors is a guest member of the editorial board of *Proceedings of IAHS* for the special issue "ICFM9 – River Basin Disaster Resilience and Sustainability by All". The peer-review process was guided by an independent editor, and the authors also have no other competing interests to declare.

Disclaimer. Publisher's note: Copernicus Publications remains neutral with regard to jurisdictional claims made in the text, published maps, institutional affiliations, or any other geographical representation in this paper. While Copernicus Publications makes ev-

ery effort to include appropriate place names, the final responsibility lies with the authors.

Special issue statement. This article is part of the special issue "ICFM9 – River Basin Disaster Resilience and Sustainability by All". It is a result of The 9th International Conference on Flood Management, Tsukuba, Japan, 18–22 February 2023.

Review statement. This paper was edited by Mamoru Miyamoto and reviewed by two anonymous referees.

References

Floodlist News in Africa: Breaking News 3 days ago – Guatemala – 25 000 Affected by Heavy Rain, Floods and Landslides West Africa – Floods Hit Burkina Faso and Northern Ghana, http://floodlist.com/africa/west-africa-burkinafaso-ghana-september-2020 (last access: 22 April 2024), 8 September 2020.

Kubota, T., Aonashi, K., Ushio, T., Shige, S., Takayabu, Y. N., Kachi, M., Arai, Y., Tashima, T., Masaki, T., Kawamoto, N., Mega, T., Yamamoto, M. K., Hamada, A., Yamaji, M., Liu, G., and Oki, R.: Global Satellite Mapping of Precipitation (GSMaP) Products in the GPM Era, in: Advances in Global Change Research, 355–373, Advances in Global Change Research, vol. 67, Springer, https://doi.org/10.1007/978-3-030-24568-9_20, 2020.

Rasmy, M., Sayama, T., and Koike, T.: Development of Water and Energy Budget-based Rainfall Runoff-Inundation Model (WEB-RRI) and Its Verification in the Kalu and Mundeni River Basins, Sri Lanka, J. Hydrol., 579, 124163, https://doi.org/10.1016/j.jhydrol.2019.124163, 2019.

Rasmy, M., Yasukawa, M., Ushiyama, T., Tamakawa, K., Aida, K., Seenipellage, S., Hemakanth, S., Kitsuregawa, M., and Koike, T.: Investigations of Multi-Platform Data for Developing an Integrated Flood Information System in the Kalu River Basin, Sri Lanka, Water, 15, 1199, https://doi.org/10.3390/w15061199, 2023.

Proc. IAHS, 386, 271–276, 2024
https://doi.org/10.5194/piahs-386-271-2024

Developing an Integrated Flood Management Plan under climate change for Grand River North West River basin, Mauritius

Akshay Kowlesser[1], **Mohamed Rasmy**[2], **and Toshio Koike**[2]
[1]GERS-LEE, Univ. Gustave Eiffel, IFSTTAR, 44344 Bouguenais, France
[2]ICHARM, PWRI, Tsukuba, Ibaraki, Japan

Correspondence: Akshay Kowlesser (akshay.kowlesser@univ-eiffel.fr)

Received: 18 June 2023 – Revised: 17 March 2024 – Accepted: 23 May 2024 – Published: 21 June 2024

Abstract. In this article, the catchment of Grand River North West (GRNW) in Mauritius was selected as a case study to develop an Integrated Flood Management Plan to address the flood risks that Mauritius is and will be facing due to climate change. The study contains three main components: (i) a climatology analysis for the past (2003–2018) and future climate (2025–2040) to assess the trends in floods under climate change, (ii) the development of a hydrological model to study the hydrological responses of the basin for extreme flooding events, and (iii) a GIS-based flood risk model to develop a risk map. The results show that extreme rainfall events exceeding $50\,\mathrm{mm\,d^{-1}}$ have a high likelihood of increasing. A flood vulnerability map for the basin was developed based on nine indicators to aid decision making at the village council area (VCA) level and guide financial investments flood mitigation measures. The study established a novel end-to-end approach to develop an Integrated Flood Management Plan based on a case study for a river basin in Mauritius.

1 Introduction

From 2000 to 2019, floods affected 1.6 billion people worldwide and accounted for 44 % of all disaster events, the highest figure for any disaster type (CRED and UNDRR, 2020; UNDRR, 2020). Floods also killed 104 614 people and caused USD 651 billion in economic damage during this period (Toledo-Gallegos et al., 2022). Flood impacts are, however, felt differently in various parts of the world due to the physical, socio-economical, environmental, and geographical specificities and this rings even truer for Small Island Developing States (SIDS) (Scandurra et al., 2018) like Mauritius. With the limited space for land use development coupled with an economic strategy driven to support real estate development, Mauritius, is undergoing consequential urbanization (Baguant-Moonshiram et al., 2013), and as a result, Mauritius is becoming increasingly exposed to flood risks. The second most significant hydroclimatic hazard that Mauritius faces after cyclones, is flooding. It leads to direct losses of about USD 22 million annually, which accounts for 20 % of the country's economic losses related to natural disasters(Simpson et al., 2016). This article therefore aims at developing an end-to-end scalable framework to propose an integrated flood management plan based on the case study of the second largest river basin in Mauritius, the Grand River North West Basin.

2 Materials and Methods

2.1 Study Area

The Grand River North West (Fig. 1), has an area of $114\,\mathrm{km^2}$ and consists of young lava series geology for the most part. The GRNW basin has a mixed land use; 35 % for urban areas, 34 % for agricultural land (consisting mainly of sugarcane fields), and 24 % for forest land. The catchment of GRNW has undergone significant land use changes in the past decade alone. The catchment of GRNW contains the Moka Smart City, currently under construction of an extent of around $19\,\mathrm{km^2}$ taking over sugarcane fields. This large-scale development will further accentuate the very present

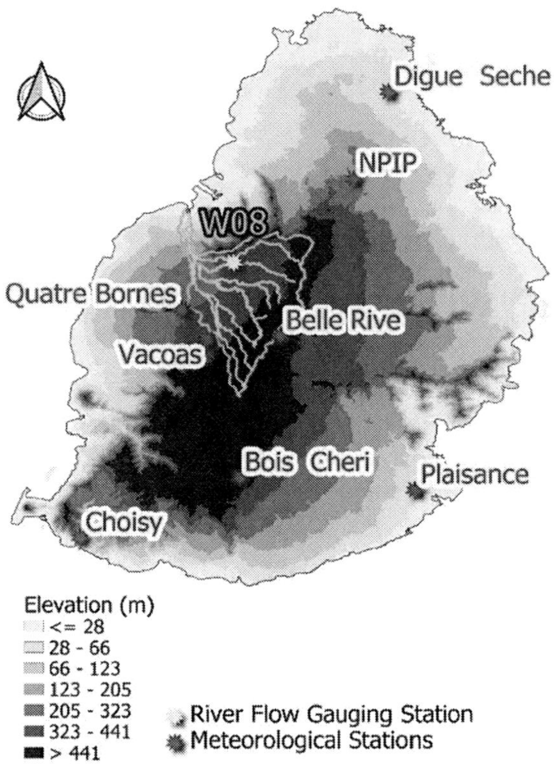

Figure 1. Location of GRNW Basin.

flooding issues in the catchment if not integrated with the overall water management and land use strategies.

2.2 Input data collection and preparation

For the purpose of this study, meteorological data and river discharge data were used for climate change impact assessment and robust hydrological model development. The respective stations are shown in Fig. 1. A digital elevation model (DEM) with a resolution of 60 m was also used for the model construction as well as MODIS (MCD15A2H), Japanese Reanalysis 55 years Data (JRA-55) 3 h temporal resolution, and 0.125° spatial resolutions for air temperature, wind speed, specific humidity, and surface pressure, and 0.56° spatial resolution for downward radiations, respectively. These data were interpolated to the model grid resolution of 60 m and a model temporal resolution of 1 h using a linear interpolation method.

2.3 Climate Analysis

To enable addressing island-scale rainfall trends for future climate analysis and ensuing flood risks, general circulation model (GCM) outputs were used. GCMs are important tools, but their outputs, although reasonable, are often not adapted to regional climate characteristics, which are critical for basin- and national-scale studies (Kawasaki et al., 2017). The Coupled Model Intercomparison Project Phase 5 (CMIP5) (Taylor et al., 2012) Data Integration and Analysis System (DIAS) tool is a state-of-the-art multimodel dataset comprised of 44 GCMs and tools to reduce this uncertainty associated with GCMs so as to better represent regional climate conditions. A methodology developed by Koike et al. (2014) proposed the selection of the best suited GCMs to best represent regional climatology based on 7 key meteorological elements which are:

a. Precipitation-Small scale

b. Air Temperature-Medium Scale

c. Outgoing Long Wave Radiation- Medium Scale

d. Sea Level Pressure- Medium Scale

e. Zonal Wind- Medium Scale

f. Meridional Wind- Medium Scale

g. Sea Surface Temperature-Large Scale

A small domain was chosen for the precipitation and a larger domain for the sea surface temperature while a medium domain was chosen for the other parameters. The larger domain was selected for the sea surface temperature due to the effects of the El Nino Southern Oscillation, which affects the climate in the Indian Ocean region.

The spatial correlation (Scorr) and the root mean square were used to compare the variances between the models and the observed global datasets. The Global Precipitation Climatology Project (GPCP) dataset was used to analyze the mean monthly rainfall, while the Japan reanalysis data (JRA55) was used to differentiate other meteorological variables.

The Scorr and absolute values were thus used as indices to evaluate the best GCMs.

Based on the past climate records (1 January 2003–31 December 2018), the future climate projections and trends can be analyzed using these GCMs. Four rain gauge stations were chosen for that purpose since they contained relatively complete daily data, and wherever data was missing, the data was filled using nearby points. The four selected stations were: Belle Rive, Bois Cheri, Digue Seche and Plaisance.

The precipitation data was then downscaled and bias-corrected. The CMIP5-DIAS adopts a three-step statistical bias correction (Nyunt et al., 2016). For each of the four gauge stations, the GCM data was extracted and divided into extreme rainfall, normal rainfall, and no rainfall thresholds. The Generalized Pareto Distribution was used for extreme-rainfall correction, the gamma distribution was used for normal-rainfall correction, and the statistical ranking order was used for no-rain days correction (Koike et al., 2014). The RCP 8.5 radiative forcing scenario was used to investigate the effects of climate change on precipitation. The corrected and downscaled rainfall data for each of the four stations was then obtained for future projections for the years 2025–2040.

For the purpose of this study, and based on the aforementioned methodology, the best suited GCMs were identified to be:

i. CMCC-CESM@ens_mean (Scoccimarro et al., 2011)

ii. ACCESS1.3@ens_mean (Lewis et al., 2014)

iii. CESM1(CAM5)@ens_mean (Neale et al., 2010)

iv. CSIRO-Mk3.6.0@ens_mean (Gordonet al., 2002)

2.4 Hydrological Model Development

The Water and Energy Budget-based Rainfall-Runoff-Inundation (WEB-RRI) model (Rasmy et al., 2019) was used for the simulation of catchment-scale water budgets as well as rainfall-runoff-inundation processes so as to perform operational applications such as hydrological extremes forecasting, land use changes, and water-cycle variability.

The WEB-RRI model combines physical processes such as vertical transfer of moisture and energy fluxes, soil-vegetation-atmosphere interactions, soil structure, soil-moisture dynamics, lateral movements of water flows, and interaction between vertical and horizontal water flows. The model structure is comprised of four components:

i. The Simple Biosphere (SiB) Model 2 module

ii. The unsaturated zone vertical flow

iii. The 2D diffusive wave lateral flow module

iv. The 1D diffusive wave river flow module

2.5 Flood Vulnerability Assessment

The vulnerability index for each VCA (the smallest administrative unit for infrastructure management) was computed to enable local management of the vulnerability at an urban scale while taking into consideration a larger basin scale effect.

Vulnerability can be defined as the "extent of harm that can be expected under certain conditions of exposure, susceptibility and resilience." (UNESCO IHE, 2021).

The flood vulnerability equation can therefore be summarized as follows:

$$\text{Vulnerability} = \text{Exposure} + \text{Susceptibility} - \text{Resilience} \quad (1)$$

where exposure is defined in this context as the cost of damage to buildings and agricultural land and risk of loss of life as well as access to main roads and highways; susceptibility, on the other hand, relates to the system characteristics and are elements exposed within the system that influence the damage probability of flooding (UNESCO IHE, 2021) and resilience can be expressed as any system's ability to withstand and come back from a significant perturbation such as floods while maintaining a certain level of efficiency and functionality in its social, environmental, economic, and physical components (Kerner and Thomas, 2014) with indicators such as shelters and flood defence implementation.

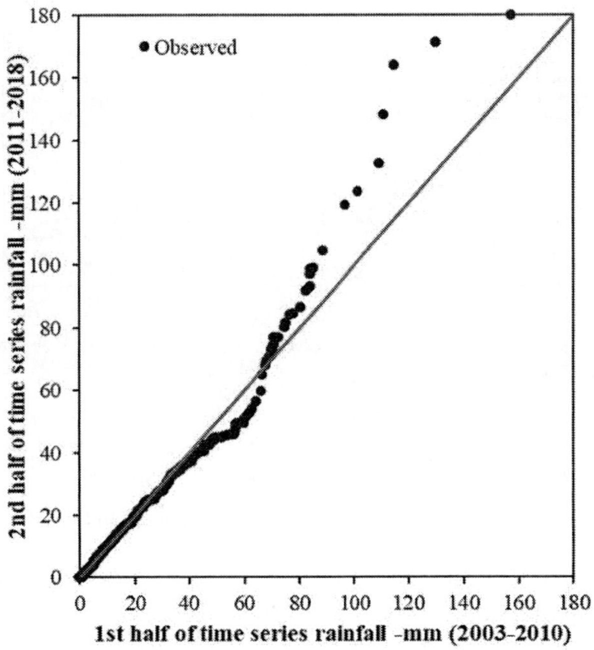

Figure 2. Innovative Trend Analysis (Past)-daily frequency.

3 Results and Discussion

3.1 Extreme Rainfall Analysis

The Innovative Trend Analysis (Sen, 2012) was used to study the climatic trends of the past recorded climate by looking at changes in precipitation occurring at two equal historical consecutive scales. The results in Fig. 2 illustrate that during the last 8 years, precipitation exceeding 100 mm became more common. This increasing trend of extreme rainfall events suggests potentially significant flood damage in the past decade.

To assess the level of confidence in the future simulations, a simple methodology was adopted:

a. If 4 out of the 4 models agree on their output, the event is deemed very likely to happen (100 %).

b. If 3 out of the 4 models agree, the event is deemed likely to happen (75 %).

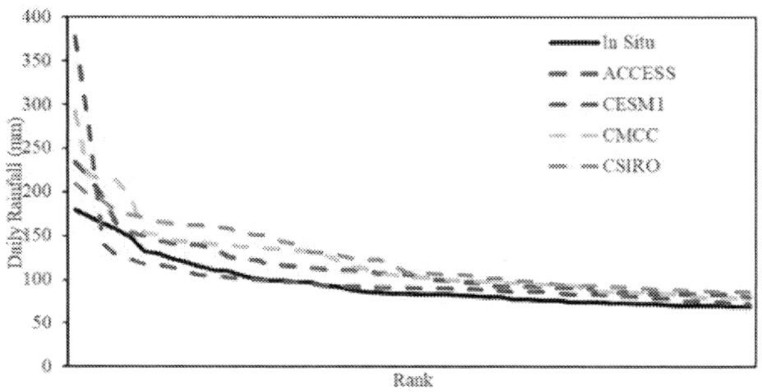

Figure 3. Extreme Rainfall comparison between in-situ and 4 GCMs.

c. If 2 out of the 4 models agree, the event is deemed uncertain. (50 %).

The trends in extreme precipitation obtained from the GCM simulations were analysed to investigate the possible increase in hazard caused by heavy rainfall. The highest 30 daily rainfall events for the simulations for 2025–2040 were compared to the previous observed records for the past 16 years (Fig. 3). It can be observed that all the models agree that there will be an increase in the extreme daily rainfall intensity, with the CESM1 model even showing a twofold increase. There is therefore a very high likelihood that the extreme rainfall events will increase in their intensity and that record breaking events will occur.

This was further demonstrated using the Innovative Trend Analysis comparing past and future simulations as shown in Fig. 4, where the values of in-situ rainfall were compared to the GCMs. It was observed that the models predict an increasing trend in extreme events (for a threshold of 100 mm).

Therefore, it can be inferred from the above that extreme rainfall events of above 100 mm d^{-1} will become more frequent. Moreover, record-breaking events will also become more common. As such, it is expected with a very high level of confidence that in the next 16 years, flood events will become more common in Mauritius, as has been the trend in the past.

3.2 WEB RRI Simulation

The WEB-RRI model was calibrated reasonably well for high flows and well-calibrated for low flows. The model performance was evaluated using the MBE, RMSE, and NSE. It should be noted that the W08 station, situated in Riviere Profonde as shown in Fig. 1 drains only a section of the GRNW catchment. This explains the relatively low flows in the order of 6 m^3 s^{-1} for flood peaks and 1 m^3 s^{-1} for the base flow at the river discharge measurement location. As such, since the difference between the peak and the base flow was not very discernible within the model, the calibration to model the low flows had to be carefully done so as not to overestimate the base flows and to capture the peak flows and vice versa. The model managed to capture the base flows reasonably well while the peaks were slightly underestimated by a margin of 2–4 m^3 s^{-1}. The model was calibrated with an NSE = 0.42 and MBE = −0.13 m^3 s^{-1}, and RMSE = 3.40 m^3 s^{-1} for the years 2013–2014, as shown in Fig. 5.

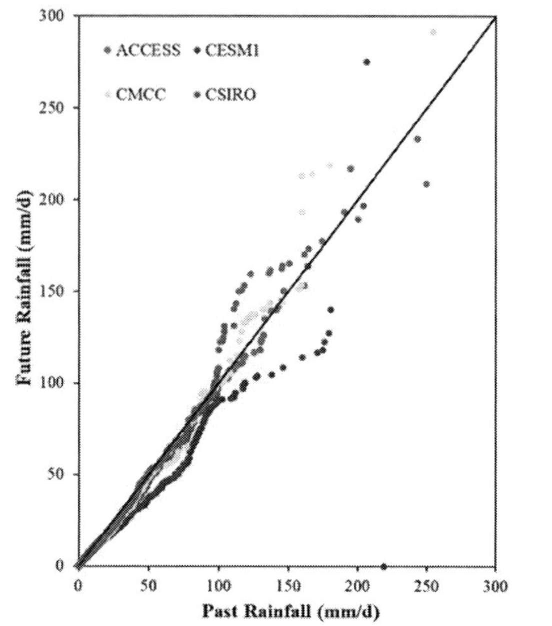

Figure 4. Innovative Trend Analysis- Future v/s past simulations.

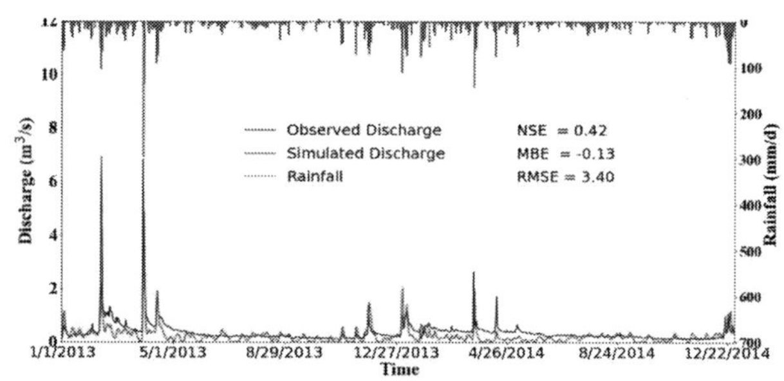

Figure 5. Calibration of WEB RRI Model.

For the validation, the model parameters were kept constant, and the model was run from 1 January 2015 to 31 December 2016 for the same location. The model was validated with a satisfactory NSE = 0.41, MBE = −0.08 m^3 s^{-1}, and RMSE = 2.06 m^3 s^{-1} for the years 2015–2016, as shown in Fig. 6. It was noted that the floods of February and March 2015 were well reproduced by the model, but that of January 2016 was not captured. The base-flow values were seen to be generally well simulated.

It was noted that the dry period which ranges from May to September gives a base flow for Riviere Profonde of less

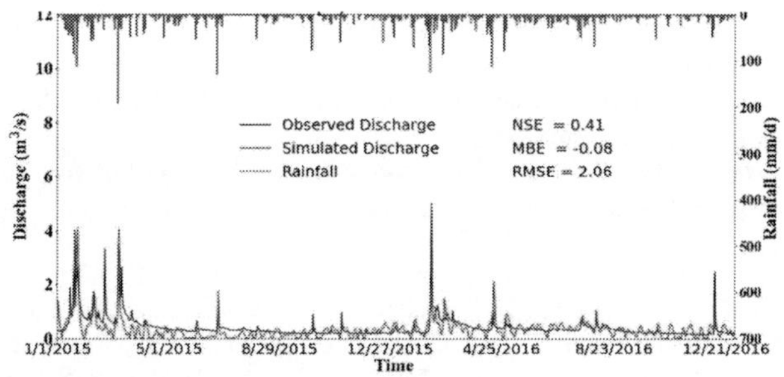

Figure 6. Validation of WEB RRI Model.

than $1\,m^3\,s^{-1}$, and the wet period only shows singular peaks for specific flood events.

3.3 Flood Vulnerability Assessment

The vulnerability assessment demonstrated that one VCA, PL2, presented a very high risk of flooding. The PL2 VCA is at the downstream end of the basin, and with the basin being "funnel-shaped", the high peak flow arrives and accumulates quickly at the outlet. Moreover, the low-lying topography of the PL2 VCA offers no gradient to discharge the flood efficiently to the sea. If a flood event coincides with a high tide or a storm surge, it is expected that the impacts of the flood at PL2 will significantly worsen. The Moka, St Pierre and VP5 VCAs are next in terms of their vulnerability due to the high concentration of assets (buildings and agricultural land) and a concentration of watercourses as well as new developments such as the Moka Smart city impacting on the basin run-off increase and decrease in lead times. The medium, low, and very low vulnerable VCAs were located mostly in the upstream sections where the majority of the land use consists of forested and agricultural land. The vulnerability assessment yielding a flood vulnerability map at a VCA level consists of an important tool in demonstrating to decision makers the risk of inaction and also the benefits of urgent and targeted actions using scientific evidence.

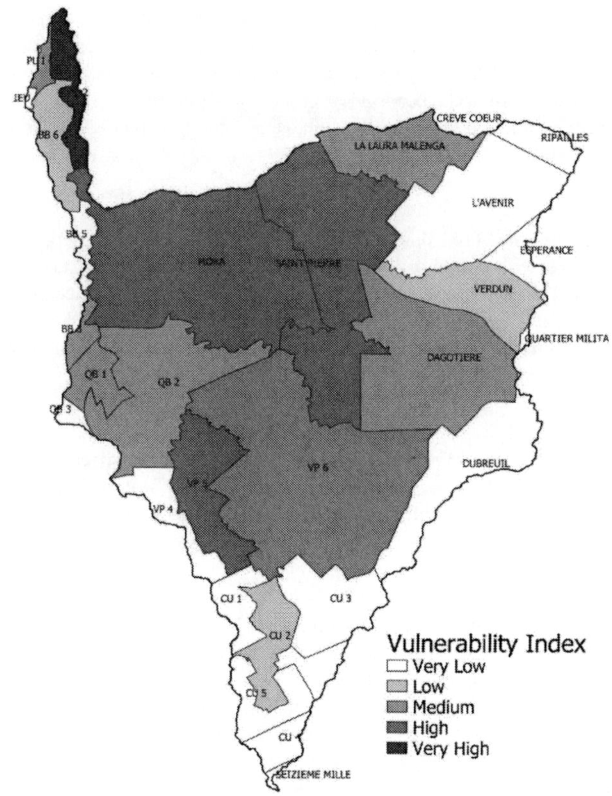

Figure 7. Flood Vulnerability Index per VCA for GRNW.

4 Conclusions

Mauritius, as a small island developing state, is exposed to a number of pressing water related hazards which have rendered Mauritius vulnerable in terms of its disaster management as well as posed threats to its water resources. This research work has aided in understanding the past climatic trends in Mauritius and their impacts on flood hazards. Based on the past climate trends, future climate trends were inferred based on GCMs that were downscaled and bias-corrected. It was inferred that it is very likely that in the near future, extreme rainfall exceeding 100 mm will become more common. By implementing hydrological modelling techniques coupled with climate change impact analyses and a GIS-based risk model, a novel technique that has been adopted on a basin scale in Mauritius, the relationship among rainfall, runoff, inundation, damage and vulnerability were identified. A vulnerability map with qualitative vulnerability indices has been developed to guide decision making in flood management strategies at a local level.

This research however provides findings based on 16 years of meteorological data for 8 stations. In order to increase the reliability and robustness of the developed models, longer time series and a higher number of stations should be included in the computation together with a higher resolution DEM.

Data availability. The data is not publicly available and has been obtained from the Mauritius Meteorological Services, the Water Resources Unit and the Ministry of Housing and Land Use Planning of Mauritius.

Author contributions. MR conceptualized the study. AK developed the methodology, wrote the initial draft, and prepared the figures. TK supervised the work.

Competing interests. At least one of the (co-)authors is a guest member of the editorial board of *Proceedings of IAHS* for the special issue "ICFM9 – River Basin Disaster Resilience and Sustainability by All". The peer-review process was guided by an independent editor, and the authors also have no other competing interests to declare.

Disclaimer. Publisher's note: Copernicus Publications remains neutral with regard to jurisdictional claims made in the text, published maps, institutional affiliations, or any other geographical representation in this paper. While Copernicus Publications makes every effort to include appropriate place names, the final responsibility lies with the authors.

Special issue statement. This article is part of the special issue "ICFM9 – River Basin Disaster Resilience and Sustainability by All". It is a result of The 9th International Conference on Flood Management, Tsukuba, Japan, 18–22 February 2023.

Acknowledgements. The authors would like to extend their gratitude to the Mauritius Meteorological Services and the Water Resources Unit and the Ministry of Housing and Land use Planning for their help in the data acquisition. The authors would also like to extend their thanks to the Japan International Cooperation Agency (JICA) for supporting this cooperation.

Review statement. This paper was edited by Christophe Cudennec and reviewed by two anonymous referees.

References

Baguant-Moonshiram, Y., Samy, M., and Thomas, K., The Challenges of Building Sustainable Cities: A Case Study of Mauritius, THE SUSTAINABLE CITY 2013, Putrajaya, Malaysia, 247–253, https://doi.org/10.2495/SC130211, 2013.

Centre for Research on the Epidemiology of Disasters (CRED) and UN Office for Disaster Risk Reduction, Human Cost of Disaster (UNDRR): An Overview of the Last 20 Years 2000–2019, 2020.

Gordon, H. B., Rotstayn, L. D., McGregor, J. L., Dix, M. R., Kowalczyk, E. A., O'Farrell, S. P., Waterman, L. J., Hirst, A. C., Wilson, S. G., Collier, M. A., and Watterson, I. G.: The CSIRO Mk3 Climate System Model, Aspendale, CSIRO Atmospheric Research technical paper, 60, 130 pp., http://www.dar.csiro.au/publications/gordon_2002a.pdf (last access: 15 August 2021), 2002.

Kawasaki, A., Yamamoto, A., Koudelova, P., Acierto, R., Nemoto, T., Kitsuregawa, M. and Koike, T.: Data Integration and Analysis System (DIAS) Contributing to Climate Change Analysis and Disaster Risk Reduction, Data Sci. J., 16, 41, https://doi.org/10.5334/dsj-2017-041, 2017.

Kerener, A. and Thomas, J. S.: Resilience Attributes of Social-Ecological Systems: Framing Metrics for Management, Resources, 3, 672–702, https://doi.org/10.3390/resources3040672, 2014.

Koike, T., Koudelova, P., Jaranilla-Sanchez, P. A., Bhatti, A. M., Nyunt, C. T., and Tamagawa, K.: River Management System Development in Asia Based on Data Integration and Analysis System (DIAS) under GEOSS, Earth Sci., 58, 76–95, 2014.

Lewis, S. C. and Karoly, D. J.: Assessment of Forced Responses of the Australian Community Climate and Earth System Simulator (ACCESS) 1.3 in CMIP5 Historical Detection and Attribution Experiments, Aust. Meteorol. Oceanogr. J., 64, 87–101, 2014.

Neale, R. B., Chen, C. C., Gettelman, A., Lauritzen, P. H., Park, S., Williamson, D. L., Conley, A. J., Garcia, R., Kinnison, D., Lamarque, J. F., and Marsh, D.: Description of the NCAR Community Atmosphere Model (CAM5.0), NCAR Tech. Rep. NCAR/TN-4861STR, 268 pp., 2010.

Nyunt, C. T., Koike, T., and Yamamoto, A.: Statistical bias correction for climate change impact on the basin scale precipitation in Sri Lanka, Philippines, Japan and Tunisia, Hydrol. Earth Syst. Sci. Discuss. [preprint], https://doi.org/10.5194/hess-2016-14, 2016.

Rasmy, M., Sayama, T. and Koike, T.: Development of Water and Energy Budget-based Rainfall Runoff-Inundation Model (WEB-RRI) and Its Verification in the Kalu and Mundeni River Basins, Sri Lanka, J. Hydrol., 579, 124163, https://doi.org/10.1016/j.jhydrol.2019.124163, 2019.

Scandurra, G., Romano, A. A., Ronghi, M., and Carfora, A.: On the Vulnerability of Small Island Developing States: A Dynamic Analysis, Ecol. Indic., 84, 382–392, https://doi.org/10.1016/j.ecolind.2017.09.016, 2018.

Scoccimarro, E., Gualdi, S., Bellucci, A., Sanna, A., Fogli, P. G., Manzini, E., Vichi, M., Oddo, P. and Navarra, A.: Effects of Tropical Cyclones on Ocean Heat Transport in a High Resolution Coupled General Circulation Model, J. Climate, 24, 4368–4384, https://doi.org/10.1175/2011JCLI4104.1, 2011.

Sen, Z.: Innovative Trend Methodologies in Science and Engineering, Springer, J. Hydrol. Eng., 17, 1042–1046, https://doi.org/10.1061/(ASCE)HE.1943-5584.0000556, 2017.

Simpson, A., Phillips, E., Balog, S., Murnane, R., Deparday, V., Fraser, S., Jongman, B., and Parmelee, L. F.: Southwest Indian Ocean Risk Assessment and Financing Initiative, Washington, D.C., World Bank Group, 2016.

Taylor, K. E., Stouffer, R. J., and Meehl, G. A.: The CMIP5 Experiment Design, B. Am. Meteorol. Soc., 93, 485–498, 2012.

Toledo-Gallegos, V. M., My, N. H. D., Tuan, T. H., and Börger, T.: Valuing Ecosystem Services and Disservices of Blue/Green Infrastructure: Evidence from a Choice Experiment in Vietnam, Econ. Anal. Policy, 75, 114–128, https://doi.org/10.1016/j.eap.2022.04.015, 2022.

UNDRR: The Human Cost of Disasters: An Overview of the Last 20 Years (2000–2019), https://www.undrr.org/publication/human-cost-disasters-overview-last-20-years-2000-2019 (last accessed: 28 December 2023), 2020.

UNESCO IHE: Flood Vulnerability Index, FVI, http://www.unesco-ihe-fvi.org/ (last access: 7 August 2021), 2021.

Proc. IAHS, 386, 277–283, 2024
https://doi.org/10.5194/piahs-386-277-2024

Open Access PIAHS

Flood risk management in planning and construction of city: the Guangzhou experience

Minyi Kuang[1,2], Yanyan Zheng[1,2], Xingdong Deng[1,2], Yukui Yang[1,2], Jingchen Wang[1,2], Xintian Sui[1,2], and Yuan Peng[1,2]

[1]Guangzhou Urban Planning & Design Survey Research Institute, Guangzhou, 510060, China
[2]Guangdong Enterprise Key Laboratory for Urban Sensing, Monitoring and Early Warning, Guangzhou, 510060, China

Correspondence: Minyi Kuang (270925174@qq.com)

Received: 20 July 2023 – Revised: 12 December 2023 – Accepted: 8 June 2024 – Published: 4 July 2024

Abstract. Recently, urban flood disasters caused by rainstorm has happened more and more frequently. Chinese government has proposed that the urban flood disasters problem should be solved by integrated and systematic approach. The analysis of the historical rainstorm flood events of Guangzhou in the past 10 years showed that, the urban rainstorm flood has happened more frequently in lowland, hardening underlying surface and drainage system jacked by high water level. Guangzhou has innovatively delimited high-risk regions by watershed units, using Remote Sensing (RS) data and GIS data, which become the basis of flood risk management. The vertical planning, the storage capacity and river managing boundary have become three important conditions, which should be taken into consideration in the construction projects and serve as the basis for the whole life cycle. As a pioneer, Guangzhou has provided experience for metropolis to reduce flood risk in the construction stage. This paper analyzes the changes and causes of the overall flood safety situation in Guangzhou, and introduces the practice and exploration in Guangzhou from the aspects of promoting the accurate identification of flood risk areas, guiding the spatial control of flood control and drainage safety with planning, establishing the planning transmission mechanism, implementing the flood risk assessment system, and ensuring the transformation of resilient city goals.

1 Background

On the study of urban waterlogging, the industry and academia have accumulated many research results. In the 1980s, mainly studying urban waterlogging caused by rainstorms, focusing on urban flooding caused by river basin floods. In the 1990s, the main research was on the influence of human factors such as urban construction on urban waterlogging. Since 2000, the drainage model began to simulate, summarize the causes of urban waterlogging or conduct in-depth research on a cause, and study foreign advanced concepts and technical means (Xia et al., 2018).

At present, the causes of waterlogging are all single factor analysis. There are many studies on single factors such as drainage system, underlying surface, depression, and temporal and spatial characteristics of rainfall, but there is still a lack of relevant research content on the coupling of these factors to identify potential waterlogging prone areas (Wang and Wang, 2018). Therefore, it is of great significance in planning, construction and management to study the causes of waterlogging and identify potential flood-prone areas by integrating natural rainfall factors and ground hazard factors.

In April 2021, Chinese General Office of the State Council issued the implementation opinions on enhancing urban rainstorm flood. "According to the requirements of building sponge cities and resilient cities, measures should be taken according to local conditions and cities to improve urban flood control and drainage capacity, and urban rainstorm flood problems should be solved in a coordinated and systematic way." (Suggestions on strengthening urban waterlogging control, 2021) has been put forward. Guangzhou is located at the Pearl River Estuary, with the geographical char-

Published by Copernicus Publications on behalf of the International Association of Hydrological Sciences.

acteristics of coastal area. The north of Guangzhou is vulnerable to torrential flood, the middle is prone to rainstorm flood and tropical cyclones and storm surges in the south. In recent years, due to the influence of extreme weather, extraordinary storm happened in Guangzhou frequently (Ai and Wu, 2018; Wang et al., 2020). Rapid urban development leads to the expansion of hardened underlying surface and the increase of rainstorm flood risk (Guo and Deng, 2011; Cheng and Li, 2015). Guangzhou has tried to solve the problem of rainstorm flood management in the source and system (Cheng, 2022; Zhan et al., 2021). This paper analyses the changes of flood risk in Guangzhou, and introduces the practice and exploration of Guangzhou in promoting the accurate identification of flood risk areas, which can guide the spatial flood control in the stage of planning, and implementing the flood risk assessment system.

2 Reasons of rainstorm flood in Guangzhou

Combination of drainage and storage has been basically formed in Guangzhou. In recent years, Guangzhou improved disaster emergency management by the optimizing channel management, building water conservancy facilities and improving the construction standard of drainage system.

However, affected by extreme weather changes and urban development, urban flood control has faced with new challenges (Huang et al., 2021). By 2020, there have been 639 rainstorm flood points in the whole city. Compared with 2010, the situation of rainstorm flood has increased outside the central area (Fig. 1).

2.1 Meteorology and hydrology

According to the statistics of rainfall data from 1955 to 2010, the frequency of rainstorms in the central area of Guangzhou was 8.7 d, the average contribution rate of rainstorms was 32.8 %, and the average intensity of rainstorms was 79.5 mm d^{-1}. In recent years, the heavy rainfall showed the characteristics of greater intensity, more frequent and wider range. The highest tide level of the Pearl River in Guangzhou shows an upward trend. Since 2005, there have been four high tide levels exceeding the design standard (200-year return period) of the embankment.

2.2 Underlying surface

With the intensification of urban development, the total built-up area of the city increased from 952 km^2 in 2010 to 1324 km^2 in 2019, with a net increase of 372 km^2. In recent ten years, the ground impervious rate has increased from 20.18 % to 26.85 % (Fig. 2). The expansion of impervious surface causes the increase of runoff, which is easy to cause rainstorm flood.

2.3 Stormwater storage

The water surface rate of the whole city is 10.15 %. However, the water surface rate is insufficient in the northern regions, and is large in the southern river network regions. During the construction of city, rainwater storage space and drainage channels are crowded, so that the drainage and flood prevention ability of the river basin is sharply reduced. At the same time, due to the large increase of impervious area, the peak flow and total flow of rainstorm flood increase, which causes the drainage system greatly impacted.

2.4 Vertical design

According to the analysis of the digital elevation model, the ground elevation of most areas in Panyu and Nansha in the south of Guangzhou is below 8.00 m (Fig. 3), while the highest tide level of the Pearl River in the past two years has exceeded 8.00 m, posing a high risk of flood.

There are three vulnerabilities in the systematic control of ground elevation in Guangzhou. Firstly, the elevation connection between the old and new construction areas is not reasonable. The terrain of the old residential areas and villages is often lower than the surrounding municipal roads and new development plots, which are prone to rainstorm flood disasters during heavy rains. Secondly, in the process of urban planning and construction, the ground elevation is not fully connected with the drainage system, resulting in local depressions, leading to systemic flood risk. Thirdly, improper vertical elevation control of major construction projects such as bridges, tunnels and culverts, and rail transit forms local depressions, which become the weakness of regional drainage system. According to the analysis and statistics of causes of flood points in 2020, 20 % of these points are caused by insufficient connection between ground elevation and drainage system.

In the face of uncertain rainfall exceeding the standard, the flood risk management in Guangzhou has shifted from focusing on engineering defence to urban planning control. The following are the main practices and exploration of flood risk management in Guangzhou in recent years.

3 Accurate identification of high-risk area

In recent years, Guangzhou has been exploring the experience of rainstorm flood management in megacities, trying to solve the problem of flood risk management with remote sensing means. Through the innovative use of remote sensing technology and LIDAR digital elevation model and other geographic information technology, we identified the impervious ground and low-lying land, and reveal the relationship between impervious ground, depression and urban rainstorm flood.

Firstly, we extracted impervious ground from 105 catchments area in the city. Using remote sensing technology, we

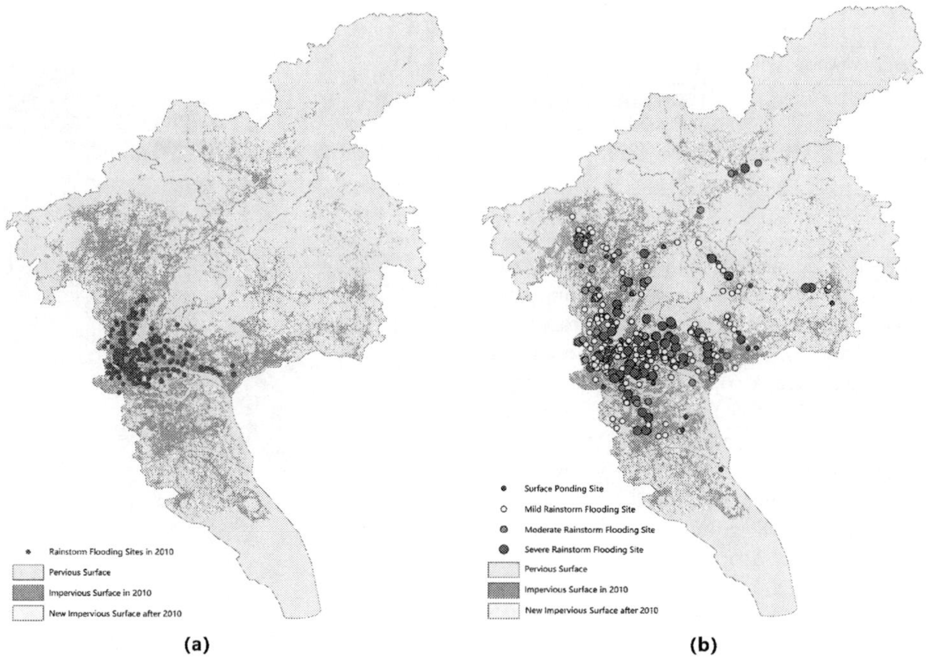

Figure 1. Distribution of rainstorm flooding sites in Guangzhou. **(a)** Analysis in 2010. **(b)** Analysis in 2020.

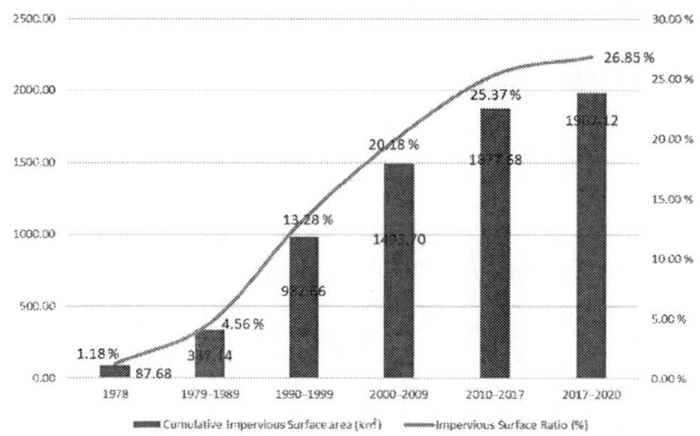

Figure 2. Analysis of impervious surface in Guangzhou.

found that, the impervious surface rate of the central urban area is 80 %–99 %, and the impervious ground area has been basically stable (Fig. 4). In other urban expansion areas, the impervious surface rate is 60 %–80 %, and the impervious surface rate increases by more than 20 % in recent years, indicating a high risk of rainstorm flood. The catchment area, which has a low average impervious surface rate in the upstream, but has a high rate in the downstream, has a high risk of flooding.

Secondly, by combining the DEM (2 m accuracy) of the city's LIDAR high-resolution digital elevation model with the ArcGIS water conservancy analysis model, the maximum

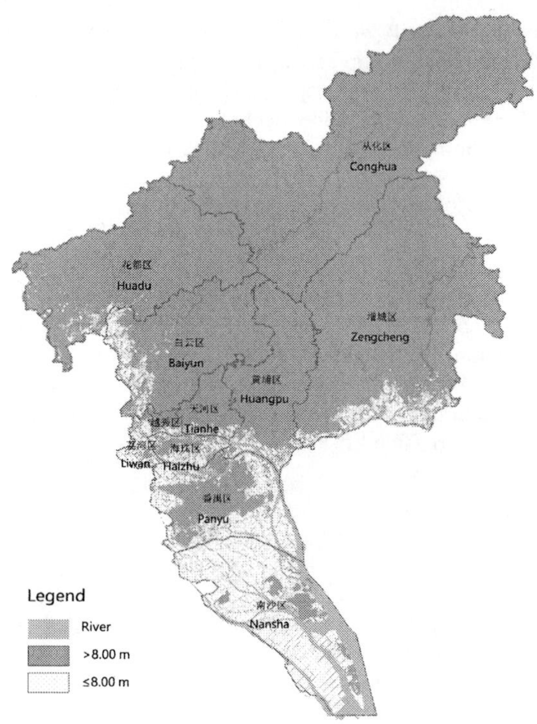

Figure 3. Digital elevation model analysis of Guangzhou.

Table 1. Classification of rainstorm flooding risk degrees.

Low-lying Depth (D)	Pervious Surface	Impervious Surface
Non-depression	Low ($F \leq 0.4$)	Lower ($0.4 < F \leq 0.6$)
$D < 0.5$ m		Median ($0.6 < F \leq 0.8$)
0.5 m $\leq D < 1.5$ m		Higher ($0.8 < F < 1.0$)
$D \geq 1.5$ m		High ($F = 1.0$)

slope method was used to accurately identify the low-lying land in the city (Fig. 5). We found that, the proportion of non-depression land in Guangzhou is 84.85 %, and the proportion of depression land is 15.15 %. Among them, the low-lying depth less than 0.5 m accounted for 6.53 % of the city's area, the low-lying depth of 0.5–1.0 m accounted for 3.67 %, the low-lying depth of 1.0–1.5 m accounted for 1.73 %, the low-lying depth of 1.5–2.0 m accounted for 0.98 %, the low-lying depth greater than 2.0 m accounted for 2.24 %.

Thirdly, dividing impermeable ground into 2 levels and low-lying land into 4 levels, the two main impact factors were superimposed and standardized, and then we calculated the environmental flooding risk index (F) of points and divided five degrees of rainstorm flooding risk (Table 1).

Fourthly, The rainstorm flooding risk degrees was divided according to the two main influencing factors of impervious ground and depth of low-lying land, and the contiguous areas with high, higher and median rainstorm flooding risk degree were preliminarily designated as high-risk areas. By comparing the relationship between the water surface line of the river surge and the topography, and considering the influencing factors of the flood tide level of the Pearl River breaking 8.00 m, the boundary of the preliminarily drawn waterlogging prone risk areas was revised, and the modified 62 high-rainstorm-flood-risk areas of the city were finally determined (Fig. 6).

The area prone to rainstorm flooding is totally 772.84 km^2, affecting about 4.8 million people. Rainstorm flood points in 2010 and 2020 were compared with the prone areas for verification. We found that, except the points caused by pipe plugging, inappropriate management and other reasons, rainstorm flood points are highly consistent with the designated high-rainstorm-flood-risk areas.

4 Countermeasure

Guangzhou regards the delineation of rainstorm flood risk prone areas as the base of risk management and control, and regards the spatial requirements for storage, vertical design and facilities layout related to flood risk control as the rigid constraints of urban construction and the core elements of urban safety (Kuang, 2004).

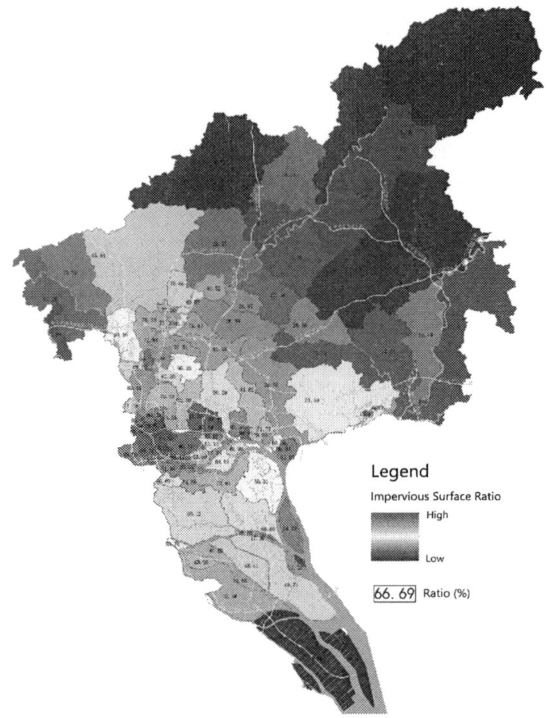

Figure 4. Analysis of ground im-permeability in 105 catchments of Guangzhou.

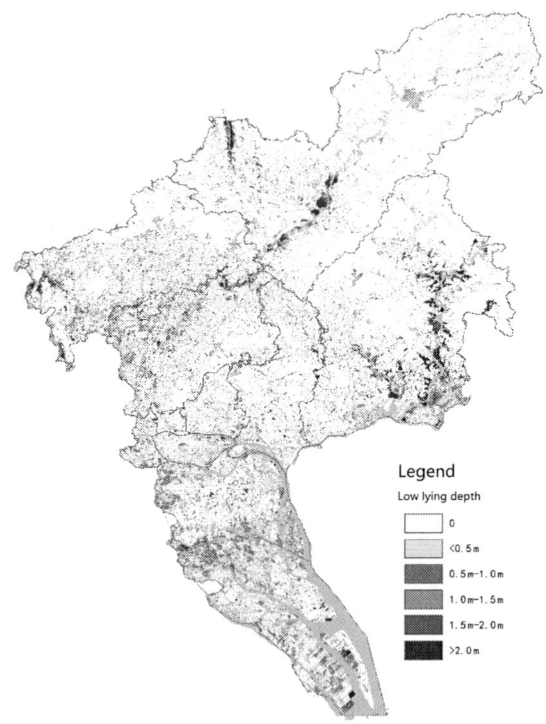

Figure 5. Distribution of low-lying land in Guangzhou.

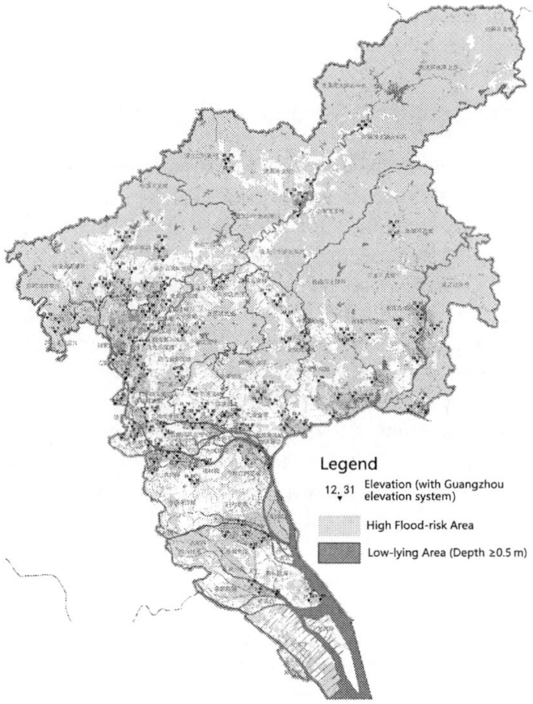

Figure 6. Distribution of flood-risk areas in Guangzhou.

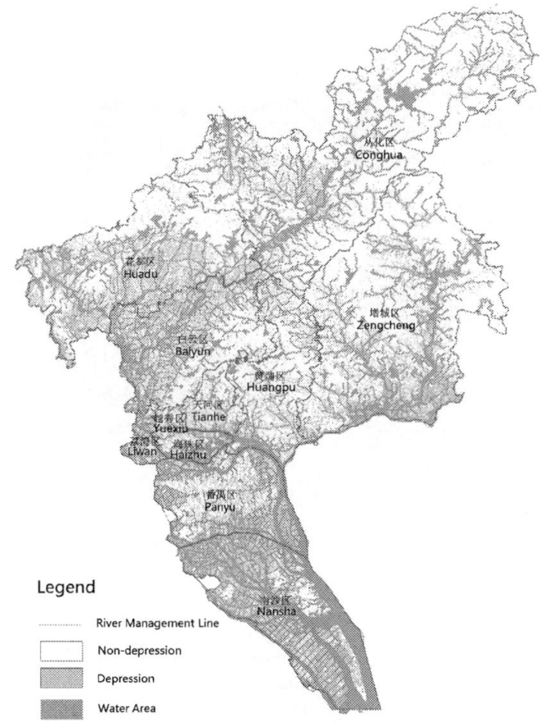

Figure 7. Guangzhou River System.

4.1 Blue and green space management

The concept of blue and green space management has been promoted. Blue and green spaces have been reasonably delineated. Flood and drainage channels and rainstorm storage space has been reserved. The designated ecological control zone covers an area of 2680 km^2, accounting for about 36 % of the city's total area. 38 blue lines of municipal main rivers and 17 protection lines of large and medium-sized reservoirs have been defined, and natural ecological spaces such as river and lake systems, woodland, grassland, wetland, waterfront space, green space and ecological corridor have been strictly protected, while the natural flood discharge channel and flood storage and detention area have been reserved.

4.2 River system management

According to the grade of rivers and lakes, the management scope of rivers and lakes are between 6 and 20 m. Strictly implement water requisition-compensation balance to ensure effective stormwater storage space. The planning department and the water department jointly delimit more than 1300 river management lines, take the delimited management area as the legal basis, and manage water in different levels (Administrative Measures for the Scope of the Urban Blue Line, 2006). New projects are not allowed to enter the management scope of rivers and lakes, and sufficient stormwater storage space is reserved (Fig. 7).

4.3 Vertical elevation management

Vertical elevation is managed in concert with flood control systems. According to the designed flood level of the river, the vertical design requirements based on flood safety are clarified, and the vertical elevation control of the construction land is given priority in accordance with the principle that is conducive to rainwater removal.

For areas with ground elevation lower than the maximum flood level of 8.00 m along the Pearl River, different ground elevation control strategies should be adopted in different areas. In the southern coastal wetland area, the terrain is relatively low-lying, and the drainage method is combined with storage and drainage, and super levees are constructed to cope with the superposition of rainstorm and storm surge. In the central urban area, the drainage mode is mainly gravity flow.

4.4 Low impact development

Guangzhou divides the city into many cells, and reduce runoff of every cell by low impact development. A special section on low impact development is set up in regional planning. In the stages of land transfer and construction project planning permits, the requirements of low impact development should be clearly stated. The planning of each cell must reserve enough rainwater storage space to ensure the runoff

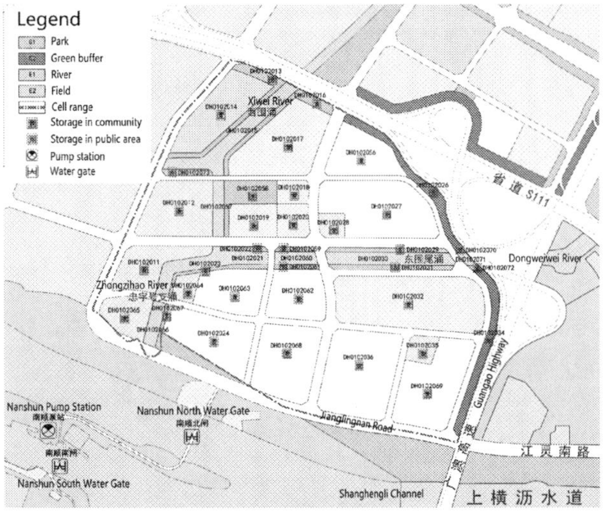

Figure 8. An example of rainwater storages layout map.

control capacity (Zheng and Qin, 2022). The layout of rainwater storage space shall be shown on the planning drawing, which will become a legal basis map (Fig. 8).

In the pross of creating Lingnan eco-city with mountains and water, Guangzhou has controlled the source of urban flood and improved the rainstorm flood prevention capacity of the watershed.

5 Discussion

5.1 Focus on the whole basin system flood risk assessment

Based on the flood risk assessment and analysis of plots, the subsequent focus should be on the flood risk assessment of river basins, and the water security content should be analyzed from the perspective of river basins, so as to improve the flood security of the region more systematically.

5.2 More closely coordinate the relationship between river management and land development

In urban planning and construction, the blue-green ecological space should be reasonably delineated, the use control of potential high-risk areas for waterlogging should be reasonably adjusted in combination with land use planning, and the relationship between river system and land development should be coordinated and optimized.

5.3 Detailed flood risk assessment on major infrastructure

Carrying out refined flood risk assessment in the early stage of major infrastructure such as tunnels, subways and culverts, identifying potential flood risk points, and improving the flood risk resistance of major infrastructure.

6 Conclusions

This paper mainly analyses Guangzhou rainstorm flood risk management experience in recent years. In the face of uncertain excessive rainfall, Guangzhou rainstorm flood governance practice innovation mainly embodied in:

Firstly, change the point from constructing engineering defence to building the flood risk control system, through the planning, construction control and improve management, which reduce the risk of rainstorm flood.

Secondly, institutional innovation. Guangzhou took the lead in establishing a transmission mechanism for rainstorm flood control in China, and implemented a flood risk assessment system at the stage of planning. Strictly protect the blue-green space, rationally control the intensity of urban development, use ecological rainstorm flood control measures in low impact development, and coordinate vertical elevation management with drainage system.

Thirdly, use the GIS and other technical means, and take watershed as the unit to identify the high rainstorm flood risk area accurately, which becomes the base of risk management and control.

This planning method has supported the Security and Resilience of Guangzhou's Territorial Spatial Planning, Urban Renewal, Guangzhou Sponge City Construction and other major planning and construction programs, and has been already applied and practiced to more than 100 flood risk assessment projects. Since the launch of the flood risk assessment system, a total of $17.66 \times 10^6\,m^3$ of storage capacity has been implemented, equivalent to more than 170 small reservoirs.

Data availability. The data are not publicly accessible. It is required in the analysis of rainstorm flood events of Guangzhou.

Author contributions. MK is responsible for the methodology development, project administration and supervision. YZ is responsible for the formal analysis, writing – original draft preparation and writing – review and editing. XD and YY are responsible for the conceptualization and funding acquisition. JW and XS are responsible for formal analysis. YP is responsible for the writing – review and editing.

Competing interests. The contact author has declared that none of the authors has any competing interests.

Disclaimer. Publisher's note: Copernicus Publications remains neutral with regard to jurisdictional claims made in the text, published maps, institutional affiliations, or any other geographical representation in this paper. While Copernicus Publications makes every effort to include appropriate place names, the final responsibility lies with the authors.

Special issue statement. This article is part of the special issue "ICFM9 – River Basin Disaster Resilience and Sustainability by All". It is a result of The 9th International Conference on Flood Management, Tsukuba, Japan, 18–22 February 2023.

Financial support. This research has been supported by the Guangdong Enterprise Key Laboratory for Urban Sensing, Monitoring and Early Warning (grant no. 2020B121202019) and The Science and Technology Foundation of Guangzhou Urban Planning and Design Survey Research Institute (grant no. RDI220200205025).

Review statement. This paper was edited by Miho Ohara and reviewed by two anonymous referees.

References

Administrative Measures for the Scope of the Urban Blue Line: http://www.gov.cn/fefg/2006-02/15/content_191814.htm (last access: 15 February 2006), 2006.

Ai, H. and Wu, X.: Variation Characteristics of Rainstorm in Guangzhou, Guangdong Meteorology, 40, 20–23, https://doi.org/10.3969/j.issn.1007-6190.2018.04.005, 2018.

Cheng, X.: Lessons and reflections from the case of Guojiazui Reservoir disaster caused by "7 · 20" extreme rainstorm disaster in Zhengzhou in 2021, China Flood and Drought Management, 32, 32–36, https://doi.org/10.16867/j.issn.1673-9264.2022043, 2022.

Cheng, X. and Li, C.: Evolution trend, important characteristics and countermeasures of urban flood risk, J. China Flood and Drought Management, 25, 6–9, https://doi.org/10.16867/j.cnki.cfdm.2015.03.002, 2015.

Guo, C. and Deng, L.: Analysis and Countermeasure of Waterlogging in Central Urban Area of Guangzhou, China Water and Wastewater, 27, 25–28, 2011.

Huang, H., Wang, X., and Liu, L.: A review on urban pluvial floods: Characteristics, mechanisms, data, and research methods, Progr. Geogr., 40, 1048–1059, https://doi.org/10.18306/dlkxjz.2021.06.014, 2021.

Kuang, M.: Forming a Natural Water System of River and River Branches in Urban Planning–Taking for Example the Sectional Design of the Rivers and River Branches in Guangzhou, Planners, 2004, 33–36, https://doi.org/10.3969/j.issn.1006-0022.2004.08.009, 2004.

Suggestions on strengthening urban waterlogging control: http://www.gov.cn/zhengce/content/2021-04/25/content_5601954.htm (last access: 25 April 2021), 2021.

Wang, M. and Wang, S.: Implication and Experience of Resilient City: A Case Study of New Orleans in the United States, Urban Development Studies, 25, 145–150, https://doi.org/10.3969/j.issn.1006-3862.2018.11.014, 2018.

Wang, T., Hu, L., and Chen, Z.: Causes and countermeasures of Guangzhou subway flooding caused by heavy rain on May 22 in 2020, Guangdong Meteorology, 42, 52–55, https://doi.org/10.3969/j.issn.1007-6190.2020.04.013, 2020.

Xia, J., Zhang, Y., Liang, C., and Liu, J.: Review of urban stormwater models, Engineering Journal of Wuhan University, 51, 95–105, https://doi.org/10.14188/j.1671-8844.2018-02-001, 2018.

Zhan, M., Wei, Z., Wang, J., Zhou, X., and Peng, D.: Urban physical examination evaluation methods and improvement strategies for territory space security: The case study of Guangzhou, Journal of Natural Resources, 36, 2382–2393, https://doi.org/10.31497/zrzyxb.20210915, 2021.

Zheng, Y. and Qin, H.: Hydrological Modeling of Low Impact Development Combined Facilities and Its Uncertainty Analysis, China Water & Wastewater, 38, 114–121, https://doi.org/10.19853/j.zgjsps.1000-4602.2022.01.018, 2022.

Proc. IAHS, 386, 285–290, 2024
https://doi.org/10.5194/piahs-386-285-2024

Working Together on Civil-Engineering Structures within the Dutch Flood Protection Programme: challenges and opportunities

Ellen Tromp[1], Joost Breedeveld[1], Johan Apperloo[2], and Jaap Kottier[2]

[1]Deltares, Delft, P.O. Box 177, 2600 MH, the Netherlands
[2]Hoogheemraadschap Hollands Noorderkwartier, Heerhugowaard, P.O. Box 250, 1700 AG, the Netherlands

Correspondence: Ellen Tromp (ellen.tromp@deltares.nl)

Received: 3 July 2023 – Revised: 13 March 2024 – Accepted: 8 June 2024 – Published: 11 July 2024

Abstract. Over 400 civil-engineering structures (e.g. sluices, pumping stations) are listed in the Dutch Flood Protection Programme (DFPP). That is a daunting task for the next 30 years, meaning that every three weeks one structure should be reinforced to meet the statutory standards. In an innovation project, the regional water authorities, Rijkswaterstaat, the Dutch Flood Protection Programme, research institutes and the consultancy agencies joint forces. In the past period, we have identified numerous obstacles, using the critical success factors framework, that hamper the reinforcement of these civil engineering structures. To tackle the barriers, we developed a roadmap that gives more insight in the steps for strengthening these structures. In addition, lessons learned from current strengthening projects on civil engineering structures are used to further optimize the overall programming of these structures within DFPP, and to enhance the knowledge transfer and uptake within the triple helix.

1 Introduction

Two-thirds of the Netherlands is below the current sea level, and this amount is likely to increase due to climate change. In the past, the Dutch responded reactively to floods. However, after the catastrophic floods of 1953, the Dutch took measures to prevent a similar disaster: a system of dike rings with flood protection levels was developed for flood risk management (Most et al., 2014), and statutory standards were set out in the Flood Protection Act of 1996 to maintain this system. Those standards related to the primary flood defences: those on the coast, major rivers and lakes. In 2008, the second Delta Commission advised changes to take future uncertainties, such as climate change and land subsidence, into account to sustainably protect the Netherlands. The Dutch national government adopted a risk-based approach (Most et al., 2014) for the Dutch flood risk management policy. This is based on new knowledge about the safety of dikes and the impact of serious floods because of the failure of dikes and/or other structures. This is a proactive approach, in which protection standards are based on both the probability and the impact of flooding in 2050, considering climate change and socio-economic developments. The Environment Act sets out the statutory standards for the flood protection structures, such as dikes, dams, and other civil-engineering structures (e.g. locks, coupures, siphons). Every 12 years, the regional water authorities are required to conduct assessments to ensure that the flood defences still meet the statutory standards. Where the safety standards are not met, the responsible body can apply for funding from the Dutch Flood Protection Programme (DFPP) (Jorissen et al., 2016). Over 1500 km of dikes and 400 civil-engineering structures will have to be upgraded between now and 2050.

In this paper, we focus primarily on the civil-engineering structures that often consist of a location-specific combination of components and materials and are required to perform other functions too. For these structures, multiple parties with their own individual interests and responsibilities - play a role, and they seek to minimise trade-offs. This complexity further emphasises the importance of knowledge management and continuous learning around these structures. In addition, within the DFPP programme board it is felt that these structures seem to get limited attention, whereas the investment needed to strengthen these structures to meet the statutory standards is high. The number of structures that are currently on the programme requires that every three weeks a structure is strengthened. This sense of urgency led the programme board, together with other parties, to initiate an innovation project dedicated only to these civil-engineering structures to develop instruments. This paper aims to provide an overview of current activities to gain more attention to the DFPP-listed civil-engineering structures.

The paper is structured as follows: Sect. 2 presents the methods, looking at the Dutch Flood Protection Programme and specifically to the civil-engineering structures, followed by a description of the methodology used in our study; in Sect. 3, we describe how the framework helped to identify challenges and opportunities, and how they are faced. Finally, Sect. 4 presents the discussion and conclusions of this paper.

2 Materials and methods

2.1 Dutch Flood Protection Programme: context

Currently, the DFPP faces the task to upgrade about 1500 km of dikes and about 400 civil-engineering structures before 2050. The exact numbers depend on the results of the statutory assessment of flood defences by the regional water authorities; first round after the new implementation of the risk-based approach finishes in 2023. To ensure that the reinforcement task of DFPP is realized within time and available budget, they adopted a knowledge and innovation strategy. In Tromp et al. (2022b) more information is available on this learning-while-doing strategy. In 2018, representatives of Rijkswaterstaat, the regional water authorities, knowledge institutes and consultancy agencies together underline the necessity to start an innovation project focused on the civil engineering structures that are part of the primary flood defence system in the Netherlands (DFPP, 2018). This includes locks, sluices, culverts, pumping stations and storm surge barriers. The DFPP uses the definition for civil-engineering structures according to The Dutch Environment Act (2016): a construction that is part of a flood defence and, over a limited length, takes over all or part of the flood protection function of the body of land, but is constructed for the purpose of another (utilitarian) function that crosses the flood defence (such as draining and facilitating navigation). In connection with this utilitarian function, these hydraulic structures are usually equipped with one or more valves. For the DFPP, the stricter requirement is that the flow surface of the water retaining structure must be $>0.5\,\text{m}^2$ (DFPP, 2016).

In 2021, Regional water Authority Hollands Noorderkwartier initiated the innovation project "Working jointly together on Civil-engineering structures". This specific regional water authority already had experience with strengthening structures, as they carried out a pilot where seven sluices were strengthened. Aim was to gain more insight in the current sense of urgency for the civil-engineering structures within the Dutch Flood Protection Programme. For the first phase, the core team identified four pillars: (1) Inventory on the overall reinforcement scope within DFPP, (2) Inventory on the challenges and opportunities when starting a reinforcement project, (3) Inventory on knowledge gaps around hydraulic structures, and (4) Identification of the current sense of urgency around these civil engineering structures, both at the administrative and executive level.

2.2 Methodology

The core team used the methodology and framework that was developed by the DFPP programme board, as laid down in the knowledge and innovation strategy; more specifically the insights gained on knowledge management and the critical success framework, as shown in Tromp et al. (2022b). This framework (Fig. 1) identifies six critical success factors (CSFs):

1. *People involved*: is sufficient knowledge and expertise available for the project team, and does the involved people have the required competences?
2. *Instruments*: are the knowledge and tools available to design the structures in its surroundings?
3. *Organisation*: is there political willingness to take risks, and what boundary conditions have been given to the project team?
4. *Governance*: how is decision-making organised, and how are organisations such as contractors and local stakeholders involved?
5. *Legislation*: how can the project comply with formal and informal regulations?
6. *Discourse*: is there enough support in society for strengthening the structures?

The three success factors on the lower row of Fig. 1 are preconditions located outside the boundaries of influence of flood defence reinforcement projects. The remaining three success factors can be controlled by either the regional water authorities or the project team members. The traffic lights in Fig. 1 show that if one of the CSF's is at red or orange a project stagnates in its implementation. This specific framework helps the DFPP to formulate activities to enhance, and possibly steer, the development of knowledge and innovations.

While the focus in Tromp et al. (2022b) was on an enhanced understanding of which elements of the knowledge and innovation strategy work and others fail, the focus here was on finding out why regional water authorities do have not yet undertaken water retaining structures, using the CSF

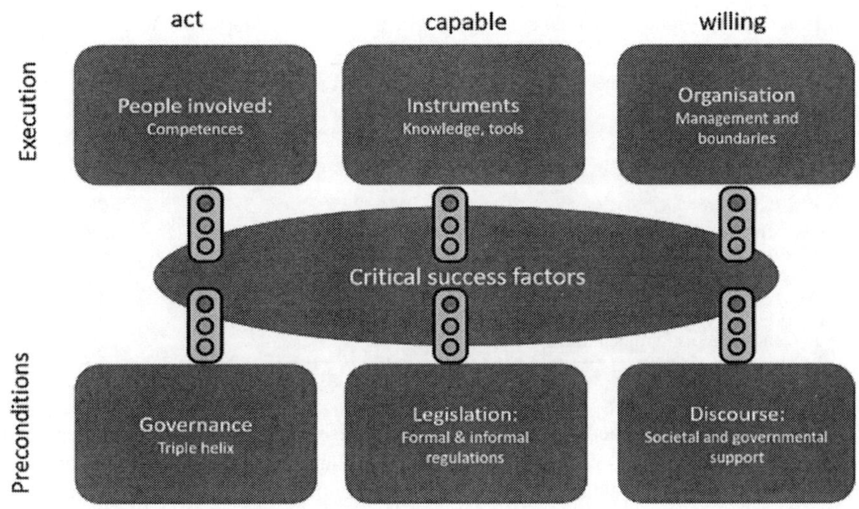

Figure 1. The six critical success factors framework (Tromp et al., 2022b).

framework. However, this framework has never been used from the onset of an (innovation) project to explore, for the different phases and at so many different organisations, at what factors caused, in this case the non-reinforcement of civil engineering structures within DFPP. With this approach, we illustrate how knowledge development, transfer and uptake can be enhanced by focusing on one or more CSFs in particular way.

In the next section, we describe in more detail the scope and challenges for the civil-engineering structures within the DFPP.

3 Results: opportunities and challenges

Within the DFPP project "Working jointly together on Civil-engineering structures", further insight into the scope of the task, a uniform approach and available knowledge is gained. In this section we describe these results of the research.

3.1 Scope of the civil-engineering structures within DFPP

In the Netherlands a total of 1824 civil-engineering structures are in the primary flood defence system, this includes culverts, sluices, storm surge barriers, movable barriers (see Fig. 2). All regional water authorities and Rijkswaterstaat were approached to share information already available from their running statutory assessments. This was mostly mid-term information because the organisations had not all finished. By summer 2021, about 40 % had been assessed (total of 685). Of these, 547 structures were found to meet the specified lower limit of the safety requirement and 138 structures did not meet the lower limit. Of the 138 structures that failed the statutory assessment, the authorities have indicated a wish to include 91 in the DFPP. Of the other remaining structures, this was unknown.

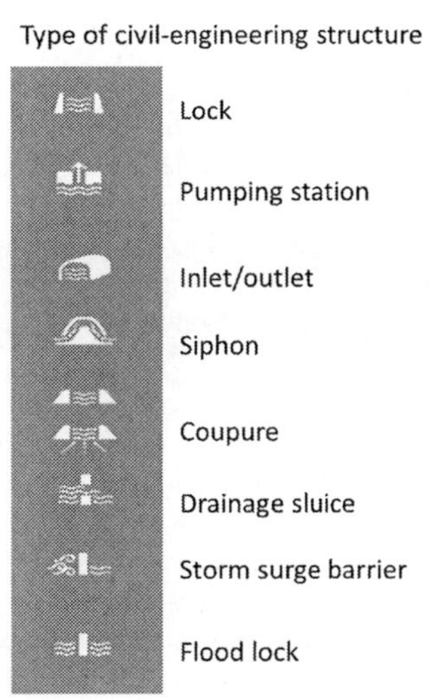

Figure 2. Artists impressions of civil-engineering structures in the primary flood defence system of the Netherlands (credit: project Working jointly together on Civil-engineering structures).

The team (IV Infra, 2021) made an extrapolation to gain insights how many structures could potentially be on the DFFP programme. This led to the insight that approximately 370 civil-engineering structures, of which 50 large scale structures must be strengthened within the Dutch Flood Protection Programme. Most structures fail to meet two dominant safety requirements: sufficient height and a sufficiently reliable closure process; less dominant are piping and sta-

Table 1. Overview of civil-engineering structures that do not comply with the Dutch statutory standard.

Type of civil-engineering structure	# of structures that does not comply with standard	# through extrapolation
Lock	15	41
Flood lock	14	38
Coupure	37	100
Inlet/outlet pipe	47	127
Pumping station	9	24
Drainage sluice	16	43
Civil engineering structure does not comply with standard	138	~370

bility. In the study the team also distinguished between the different types, as shown in Table 1.

This scoping should be updated in 2024 when all the results of the statutory assessment are available. This will undoubtedly lead to changes in the overall scoping.

3.2 Uniform approach: opportunities and challenges

For the uniform approach, we have created the so-called Civil-engineering Roadmap (Fig. 3). This roadmap describes the set of steps taken during the life cycle of a Civil-engineering structure in a primary flood defence. Previous analyses and discussions revealed the need to visualise this process. The drafted roadmap distinguishes the following three steps: (1) Obligation to provide care on a daily basis, (2) Initiation; and (3) (DFPP) Project.

For each of these steps, we researched via semi-structured interviews and desk study whether there were any points of attention and possible showstoppers leading to a delay in the reinforcement of a civil engineering structure (as well as maintenance (and replacement)). The respondents of the regional water authorities and Rijkswaterstaat indicated that, until now, they did not experience showstoppers. However, points of attention did emerge within each step of the roadmap. The raised concerns were classified by the six critical success factors, as described earlier. This research (INFRAM, 2021) led to several key focus points for each critical success factor, as shown in the Table 2.

In several interviews with representatives of the private sector, regional water authorities, Rijkswaterstaat and other governmental organisations, we found that civil-engineering structures still not get the required attention. Often, asset managers are unaware that these structures require a slightly different approach. In Dutch, the word for civil engineering structure is similar to the word that represents paintings and statues. So often a cognitive barrier (Tromp, 2019) occurs, thus hampering the knowledge uptake and transfer. The interviews highlighted that professionals at the water authorities sometimes lack insight in the available knowledge and how to use it. For this purpose, we analysed the available knowledge and highlighted relevant knowledge gaps for further research. We presented the available knowledge using the failure path approach (Tromp et al., 2022a). A failure path is an entire chain of successive events that together lead to a flood according to the definition in the current Water Act. The failure path with the highest probability of occurrence for a specific flood defence structure strongly depends on the design and geometry of the structure and the environment in which it is located. In an in-depth analysis, a distinction is made between the events (i.e. processes of change from one condition to a new, successive condition) that, given a flood situation, may lead to flooding on the one hand, and the condition (based on scenarios) of parts of the flood defence prior to the flood situation on the other.

For the failure paths, we used the four distinctive failure mechanisms used for civil-engineering structures in BOI as input, namely:

- Reliability of closure process of the structure, probability of non-closure per closure demand;
- Height of hydraulic structure, probability of excessive wave overtopping and/or overflow;
- Piping at structure, probability of insufficient resistance to piping;
- Structural strength and stability, probability of structural failure.

This overview (Tromp et al., 2022a) provides insights where additional knowledge is required and steers the knowledge agendas of the Ministry of Infrastructure and Water Management, Rijkswaterstaat and the research institutes.

4 Discussion and conclusions

The first phase of the 'Working jointly together on Civil-engineering structures' project, has provided insights that civil-engineering structures do not yet always get the attention they deserve. This not only affects the stability of the DFPP programme, but also the projects themselves. An update on the scope and number of civil-engineering structures

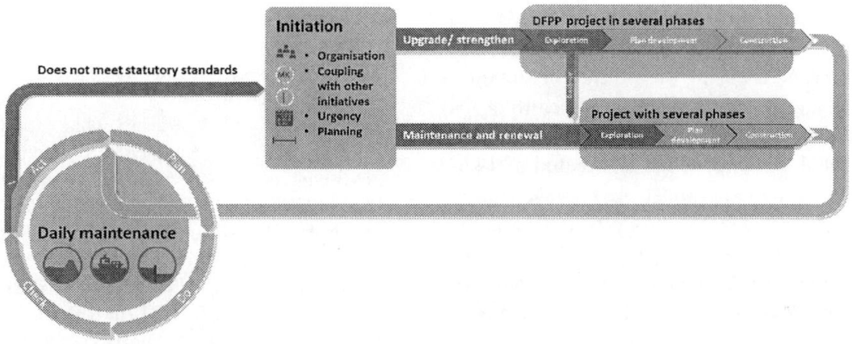

Figure 3. The Civil-engineering roadmap highlighting the different steps during the life cycle of a structure in a primary flood defence (INFRAM, 2021).

Table 2. Key focus points for each Critical Success factor with link to step in the Civil-Engineering roadmap (INFRAM, 2021).

CSF	Focus point(s)	Step
People	– Limited capacity and knowledge. – Lack of instruments to deal with uncertainties. – Overview of the task is lacking	– Obligation to care – Initiation – Project
Instruments	– Data is out of date, not of the appropriate level or unavailable. – Need for an overview of all available knowledge and know-how on innovations. – Knowledge gaps are not always clear	– Obligation to care – Initiation
Organisations	– A strong commissioner with adequate expertise daring to make decisions. – No scenarios available (regarding the approach)	– Obligation to care – Initiation – Project
Governance	– Depending on knowledge of private parties. – Unclear decision-making	– Obligation to care – Initiation – Project
Legislation	– No clear distinction between maintenance, replacement and reinforcement – Grant experiences	– Obligation to care – Initiation – Project
Discourse	– Attention to executive interest, urgency and action	Initiation

Figure 4. Several meetings, held by Hollands Noorderkwartier to transfer and uptake knowledge around seven sluices (Photocredit: Regional Water Authority Hollands Noorderkwartier, 2022).

in DFPP is required in the second phase. The regional water authorities need handholds to help them get started with civil engineering structure.

Fortunately, some projects have already been taken up within the DFPP. Regional water authority Hollands Noorderkwartier has worked on the pilot in which seven locks, predominantly have been strengthened in the past years. The knowledge gained has been disseminated in several meetings, but still requires further knowledge transfer to all water authorities having a task (Fig. 4).

In the second phase, allowing knowledge to flow is key. Projects such as those carried out at Regional water authority Hollands Noorderkwartier must actively share their knowledge with similar reinforcement projects on flood defence structure to prevent the traffic lights turning red or orange.

The innovation project will make the available knowledge accessible via a knowledge repository, but will also organise frequent meetings, each focusing on a different step and experience of the civil engineering roadmap. A wealth of information and knowledge needs has already been identified from the interviews held. As knowledge is situated and socially constructed, this knowledge must be actively shared and restated after each change in the group of participants. At the same time, a constant awareness of the impact of change on the programme's implementation will be required, as well as additional knowledge. This could also affect how knowledge is shared and absorbed in the future.

Data availability. No data sets were used in this article.

Author contributions. Conceptualization: ET. Research on Scope of the civil-engineering structures within DFPP: JB and JK. Research on the uniform approach and critical success factors: ET and JA. All authors contributed to the final version of the manuscript and have read and agreed to the published version.

Competing interests. The contact author has declared that none of the authors has any competing interests.

Disclaimer. Publisher's note: Copernicus Publications remains neutral with regard to jurisdictional claims made in the text, published maps, institutional affiliations, or any other geographical representation in this paper. While Copernicus Publications makes every effort to include appropriate place names, the final responsibility lies with the authors.

Special issue statement. This article is part of the special issue "ICFM9 – River Basin Disaster Resilience and Sustainability by All". It is a result of The 9th International Conference on Flood Management, Tsukuba, Japan, 18–22 February 2023.

Acknowledgements. The authors would like to thank the reviewers for their critical and targeted comments.

Financial support. This research has been supported by Rijkswaterstaat and the Regional Water Authorities (Dutch Flood Protection Programme).

Review statement. This paper was edited by Kensuke Naito and reviewed by two anonymous referees.

References

DFPP: Memo waterkerende kunstwerken binnen het HWBP: een definitie, 2016 (in Dutch).

DFPP: Startnotitie POS Kunstwerken, report, 2018 (in Dutch).

INFRAM: Aandachtspunten Kunstwerkroute (focus point civil-engineering structures roadmap), report, Project number: 21i861, version 2.0, https://publicwiki.deltares.nl/display/HWBPKunstwerken/Over+Samenwerken+aan+Kunstwerken?preview=/292061954/292061962/21.1065099RapportageHWBPSKKunstwerkroute_DEF.pdf (last access: 4 July 2024), 2021 (in Dutch).

IV Infra: Inventarisatie kunstwerkenopgave Samenwerken aan Kunstwerken (Inventory scope civil-engineering structures), Reference INFR210556-R02 Revision: 1, https://publicwiki.deltares.nl/display/HWBPKunstwerken/Over+Samenwerken+aan+Kunstwerken?preview=/292061954/292061960/21.1065130RapportageInventarisatiekunstwerkopgave_rev1.pdf (last access: 4 July 2024), 2021 (in Dutch).

Jorissen, R., Kraaij, E., and Tromp, E.: Dutch flood protection policy and measures based on risk assessment, in: Proceedings of the 3rd European Conference on Flood Risk Management (FLOODrisk 2016), Lyon, France, 17–21 October 2016, https://www.e3s-conferences.org/articles/e3sconf/abs/2016/02/e3sconf_flood2016_20016/e3sconf_flood2016_20016.html (last access: 4 July 2024), 2016.

Most, H., van der Tánczos, I., Bruijn, K. M., and de Wagenaar, D.: New Risk-based standards for flood protection in the Netherlands, in: Proceedings of the 6th International Conference on Flood Management, Sao Paulo, Brazil, 16–18 September 2014, https://www.researchgate.net/publication/266758851_New_risk-based_standards_for_flood_protection_in_the_Netherlands (last access: 4 July 2024), 2014.

Tromp, E.: Enhancing Knowledge Transfer and Uptake in the Design Processes of Flood Defences, PhD Thesis, Delft University of Technology, Delft, the Netherlands, https://doi.org/10.4233/uuid:df56c11b-5366-4720-9f58-015c9bf0316e, 2019.

Tromp, E., Verdijck, N., and Breedeveld, J.: Rode Draad Waterkerende Kunstwerken. Ontwikkeling kennis en kunde voor overstromingskansanalyse waterkerende kunstwerken, version 1.0, Green version, 22 September 2022, https://publicwiki.deltares.nl/display/HWBPKunstwerken/Over+Samenwerken+aan+Kunstwerken?preview=/292061954/292061961/20220920EindrapportageGroeneversieRodedraadKunstwerken_v1.pdf (last access: 4 July 2024), 2022a (in Dutch).

Tromp, E., te Nijenhuis, A., and Knoeff, H.: The Dutch Flood Protection Programme: Taking Innovations to the Next Level, Water, 14, 1460, https://doi.org/10.3390/w14091460, 2022b.

Proc. IAHS, 386, 291–297, 2024
https://doi.org/10.5194/piahs-386-291-2024

Significance of Having Integrated Water Resource Plan in a Complex Watershed System for Better Water Management during Covid Pandemic: The Case of Mahaweli River Basin

Roshan Indika Jayasinghe[1], **Mohamad Rasmy**[2], **and Toshio Koike**[2,3,☆]

[1]Chief Resident Engineer's Office (Ellewewa Reservoir Project), Irrigation Department, Embilipitiya, Sri Lanka
[2]International Centre for Water Hazard and Risk Management (ICHARM), Public Works Research, Tsukuba, Japan
[3]Department of Civil Engineering, University of Tokyo, Tokyo, Japan
[☆]retired

Correspondence: Roshan Indika Jayasinghe (roshan.jaye@gmail.com)

Received: 20 July 2023 – Revised: 14 March 2024 – Accepted: 8 June 2024 – Published: 12 July 2024

Abstract. Water scarcity and disasters, particularly floods and droughts, are the most devastating disasters in Sri Lanka, causing severe damage to agriculture, lives and property. This condition has been worsen further during the Covid19, due to isolation and travel restrictions. Collaboration and effective watershed management among the stakeholders is one of the key solutions for effective operation, but watershed management is more complex and complicated due to its physical processes, and conflicts among stakeholders. This study attends to attract stakeholders and develop Decision Making System by integrating complex hydrological model with dam operation and diversion, based on topography, rainfall, evaporation with bias-corrected satellite data. Finally, we developed an Integrated System for decision makers for effective water resource management in River basins. The proposed approach was applied to the complex Mahaweli watershed in Sri Lanka, to address prevailing issues.

1 Introduction

A watershed is area, which drained the each drop of water to single outlet that demarcated topographically. This watershed can be vary in size from small to thousands of square kilometres. Among these watersheds, huge type of natural disasters are common, such as flood, debris flow, landslide, and drought. Understanding and managing the watershed play the crucial role to minimize the damage and vulnerability. However, understanding and managing watershed is complex due to its hydrological processes. The watersheds are not simple or linear systems due to several reasons, e.g., spatial and temporal changes of precipitation, evaporation, infiltration, snowmelt process, water storage and diversion, vegetation type and its penology, land-use and soil characteristics, topography, and river morphology. Particularly, human-made changes such as water storage and their usage for power generation, irrigation water supply, diversion to other basins, and drinking water requirement, make the hydrological processes complicated. In addition, the frequent confliction between the users and stakeholders, become more complex, while managing available resources during the wet and dry period.

Focusing the Mahaweli river basin will get the overall model for the complex watershed management. Because, this river basin is largest in Sri Lanka and flows through wet, intermediate and dry zone. Throughout the journey, it meet different activities, such as flood control and storages in reservoirs, trans-basin diversion, and cultivation more than 165 000 ha of agricultural lands, and hydropower generation, which produce 25 %–30 % of country demand (Annual Reports of Mahaweli Authority of Sri Lanka, 2018). Sharing these activities by multiple organizations will causes the confliction among the stakeholders. During the Covid pandemic, it further worsen, due to isolation of people and limited working capacity in the organizations. Developing integrated water resource plan is essential to manage the available resources and disasters within such complex watershed effectively.

Dynamic system approach is one of the best method to overcome the complexity of the watershed (Khan et al., 2007). However, it will not perform well in extreme events and long-term processers due to systems uncertainties and multipurpose behaviours. There are hydrological models developed to understand the hydrological behaviour, but those are simpler and difficult to apply for complex behaviour. Due to this problem science-based system approach (Xin et al., 2018) combined with the complex system for Integrated Water Resources Management (IWRM), by considering the complicated physical activities, which limited understanding of physical processes, and limited facilities to introduce human-made activities. During the our study, Rainfall-Runoff Inundation Model (Sayama, 2013) selected to understand the hydrological processes in the basin, due to its flexibility and diffusion wave approximation of the stream, and incorporate the human-made changes to this model and developed Complex System Approach Model (CSAModel).

This CSAModel applied to complex Mahaweli river basin to identify the capabilities by using its physical behaviours. Then reservoirs and diversions introduced to complete the complex behaviour (Sajjad et al., 2000). Then, dam optimization technics were introduced to calculate the hydropower generation, water distribution, and flood management. The outcome shows the realistic results to stakeholders to manage the system even in the pandemic period as one intergraded system.

2 Theory and Methodology

2.1 Complex system approach

The approach developed to understand the complexity of the watershed, by dividing the watershed to different sections, which hydrological system, geological system, and man-made system, for better performance. The final combined product used for its applications of combined complex hydrological model, to understand the present condition, and policy development for effective basin management.

2.1.1 Hydrological system

Hydrological system include the basic hydrological model considering topography, and spatial and temporal variation of rainfall, infiltration, and evapotranspiration.

2.1.2 Geological system

Geological behaviour in the watershed considered with sub division of the main tributaries. This helped to include the different behaviours of the watershed.

2.1.3 Man-made system

Man-made structures was changed the natural behaviour of the watershed. Such components like barrage, reservoir and diversions included to the system.

2.2 Methodology

The complex watershed model developed using Rainfall-Runoff and Inundation (RRI) Model (Sayama, 2013) by combining:

- Observed data used for physical activities like, Rainfall, Topography, infiltration, and MODIS (or Moderate Resolution Imaging Spectroradiometer) satellite data for Evapotranspiration. For Non availability of rainfall data JAXA (Global Satellite Mapping of Precipitation) GSMaP data used after biased correcting.

- Reservoir operations, based on monthly-based flood storage concept for different reservoirs, according to its climatic behaviour, like seasonal rainfall characteristics, to reduce floods in lower segment of the basin.

- Four-stage of reservoir operations, such as pre-flood operation stage, flood operation stage, pre-emergency operation stage, and emergency operation stage. Spill operation combined with fuzzy logic technic with linear operation.

- Monthly-based water diversion, which calculating the demand by using water requirement in the dry zone.

Two FORTRAN Modules, which is Dam Operation Module (DOM), that developed to include the monthly-based reservoir operation technic and four-stage reservoir operation technic, and Diversion Module (DM), that developed to include the monthly based water requirement in dry zone. Additionally CROPWAT 8.0, which developed by Food and Agriculture Organization (FAO), used to find out the irrigation water requirement of paddy and other field crops (Surendran, 2015). Finally, validated Mahaweli basin RRI model coupled with the two modules, that used for the complex watershed management as Complex System Approach Model (CSAModel).

2.3 Case Study: Mahaweli river basin in Sri Lanka

The developed CSAModel applied to Mahaweli river basin to verify the results and check its performance of the model, because Mahaweli basin complex due to:

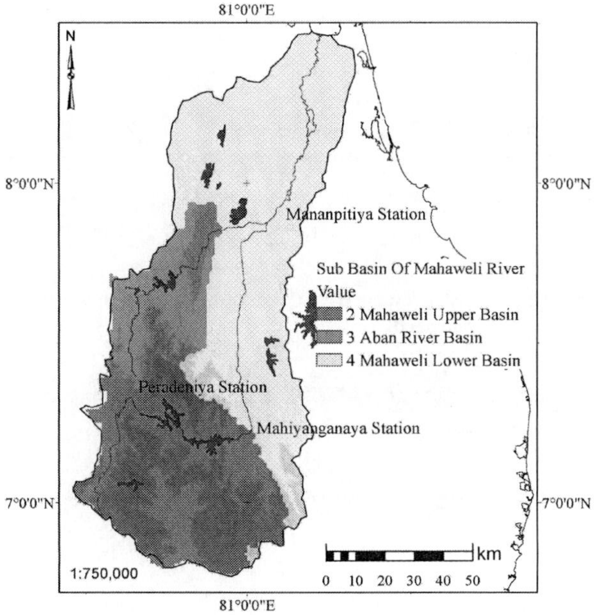

Figure 1. Mahaweli river basin with Sub basins.

- River flows through different climatic zones.
- Existing Reservoirs and water Diversions.
- Inundation in lower part with huge property and agricultural damage.
- Dry zone water scarcity.
- Frequent confliction between the users and stakeholders.

The developed CSAModel applied to overcome the above complex behaviour in watershed.

2.3.1 CSAModel for Mahaweli river basin

Basic hydrological model prepared using RRI Model and 30Arcsec grid Digital Elevation Model (DEM) data. Model coupled with hydrological processes, spatial and temporal variation of rainfall distributions (Hydrological System), topographic change, sub-basin distribution and water contribution to mainstream (Geological System), water diversion locations, reservoir locations and flood situation (Man Made System). Considering above factors system divided to upper river basin (Eq. 1), Aban river basin (Eq. 2) and lower river basin (3) as shown in Fig. 1.

40 years (1975–2015) ground rainfall data used as temporal data in 15 stations, those data spatially interpolated by using Inverse Distance Weighting (IDW) method "Eq. (1)" this method gave the additional advantage to system for better results. Satellite based Evaporation data used after bias correcting with available ground data as in "Eq. (2)"

$$Z_i = \frac{\sum_{i=0}^{n} \frac{Z_j}{d_{ij}^P}}{\sum_{i=0}^{n} \frac{1}{d_{ij}^P}}. \tag{1}$$

Where Z_i Interpolation value in an unknown cell, d_{ij} distance between the known cell and unknown cell, Z_j Known cell Value, P power function for weighting of the distance for this it takes as 6.

$$P_{\text{corr}} = P_{\text{sat}} \frac{\mu(P_{\text{ob}})}{\mu(P_{\text{sat}})}. \tag{2}$$

Where, P_{sat}, satellite-observed data, $\mu(P_{\text{sat}})$ mean of the satellite data and, $\mu(P_{\text{ob}})$ mean of the observed data.

The developed basic Hydrological Model calibrated and validated for several flood events within 1969–1972 period, without considering reservoir and water diversion, because basin development started after 1972. During the model preparation stage it was used the three stage modelling, which considering Peradeniya, Mahiyanganaya and Mananpitiya Stations (Fig. 1) step by step from upper basin to lower basin. This was very effective for better performance of the model, because physical behaviours like river profile, multiple land use patterns, and soil parameters vary from the upper basin to lower basin. The Nash-Sutcliffe Efficiency (NSE) (Nash and Sutcliffe, 1970) value used for check the efficiency of Model.

2.3.2 CSAModel with Dam and Diversion

The developed basic Hydrological Model converted to CSAModel by combining Dam Operation Module (DOM) and Diversion Module (DM), which represent present condition of the Mahaweli watershed, this CSAModel further analysed and validated using existing operation sequence of the reservoir and diversion. This used to develop of new set of rules to achieve better results by considering following watershed characteristic and conditions.

- The way of reservoir operation is key task for controlling flood in lower basin (Hoang, 2017) which can control the loss of property, economic and agricultural damage.
- Analysed and identified the significance of spatial (Fig. 2) and temporal (Figs. 3 and 4) variation of the rainfall in watershed considering past 40 years data (1975–2015).
- Identified the required flood storage capacity in the reservoir in monthly wise, by studying of annual spatial variation of rainfall (Fig. 2).

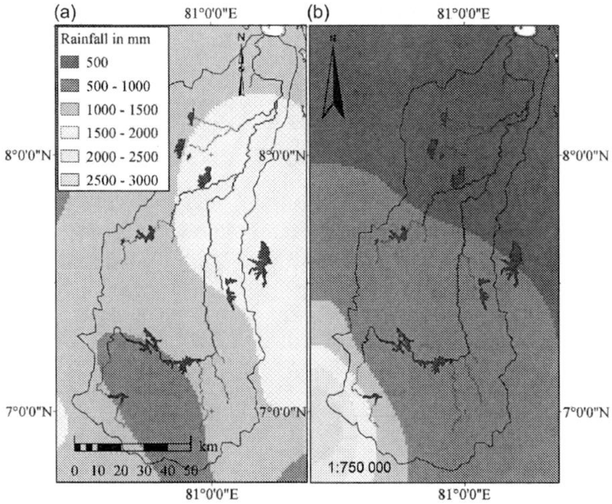

Figure 2. Average spatial variation of rainfall, North East Monsoon (NEM) at left, South West Monsoon (SWM) at right.

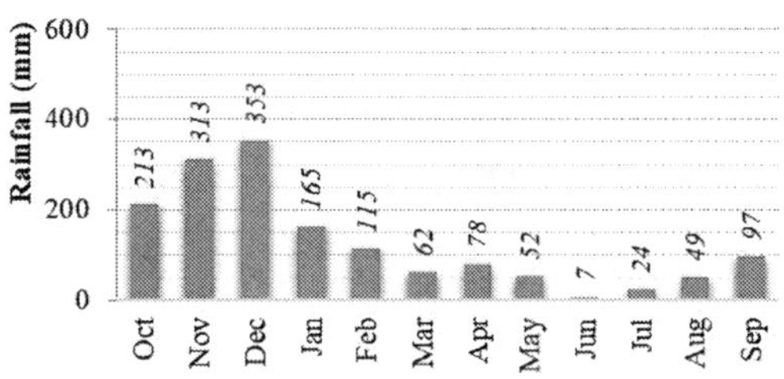

Figure 3. Average temporal variation of the rainfall in lower river basin of the watershed.

2.3.3 Flood Operation characteristic during monsoon

Lower basin scatted with dry zone reservoirs, which need water from wet zone, to fulfil the dry condition and upper basin consist with major reservoirs to control flood, water diversion and hydropower generation. The study identified that, spatial and temporal variation of the rainfall throughout the watershed, then it analysed, and categorized the reservoirs to two groups as follows.

Upper part get more rainfall during the June, July, August and September months (Fig. 4), due to South West Monsoon (SWM) (Fig. 2b). Therefore, Kothmale Reservoir selected as SWM flood reservoir and decided to maintain the flood storage as 15 %, 20 %, 20 % and 15 % from the gross storage, during these months. However, the lower part gets high rainfall from October, November and December months (Fig. 3), during North East Monsoon (NEM) (Fig. 2a), therefore lower part reservoirs were selected as NEM flood reservoirs, during the October, November, and December months and kept 20 % of flood storage in Victoria, Randenigala, Rantembe, and Bowathenna reservoirs.

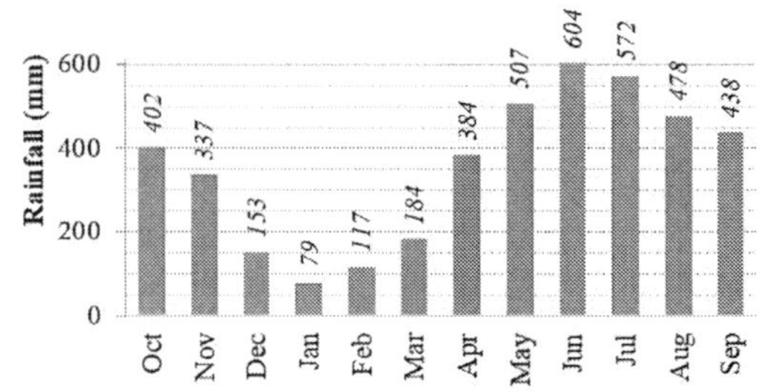

Figure 4. Average temporal variation of the rainfall in upper river basin of the watershed.

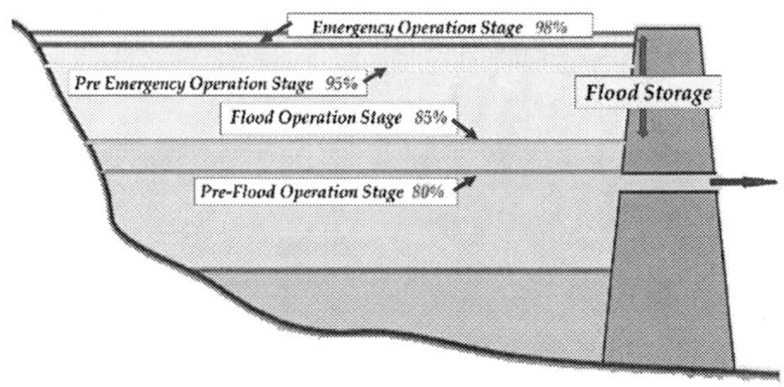

Figure 5. Stages of flood operation in reservoir.

The modified monthly-based flood storage used for reservoir operation, as mentioned above from January to December.

This flood storage divided to main four sections for control the reservoir, by considering the reservoir operation technics, then four-stage reservoir operation technic was introduced during the identified flood period as follows (see Fig. 5).

- Pre-Flood Operation Stage, when reservoir water level fill to this level, reservoir operation will activate.
- Flood Operation Stage, controlled spilling discharge to minimize the damage to lower basin.
- Pre Emergency Operation Stage, indicate the to lower basin minor flood during the operation.
- Emergency Operation Stage, the out flow will control for safety of the dam, with the operation rules.

This optimization technic introduced to the five major reservoirs in Mahaweli watershed, which Kothmale, Victoria, Randenigala, Rantembe, and Bowathenna as flood controlling reservoirs.

Diversion Module (DM) used for Trans-basin water diversion, which fulfil the dry zone water requirement. In our case study, actual diversion points in the stream introduced to module for dry zone diversion that started from Polgolla barrage to four major watersheds, which is outside of Mahaweli watershed, Kala Oya, Malwathu Oya, Yan Oya and Maduru

Oya basin, altogether this diversion provide water to cultivate more than 165 000 ha agricultural lands in the dry zone. Other diversion locations are Minipe, Elahera and Mawilaru.

2.4 Data used

- Rainfall data from 1965 to 2015 used
- Ground Observed Discharge data (1965–2015) in Peradeniya, Mahiyanganaya and Mananpitiya station (see Fig. 1).
- MODIS satellite evapotranspiration data (2014–2015), and ground observed evaporation data (2014–2015) in Palugasdamana and Mahailippallama (Hydrological Annual, 2009–2015).
- Observed Reservoir operation data (2014–2015) in Bowathenna, Victoria, Randenigala, Rantembe, and Kothmale.

3 Results and discussion

Basic hydrological model was calibrated (NSE value 0.61) and validated (NSE value 0.76) by using data 1969–1972, which is no dam and diversion in the watershed. Pre-optimized CSAModel simulated and check the results, one by one in each reservoirs with actual operation data with the real flood situation in lower basin for validation the CSAModel.

3.1 Optimized dam operation for flood

Monthly based flood storage and four-stage dam operational rules applied to CSAModel and reservoir water discharge and storage change observed with the lower part flood in 2014 event. During the simulation flood storage, diversion capacity was adjusted for minimum inundation of the lower basin and keep the maximum storage after the flood events.

Figure 6a and b curves shows that observed discharge and the modelled discharge with original operation in lower part. Figure 6c shows that performance of newly proposed four-stage dam operational rules. When comparing the both results, it shows that flood peak in Mananpitiya area reduced from 2800 to 1450 $m^3 s^{-1}$ according to newly introduced flood storage system. That reduced the actual flood level by 1.5 m in lower basin. Figure 7 shows the reduction of the inundation area. Figure 8 indicate that how new flood storage applied and varied during the months, and reservoir capacity variation according to new CSAModel in Victoria reservoir. This was applied for all five reservoirs, which I mentioned earlier.

3.2 Optimized diversion for drought

The developed CSAModel diversion function used to optimize the water diversion in dry zone. In Mahaweli watershed,

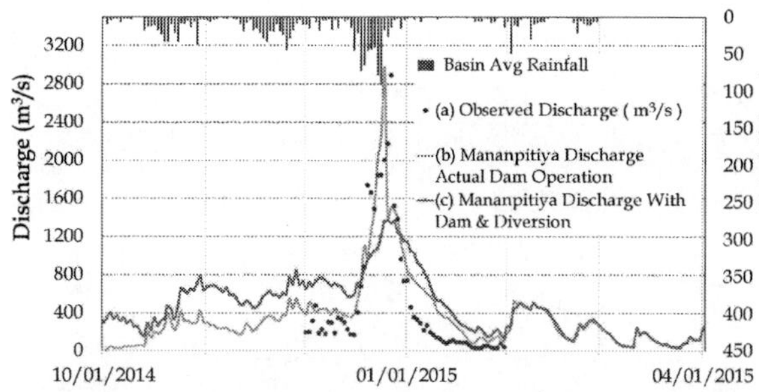

Figure 6. Discharge change **(a)**. Actual Dam Operation **(b)**. Observed data **(c)**. Application of new set of rules for dam operation.

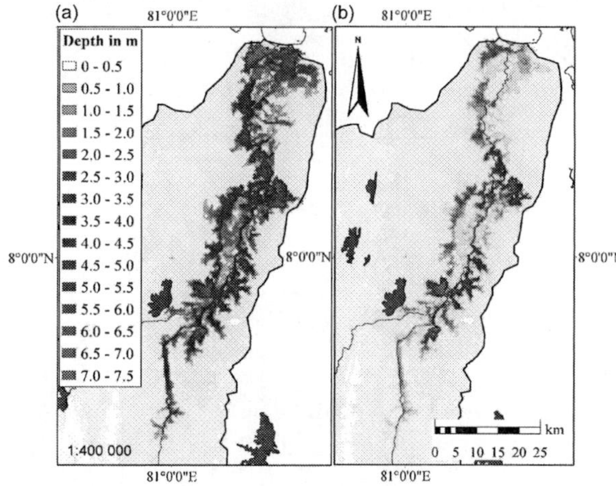

Figure 7. (a) Actual Inundation, **(b)** Change of inundation in Lower Basin, due to new set of Optimization Rules.

trans-basin diversion (Fig. 9) stared from Polgolla barrage to dry zone through tunnel. Water diversion is not enough for dry zone. Therefore, during the study two different scenario compared to optimize the maximum diversion, by changing tunnel capacity (S01) and using same tunnel (S02). Model output shown that 21 % of cultivation lands could have increased in 2014 Yala, and 17 % in 2015 Yala, for S01, and. 10 % of cultivation lands could have increased in 2014 Yala and 6 % in 2015 Yala for S02 in dry zone. This diversion optimization shows that rice production increment in dry zone, if used such optimization, and it will save USD 15–20 million per year.

4 Conclusion & Recommendation

Watersheds are more complex to predict, even it more complex due to man made changes in the system, during the COVID period it was more difficult due to unavailability of proper system. During the several decades, different studies carried out to find the watershed management tool, which can used for forecast and reduce the disasters in watershed

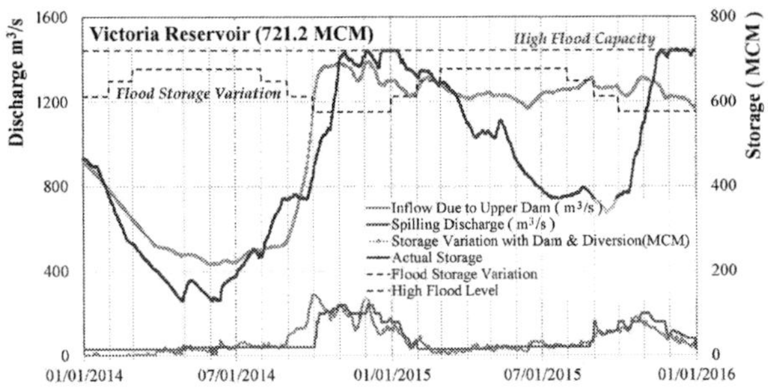

Figure 8. Before and after optimization of Victoria reservoir.

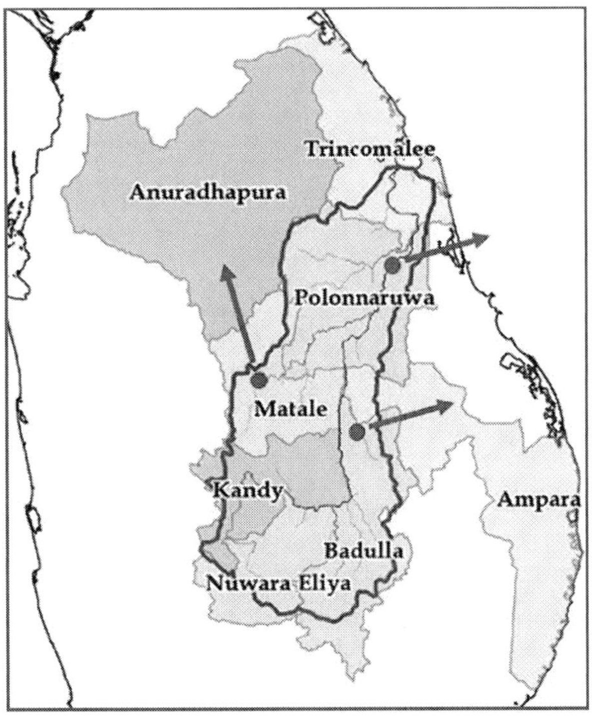

Figure 9. Trans basin Water Diversion from Mahaweli River Basin.

as discussed earlier. In this study, CSAModel developed to address this by using following steps, (a) identifying of Natural processes such as hydrological processes, topography, spatial and temporal variation of rainfall distributions and evapotranspiration, (b) simplification of complex watershed, (c) introducing physical processes and water storage and diversion facilities and, (d) application of the combined complex hydrological model.

The CSAModel applied to complex Mahaweli watershed, which is largest in Sri Lanka, to check the efficiency of the system and take the output for reduce the disasters, efficiency management, and decision making for stakeholders. The developed CSAModel performed well by selecting the spatially varied flood storage and four-stage dam operation rules. Finally model output shows that considerable amount of flood reduction in lower basin. The capacity of flood storage reduction after the event is very effective to increase the water diversion to dry zone and it shows that increase of paddy cultivation in Yala season.

Further, the study identified that, flood reservoirs for flood management, and some reservoirs for drought management according spatial variation and temporal variation of the rainfall in watershed. The model flexible for use the optimum hydropower generation with maximum cultivation performance after some modification. Finally, CSAModel helps for integrated water resource management (IWRM) in watershed by sharing data among the stakeholders for decision-making.

The study recommend that to form new committee among the shareholders e.g., Hydropower, Irrigation and drinking sector. Additionally, coupling of seasonal prediction models and checking of water availability in the system is added advantage of this model. Therefore, study recommend for coupling CSAModel with seasonal rainfall forecasting model for better forecast and manage the whole system as one unit with stakeholders during this type of Pandemic situation.

Code availability. Code developed based on the Open Source software called RRI Modeling developed by ICHARM (https://www.pwri.go.jp/icharm/research/rri/rri_top.html, Sayama, 2013), modified the existing version of the code to include the new Dam Operation Module (DOM) and Diversion Module (DM) to it.

Data availability. The hydrological data, including river flow, rainfall, and evaporation records, is officially owned by the Department of Irrigation and the Meteorology Department. This data, covering the period from 1965 to 2015 within the Mahaweli Watershed, is designated for research purposes and can be accessed through a formal written request to the respective departments.

Author contributions. The authors confirm contribution to the paper as follows: study conception, data collection, analysis, interpretation of results and draft manuscript preparation: RIJ. MR, TK: reviewed the results and approved the final version of the manuscript.

Competing interests. At least one of the (co-)authors is a guest member of the editorial board of *Proceedings of IAHS* for the special issue "ICFM9 – River Basin Disaster Resilience and Sustainability by All". The peer-review process was guided by an independent editor, and the authors also have no other competing interests to declare.

Disclaimer. Publisher's note: Copernicus Publications remains neutral with regard to jurisdictional claims made in the text, published maps, institutional affiliations, or any other geographical representation in this paper. While Copernicus Publications makes ev-

ery effort to include appropriate place names, the final responsibility lies with the authors.

Special issue statement. This article is part of the special issue "ICFM9 – River Basin Disaster Resilience and Sustainability by All". It is a result of The 9th International Conference on Flood Management, Tsukuba, Japan, 18–22 February 2023.

Acknowledgements. The authors would like to thank the constructive comments from reviewers, and special thanks to coauthors, Mohamad Rasmy (Senior Researcher, ICHARM), and Toshio Koike (Director of ICHARM/Professor Emeritus of the University of Tokyo) for guiding to research. Further, I would like to express my gratitude to Director General of Irrigation and his staff for supporting the research by allowing me to use relevant data. Also, I am deeply grateful to the Japan International Cooperation Agency (JICA), Graduate Institute of Policy Studies (GRIPS), Public Work Research Institute (PWRI) and International Center for Water Hazard and Risk Management (ICHARM) for their support. Finally, I wish to extend my gratitude to my parent, loving wife and all others for their noble supports.

Review statement. This paper was edited by Kensuke Naito and reviewed by two anonymous referees.

References

Hoang, N. V.: Analysis of Effective Reservoir Operation for Mitigating Flood Damages in a Vuong Dam Basin, ICHARM Master Paper No. 119, https://www.pwri.go.jp/icharm/training/master/img/2017/synopses/mee16728.pdf (last access: 6 July 2024), 2017.

Hydrological Annual: Irrigation Department of Sri Lanka, Irrigation Department, https://www.irrigation.gov.lk/web/index.php?option=com_content&view=article&id=115&lang=en (last access: 6 July 2024), 2009–2015.

Khan, S., Yufeng, L., and Ahmad, A.: Analysing complex behaviour of hydrological systems through a system dynamics approach, Environ. Modell. Softw., 24, 1363–1372, https://doi.org/10.1016/j.envsoft.2007.06.006, 2007.

Nash, J. E. and Sutcliffe, J. V.: River Flow Forecasting through Conceptual Models Part I – A Discussion of Principles, J. Hydrol., 10, 282–290, https://doi.org/10.1016/0022-1694(70)90255-6, 1970.

Sajjad, A. and Slobodan, P. S.: System Dynamics Modelling Of Reservoir Operations For Flood Management, J. Comput. Civil Eng., 14, 190–198, https://doi.org/10.1061/(ASCE)0887-3801(2000)14:3(190), 2000.

Sayama, T.: Rainfall-runoff-inundation (RRI) model ver 1.4.1, International Center for Water Hazard and Risk Management (ICHARM) and Public Works Research Institute (PWRI) [code], https://www.pwri.go.jp/icharm/research/rri/rri_top.html (last access: 6 July 2024), 2013.

Surendran, U.: Modelling the crop water requirement using FAO-CROPWAT and assessment of water resources for sustainable water resource management: A case study in Palakkad district of humid tropical Kerala, International Conference on Water Resources, Costal and Ocean Engineering (ICWRCOE 2015), Mangalore, India, https://doi.org/10.1016/j.aqpro.2015.02.154, 2015.

Xin, L.: Watershed system model: The essentials to model complex human-nature system at the river basin scale, J. Geophys. Res.-Atmos., 123, 3019–3034, 2018.

Proc. IAHS, 386, 299–306, 2024
https://doi.org/10.5194/piahs-386-299-2024

Assessment of flood risk in a coastal city considering multiple socio-economic vulnerability scenarios

Mousumi Ghosh[1], **Subimal Ghosh**[1,2], **and Subhankar Karmakar**[1,3]

[1]Interdisciplinary Program in Climate Studies, Indian Institute of Technology Bombay, Mumbai 400076, India
[2]Department of Civil engineering, Indian Institute of Technology Bombay, Mumbai 400076, India
[3]Environmental Science and Engineering Department, Indian Institute of Technology Bombay, Mumbai 400076, India

Correspondence: Subhankar Karmakar (skarmakar@iitb.ac.in)

Received: 25 July 2023 – Revised: 20 May 2024 – Accepted: 8 June 2024 – Published: 15 July 2024

Abstract. The Intergovernmental Panel on Climate Change stresses the importance of vulnerability and exposure along with hazard in defining flood risk. Therefore, a novel approach for flood risk quantification is proposed that evaluates the impact of various realistic socio-economic scenarios on risk reduction. The flood hazard is derived from a hydrodynamic flood modelling framework over Mumbai, India's highly flood-prone coastal megacity. The socio-economic vulnerability is assessed by a multivariate approach based on principal component analysis and data envelopment analysis. Finally, the flood risk is quantified and mapped by aggregating hazard and vulnerability for different socio-economic scenarios, also key indicators contributing to a significant reduction in vulnerability and risk are identified. This non-structural long-term flood risk management approach will benefit densely populated urban areas, especially in developing, and underdeveloped countries.

1 Introduction

The assessment of flood risks is the most critical non-structural flood management strategy that requires global attention and is of utmost importance; Civic bodies, decision-makers, and the society alike have found it imperative (Shao et al., 2016). Flood risk primarily consists of two major components, flood hazard (H) and flood vulnerability (V), and is characterized in an univariate representation as a combination of both factors (Karmakar et al., 2010) i.e., a product of H and V. In other words, it evaluates the likelihood of flooding, and the degree of risk is principally characterised in terms of probability of occurrence (Kron, 2005), depth, velocity or combination of both and residence time of flood water. However, very less importance has been given to socio-economic vulnerability in the previous studies in terms of flood risk mapping. It has become important to recognize the significance of socioeconomic aspects of vulnerability, since floods bring substantial losses in society like death of family member, loss of temporary or permanent jobs, destruction of property, while concurrently quantifying the flood damage in terms of hazard. In these lines, this study aims to develop a comprehensive vulnerability mapping framework addressing the socio-economic aspects, by utilising a nonparametric data envelopment analysis (DEA). Simultaneously, a severe flood hotspot of Mumbai, the Mithi catchment, is considered to assess the overall changes in flood risk maps, an amalgamation of flood hazard (developed with a hydrodynamic flood modelling framework) and vulnerability maps, under various socio-economic scenarios.

2 Methodology

Mumbai, one of the most crowded cities in India with a population of around 12 million as per the most recent Census of India 2011 data and has now increased to around 20 million (Ghosh et al., 2021b), is the focus of this study. The average population density of 20 000 per km^2; where more than half of the population resides in slums (Census of India, 2011). As the city grew, many areas were reclaimed to meet the demand for space. Originally, the city was made up of a cluster of islands. This has led to its inherent vulnerability to floods especially in the low-lying regions at or below mean sea level. The city primarily comprises of two revenue districts under the Municipal Corporation of Greater Mumbai (MCGM) – Mumbai city district (67.79 km^2) and Mumbai Suburban district (370 km^2). These are further divided into six zones, each consists of three to five wards – accounting to a total of twenty-four wards alphabetically named by MCGM. There are nine wards in Mumbai city district and fifteen wards in Mumbai suburban district. The city has been further divided into eighty-eight sections for vulnerability assessment in this study to efficiently manage the city's huge human population. As fine resolution DEM data are available for the flood-prone Mithi catchment of Mumbai, the hazard assessment was conducted over this catchment.

2.1 Vulnerability Assessment

The social and socioeconomic theme is considered for vulnerability assessment by utilising an indicator-based approach using the framework described by Sherly et al. (2015) (Fig. 1). A list of indicators representing socio-economic conditions were selected and categorised into positive (increasing vulnerability) and negative indicators (decreasing vulnerability) categories based on a comprehensive literature review. Since units of different indicators vary, an initial standardisation process was conducted as per the Eq. (1):

$$V_i^{\text{std}} = \frac{V_i - V_i^{\min}}{V_i^{\max} - V_i^{\min}} \tag{1}$$

where, V_i^{std} is the standardised vulnerability = of the ith section/grid; $V_i^{\max}$ and $V_i^{\min}$ are the maximum and minimum counts of vulnerability indicator respectively; and V_i is the count pertaining to the ith grid. After the identification of vulnerability indicators in each of these grids, data envelopment analysis (DEA) is performed to give an optimal weight for each indicator from which the vulnerability is quantified (Van Rossum and Drake, 1995). Further to decorrelate the higher correlated data sets, increase explicability, and computational efficiency, principal component analysis (PCA) is performed (Jolliffe, 2002). The PCA aids to decrease the sensitivity to overfitting in DEA, by reduction of dimensions of the problem, i.e., the 11 socio-economic indicator values obtained for all the grid points over the spatial domain. The indicators are linearly transformed to orthogonal variables in order of explanation of variance known as principal components (PCs). A considerable variance is explained by the PCs with higher eigenvalues which reduces the dimension of the dataset.

DEA is a method for estimating the relative efficiency of units that are called decision making units (DMU), when it is difficult to identify absolute measures of efficiency. Here each grid or spatial unit may be considered to be a DMU. This non-parametric technique makes no assumptions about the form of the function. It is a non-stochastic approach in which all observations are treated as deterministic. The DEA aims to evaluate how well the DMUs are performing using a data-oriented approach, which communicates multiple inputs into multiple outputs, based on a data-driven nonparametric approach. Unlike other methods, it doesn't require expert elicitation to determine how to assign weightages to various vulnerability indicators (Cooper et al., 2011). The efficiency of each DMU is measured by the distance of its input output vectors to a piece wise linear frontier.

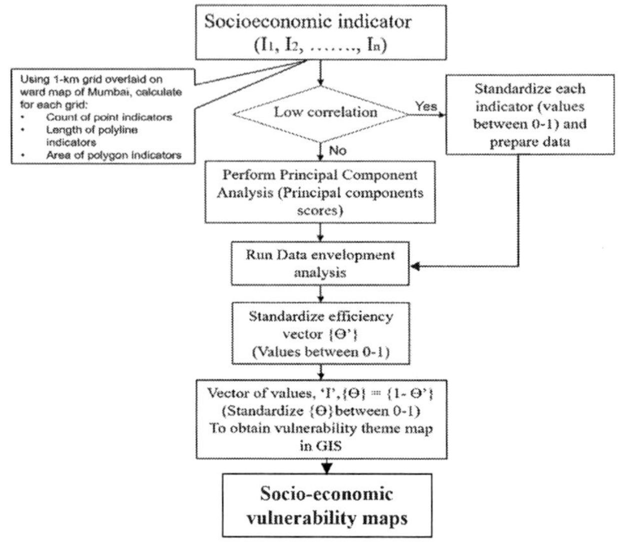

Figure 1. Proposed framework of indicator-based quantitative spatial socio-economic vulnerability assessment.

In DEA model, each DMU converts inputs to outputs. The DEA model optimises each observation with the objective of calculating a discrete piecewise frontier determined by the set of Pareto efficient DMUs. The number of DMUs must be no less than the number of inputs and outputs, and the variables must have low correlation, otherwise, the capacity of the DEA model to calculate the efficiency will be reduced.

In this study, the slack based input-oriented BCC model which considers variable returns to scale (VRS) was adopted, given by:

$$\text{minimise}\left[\theta - \epsilon\left(\sum\nolimits_{i=1}^{m} s_i^- + \sum\nolimits_{r=1}^{s} s_r^+\right)\right] \tag{2}$$

$$\sum\nolimits_{j=1}^{n} \lambda_j x_{ij} + s_i^- = \theta x_{ip}, i = 1, \ldots, m \forall p \tag{3}$$

$$\sum \%_{j=1}^{n} \lambda_j y_{rj} - s_r^+ = y_{rp}, r = 1, \ldots, s \forall p \tag{4}$$

$$\sum\nolimits_{j=1}^{n} \lambda_j = 1 \tag{5}$$

$$\lambda_j, \epsilon \geq 0 \tag{6}$$

where p is the grid being evaluated; θ ($0 < \theta \leq 1$) is the technical efficiency of the grid; y_{rj} is the amount of output r provided by grid j, x_{ij} is the amount of input I used by grid j; λ_j is the weight assigned to grid j; s_i^- and s_r^+ are the slack and remnant variable respectively; and ϵ is the non-Archimedean infinitesimal, generally taking $\epsilon = 10^{-6}$. Following this, the DMU's technical efficiency, θ, are utilised to get the respective vulnerabilities which is calculated as $v_i = 1 - \theta_i$ which differs between 0–1, here 0 is the minimum and 1 is the maximum values of vulnerability.

The vulnerability indicators chosen should be appropriately representative, relevant, justifiable and feasible with respect to the system (European Environment Agency, 2005; Vittal et al., 2020). The numerical data required for the development of social and economic indicators for the study are collected from Municipal Corporation of Greater Mumbai MCGM, Google Earth, and the Census of India (2001, 2011). The census data are available at the section level. The set of indicators considered are enumerated in Table 1.

Several efforts have been taken towards flood risk management by adapting strategies for reduction of flood hazard. However, the implementation of measures to reduce the vulnerability of a population to flood which may be beneficial in reduction of flood risk has been less discussed in literature. This can be an effective long-term measure for alleviation of flood risk. The non-working and illiterate section in the society especially female population are affected the most during a heavy flooding event. Hence taking measures towards reduction of their vulnerability to floods by educating them and providing them employment may help in reduction of flood risk and overall upliftment of society. In these lines, for the current study, vulnerability assessment is performed considering these 11 socio-economic indicators for five different socio-economic scenarios to identify the key indicators which contribute to a significant change in vulnerability. Certain sections considered to examine the reduction in vulnerability in response to the change in values of indicators based on the feasibility of change in indicators. The scenarios considered are as follows:

1. 2001 – Baseline scenario
2. 2011 – Baseline scenario
3. 2011 – Total reduction in the number of non-workers by 10 %.
4. 2011 – Total increase in female literacy by 10 %.
5. 2011 – Total increase in total literacy by 10 %.

The vulnerability maps are derived for these scenarios following the framework given in Fig. 3 in GIS interface and are compared to identify the key indicator responsible for the reduction in vulnerability.

2.2 Hazard Assessment

A flood hazard is a measure of the likelihood that a flood will occur at a specific location and time, and is represented as flood inundation depth, velocity, extent, onset, critical response time to certain depth of flooding, and other characteristics in previous studies (Ghosh et al., 2023; Mohanty et al., 2020; Russo et al., 2013). In the present study a framework is developed for flood modelling through a 3-way coupled approach for the Mithi catchment to estimate the flood hazard as illustrated in Fig. 2.

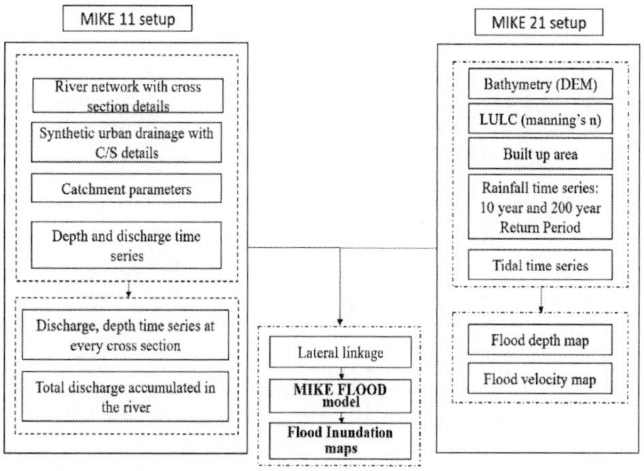

Figure 2. Proposed framework of indicator-based quantitative spatial socio-economic vulnerability assessment.

The MIKE 11 model with drainage network has been coupled with the MIKE 21 model in MIKE FLOOD (Danish Hydraulic Institute (DHI), 2020) for the Mithi catchment and flood inundation scenarios have been generated for 10 and 100 years return periods of rainfall (India Meteorological Department, 2023), discharge, and tidal elevation data (Indian National Centre for Ocean Information Services, 2023), the major influencers of flooding. Both the models are individually setup and then MIKE 11 along with stormwater drainage network along with MIKE 21 are linked in MIKE FLOOD as represented in Fig. 2. A detailed illustration for the holistic hydrodynamic framework can be found in Ghosh et al. (2021a). The momentum and continuity equations are solved in the MIKE 11 model for simulation of streamflow and the 2-D Navier-Stokes equations are solved in the MIKE 21 model for simulation of overland flow, and finally the flood inundation depth (m) and extent obtained is utilized to represent the flood hazard for the catchment. Finally, the flood inundation and hazard maps are derived for the various rainfall scenarios. The *Australian Flood Hazard Classification* is considered to classify flood inundation and hazard into various classes (AIDR, 2017). The inundation is classified into seven categories based on depth only and hazard into six categories based on both depth and velocity.

2.3 Risk Assessment

Risk is expressed as combination of two major components, i.e., hazard and vulnerability (Knighton et al., 2017). Flood risk maps indicate the potential adverse consequences of floods under a variety of probabilities, such as the number

Table 1. List of social and socio-economic vulnerability indicators considered in the study.

Sl. No.	Indicators	Justifications
1	Total population	The evacuation process during floods becomes difficult with increase in population
2	Total number of households	The potential loss and recovery increases
3	Female population	Higher responsibility for family and children makes their movement difficult; higher susceptibility to health complications from hazards
4	Number of children below 6 years	Physically weak, incapable of response to disaster and decision making; require special attention
5	Total number of Scheduled caste and Scheduled tribe population	Categorized as socially backward community by government
6	Total number of female Scheduled caste and Scheduled tribe population	Categorized as socially backward community by government; Higher responsibility for family and children make their movement difficult
7	Total illiteracy (%)	May have issues in communication, face difficulty to understand, access and respond to recover from disaster.
8	Female illiteracy	Face difficulty to follow evacuation warning and take care of family during disaster
9	Total number of marginal workers	Marginal workers work for less than six months per year. There may not always be a stable or continuous employment and income for these workers throughout the year because they are often employed irregularly or on a seasonal basis. Landless laborers who will suffer most after disaster as they would be jobless
10	Total number of female marginal workers	Landless female laborers who will suffer most after disaster as they would be jobless; higher responsibility towards family
11	Total number of non-workers	Dependent members in their family, which increase vulnerability
12	Extent of slums	Socially economically most backward

of inhabitants potentially affected, the type of economic activity affected, and the installations that could cause accidental pollution during flooding. Various approaches have been adapted in literature to quantify risk such as univariate flood risk mapping, in which risk is expressed as product of hazard and vulnerability (Vojinovic et al., 2016; Arrighi et al., 2019); and bivariate flood risk mapping, where hazard and risk are mapped as two different but spatially correlated variables in same map using a choropleth representation (Mohanty et al., 2020; Vittal et al., 2020; Ghosh et al., 2023), among many others. For the current study the univariate risk mapping approach is used for flood risk assessment. Here, the assessment of flood risk requires hazard and vulnerability estimations separately, which are subsequently aggregated. Univariate flood risk is used to calculate the magnitude of risk from hazards and vulnerabilities in the current study. By multiplying the hazard values with the vulnerability values, a risk value is generated that is further standardized between 0 and 1 and divided into *very low*, *low*, *medium*, *high*, and *very high-risk* categories (AIDR, 2017), and flood risk maps for each scenario for the area under consideration are developed.

3 Results and Discussion

Figure 3 shows the socio-economic vulnerability maps using DEA for the different scenarios considered. The framework developed proposes an indicator-based approach in which relevant attributes related to socio-economic vulnerability are brought into a spatial database. Five socio-economic scenarios are considered, namely the baseline scenario of 2001 and 2011 and three other scenarios where the indicators like percent of literate population and female population are increased each by 10 % and non-worker population are reduced by 10 % for different scenarios to depict the reduction in vulnerability. As expected, the vulnerability increased in 2011 than that of 2001, which can be attributed to the increase in vulnerable population in the study area as a result of unplanned expansion, slum dwelling population resulting in increase in non-working population, illiterate population, etc., among many other factors. Due to the presence of a large proportion of slum areas and large population density along the Mithi riverbanks making them more susceptible to the impacts of flooding, high vulnerability is found in those regions

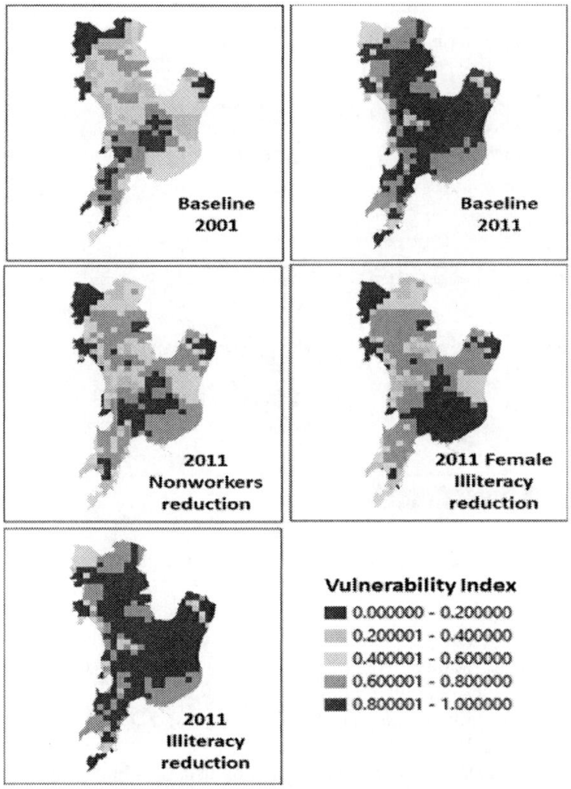

Figure 3. Socio-economic vulnerability mapping for different scenarios considered.

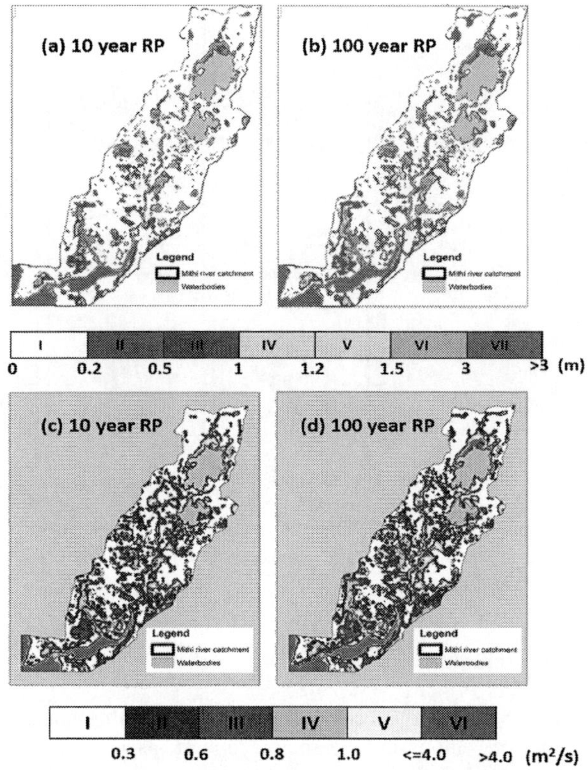

Figure 4. Inundation maps for **(a)** 10 year and **(b)** 100 year return period and hazard maps for **(a)** 10 year and **(b)** 100 year return period.

which can also be visualized from the vulnerability maps. When compared with the baseline scenario of 2011, the scenario where the percentage of non-workers was reduced contributed the most to overall vulnerability reduction among the five scenarios. Female literacy has a greater impact on overall vulnerability than total literacy. The increase in total literacy contributes almost nothing in decrease in vulnerability. The percentage of illiterate female population is high in south-eastern, eastern, and central parts when compared with other areas of the city. Correspondingly, the percentage distribution of total illiterate population across the city is not as uneven as the female illiterate population. Hence, the scenario of female illiteracy reduction results in greater reduction in vulnerability values for obvious reasons. The figures also clearly demonstrate the decadal change in vulnerability. As a result, this scenario-based socio-economic approach can be useful in reducing socioeconomic vulnerability.

Inundation and hazard maps derived with a 3-way coupled hydrodynamic flood modelling method are shown in Fig 4. The flood inundation is represented in terms of flood depth and the hazard is quantified by taking into consideration both inundation and velocity as per Australian Flood Hazard Classification (AIDR, 2017). As reflected in the flood maps, with an increase in rainfall return periods, flood inundation areas have increased. In areas with higher rainfall, depth, discharge, and tide, there is a greater chance of flooding. Likewise, for hazard maps, high intensity of rainfall and discharge leads to more inundation as well as increased flow velocities.

The risk maps are derived which are combination of vulnerability and hazard using univariate approach represented in Fig. 5. The risk maps also show similar trend, i.e. the risk increases in increase in return period of rainfall, and with increase in vulnerability the risk also gets magnified. The 2001 and 2011 baseline scenarios are marked under blue and red borders respectively while the different scenarios are represented under orange borders. It is observed that the socio-economic scenario in which the total literacy is increased has almost no effect on overall reduction of risk. The other two scenarios, i.e., 10 % reduction in non-workers and 10 % increase in female literacy with respect to 2011 census data reduces the overall risk with respect to the baseline 2011 scenario.

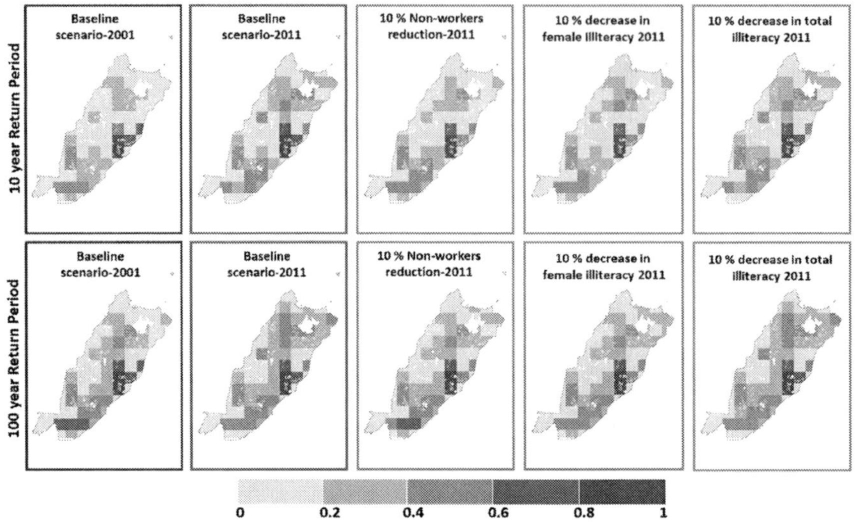

Figure 5. Demonstrative risk maps derived for 10 year and 100-year return periods for different socio-economic vulnerability scenarios for Mithi river catchment.

4 Conclusion

In the present study, a comprehensive vulnerability mapping framework addressing the socio-economic aspects (e.g., total population. illiterate population, households, elderly and children population, etc.) using a nonparametric data envelopment analysis (DEA) is developed. Further vulnerability assessment under different socio-economic scenarios is performed to identify the key indicators which contribute to a significant change in vulnerability. A 3-way coupled hydrodynamic flood model is implemented for flood modelling. The inundation and hazard maps for 10 and 100 years of return period of rainfall are derived. The results indicate that with the alteration in the feasible socio-economic indicators the overall vulnerability maybe reduced. The inundation and hazard increase with increase in return period. In addition to implementation of various methods to alleviate the flood hazard, the overall risk can be reduced by reducing the vulnerability as described in various scenarios in this study. The study may be further extended by considering more socio-economic vulnerability scenarios and for deriving bivariate flood risk maps and hence the marginal contribution of both hazard as well as vulnerability can be deduced. The retention and recession time can be considered to evaluate flood resilience of an area to flooding as a future scope of the study. The current study primarily focusses on evaluation on long term flood management options by adapting various vulnerability scenarios for flood risk alleviation, therefore the flood inundation depth and extent (major indicators) has been considered for flood hazard assessment. In future, this study may be extended to include other descriptors of flood hazard and resilience and with updated census of India data when available.

Code availability. The MIKE FLOOD platform has been utilized to develop the hydrodynamic flood model (https://www.mikepoweredbydhi.com/products/mike-flood) (Danish Hydraulic Institute (DHI), 2020). The rainfall prediction, vulnerability and risk assessment, has been performed by using Python software using packages such as numpy, pandas, and scipy (Van Rossum and Drake, 1995)).

Data availability. The census of India demographic data for vulnerability assessment can be obtained from https://censusindia.gov.in/census.website/data/census-tables (Census of India, 2001, 2011). The rainfall data can be obtained through request from India Meteorological Department (IMD), India (https://dsp.imdpune.gov.in/, India Meteorological Department, 2023). The tidal data can be obtained from Indian National Centre for Ocean Information Services (INCOIS), Hyderabad, India, used in this study can be obtained from https://incois.gov.in/portal/datainfo/drform.jsp (INCOIS, 2023).

Author contributions. SK and MG designed the problem. MG performed the analysis with inputs from SK. MG prepared the figures with inputs from SK and SG. MG wrote the original draft of the manuscript with inputs from SK. SK and SG did the funding and data acquisition for the work. All authors reviewed and edited the manuscript.

Competing interests. The contact author has declared that none of the authors has any competing interests.

Disclaimer. Publisher's note: Copernicus Publications remains neutral with regard to jurisdictional claims made in the text, published maps, institutional affiliations, or any other geographical representation in this paper. While Copernicus Publications makes every effort to include appropriate place names, the final responsibility lies with the authors.

Special issue statement. This article is part of the special issue "ICFM9 – River Basin Disaster Resilience and Sustainability by All". It is a result of The 9th International Conference on Flood Management, Tsukuba, Japan, 18–22 February 2023.

Acknowledgements. This work is supported by the Department of Science & Technology (SPLICE – Climate Change Programme), GoI, Project reference numbers DST/CCP/CoE/140/2018), and the Industrial Research & Consultancy Centre (IRCC) Award grant, IIT Bombay, Project reference numbers RI/0221-10000963-001, and the project under GISE hub, reference number RD/0121-DST0000-011.

Financial support. This research has been supported by the Department of Science and Technology, Ministry of Science and Technology, India (SPLICE – Climate Change Programme, GoI, grant no. DST/CCP/CoE/140/2018), the Industrial Research and Consultancy Centre (grant no. IRCC Award grant, IIT Bombay, Project reference numbers RI/0221-10000963-001) and the grant by the Industrial Research and Consultancy Centre (IRCC), IIT Bombay under GISE hub, reference number RD/0121-DST0000-011.

Review statement. This paper was edited by Christophe Cudennec and reviewed by Rajendra Prasad and one anonymous referee.

References

Arrighi, C., Pregnolato, M., Dawson, R. J., and Castelli, F.: Preparedness against mobility disruption by floods, Sci. Total Environ., 654, 1010–1022, 2019.

Australian Institute for Disaster Resilience (AIDR): Guideline 7-3, Supporting document for implementation of Australian Disaster Resilience Handbook 7 Managing the Floodplain: A Guide to Best Practice in Flood Risk Management in Australia, https://knowledge.aidr.org.au/media/3518/adr-guideline-7-3.pdf (last access: 10 January 2023), 2017.

Census of India: Provisional population totals, Census of India, New Delhi: Office of the Registrar General and Census Commissioner, India, https://censusindia.gov.in/census.website/data/census-tables (last access: 10 January 2023), 2001.

Census of India: Provisional population totals, Census of India, New Delhi: Office of the Registrar General and Census Commissioner, India, https://censusindia.gov.in/census.website/data/census-tables (last access: 10 January 2023), 2011.

Cooper, W. W., Seiford, L. M., and Zhu, J.: Data envelopment analysis: History, models, and interpretations, in: International Series in Operations Research and Management Science, 164, https://doi.org/10.1007/978-1-4419-6151-8_1, 2011.

Danish Hydraulic Institute (DHI): MIKE FLOOD 1D-2D Modelling User Manual, https://www.mikepoweredbydhi.com/products/mike-flood (last access: 10 January 2023), 2020.

Environment Agency: EEA core set of indicators: Guide, EEA Technical Report No. 1/2005, Copenhagen, Denmark, https://www.eea.europa.eu/publications/technical_report_2005_1/file (last access: 10 January 2023), 2005.

Ghosh, M., Mohanty, M. P., Kishore, P., and Karmakar, S.: Performance evaluation of potential inland flood management options through a three-way linked hydrodynamic modelling framework for a coastal urban watershed, Hydrol. Res., 52, 61–77, https://doi.org/10.2166/NH.2020.123, 2021a.

Ghosh, M., Singh, J., Sekharan, S., Ghosh, S., Zope, P. E., and Karmakar, S.: Rationalization of automatic weather stations network over a coastal urban catchment: A multivariate approach, Atmos. Res., 254, 105511, https://doi.org/10.1016/j.atmosres.2021.105511, 2021b.

Ghosh, M., Shastri, H., Ghosh, S., and Karmakar, S.: A Novel Response Priority Framework for an Urban Coastal Catchment Using Global Weather Forecasts-Based Improved Flood Risk Estimates, J. Geophys. Res.-Atmos., 128, e2023JD038876, https://doi.org/10.1029/2023JD038876, 2023.

Indian National Centre for Ocean Information Services (INCOIS): Tidal elevation data, INCOIS [data set], https://incois.gov.in/portal/datainfo/drform.jsp, last access: 10 January 2023.

India Meteorological Department: Stationwise rainfall data, https://dsp.imdpune.gov.in/, last access: 10 January 2023.

Jolliffe, I. T.: Principal Component Analysis, Springer-Verlag, New York, 2nd edition, Springer Series in Statistics, http://cda.psych.uiuc.edu/statistical_learning_course/JolliffeI.PrincipalComponentAnalysis(2ed.,Springer,2002)(518s)_MVsa_.pdf (last access: 10 January 2023), 2002.

Karmakar, S., Simonovic, S. P., Peck, A., and Black, J.: An Information System for Risk Vulnerability Assessment to Flood, Journal of Geographic Information System, 2, 129–146, https://doi.org/10.4236/jgis.2010.23020, 2010.

Knighton, J., Steinschneider, S., and Walter, M. T.: A vulnerability-based, bottom-up assessment of future riverine flood risk using a modified peaks-over-threshold approach and a physically based hydrologic model, Water Resour. Res., 53, 10043–10064, 2017.

Kron, W.: Flood Risk = Hazard · Values · Vulnerability, Water Int., 30, 58–68, https://doi.org/10.1080/02508060508691837, 2005.

Mohanty, M. P., H, V., Yadav, V., Ghosh, S., Rao, G. S., and Karmakar, S.: A new bivariate risk classifier for flood management considering hazard and socio-economic dimensions, J. Environ. Manage., 255, 109733, https://doi.org/10.1016/j.jenvman.2019.109733, 2020.

Russo, B., Gómez, M., and Macchione, F.: Pedestrian hazard criteria for flooded urban areas, Nat. Hazards, 69, 251–265, https://https://doi.org/10.1007/s11069-013-0702-2, 2013.

Shao, W., Xian, S., Lin, N., Kunreuther, H., Jackson, N., and Goidel, K.: Understanding the effects of past flood events and perceived and estimated flood risks on individuals' voluntary flood insurance purchase behaviour, Water Res., 108, 391–400, https://doi.org/10.1016/j.watres.2016.11.021, 2016.

Sherly, M. A., Karmakar, S., Parthasarathy, D., Chan, T., and Rau, C.: Disaster vulnerability mapping for a densely populated coastal urban area: An application to Mumbai, India, Ann. Assoc. Am. Geogr., 105, 1198–1220, https://doi.org/10.1080/00045608.2015.1072792, 2015.

Van Rossum, G. and Drake Jr., F. L.: Python reference manual, Centrum voor Wiskunde en Informatica Amsterdam, https://dl.acm.org/doi/book/10.5555/869369 (last access: 10 January 2023), 1995.

Vittal, H., Karmakar, S., Ghosh, S., and Murtugudde, R.: A comprehensive India-wide social vulnerability analysis: highlighting its influence on hydro-climatic risk, Environ. Res. Lett., 15, 014005, https://doi.org/10.1088/1748-9326/ab6499, 2020.

Vojinovic, Z., Hammond, M., Golub, D., Hirunsalee, S., Weesakul, S., Meesuk, V., Medina, N., Sanchez, A., Kumara, S., and Abbott, M.: Holistic approach to flood risk assessment in areas with cultural heritage: apractical application in ayutthaya, thailand, Nat. Hazards, 81, 589–616, 2016.

Proc. IAHS, 386, 307–312, 2024
https://doi.org/10.5194/piahs-386-307-2024

Global experiences in flood management – perspectives through ICFM Webinar Series

Cheng Zhang and Wuxia Bi

China Institute of Water Resources and Hydropower Research (IWHR), Beijing, 100038, China

Correspondence: Wuxia Bi (biwx@iwhr.com)

Received: 20 July 2023 – Revised: 11 December 2023 – Accepted: 8 June 2024 – Published: 17 July 2024

Abstract. It is a major issue of concern for international organizations and governmental departments in supporting global climate governance. As floods account for 44 % of all natural disasters in recent 20 years, it is of great significance to propose countermeasures by reviewing global extreme flood events in recent years. The International Conferences on Flood Management (ICFM) and Flash Flood Program (FFP), have hosted a series of international academic webinars since 2020. Drawing upon the outcomes of the webinars and reviewing the latest flood mitigation strategies of major countries advanced in water management, this paper focuses on the global experience in response to watershed flood, flash flood, urban flood, extreme flood under climate change, and comprehensive mitigation measures. It can be learned that flood control strategies need combine both structural and non-structural measures, and draw upon nature-based solutions that integrate green, blue and gray spaces within basins or cities through multi-purpose and targeted programs, with a view to achieving flood management goals and promoting social, economic and environmental sustainability in a holistic way.

1 Introduction

Climate change has reshaped the spatial and temporal patterns of water cycle, posing a continuous threat to human survival and development in a way that modifies water availability and endangers water security. It also has multi-directional interactions with water-sediment processes, hydro-chemical processes and hydro-ecological processes. Water cycle variability has affected global change in a very obvious way. As the footprint of climate change and human activities increase, water cycle processes and fluxes have been altered, with significant impact on water resources, droughts and floods, and other extreme hydrological events. Researches revealed that for every degree increase in global average temperature, the water circulation rate will accelerate by 2 %–4 % (Guangming Net, 2022).

Observations found that hot days extend by 2–8 d per decade in most areas of the world, with the average number in recent years tripling compared to 1961–1990, while extreme precipitation adds by 1 % per decade. Since the 1990s, the rate of sea level rise has accelerated, reaching 4 mm per year, which has doubled compared to before (Sohu Business, 2022). Related data shows that between 1998 and 2017, high-income countries suffered an average loss of 0.41 % of GDP due to climate related disasters, while low-income countries accounted for 1.77 % of GDP. And the loss was from 751 million in 1950 to 4.2 billion in 2018. Currently, 55 % of the world's population lives in urban areas. It is expected that this proportion will increase to 68 % by 2050. According to the report of the United Nations Office for Disaster Risk Reduction (UNDRR) (CRED, 2020), there were 3254 floods worldwide reported since ushering into the 21st century (2000–2019), accounting for 44 % of all natural disasters, the highest percentage among all types of natural hazards; and increasing by 1.34 times in comparison to the 1389 floods during 1980–1999.

The frequency and severity of floods increased due to climate change in the past few years, yet the risk of floods has been underestimated. Based on our statistical analysis on the typical flood events occurring during 2020 to 2022 (Fig. 1), floods have the most impact on Asia, accounting for over 35 % among total flood events worldwide (Fig. 2). The most affected country is China. According to a UNDRR report, Asia is currently exposed to the highest flood risks as it experienced 41 % of all the flood events worldwide (CRED,

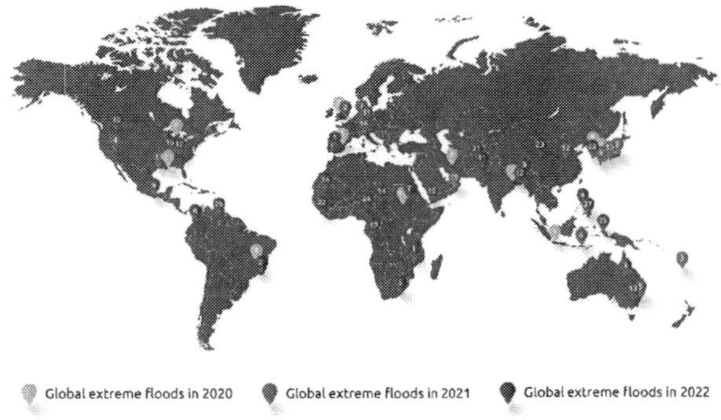

Figure 1. Distribution map of global extreme flood events from 2020 to 2022. (Source: https://www.icfm.world/Statistics/, last access: 10 July 2024).

2020). The conclusion that Asia suffered from the highest percentage of flood events in this report is consistent with the UNDRR report.

Global extreme floods show diversity in time and space in recent years. In 2021, Germany Flood resulted in 180 deaths, and have caused great economic losses of EUR 3 billion. The extreme rainstorm in Zhengzhou China is currently the most severe urban flood, which caused 398 deaths, and the economic losses reached CNY 20 billion. Rare desert floods occurred in Oman, with over 3 years of rainfall accumulated within 24 h. In 2022, rainstorms in Eastern South Africa lasted over months, and had 448 deaths. In Congo, the rainfall amount was over 100 mm in 24 h, and the death reached more than 400 people. The extreme flood in Pakistan caused over 1700 deaths, affecting over two thirds of the country, and over 3.3 million people were affected. The rainstorm in Yemen destroyed dozens of historic buildings in its ancient capital.

Under the circumstances, more attention has been paid to extreme events globally. The 2023 United Nations Water Conference passed a landmark "Water Action Agenda", which calls for strengthening institutional capabilities and integrating risk reduction measures to respond to extreme weather events.

To take joint actions in response to the increasing risk of more extreme floods, the international Conferences on Flood Management (ICFM) secretariat was hosted by China's leading water research institute, China Institute of Water Resources and Hydropower Research (IWHR), in Beijing since 2017. Main work include organizing ICFM conferences and webinar series, writing scientific reports, operating global flood database, public education, etc. Since 2020, ICFM organized international academic webinar series to offer a platform for exchanging experience and practices on flood management. So far, 10 webinars were hosted with more than 20 global experts discussing trending topics. The visitors were from 83 countries, the live streaming had over 10 000 clicks, and our expert database had more than 2900 experts. The official website since then with over 20 thousand visits in the past year.

The following sections will summarize and refine the concept, research projects, methods, insights proposed by the panelists at the webinars.

2 Global perspectives

2.1 Watershed flood

For watershed flood, both structural and non-structural measures should be applied. For structural measures, the joint scheduling of cascade reservoirs was implemented in a refined manner, this manner impounded 24.6 billion m^3 of flood and reduced the outflow by 14.8 % in Yangtze River Flood for 2020 flood (ICFM, 2020). As for non-structural measures, the forecasting system can be applied in collaboration with the decision-supporting system for flood management (ICFM, 2020). Multi-party coordination and collaboration should be strengthened because emergency response is crucial. We should also rationally use flood detention basins based on scientific assessment, and take the initiatives to flood so as to further protect life and property. The dam failures of Michigan flooding in 2020 indicate the significance to normalize safety inspections of flood control projects (ICFM, 2020).

Severe rainfall in several areas of Shanxi Province, China has caused floods. Chinese media pointed out that floods have caused over 1.75 million people to be affected, 120,000 people to be urgently relocated, and 3.58 million acres of crops have been damaged. To manage the watershed flood, we should guarantee that there are no casualties, no dam collapse of reservoirs, no breach of important embankments, and no impact on important infrastructure.

2.2 Flash flood

The Flash Flood Program (FFP) was established, in collaboration with IWHR, which mainly focuses on the floods occurred in mountain areas and emphasizes the speed of the flood. Flash floods disasters account for 70 % of total global water-related mortality. In 2022, 119 people died and disappeared due to disasters in China, which is the lowest year since statistical data is available. In recent years, 70 % of the deaths and missing people in China due to natural disasters have been caused by floods. As there were 171 people died and disappeared due to floods in 2022, coincidentally, it happened to be 70 % of the deaths caused by floods. It can be inferred that flash floods pose a huge threat to the lives and safety of the people (ICFM, 2022).

In order to save money, some bridges built on small and medium-sized rivers had insufficient height and span, which ultimately leads to problems after the formation of flash floods. This situation is a prominent issue in several disas-

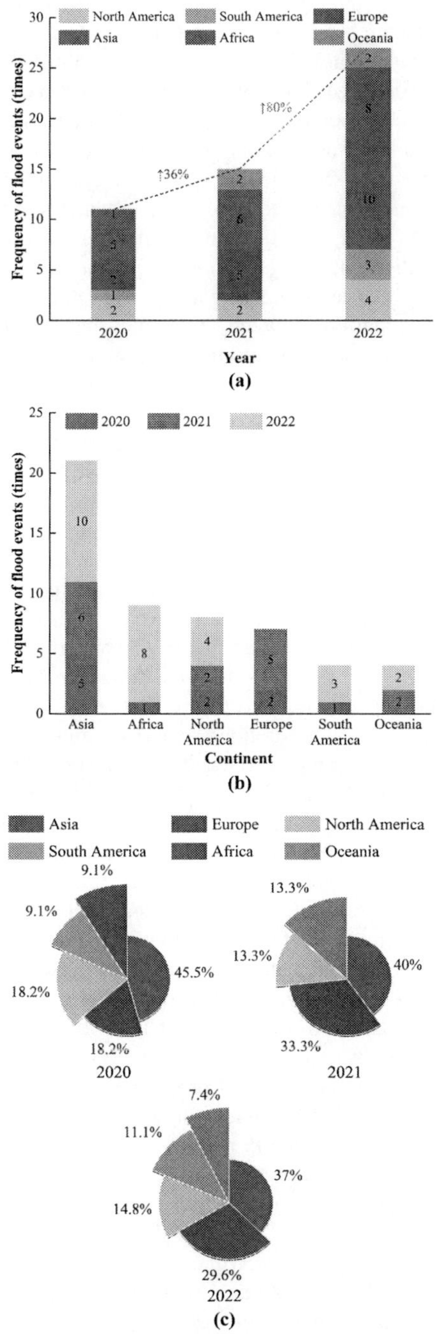

Figure 2. Statistical analysis charts of global typical flood events from 2020 to 2022. **(a)** Flood events in recent 3 years. **(b)** Flood events on each continent. **(c)** Proportion of annual flood events on each continent.

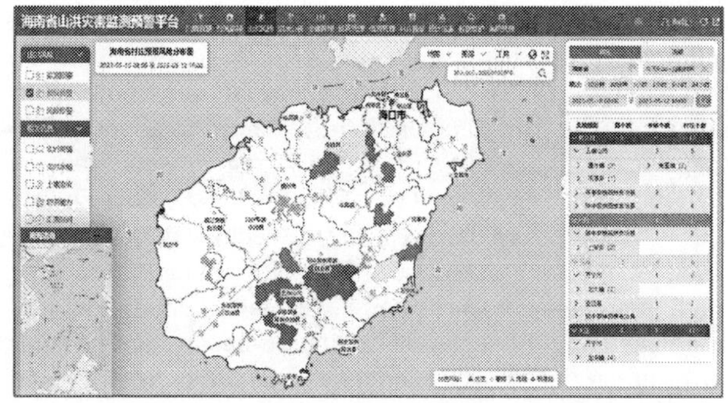

Figure 3. Flash flood forecast and early warning system developed in China. Publisher's remark: please note that the above figure contains disputed territories.

ters occurring in recent years that we have summarized. Of course, there are also bridges in some areas were not directly threatened by flash floods, but due to bridge being blocked, the water flow has been diverted. After the diversion, towns and villages that were not within the defence range would also be threatened.

We should strengthen water-related resilience through "basin-wide flood management" and "rebuilding flood-conscious societies", which coincides with River Basin Disaster Resilience and Sustainability by All proposed by Japan. Country and local government should improve monitoring and early warning technology and strengthen the education of local residents in disaster preparedness, response, and recovery. Such idea is consistent with the International Day motif of Disaster Reduction: Early Warning, Early Action. Notably, the public warning may send messages about how much rain it will rain, for example, 50 and 80 mm, but local people may not understand. Thus, the messages should point out the rainfall amount is large or small, and what measures should be taken. In addition, many transfers are in night, the transfer measures should be refined, and the practical exercises are necessary to help local people memorize the transfer route, especially the places where has a risk of congestion and congestion, such as crossing ditches, bridges and culverts (ICFM, 2022a).

Water conservation measures can be applied in flash flood ditch. Specifically for severe flash flood, it is crucial to build multi-stage progressive flash flood forecast and early warning system. The system developed in China reduced life losses by 71 % from 2013 to 2022 (ICFM, 2022a).

2.3 Urban flood

While the world is rapidly urbanizing, it is worth noting that most of the global risks caused by climate change are concentrated in cities. We are undergoing rapid urbanization, China's urbanization rate varied from 30 % in 2000 to 59.6 % in 2018. About 55 % of the world's population lives in cities and is expected to grow to 68 % by 2050, there is small space for river regulation and storage. Since there is limited space for river regulation and storage, more attention should be paid to urban flood. We need to adopt a long-term vision and conduct coastal risk assessments facing to extreme events, as about 40 % of the world's population lives within 100 km from the coast. From a technical aspect, cities should establish a data-model-forecast-command integrated intelligent management system. Meanwhile, municipal governments should watch out for the co-occurrence possibility of urban flood, river flood, and coastal risk, improve city resilience by applying urban flood model with AI and big data, and apply engineering allocation in the rivers and plains (ICFM, 2021).

2.4 Extreme flood under climate change

Statistics indicate that 50 %–70 % of the increase in extreme floods are resulted by human activities under climate change. Facing floods like 2022 Pakistan Flood, we should reduce poverty worsened by flooding (10 % increase due to flooding), reduce vulnerabilities, and increase resilience through housing standards and agriculture assistance (ICFM, 2022b), etc. Local governments should also develop overall planning and make good use of funds for flood mitigation and prevention. Forecasting models can be used in operational flood forecasting systems and early warning systems, challenges ahead mainly lie in fundamental datasets, such as, precipitation, DEM, etc. Moreover, we should pay attention to aging water conservancy projects, which triggered the flood disaster in Western Europe in 2021. For the floods in Brazil, we should focus on issues related to floods and diseases, such as arbovirus and dengue fever (ICFM, 2023), etc. The drought-flood abrupt alternation in Italy in 2023 had extreme rainfall lasting 36 h and the rainfall amount reached the precipitation amount in half a year in average years.

2.5 Comprehensive mitigation measures

Comprehensive mitigation measures should be carried out to cope with increase in extreme flood events. Lowering risk by increasing resilience, including preparedness & prevention, reconsider adaptive planning approaches. For inland floods, managed retreat (flood-plain zoning, room for the river) is a sustainable approach. New tools should be used for flood management such as big data and digital twin. The flood rescue team (personnel, equipment) should improve professionalism through equipment upgrade and efficiency of emergency response. Governments can enhance public awareness through education and media publicity, and ensure even distribution and coverage of social flood control facilities. The experience gained from epidemic prevention and control can also provide reference for flood management.

The latest flood mitigation strategies of major countries advanced in water management around the world were reviewed. Due to factors such as unique geographical location and climate change, the Dutch efforts on flood resilience began in 2008 through the multi-layer safety (MLS) approach by constructing a three-layer resilience system (Cao, 2019). The three layers include: (1) Protection, i.e. the primary dike system (10 000-year flood protection standard). Standards for flood control structures are established based on flood risk and disaster loss analysis to foster a system of flood control structures; (2) Sustainable spatial planning. The national space is zoned by secondary dikes and water structures for multiple purposes, with the concept of "Room for the River" being highlighted; and (3) Emergency response and disaster relief, which mainly deal with extreme floods, including better coordination between emergency service providers, managerial decisions, communication models and evacuation plans, building more flood resilient cities such as Dordrecht.

Japan has experienced floods multiple times since 2013. Drawing upon the lessons learned from frequent floods, the country continues to revise its flood management measures accordingly. In 2020, the Ministry of Land, Infrastructure, Transport and Tourism (MLIT) of Japan undertook a large-scale flood management reform by introducing a new policy of River Basin Disaster Resilience and Sustainability by All (Ministry of Land, Infrastructure, Transport and Tourism, Japan, 2020). In the past, Japan's flood control relied on structural measures with clear role allocation, implemented mainly in river basins and floodplains by administrators from divisions involving rivers, sewage, erosion and sediment control, and coastal management. The new policy considers river basins as spaces that include watershed and floodplains, and takes comprehensive, multi-layered actions, including flood prevention, exposure reduction and disaster resilience. It calls on all stakeholders in river basins, including the national government, prefectures, municipalities, private enterprises, residents, and water users, to take actions for enhancement in disaster resilience and sustainability.

Over the past two decades, the United Kingdom has experienced a number of floods that have posed a grave threat to individual and commercial properties across the country. To reduce flood-caused losses, the government began to put more emphasis on property resilience to flood. The Property Flood Resilience Roundtable in 2015 suggested that the government is responsible for helping the public protect their own properties from flooding, and therefore proposed an action plan to improve property flood resilience. After one year, the Property Flood Resilience Action Plan was published by the UK Department for Environment, Food and Rural Affairs, which guides the implementation of measures to improve the flood resilience of private and commercial properties.

Also, the United States has developed a set of plans and strategies regarding climate change and water security to mitigate and tackle climate-driven disasters in recent years, such as, Climate Adaptation Plan 2022 Progress Report (DOD,

2022), White House Action Plan on Global Water Security (The White House, 2022).

3 Conclusion

The frequency and severity of floods have increased due to climate change in the past few years, alarming a global challenge that needs to be responded jointly by the world. From 2000 onwards, the proportion of the flood-hit population has risen by 24 %, with the increase by more than 20 % in 40 countries. It is evident that the frequency and coverage of floods and the proportion of the affected population have all expanded since the 21st century as the global water cycle has accelerated over time. The severity of flood increased as well; for example, the 2022 Pakistan floods caused economic loss more than 75 % of the total economic losses by floods over the past 70 years. In the face of more frequent flooding events, countries have adopted a series of strategies and measures accordingly, specifically both in structual and non-structual way. Structural methods include dike safety management measures. Non-structural methods require new policies and actions incorporating flood resilience building and public awareness raising, etc. In a word, flood management should follow the law of nature, applying the nature-based solutions, integrating green, blue and gray spaces within basins. Each river has its own "genes", and there are no standardized flood management solutions, so optimization measures should be proposed as far as possible. We need to improve monitoring and early warning technology and establish intelligent management system. Global cooperations and partnerships, social cooperations on flood management should be strengthened to respond to the climate change.

Data availability. All raw data can be provided by the corresponding authors upon request.

Author contributions. CZ planned the campaign; WB analyzed the data; WB wrote the manuscript draft; CZ reviewed and edited the manuscript.

Competing interests. The contact author has declared that neither of the authors has any competing interests.

Disclaimer. Publisher's note: Copernicus Publications remains neutral with regard to jurisdictional claims made in the text, published maps, institutional affiliations, or any other geographical representation in this paper. While Copernicus Publications makes every effort to include appropriate place names, the final responsibility lies with the authors.

Special issue statement. This article is part of the special issue "ICFM9 – River Basin Disaster Resilience and Sustainability by All". It is a result of The 9th International Conference on Flood Management, Tsukuba, Japan, 18–22 February 2023.

Acknowledgements. We would like to express our gratitude to the organizers of the ICFM9 conference and anonymous reviewers.

Financial support. This research has been supported by the National Key Research and Development Program of China (grant no. 2021YFC3001404).

Review statement. This paper was edited by Miho Ohara and reviewed by two anonymous referees.

References

Cao, Z.: Exploration of Water Management System Reform in the Dutch Room for the River Program, https://mp.weixin.qq.com/s?__biz=MzIwMjAxNTIwOQ==&mid=2247495399&idx-=3&sn=5f63ddf94f85d6bf52fb7d680be656ed&chksm=96e78661a1900f7750f290f8d2af72de8e2254798eb424870c8c104f0f69be0aa37d241bf5f1&scene=27 (last access: 19 July 2023), 2019.

CRED: Human cost of disasters: An overview of the last 20 years (2000–2019), UN Office For Disaster Reduction, 2020.

DOD: Other Agencies Release Climate Adaptation Progress Reports, https://www.defense.gov/News/News-Stories/Article/Article/3182522/dod-other-agencies-release-climate-adaptation-progress-reports/ (last access: 19 July 2023), 2022.

Guangming Net: Global Ocean Salinity Differences Exacerbated, Water Cycle Accelerated: https://m.gmw.cn/baijia/2020-09/11/34178034.html (last access: 19 July 2023), 2022.

ICFM: The Flood Challenge to Resilience, https://www.icfm.world/Webinars/webinars-no-2.html (last access: 11 December 2023), 2020.

ICFM: Real-time Urban Flood Forecasting: Current Status and Future Challenges, https://www.icfm.world/Webinars/webinars-no-6.html (last access: 11 December 2023), 2021.

ICFM: Managed Retreat – From managing unavoidable to avoiding unmanageable, https://www.icfm.world/Webinars/webinars-no-8.html (last access: 11 December 2023), 2022a.

ICFM: 2022 Pakistan Floods, https://www.icfm.world/Webinars/webinars-no-10.html (last access: 11 December 2023), 2022b.

ICFM: 2023 Floods in Brazil, https://www.icfm.world/Webinars/webinars-no-11.html, last access: 11 December 2023.

Ministry of Land, Infrastructure, Transport and Tourism, Japan: River Basin Disaster Resilience and Sustainability by All, Ministry of Land, Infrastructure, Transport and Tourism, Japan, 2020.

Sohu Business: Green Economy under Climate Change is the Megatrend for the future – Zhang Xingyin, Deputy Director of Department Science, Technology and Climate Change, China Meteorological Administration, https://m.sohu.com/a/

606485721_100001551/?_trans_=010005_pcwzywxewmsm (last access: 19 July 2023), 2022.

The White House: WHITE HOUSE ACTION PLAN ON GLOBAL WATER SECURITY, 2022.

Proc. IAHS, 386, 313–319, 2024
https://doi.org/10.5194/piahs-386-313-2024

Open Access
PIAHS

How to avoid heavy life losses in Liulin-like flash flood events

Changzhi Li, Miao Zhang, Qing Li, and Bingshun He

China Institute of Water Resources and Hydropower Research (IWHR), Beijing, 100038, China

Correspondence: Changzhi Li (lichangzhi@iwhr.com)

Received: 21 July 2023 – Revised: 11 December 2023 – Accepted: 8 June 2024 – Published: 23 July 2024

Abstract. From 11 to 12 August 2021, an extreme rainstorm attacked the Liulin Town, a very common town in Hubei Province, China. The rainstorm triggered flash flood and caused heavy live losses. This study first conducted an analysis on the causes of the flash flood disaster, then, performed a simulation to review the flash flood process, and finally discussed the technical strategies and approaches to avoid life losses in similar cases. The main understandings include: (1) accumulative rainfall of shorter interval and time-sliding is significantly helpful to improve warning accuracy; (2) flash flood risk needs to be evaluated at catchment scale and making room for floods is necessary; (3) residential buildings threatened by flash floods need to rebuild some structures for live-saving in emergency circumstances; and (4) early warning system needs to have capabilities to collect real basin-wide rainfall and flood information, especially in case of rainstorm, and to be smart to judge hazard level.

1 Introduction

Flash flood is one of the main flood disasters. Globally, the mountain and hill cover about 30 % of the land area and heavy casualties and property losses result from flash flood each year with climate change (Edenhofer et al., 2012). From 1949 to 2021, more than 53,600 flash flood events attacked China. The Chinese government attaches great importance to flash flood disaster mitigation, and by years of persistent works, especially a nation-wide flash flood disaster management project implemented from 2013 to 2016, flash flood disaster mitigation systems have developed including main components of flash flood monitoring and early warning at national and county levels, accountability system, mass detection and prevention, and suitable structural works, which have greatly reduced the life losses due to flash floods at national level. However, On 12 August 2021, a heavy rainstorm flash flood attacked the Liulin Town, Hubei Province, China, resulting in heavy life losses (YRWG, 2021), which exposed some weaknesses in current flash flood mitigation. This article replayed this flash flood event, with objective of reviewing lessons and experiences, to avoid losses caused by similar disasters.

2 Catchment delineation

2.1 Langhe River Basin

Liulin Town, attacked by the flash flood, locates in the Langhe River Basin. Langhe River originates at Dayanjiao, flows through Liulin Town, Hedian Town, Zengdu District, Xihe Town, Suizhou High-tech Zone, and finally pours into the Fuhuan River. This catchment takes a fan shape in the upper reaches of Liulin Town, with drainage area 26.4 km^2, basin slope 0.225, basin relief 376 m, relief ratio 0.035, elongation ratio 0.542, and drainage density 1.99 $km\,km^{-2}$. The main channel is 10.7 km long with channel slope 2 %. The Langhe River can be roughly divided into the following three sections: (1) section in the upper reach of Liulin Town, with 150–250 m wide valley and mild channel slope; (2) section from Dayanjiao to Liulin Town, with several tributaries perpendicularly pouring into, mainly, Jinyin Creek, Lianhua Creek at the left bank, Taiping Creek and Baishuwan Creek at the right bank, and other 15 creeks into the mainstream or the tributaries; and (3) section from the lower reach of Liulin Town to the tail of Baiguohe Reservoir, with 70–100 m wide valley, obviously, narrower than that in the upper reach areas.

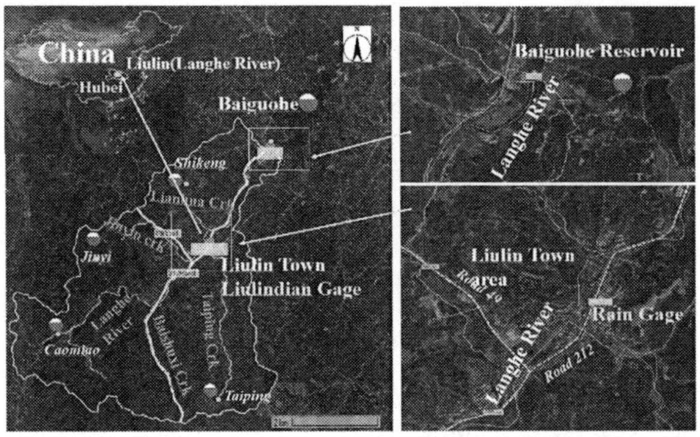

Figure 1. Schematic of the Langhe River Catchment.

Therefore, the Liulin Town lies in a slightly closed storage area, as shown in Fig. 1.

2.2 Liulin Town

Liulin Town, with a total population over 22 000 in 2020, is a low-mountain-hilly area. This Town is located at the bottom of a slightly closed wide and gentle valley in the Langhe River basin, surrounded by mountains on three sides, with good conditions of water conferencing, valley downstream narrowing, and low flood conveyance capacity. According to outputs of this Town of national flash flood disaster management project, implemented during 2013 to 2016, the Daqiao Street belongs to the area of very high-level flash flood risk (high population density without structures) while the Dayanjiao village is the area of high-level flash flood risk (moderate population density without structures).

2.3 Flood control structures

There are four small-sized reservoirs, Jinyin, Caomiao, Taiping and Shikeng, in the upper reaches of Liulin Town in this catchment, and one medium-sized reservoir, Baiguohe Reservoir, in the lower reaches of the town. Two reservoirs, Caomiao and Baiguohe, are in the main stream of the River, and the others in the tributaries (see Fig. 1).

Moreover, according to the local flood control plan, levees with standard of 10 years return period should be built on both sides of the Long River flowing through the Liulin Town. However, only part of the river dredging was completed when the flash flood event occurred.

2.4 Flash flood facilities

A county level flash flood early warning system has been developed in the Suixian County which has the capabilities of collecting real-time rainfall amount from rain gages within this county and releasing early warning once the rainfall depth exceeds the rainfall thresholds. As for the Liulin Town, the rain gage in the Dayanjiao Village and a warning broadcast in the Daqiao Street is component of this system. Moreover, each of the 4 small reservoirs mentioned above has a rain gage, but none of them transmits rainfall data to the early warning system. Totally, there are 5 rainfall gages in this catchment.

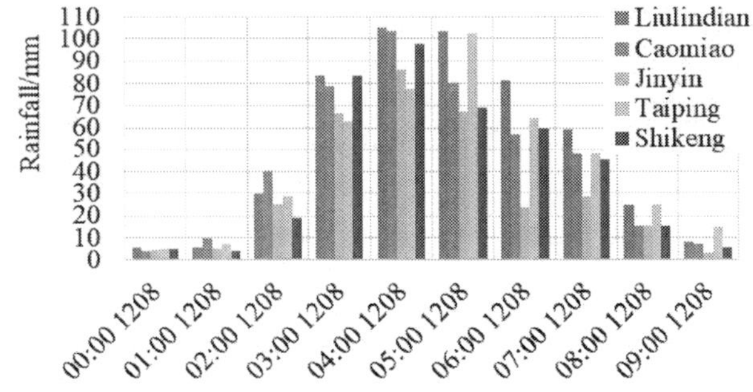

Figure 2. Catchment-wide rainstorm during the flash flood.

3 Flash flood process

3.1 Rainstorm

According to the rainfall records from the five rain gages, from 20:00 LT, Beijing Time, 11 August to 20:00 LT, 12 August, the maximum rainfall depths for 10 min, 1, 3, 6 and 24 h reached 23.0, 105.4, 277.5, 439, and 485.0 mm, respectively. And all of maximal rainfall depths of these typical durations occurred at the Liulindian Gage. Specially, Liulindian Gage has 2 h (03:00–05:00 LT) of rainfall depths exceeding 100 mm (YRWG, 2021). The rainfall hydrograph of the 5 rain gages indicates that a heavy rainstorm covered the whole catchment in the upper reach of the Baiguohe Reservoir (see Fig. 2) during this period.

Given a small drainage area of this catchment, and nearest to the centre of the catchment, the rainfall hydrograph of the Liulindian Gage has been significantly representative in the whole basin, hence, rainfall processes of 1 h interval and 10 min interval of this Gate were used to perform rainfall flood analysis, see Fig. 3.

3.2 Catchment terrain

Based on the watershed boundary with 100 m buffer, DEM data (5 m resolution) was clipped using GIS software to obtain the data for creating catchment terrain model. In the terrain model, no special considerations were taken for the four small-sized reservoirs. The reason is that the spillways of these reservoirs take form of free overflow, and all of them locates in the upstream areas with very small drainage area compared with the whole basin, and all of them operated normally during the event.

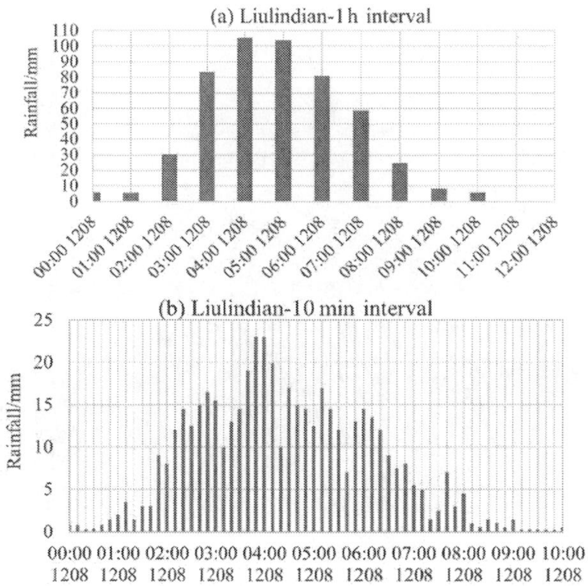

Figure 3. Rainfall hydrograph of different time intervals of Liulindian Gage.

3.3 Flash flood simulation

The simulation for flash flood was completed using 2-D shallow water equations, with rainfall hydrograph as the upper boundary condition and the whole river basin as the calculation area. The main considerations for flash flood simulation are as follows.

3.3.1 Mesh generation

In view of the valley's width in the main areas of interest in the catchment (150–250 m in the upstream valley of Liulin Town, and 70–100 m of the downstream valley), meshes of 30 m × 30 m were generated for common computation areas, and 15 m × 15 m for key computation areas, and two control sections were set up in the upper and lower reaches closing to Liulin Town to obtain flood hydrograph, a boundary line was set as lower boundary condition at the outlet of the basin (the tail of the Baiguohe Reservoir). The channel slope of this section was measured as 0.005 according to DEM data. Given the Baiguohe Reservoir having a total volume capacity of 46.22 million m^3, and can be reckoned as a large water body, normal water depth was adopted as the lower boundary in the hydrodynamic model, see Fig. 4.

3.3.2 Flood calculation

Flood calculation was performed for each grid, and its algorithms include horizontal and vertical directions as follow (USACE, 2020).

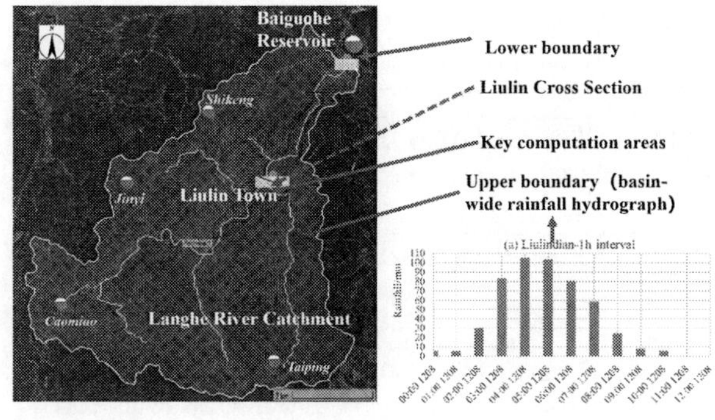

Figure 4. Catchment-wide model for rainfall-flood simulation.

Horizontal direction

Continuous equation:

$$\frac{\partial H}{\partial t}+\frac{\partial(hu)}{\partial x}+\frac{\partial(hv)}{\partial y}+q=0 \quad (1)$$

Momentum equations:

$$\begin{aligned}\frac{\partial u}{\partial t}&+u\frac{\partial(u)}{\partial x}+v\frac{\partial(u)}{\partial y}=-g\frac{\partial H}{\partial x}\\&+\frac{1}{h}\frac{\partial}{\partial x}\left(v_{xx}h\frac{\partial u}{\partial x}\right)+\frac{1}{h}\frac{\partial}{\partial y}\left(v_{yy}h\frac{\partial u}{\partial y}\right)\\&-c_{\mathrm{f}}u+fv+\frac{\tau_{s,x}}{\rho h}\\\frac{\partial v}{\partial t}&+u\frac{\partial v}{\partial x}+v\frac{\partial v}{\partial y}=-g\frac{\partial H}{\partial y}+\frac{1}{h}\frac{\partial}{\partial x}\left(v_{xx}h\frac{\partial v}{\partial x}\right)\\&+\frac{1}{h}\frac{\partial}{\partial y}\left(v_{yy}h\frac{\partial v}{\partial y}\right)-c_{\mathrm{f}}v+fu+\frac{\tau_{\mathrm{s},y}}{\rho h}\end{aligned} \quad (2)$$

in which, t: time, s; H, h: water elevation and water depth, ; u, v: velocity component in xy direction, $\mathrm{m\,s^{-1}}$; q: source and sink, $\mathrm{m^2\,s^{-1}}$; ρ: density of water, $\mathrm{kg\,m^{-3}}$; g: gravitational acceleration, $\mathrm{m\,s^{-2}}$; $v_{xx}\,v_{yy}$: coefficients of horizontal viscosity in x y direction, $\mathrm{m^2\,s^{-1}}$; c_{f}: friction coefficient at river bottom; τ_{s}: wind stress at surface $\mathrm{kg\,m^{-1}\,s^{-2}}$; f: coefficient of Coriolis force.

Vertical direction

The input at vertical direction of the grid is the rainfall while the loss is the soil infiltration. In this rainstorm flash flood event, rainfall occurred in a very concentrated and short time, and the soil type of the underlying consists mainly of loam soil and clay, soil infiltration is very minor comparing with the rainstorm depth, and was not considered in the model.

3.3.3 Friction

Friction is related to the surface roughness that usually described using land cover. This basin is covered mainly by

Table 1. Land cover class and roughness coefficient.

Code	Class	Roughness	Description
10	Cultivated land	0.065–0.080	Lands used for agriculture, horticulture and gardens
20	Forest	0.080–0.100	Lands covered with trees, with vegetation cover over 30 %
30	Grassland	0.050–0.065	Lands covered with natural grass with cover over 10 %
40	Shrubland	0.060–0.080	Lands covered with shrubs with cover over 30 %
50	Wetland	0.035–0.050	Lands covered with wetland plants and water bodies
60	Water body	0.020–0.025	Water bodies in the land area, such as lake, river, reservoir
80	Artificial surface	0.080–0.100	Lands modified by human activities

cultivated land, forest land, grassland, wetland, water body and artificial surface, correspondingly, the roughness coefficients for friction are compositely estimated according to the river channel, the composition of land cover class. Table 1 presents the values of roughness coefficient for typical land cover classes in the model.

3.3.4 Calibration

A site investigation was conducted to identify the flash flood process, in which, the peak discharge was calculated using Manning's formula and the process was deduced using instantaneous unit hydrograph approach with interval of 0.5 h based on local rainfall-flood manual, and Fig. 5 presents the flood hydrographs from the investigation and the simulation at the upper control section of the Liulin Town. Obviously, both are in good agreement with each other in terms of flood peak discharge, rising limb, and falling limb. In addition, most of the simulated water depth in the town area ranged between 3.0–3.5 m, which is also consistent with the results from site investigation. Therefore, the simulated results are reliable to be used for further analysis.

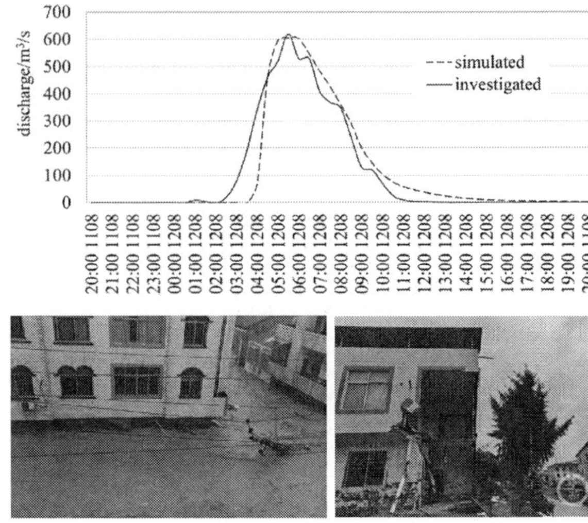

Figure 5. Simulated and investigated flood hydrographs well matched in flood peak discharge, rising limb and falling limb.

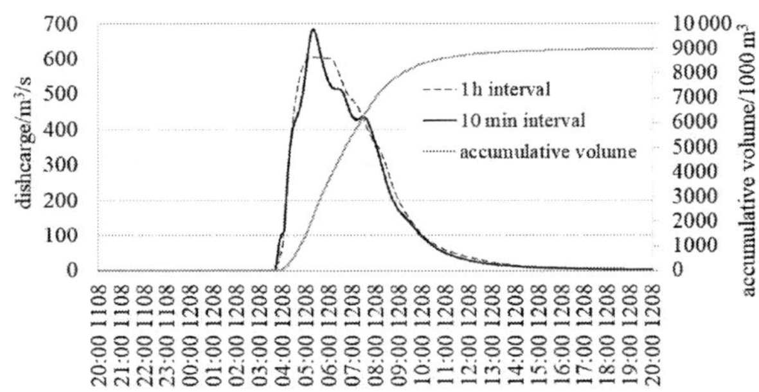

Figure 6. Simulated flood hydrographs and accumulative volume at the Liulin cross section.

4 Results and analysis

4.1 Influence of time intervals of rainstorm

Flood process was simulated according to the rainfall process of two intervals (1 h and 10 min) at Liulindian Gage, and Fig. 5 presents the results at the upper cross section of Liulin Town. Here, one can see that in this event, with same total rainfall amount, the flood peak discharge of a rainstorm with 10 min interval is $80\,\mathrm{m^3\,s^{-1}}$ larger than that with 1 h interval, and the increasement reaching 13 %. In fact, the rainfall process with 1 h interval has more homogenization than that with 10 min interval, which reduces the flood peak discharge to a certain extent. See Fig. 6.

Currently, most flash flood early warnings are issued based on rainfall thresholds at an hourly clock time, however, a very strong rainstorm within 60 min crossing an hourly clock time may be heavily cut off and averaged, which probably lead to neglecting a real rainfall depth exceeding the threshold of

Table 2. Differences between recorded and sliding.

Sliding/	Rain gage				
min	Liulindian	Caomiao	Jinyin	Taiping	Shikeng
10	−7.5	−6.7	−6.6	−14.6	−8.8
20	−17.5	−13.2	−11.2	−21.2	−17.5
30	−27.5	−19.3	−18.4	−26.4	−28.3
40	−37	−30.6	−27.9	−36.2	−42.4
50	−12	−6.6	−8.5	−9.5	−11.1

1 h, then missing early warning in a timely way. Therefore, a sliding analysis was performed for the rainfall records with various intervals (10, 20, 30, 40 and 50 min) at the 5 gages, as shown in Table 2, where, at a same hourly clock time, the data are the differences between the recorded and the sliding values. Obviously, the differences are significant, and sliding is necessary for rainfall threshold determination.

Currently, most rain gage has the capability to record rainfall each 5 min, so it is feasible to provide rainfall data of 10 min time interval for flash flood early warning, which will be significantly helpful to improve early warning.

4.2 Flood peak discharge and flood control capacity

The bankful discharge of the Langhe River section along the Liulin Town is only $250\,m^3\,s^{-1}$, which was significantly exceeded in this flash flood event, with flood peak discharge $640\,m^3\,s^{-1}$, and flood volume (24 h) about 9.2 million m^3 (YRWG, 2021). The simulation provides peak discharges of 606 and $686\,m^3\,s^{-1}$, and accumulative flood volume of 9.08 million m^3 (see Fig. 6). due to the special terrain of this catchment and the location of the Liulin Town, floods can easily form in a short time and rise rapidly at the town once heavy rainfall occurs. Moreover, some spaces for floods have been occupied step by step in this town which significantly amplified and worsen the disaster if extreme flood comes. Many researches indicate that due to climate change, the possibility of extreme rainstorm is increasing globally and will bring about more unexpected extreme floods (Abbs and Rafter, 2009; Trenberth, 2011; Bates and Westra, 2013). Therefore, for towns with similar locations like the Liulin Town, special and enough attention needs to be paid to room for floods.

4.3 Flood formation, evolution and early warning

The main process of flood formation at Liulin Town are as follows: 03:30 LT, 12 August, flood water level began to rise and peaked at 05:10 LT, about $610\,m^3\,s^{-1}$, deduced by Manning's formula; flood peaked for about an hour and started to decline at 06:20 LT. And flood rose rapidly during 03:00 to 04:00 LT, 12 August. According to the site investigation, the best time for emergency evacuation was very short, from 03:40 to 04:20 LT, because the ground floors of Liulin

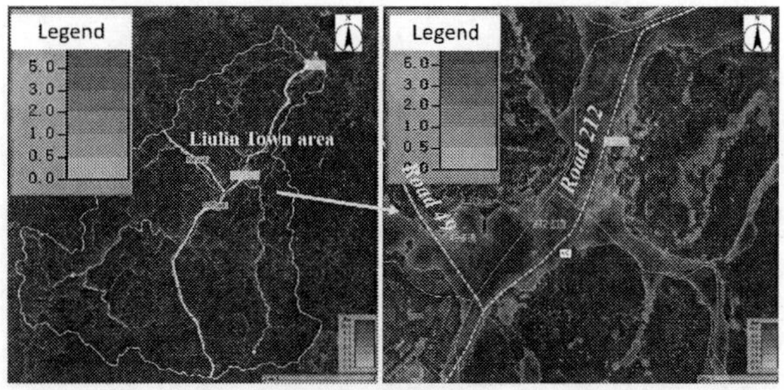

Figure 7. Flood depth during the flash flood.

Town along the river was completely flooded within less than 40 min. In fact, none of the 4 rain gages at reservoirs has been connected to the early warning system and delivered real information of catchment-wide rainstorm to the system during the flash flood event. In other words, how dangerous the rainstorm has not been conveyed to the exposures in a timely way, and the precious-lifesaving leading time was wasted in vain, even those existing handy facilities were easily available. Therefore, it is urgent and necessary to collect and integrate rainfall-flood information at the catchment scale, based on the location of towns or villages.

4.4 Water depth and flow velocity

According to the simulation, the inundation depth in most areas of Liulin Town ranges among 3.0–3.5 m. And Fig. 7 shows the simulated inundation depth scenarios. As can see that Liulin Town was almost completely inundated, except a few houses on the highland. In this figure, the yellow lines are provincial level road (Road 212 and Road 49). Road 212 runs through the Liulin Town and became a temporary conveyance way during the flood, and houses on both sides were flooded and water level reached the top of ground floor (see Fig. 5).

As for flood velocity, Fig. 8 presents the distribution of maximum flood velocity in Liulin Town, where the speeds in the flooded areas varies greatly, ranging among $1–5\,m\,s^{-1}$ in most areas, and reaching up $7\,m\,s^{-1}$ in some steep areas. Moreover, the flood speeds along Road 212 were generally among $3–5\,m\,s^{-1}$, and some areas up to $7\,m\,s^{-1}$.

According to the above analysis on flood depth and velocity, two points need to be strengthened to improve public awareness of flash flood risk: (1) studies (Smith et al., 2014; Wang et al., 2019) indicate that people or cars in flood with deep flood depth and high velocity are very dangerous (see Fig. 9), so, no one should take any chances to evacuate in flash flood, but keep away from it as far as possible; (2) in terms of permanent buildings closing to the river threatened by flood, indoor vertical evacuation passages, such as indoor upstairs and downstairs, should be considered below the third floor, and the authors strongly suggest it should be involved

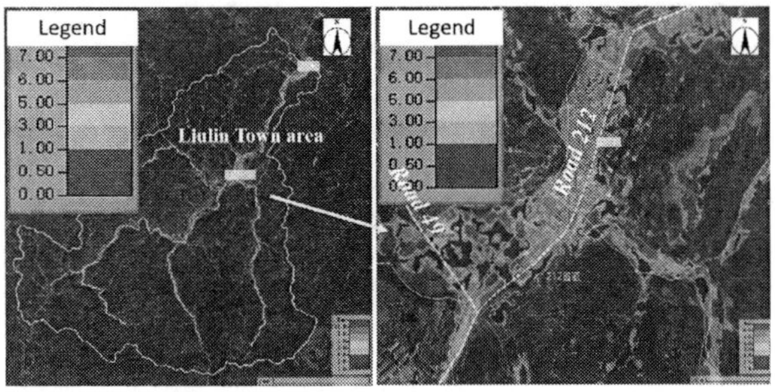

Figure 8. Flow velocity during the flash flood.

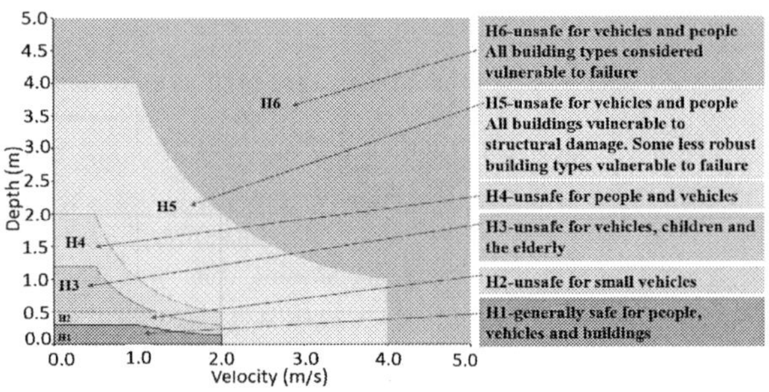

Figure 9. Combined Flood Hazard Curves (Smith et al., 2014).

in building codes. In this event, most people were killed indoor by flood water, because they have no indoor vertical evacuation channels, and could not open the rolling gates due to high water pressure, let alone reaching the outdoor staircases to move to high places.

5 Conclusions

Taking the flash flood disaster in Liulin Town as an example, this paper discussed the causes of flash flood disaster in the Liulin Town, performed simulation on the flood using shallow water equations, and then, combined with the specific situation of the Langhe River Basin, discussed the influence of time intervals of rainstorm, the flood peak discharge and flood control capacity, flood process and early warning, flood depth and velocity of flood and flash flood risk, rainfall process interval and flood peak, with purpose of finding technical strategies and approaches to mitigate disasters for similar cases. The main conclusions are as follows:

1. The early warning system should adopt a sliding accumulation rainfall with a finer time interval to improve the accuracy of early warning.

2. The Liulin Town is surrounded by mountains on three sides and located at the bottom of a slightly closed valley. The downstream valley of this town is narrow and the flood conveyance capacity is not so high, and some room for flood has been occupied by human actions. For towns at similar locations in catchment like the Liulin Town, it is necessary to evaluate flash flood risk at catchment scale and to make enough room for floods.

3. Flash floods are characterized by high depth and rapid flow, therefore, in addition to timely evacuation, residents threatened by flash floods need to rebuild some necessary structures for live-saving in emergent situations, such as indoor vertical evacuation passages, which should be encouraged in building codes and can save lives at some extreme cases.

4. Currently, most of rain gages are usually installed in or near residential sites for the sake of operability of both installation and maintenance, which often results in inadequate, even no access to rainfall amount in upstream areas of a catchment. It is necessary to collect and integrate real rainfall and flood information to the early warning system in time at catchment scale, especially in case of rainstorm; moreover, the system needs to be smart enough to give a judgement on the hazard level based on the rainfall depth and tensity.

Code availability. All of the simulation in this paper was done based on HEC-RAS software, a free and publicly available software.

Data availability. All of the underlying research data in this study were from the Report for "8.12" Flash 35 Flood Disaster in Liulin Town, Hubei Province, China, which was investigated by the Yangtze River Work Group and submitted to the Ministry of Water Resources, PRC. This report is not publicly accessible.

Author contributions. CL: conceptualization, supervision, funding acquisition, writing – review & editing; MZ: data curation, formal analysis, writing – original draft preparation; QL: validation, visualization; and BH: funding acquisition, resources.

Competing interests. The contact author has declared that none of the authors has any competing interests.

Special issue statement. This article is part of the special issue "ICFM9 – River Basin Disaster Resilience and Sustainability by All". It is a result of The 9th International Conference on Flood Management, Tsukuba, Japan, 18–22 February 2023.

Financial support. This research has been supported by the project of the Ministry of Science and Technology of the People's Republic of China (grant no. 2019YFC1510700) and IWHR Project "Theoretical and Technical Exploration on Flash Flood Mitigation and Control".

Review statement. This paper was edited by Miho Ohara and reviewed by two anonymous referees.

References

Abbs, D. and Rafter, T.: Impact of Climate Variability and Climate Change on Rainfall Extremes in Western Sydney and Surrounding Areas: Component 4 – dynamical downscaling, Report to the Sydney Metro Catchment Management Authority and Partners, CSIRO, PMB No 1, Aspendale, Victoria, 3195, 81 pp., 2009.

Bates, B. and Westra, S.: Climate Change Research Plan - Summary. Report for Institution of Engineers Australia, Australian Rainfall and Runoff Guideline: Project 1, Engineers Australia Water Engineering, https://arr.ga.gov.au/__data/assets/pdf_file/0016/40570/ARRCCS_ResPlan_Summary_17Jan2013.pdf (last access: 9 November 2022), 2013.

Edenhofer, O., Pichs-Madruga, R. Y., Sokona, V., Barros, C. B., Field, T., Zwickel, S., Schloemer, K., Ebi, M., Mastrandrea, K., Mach, C., and von Stechow, C.: Workshop Report of the Intergovernmental Panel on Climate Change Workshop on Socio-Economic Scenarios, IPCC Working Group III Technical Support Unit, Potsdam Institute for Climate Impact Research, Potsdam Germany, 51 pp., 2012.

Smith, G. P., Davey, E. K., and Cox, R. J.: Flood Hazard, UNSW Australia Water Research Laboratory, Technical Report 2014/07, 2014.

Trenberth K. E.: Changes in precipitation with climate change, Clim. Res., 47, 123–138, https://doi.org/10.3354/cr00953, 2011.

USACE (US Army Corps of Engineers): HEC-RAS Hydraulic Reference Manual, CPD-69, https://www.hec.usace.army.mil/publications/ (last access: 5 October 2022), 2020.

Wang, J., Li, N., and Wang, S.: Review on flood hazard assessment index and grade classification, China Flood and Drought Management, 29, 21–26, https://doi.org/10.16867/j.issn.1673-9264.2019213, 2019 (in Chinese).

YRWG (Yangzte River Work Group): Report for "8.12" Flash Flood Disaster in Liulin Town, Hubei Province, China, YRWG (Yangzte River Work Group), MWR, China, 48 pp., 2021 (in Chinese).

Proc. IAHS, 386, 321–326, 2024
https://doi.org/10.5194/piahs-386-321-2024

Open Access
PIAHS

The role of construction consultants in disaster recovery

Minoru Kamoto[1], **Akira Nakamura**[2], **and Masashi Ogura**[3]

[1]Infrastructur, Research Centre, Japan Civil Engineering Consultants Association (JCCA), 101-0075 Tokyo, Japan
[2]Shikoku Branch Office, Nippon Koei Co., Lit. 760-033 Kagawa, Japan
[3]Osaka Headquarters Transportation Infrastructure Dept Pacific Consultants Co, Ltd, 530-0004 Osaka, Japan

Correspondence: Minoru Kamoto (kamoto@jcca.or.jp)

Received: 16 June 2023 – Revised: 4 December 2023 – Accepted: 8 June 2024 – Published: 24 July 2024

Abstract. Disaster recovery requires speed above all, and the public and private sectors need to work together. In the event of a disaster, the activities of the construction industry play a crucial role. Construction consultants, surveying companies, and geological survey companies play an important role. Construction consultants oversee the entire process of disaster recovery, from assessing the damage immediately after the disaster to managing the restoration work. In some cases, local governments order construction management services to supervise the entire disaster recovery work, and in such cases, construction consultants are even responsible for part of placing construction procurements. In this way, construction consultants have become an indispensable part of the disaster recovery process. This report is an introduction to the activities of construction consultants in times of disaster, using the heavy rains situations as examples, and presents of challenges and recommendations.

1 Introduction

Japan has high precipitation, mountainous terrain with looming plains, and rushing rivers. As a result, the country has always been prone to flood disasters. Recent climate change has caused extreme precipitation events, increasing the frequency and intensity of disasters. Population and assets are concentrated in the plains, which account for only a small proportion of the country's total land area, and flood control measures are extremely important for the stability of the population.

Although the civil engineering departments of central and local governments have been primarily responsible for all aspects of flood control, including research, framing, planning, construction, and management, their organizations have been downsized and weakened, especially in municipalities, and public agencies are seemed to shift their work from being proactive in specific tasks to focusing on public relations and procurement. Therefore, it is more important than ever for the public and private sectors to work together to ensure the steady and robust practical operations.

At the same time, science, technology, and project operation and management methods are evolving, and it is important to create paths for further development and utilization of these through cooperation between the public and private sectors. The construction industry has been receiving procurements from public organizations and supporting social improvements for many years and has accumulated a wealth of experience in this field. Construction consultants (CC) have accumulated and honed their knowledge through many years of research, framing, and planning, and there are many talented individuals who can play an active role in the industry.

This paper describes how CC are playing an active role in disaster recovery and reconstruction, presents challenges and prospects.

2 Case study of torrential rain disaster

2.1 Response to the July 2018 torrential rain disaster, on the Oda River system (Kurashiki City, Okayama prefecture)

2.1.1 Outline of the disaster

From the end of June to early July 2018, a rainy season front stagnating near Japan and typhoon No. 7 over the South Sea caused record-breaking torrential rains over area nationwide, mainly in western Japan. As a result, rivers overflowed and mudslides occurred simultaneously over a wide area, with 52 people were killed, and 5977 houses were damaged, partially, and fully in Kurashiki city, where Mabi is located (Okayama prefecture, 2020). In addition, the damage to lifelines such as electricity, water and sewage systems, and transportation infrastructure in the urban centres caused extensive socioeconomic damage. Figure 1 shows Takahashi and Oda River Basin.

2.1.2 Survey of damage situation

The Oda River in the Takahashi River system and its tributaries were damaged by torrential rains on 5 and 6 July 2018, and a committee was established to investigate the cause of the damage and to study restoration methods, etc. The committee conducted a survey of the damage on 10 July. The CC conducted a joint on-site survey and, under the 30 direction of the procurer, conducted a trace water level survey, soil investigation, and simple sketches at the breach points to collect information for the investigation of the cause of the damage and the study of countermeasures (JCCA, 2019, 2022).

2.1.3 Investigation of the cause of the damage and basic policy for countermeasures by the Oda River Levee Investigation Committee

Based on the results of the field survey, the Oda River Levee Investigation Committee held five meetings, including a field survey, to analyse the causes of the damage and to formulate a basic policy for countermeasures at the section where the levee broke. In the committee meetings, the CC handled the field survey and geological survey, levee stability study related to seepage, composition, and verification of the general situation of flood (water level, flow volume, and plane flow), and verification of the inundation through flood analysis (setting the inner and outer water levels of the levee) under a tight schedule. Prepared materials and assisted in committee operations based on the results of the studies.

2.1.4 Examination of special projects

On 7 September, MLIT (Ministry of Land, Infrastructure, Transport and Tourism), CRDB (Chugoku Regional Development Bureau) announced that the Oda River confluence will be replaced to 4.6 km downstream of Takahashi river and that the Oda River will be excavated and its banks strengthened, and branch rivers banks will be raised and strengthened through intensive river improvement projects that will be implemented over approximately five years in order to safely carry the river flow volume caused by the future same heavy rainfall as this time. The CC will work with the MLIT CRDB and Okayama Prefecture on the project. The CC assisted in preparation of the project plan by studying the target flow volume, improvement menu, drawing, and project cost calculation under the tight schedule (Das et al., 2023).

2.2 Response to the torrential rain disaster caused by typhoon No. 19 in 2019, on the Chikuma River (upstream name of the Shinano River system)

2.2.1 Outline of the disaster

On 12 October 2019, a typhoon hit the Izu Peninsula, resulting in "Typhoon no. 19 torrential rainfall in 2019" which caused record-breaking rainfall in a wide area of Shizuoka and Niigata prefectures, the Kanto Koshin and the Tohoku region, with 99 were killed, 67 985 houses completely or partially destroyed (Fire and Disaster Management Agency, 2020). The Nagano Shinkansen Train Centre was flooded due to overtopping of the Chikuma River, which paralyzed the Hokuriku Shinkansen. Figure 2 shows Shinano and Chikuma river basin.

2.2.2 Immediate response to the disaster

CC voluntarily collected and organized meteorological and hydrological data on a real-time basis from publicly available information for basins where they might receive procurements and provided support for the prompt publication of an outline of the outflow of water and the effects of river improvement projects. In the Chikuma River, they also collected and organized data in the same way and monitored real-time images such as CCTV from offices and municipalities' websites to estimate the risk areas. These results were used to estimate flow volumes in the disaster. Based on these results, iso-rainfall maps, actual water levels, and flow volumes were prepared and used as materials for basin committees, etc., and for outlining outflows (JCCA, 2020).

2.2.3 Various studies for disaster application materials

MLIT has launched the "Emergency Flood Control Project", in which the national government, prefectures, and municipalities concerned work together to implement hard and soft measures to prevent another disaster in the entire basin for rivers that have been severely damaged by floods in recent years. On 31 January 2020, the government announced that the Chikuma River, together with the downstream Shinano River, would be included in the Shinano River System Emergency Flood Control Project, in which the national, prefectural, and municipal governments work together to implement riverine measures, basin measures, and community development and soft measures at a cost of approximately JPY 180 billion over 10 years. Under a tight schedule, CC companies assisted in the study of the target flow volume, improvement menus, preparation of drawings, calculation of project costs, and preparation of application materials. In the study of the improvement menus, the damage situation

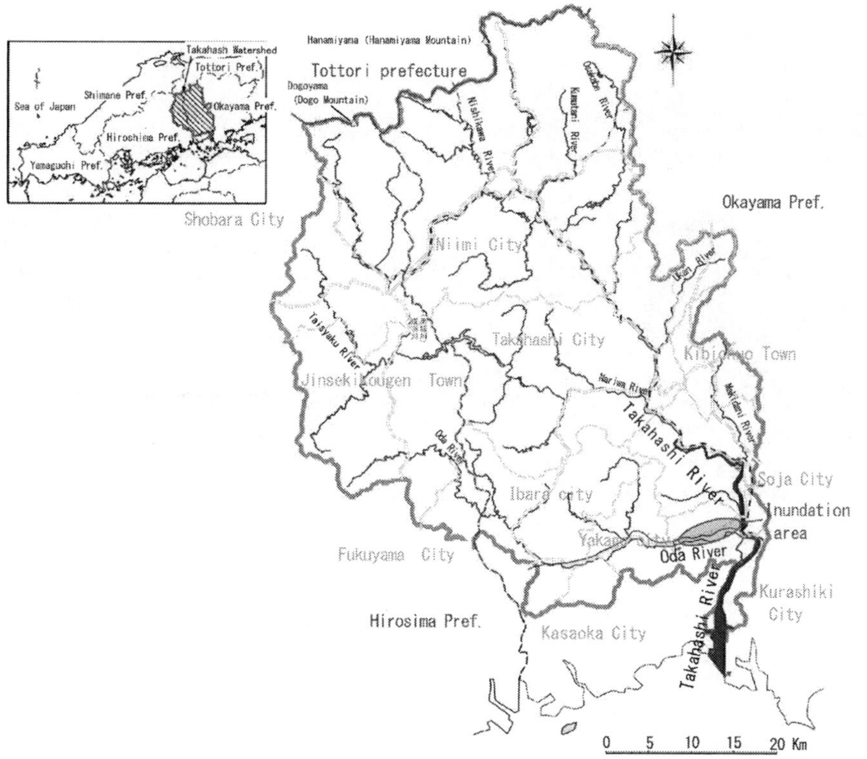

Figure 1. Map of Takahashi and Oda River basin based on the map of Ministry of Land, Infrastructure, Transport and Tourism Japan.

was assessed through on-site inspections, and runoff analysis, water level calculations, inundation analysis, etc. were conducted and the proposed improvement menus were presented to procurers.

2.3 Comparison of Oda and Chikuma Rivers' case

Due to changes in rainfall caused by climate change, the Chikuma River (Shinano River) underwent a revision of the target discharge volume in the River Improvement Plan. Shinano River had to be examined over the entire length of the river. Oda River, in comparison, was a localized study. CC contributed to them from a technical standpoint. Since both Oda and Chikuma Rivers broke their levees, nationwide studies to improve the durability of levees are continuing, and CC's technical contributions for field surveys, experiments, and simulations are indispensable.

In both rivers, many activities were conducted by CC after the flood disasters. These are introduced as next.

3 Engagements in large-scale disasters

Disaster recovery requires speed and a concerted effort by the public and private sectors. In times of disaster, the activities of the construction industry play a role from the occurrence of a disaster to the completion of restoration work. CC, surveying firms, and geological surveying firms, known as construction-related industries, play a particularly important role in disaster recovery. CC oversee the entire disaster recovery process, from the initial assessment of the damage immediately after a disaster occurs to the construction management of the recovery work. In some cases, local governments place orders for construction management (CM) services to supervise the entire disaster recovery work, and in such cases, CC may even be responsible for part of the procurement of the work. In this way, CC have become an indispensable part of the disaster recovery process. Figure 3 shows the flow of disaster recovery and the work of CC at the time of a disaster. The specific activities from (1) to (6) in Fig. 3 are written in the following, with the CC, surveying company, geological survey company, general construction company, and specialized construction company clearly identified.

(1) Confirmation of damage

"CC, Surveying companies (SC) including aerial surveying)" will conduct surveys and measurements to record the scale and situation of the damage when the topography is altered after the disaster due to emergency measures. Provide advice

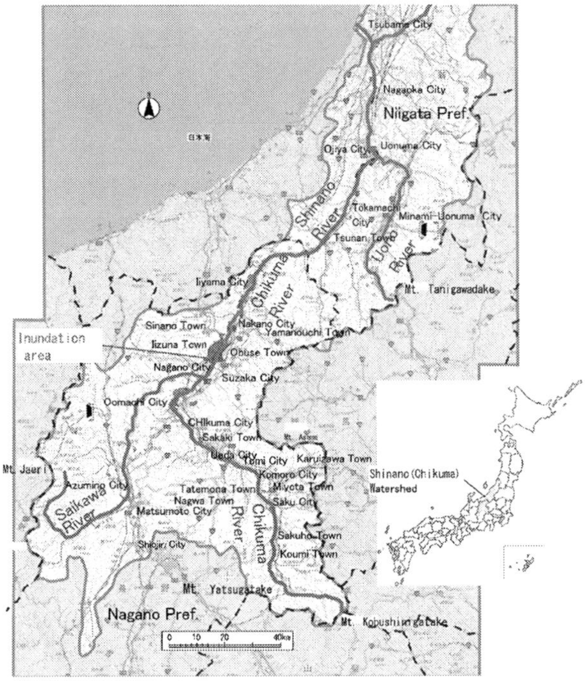

Figure 2. Map of Chikuma and Shinano River basin based on the map from Ministry of Land, Infrastructure, Transport and Tourism Japan.

on evacuation of residents and road closures. Conduct monitoring of the damaged area. Planning of emergency countermeasures, which must be done urgently and quickly so as not to delay the emergency countermeasure work. In recent years, survey use new technologies such as UAVs. In case of a large-scale disaster, aerial surveying is conducted to confirm safety during lifesaving operations.

(2) Emergency response work

"General construction companies (GCC), specialized construction companies (SCC)" perform work that requires emergency response for safety after a disaster. This includes temporary restoration of levees and road reopening. The work is usually started immediately after the disaster, and usually takes place day and night. Safety checks are conducted during emergency work and after temporary restoration.

(3) Technical review committee

"CC, Geological survey companies (GSC)" will confirm the magnitude and damage mechanisms of floods and earthquakes prior to the main restoration, such as levee breaches and facility collapse due to flood, earthquakes, tsunami, and liquefaction. If necessary, Technical Review Committee (TRC) of academics will be formed to provide technical decision and support. CC and geological survey companies will conduct surveys and analyses in response to the T RC, which will be reflected in the restoration design.

(4) Preparation for the application of gov. budget

11CC, GSC, SCC" will conduct survey and design in a short time for disaster assessment and calculate construction costs. In the design process, they also plan the level of improvement and conduct necessary geological surveys. In the case of a large-scale disaster, support engineers are assembled from nationwide to deal with many issues in a short time.

(5) Design implementation and supervision

"CC, GSC, SC" conduct the design implementation for the main restoration. In the case of a large-scale disaster, a large amount of design work needs to be done in a short time, as deadlines are set for special emergency projects for severe disasters.

"CC GCC" provide Project Management (PM) and CM when there is a shortage of engineers from local governments or others.

(6) Restoration work and construction supervision

"GCC, SCC" will conduct the main restoration work. The deadline is set for special emergency projects for severe disasters. The work must be carried out in a short time.

"GCC, CC", if any problems arise during the construction work, will consult with the parties concerned and make changes to the design as appropriate. They provide PM and CM when there is a shortage of engineers from local governments or others (JCCA, 2019; MLIT, 2022).

4 Challenges and recommendations

4.1 Decrease in the number of staff in the authority and to recommend CM method

In recent years, procurers have been required to ensure quality, reduce costs, and provide highly transparent contracts for public works projects, and their responsibilities have been increasing. In response to this, it has become necessary to make use of private sector which can appropriately perform procurement-related work, under the situation of significant decrease of the number of procurer employees in local public entities (see Fig. 4). Depending on the scope of support for procurement-related work, the "CM method" is the one in which part or all procurement-related work related to construction supervision is outsourced to the private sector, the "project promotion PPP (Public Private Partnership) method" is the another in which part of the procurement-related work is outsourced to the private sector from the investigation and design stages (MLIT, 2002).

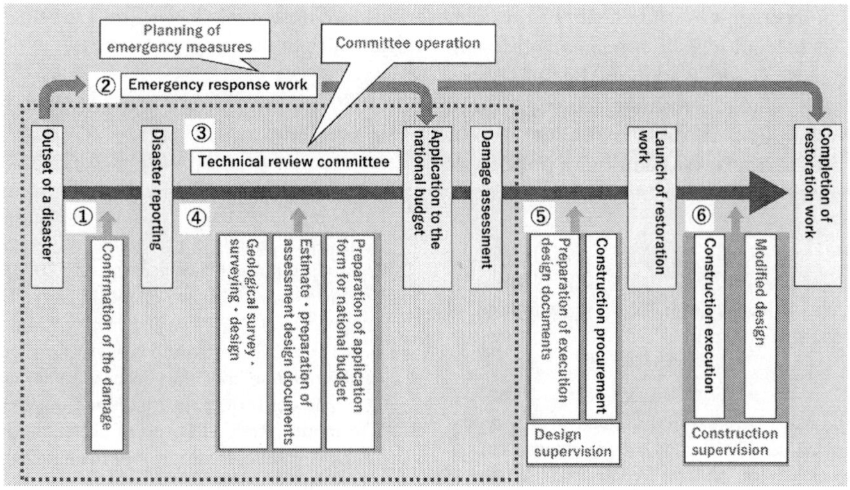

Figure 3. Role of construction sector in disaster recovery.

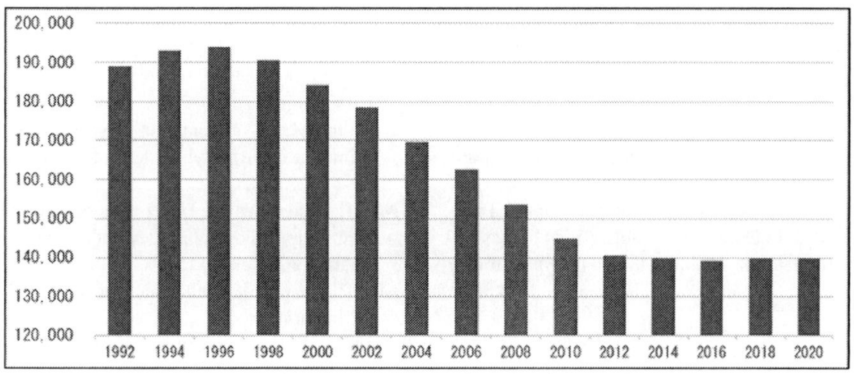

Figure 4. Number of staff at civil engineering department of local governments in Japan (Source: the Ministry of Internal Affairs and Communications' Survey of Capacity Management of Local Governments, Vertical axis is number of staff; horizontal axis is fiscal year).

The major challenge for the affected municipalities, which had no experience in procuring large-scale public works projects, was to establish a project implementation system necessary to start the project early and implement it in a short time. The affected municipalities placed the highest priority on the early start and completion of the projects. At the same time, it must be ensured transparency and fairness in the use of government funds and be fulfilled accountability to the public. Number of staff in the civil engineering department, which is responsible for public works projects in local governments, has decreased by approximately 55 000 (−28.4 %) since its peak in FY1996. For the Great East Japan Earthquake, Urban Renaissance Agency (UR) introduced CM method in the "Reconstruction and Urban Area Development Project" in 19 areas in 12 cities and towns, utilizing private-sector know-how to promote the project smoothly (MLIT, 2017). For the river improvement project on the Ouse and Natsui Rivers in Fukushima Prefecture, CM method was applied in a two-year lump-sum contract from April 2019 to March 2021 to expand the river area and reduce flood damage to houses along the rivers. CMR (construction manager) was involved in the entire process from basic planning, basic design, implementation design, site acquisition, construction procurement, and construction. As a result, procure system was secured, technical skills of young inhouse engineers were developed, the flood control safety level was improved, and coordination with many related organizations, including the relocation of obstructions and bridge attachments, was smoothly carried out.

5 Contribution to disaster prevention

CC work was like an essential job for science and technology. The disaster recovery and infrastructure could not be done without CC. CC have been increasing their competence by conducting substantial surveys and planning on a contract basis. Inhouse engineers in government are to focus on regulation (Kamoto, 2012) and logistic arrangement, losing the engineering skills these years. In addition, the number of those are decreasing. CC should substitute/cooperate more for public sector's work to lead the disaster-resilient nation. CC also themselves need to improve its ability for policy recommendations and techniques for proactive disaster management.

Data availability. Figure 4 data is from https://www.mlit.go.jp/common/001231371.pdf (MLIT, 2024).

Author contributions. MK designed and conceived this study. MK, AN and MO collected data. MK analyzed and interpreted the results and drafted the manuscript. AN and MO drafted the figures. All authors reviewed the results and approved the final version of the manuscript.

Competing interests. The contact author has declared that none of the authors has any competing interests.

Disclaimer. Publisher's note: Copernicus Publications remains neutral with regard to jurisdictional claims made in the text, published maps, institutional affiliations, or any other geographical representation in this paper. While Copernicus Publications makes every effort to include appropriate place names, the final responsibility lies with the authors.

Special issue statement. This article is part of the special issue "ICFM9 – River Basin Disaster Resilience and Sustainability by All". It is a result of The 9th International Conference on Flood Management, Tsukuba, Japan, 18–22 February 2023.

Acknowledgements. This paper is based on the one prepared for presentation at ICFM9. We would like to thank all the ICHARM members who were involved in organizing ICFM9. Shinji Egashira of ICHARM/GRIPS gave valuable advice. The flood damage situation at the Oda and Shinano Rivers and subsequent countermeasures were given by Kunihiro Yamada (Advisor to the Ministry of Land, Infrastructure, Transport and Tourism, Japan, at that time). Asao Yu and Hiroshi Matsuda, who compiled the disaster activity reports for 2018, 2019, Koichi Kuroyanagi advised PPP issues, at the Japan Civil Engineering Consultants Association. We would like to thank all those who supported us. The authors are solely responsible for the content of this paper.

Review statement. This paper was edited by Miho Ohara and reviewed by two anonymous referees.

References

Das, S., Ishiwatari, M., Kanbara, S., Komino, T., and Shaw, R.: Lessons from Mabi: Five Years of Recovery from the 2018 Western Japan Flood, https://www.cwsjapan.org/2023/07/06/report_lessonsfrommabi (last access: July 2023), 2023.

Fire and Disaster Management Agency: Damage Caused by Typhoon No. 19 in 2019 and Heavy Rainfall Caused by a Frontal Line (65th Report), https://www.fdma.go.jp/disaster/info/items/taihuu19gou67.pdf (last access: 2 February 2020), 2020.

JCCA: Record of Construction Consultants' Activities in the Torrential Rain Disaster of 2018, https://www.jcca.or.jp/files/achievement/committee_report/saigai2019.pdf (last access: May 2019), 2019.

JCCA: Record of Construction Consultants' Activities in the Typhoon No. 19 Disaster in 2019, https://www.jcca.or.jp/files/achievement/committee_report/saigai18.pdf (last access: July 2020), 2020.

JCCA: White Paper P34, 131–136, https://ofiles.jcca.or.jp/files/achievement/annual_report/pdf/2022/wp2022_all.pdf (last access: August 2022), 2022.

Kamoto M.: Systematized River Management by River Administrator, http://whrm-kamoto.com/top.html (last access: August 2022), 2012.

MLIT: Guidelines for Utilization of the CM Method, https://www.mlit.go.jp/totikensangyo/const/1_6_bt_000215.html (last access: February 2022), 2002.

MLIT: Report of the Study Group for Verification and Future Utilization of the CM Method for Reconstruction of East Japan, https://www.mlit.go.jp/common/001181972.pdf (last access: March 2017), 2017.

MLIT: Guidelines for the Smooth Implementation of Disaster Recovery Projects in Municipalities, https://www.mlit.go.jp/river/shinngikai_blog/saigai_fukkyu/pdf/gaidrain.pdf (last access: 30 October 2022), 2022.

MLIT: Promotion of Work style reforms and Securing and developing mid- to long-term staff, https://www.mlit.go.jp/common/001231371.pdf (last access: 19 July 2024), 2024.

Okayama prefecture: July 2018 heavy rain disaster record book, https://www.pref.okayama.jp/page/653529.html (last access: March 2020), 2020.

Proc. IAHS, 386, 327–332, 2024
https://doi.org/10.5194/piahs-386-327-2024

4 "R" Approach of Catastrophic Disaster of Melamchi Flood: A Sustainable Urban Planning

Tapendra Bahadur Khadka[1,a,b]

[1]freelance civil engineer/urban planner, Kathmandu, 44600, Nepal
[a]formerly at: City Planning Commission, Kathmandu Metropolitan City, Kathmandu, 44600, Nepal
[b]formerly at: Ministry of Urban Development, Government of Nepal, Kathmandu, 44600, Nepal

Correspondence: Tapendra Bahadur Khadka (tapendrabk@yahoo.com)

Received: 30 June 2023 – Revised: 18 March 2024 – Accepted: 8 June 2024 – Published: 26 July 2024

Abstract. This 4 "R" (i.e., Reclaim, Readjustment of Land, Rehabilitation and Reconstruction) concept is an appropriate solution for a resilient urban environment with sustainable development after the disaster events. On the late night of 16 June 2021, Melanchi Bazar of Nepal a flash flood of outburst of landslide dam created unexpected disaster destroying the lives and properties with the loss of 291 households and displaced more than 600 peoples, cultivated land, and major infrastructures. Land reclaim is necessary in the flood plain area with the "Built Back Better" concept. Land readjustment with new cadastral survey should be conducted to update the existing record of the privatively owned land of the citizens. Rehabilitations of urban facilities, with the function of flood plain area providing the storage to reduce stream velocities during high floods. Revitalization and recuperative treatment of the urban settlements with an enhancement of wetlands, creeks, and river corridors.

1 Background

Melamchi is one of the major emerging urban centres in the area and also the headquarter of the Melamchi Municipality, Sindhupalchok District of Bagmati Province of Nepal. It lies downstream of the confluence of the Melamchi River and the Indrawati River. The origins of these two rivers are from the Himalayan range and sources of water are snowmelt and rainfall. It is located at the 27°49′48″ latitude and 85°49′48.575″ longitude at an elevation of 940 m from the mean seal level.

Melamchi, the urban centre was the destination for the last two decades for domestic as well as international visitors due to the mega national pride infrastructure project Melamchi Water Supply Project. It is also a gateway to Helambu. Melamchi is not only an urban centre, it is developing as a growth centre of surrounding market centres providing the spillover effect to the hinterland promoting the rural-urban linkages. It also helps the adjoining rural municipalities for economic activities providing the market for their cash crop and other agriculture products. Melamchi Municipality also focuses on developing the urban infrastructure to feel their citizens they are urban dwellers.

The late night of 16 June 2021, a flash flood and the outburst of landslide dam created unexpected disaster with the debris flow destroying the lives and properties of Melamchi inhabitants and destroyed number of houses, cultivated land, and major infrastructure along the road alignment. The study reported the loss of 291 households and displaced more than 600 people.

A devastating earthquake with a moment magnitude (M_w) 7.8 hit the central Nepal on 25 April 2015. The epicenter of the earthquake located near Baarpak, Gorkha (28.1470° N, 84.7080° E) 77 km northwest of Kathmandu. Another big aftershock of M_w 7.3 occurred in the northeast of Kathmandu on 12 May 2015 (Collins and Jibson, 2015) with an epicenter in Kodari Sindhupalchowk. It was also followed by more than 424 aftershocks having local magnitude of 4 which caused additional damages.

It is estimated that the lives of eight million people, almost one-third of the population of Nepal, have been impacted by the earthquakes pushing back more than one million people below the poverty line. The adverse effects of earthquake in the geo-spatial location had tremendously affected the surrounding geology of the settlements in the Gorkha and Sindhupalchowk districts were observed and surveyed by the geologist.

Nepal is among the 20th most disaster-prone countries in the world, both natural and man induced. It ranks 4th, 11th and 30th in terms of climate change (CC), earthquake and flood risk respectively (MoHA, 2015). It is the 7th most vulnerable nation in the world for deaths related to floods, landslides and avalanches combined; 8th for flood-related deaths alone; and 23rd in terms of total natural hazard related deaths (MoHA, 2009).

Whereas nearly 25 numbers of multi-hazards are enlisted within the country, Due to the diverse topography, complex geological strata and drastic change in climatic condition due varying altitude of 80 to 8848.85 m the top of World, Nepal is exposed to the many types of man-made and natural disasters. More than 80 % of the population is exposed to the risk of natural hazards (MoHA, 2018). There are many factors that have influenced landslides and flood hazards including high intensity rainfall and changes in land use pattern without considering the objective of the sustainable development.

Landslides are one of the most prevalent multi-hazards frequently occurred in Nepal because of the active seismicity, intensive monsoon rains, and hastened erosion due to haphazard deforestation and expedite in construction. High chances of slope failures and rock falls occurred at shallow depth. Various ground fissures were found in devastated area.

2 Rationale and Uniqueness of 4 "R" Approach

This topic is unique because it is not only directly related with flood disaster mitigation. The major part is reclaimed of destroyed urban land (flood plain). Land readjustment planning also help to relocate the plots of individual land owner. Who lost their land boundary after disaster. It's uniqueness due to Reclaim, Readjustment, Rehabilitation and Reconstruction. When people talk/argue/discuss about flood management they only think about flood modelling, rainfall intensity, high flood levels, climate change, flood affected areas, inundation area, river bank erosion, raise of river beds due to land slide in upstream sides and river training work on the flood plain area. They never think from the side of urban planner and urban planning concept.

It outlines about the disaster risk reduction along the river basin area to develop the plan and policies with critical thinking for resilient community development.

It covers all the sectors of Disaster Risk Reduction including the reclaim, readjustment of land, rehabilitation, and reconstruction areas that have not been previously considered together in flood management sector.

It can be applied to all the disaster-prone area before and after the disaster events to save the life and property of people during disaster and also rehabilitate the existing livelihood too. It also helps to the local community leaders to advocate for specific changes in policies.

3 Government Policies and Program

After 2015 Gorkha earthquake, National Reconstruction Authority (NRA) carried out a geological investigation subdividing of 662 settlements in 3 different categories. There were 136 settlements identified in Category III, 175 settlements in Category II and 351 settlements in Category I respectively. Detail studies of 136 vulnerable settlements have been carried out at various stages.

As per the recommendation of geologists, the government has been initiated to relocate at least 14 vulnerable community settlements of Sindhupalchowk district to the safer places. Few settlements were safer to live in and few settlements were identified to be relocated.

After the disaster Decisions of the Seventh Meeting of National Council for Disaster Risk Reduction and Management held on Tuesday 26 July 2078. With the decision to instruct the National Disaster Risk Reduction and Management Authority (NDRRMA) to conduct a detailed study of the debris flood of the Melamchi River and the risks caused by it and the measures to solve them, including the study of all kinds of past experiences.

After the council decision the NDRRMA quickly responded to the disaster, with a series of activities including a drone survey to identify all damaged houses, map out inundation areas, and record the changes in topography caused by massive erosion and deposition (Takamatsu et al., 2021). Flood modelling has been carried for evaluating future flood risks and policy formulation regarding to fix the setback distances for flood plain.

Considering the risks, the vulnerability and the hazards, there is an urgent need to develop a risk sensitive land use planning (RSLUP) that focuses on reducing the probable disaster risks and build resilient communities. According to the National Land Use Policy, there is a provision to promote the development of safe settlements and its appropriate management by formulating a risk sensitive land use plan based on multi-hazard risk assessment of the municipal area.

Similarly, in the Local Government Establishment Act 2074, there is a provision that municipalities should formulate and implement local level land use policies, plans and programs in accordance with the duties, roles and responsibilities assigned to municipalities under the jurisdiction of federal and provincial laws. In order to achieve the main goal of creating disaster-resistant communities through the land-use planning system, a fundamental change in planning concepts and practices seems necessary.

Figure 1. Melamchi Bazar before the Flash Flood. Story Map Source: Created by Nepal Flying Labs using the Arc GIS Story Mapping tool by ESRI. A group of organizations (Nepal Flying Labs, Nepal Engineers' Association, Institute of Himalayan Risk Reduction, Geovation Nepal and NAXA).

Figure 2. Melamchi Bazar after the Flash Flood. Story Map Source: Created by Nepal Flying Labs using the Arc GIS Story Mapping tool by ESRI. A group of organizations (Nepal Flying Labs, Nepal Engineers' Association, Institute of Himalayan Risk Reduction, Geovation Nepal and NAXA).

4 "R" Approach for Sustainable Urban Planning

4.1 Reclaim of Land

On 16 June 2021 a devastating catastrophic disaster with heavy rainfall and landslide had been occurred due to slope failure in the upstream side of the Indrawati River and the total flow was overstated for the time as shown in the Figs. 1 and 2.

After the outburst of the overstate flow of the river, the downstream settlements had been washed out with an unexpected disaster destroying the life and properties. It has a major impact on the livelihood, social-economic and lifeline of the people. The loss on the economy is immeasurable and impacts on daily life activities is also unpredictable. Approximately 985 359 m^2 of land including the river flood plain area was covered with the debris (from drone survey data).

4.2 Present Land Condition

Now, it is very important to carry out the detailed geological study of the location where the river changes its original course and flow in the new path. Huge quantity of debris also spread over the entire land deleting the past demarcation of the land boundary as shown in the Figs. 3 and 4. below. There are also major impacts on the land administration system of the Melamchi area changing the existing land use pattern with new challenges; two main types of land (i) river left barren land and (ii) registered land under the new river course flow. Three levels of government (i.e., Federal, Provincial and Local) should take the responsibilities to reclaim the land with the detail hydrological and geological studies of the location. Land reclaim project shall be created for necessary river protection and other infrastructure work along the bank of the river through the "Built Back Better" concept.

5 Readjustment of Land

Land readjustment planning is very important at this juncture of time because existing land demarcation is missing due to spatial change in the existing land use pattern after a devastating flash flood. The existing scenario of the area with the occupancy also drastically changed. At this time people of the affected area and local government should cooperate and support to cope with these problems. It can be solved through land readjustment planning, where marginal land can be managed for public utility space and optimum use of private land. The local government has to take the

Figure 3. Melamchi Bazar affected by the Flash Flood and flood plains from the drone survey (20 June 2021). Photo Source: Geovation, Nepal.

Figure 4. Disaster encountered by Melamchi Bazar due to Flash Flood Photo Source: Kathmandu Post, Nepal.

municipal council decision for Urban development planning concerning the participation of landowners and tenants. After approval of the Urban planning project from the concerned local body, shall publish a detailed notice in the national daily newspapers for the information of the general public.

A new cadastral survey should be conducted to prepare the updated land administration record of the privatively owned land of the citizens of the Melamchi area. According to the existing land use pattern and the land registration certificates, all information should be updated and the land acquisition process should be applied immediately. During this land acquisition time, people should be aware of the process and compensation will be in terms of the developed plot after deducting all the contribution percentage of land. Before instigating all these processes, the local government should conduct an orientation meeting with the land and property owners about the land readjustment (Land Pooling) system. The project's major objective shall be "no one should be left behind" with this benefit and "no one should be displaced" from the area. It also helps to consolidate the fragmented land into a single unit and it will easy to allocate such land in a particular location with the optimum use.

5.1 Provision of Land Development in Town Development Act 2045

The provision of land development in Town Development Act 1988 AD, the committee may operate the land development programs to manage existing land with the basic need of residents in a planned manner and with other township activities. To conduct the land readjustment project at least fifty-one percent of landowners or tenants of the project area should agree, which is mandatory and necessary. During the course of conducting land pooling program, the contribution from the land owner or tenants shall be based on the current monetary value of that land and existing road and other provided facilities in that area, to provide services and facilities such as road, drainage, open space, drinking water and electricity. To develop the sites the entire unit of lands shall be integrated into a single unit.

After fixing land contribution percent, the committee shall return developed plots to the former landowner or tenants along with provisional certificate within the period of mentioned timeline. After returning developed plots, the Committee shall write to the concerned Land Survey and Land Reform and Management Office for preparing the final survey map, maintaining the record of the landowner and for issuing land ownership registration certificate to the concerned landowner or tenants. After the preparation of the newly developed survey map, the record of the landowner, and the landowner registration certificate shall be updated, and the previous survey and entire previous inventory of the landowners and tenants of the concerned area shall be void.

6 Rehabilitation

Urban Flood Plain Rehabilitation Project shall be launched in such a highly urbanized and subject to frequent flooding area to develop flood plain. As per the present site condition of the land regarding the vulnerability it can be relocated residents and razed existing buildings. With the proper treatment as per the engineering consideration, the disaster-prone zone can be revitalized to restore the natural flow of the river. Besides other urban infrastructure facilities, the flood plain area can be integrated into parks, greenways along the river, and public recreation space. The flood plain area will be a function to provides storage with reducing stream velocities during large flood events. It also provides buffers zone area from flooding too. The river cross-section should be designed as per the existing hydrological dataset, providing three consecutive important factors considering the future use of land as follows.

a. Annual maximum floodway

b. Regulatory floodway (for 25 years of return period high flood level) and,

c. Standard Project Flood (SPF) limit (for 100 years of return period high flood level)

According to the above categories of the river section, standard zoning regulation such as Land Use Policy at municipal level shall be implemented on the use of flood plain areas. Regulatory floodway fringe use can be permitted for a selected use or otherwise protected it should be mentioned in the Building Byelaws. Standard Project Flood (SPF) limit area subject to possible flooding by very large floods with the 100 years of the return period. These types of areas can be restricted for the particularly important buildings of emergency uses, i.e., hospital, communication, security, etc. The studies show, there is a 1 % chance of occurrence for a 100 years flood event for any given year, and similarly, there is a 4 % chance of occurrence for a 25 years flood event.

7 Reconstruction

The flash flood disaster destroyed the entire urban settlements along with the potential agricultural land on either side of the river converting into a barren flood plain filled with lots of debris. After every disaster there will be a golden period of renaissance for the spatial planning. After every disaster occurred it is very important to rescue the lives and properties of the people, then have to focus on the relief to those victims. After the relief of recent disaster events then they start on the search for loosed properties and belongings to survive for the future. Revitalization and recuperative treatment of the urban settlements are very important to invest in the reconstruction of major urban infrastructure with a planned urban development technique. Urban Flood Plain Restoration Project with a combined engineering and landscape design approach the restoration and enhancement of wetlands, creeks, and river corridors can be implemented.

Huge investment requires in the construction of raised road embankment and river levee. All types of urban utilities with proper recreational and public open spaces with public park maintaining the eco-system with a pleasant built environment needs to be incorporated. Land readjustment planning will definitely fulfill the demand for housing as well as commercial lots with a pleasant urban environment and meet the "Built Back Better" concept. Concerned authorities of the government of Nepal should give the proper attention to cope with the prevailing practice to redevelop the flash flood disaster destroyed urban settlements along the side of the river downstream. In this reference, the 4 "R" concept is an appropriate solution for a resilient urban environment with sustainable development after disaster events.

8 Conclusion and recommendation

Nepal lies within the top twenty rank of multi-hazard countries as per the world ranking. The different types of hazards caused different types of disaster in different timelines. The disaster events can't be predicted but pre-information about meteorological transformation early alarm system may announce. The different time multi-hazard events have great impacts on the livelihood with a loss of lives and properties of the people. Disaster can't be controlled but if the mitigation measure will be applied it may reduce the chances and impacts of disaster.

Rural and urban municipalities settlements area are mostly seeming vulnerable due the flood and landslides disaster. Multi hazard risk assessment should be done for the municipalities for the investment in the safer location. Bank protection, erosion control, raised embankment with resilient physical urban infrastructures should be construct. After the flood disaster events 4 "R" approach i.e.; Reclaim of land, Readjustment of land, Rehabilitation and Reconstruction should be adopted for the sustainable development. It will definitely address the adaptation of disaster risk reduction and management for the victims. Therefore, the Government of Nepal should prepare the policy and program to minimize the risk. This types approach can be replicate in the other similar emerging market and urban centres to protect from the such disaster in the future. To formulate the national policy regarding the response, rescue, relief work including the rehabilitation and the reconstruction for the victims of the affected areas.

It is very important that each of the municipalities including Rural municipalities also have to prepare Risk Sensitive Land Use Plan (RSLUP) with the proper study of multi hazard risk assessment. Political leader must realize it, Federal and Provincial government should be enforced to all the local government to prepare and implement Risk Sensitive Land Use Plan.

Code and data availability. The code and data used in this article are available at https://storymaps.arcgis.com/stories/5df39824b5604c80be8dfc429be93cfa (Nepal Flying Labs, 2021).

Competing interests. The author has declared that there are no competing interests.

Disclaimer. Publisher's note: Copernicus Publications remains neutral with regard to jurisdictional claims made in the text, published maps, institutional affiliations, or any other geographical representation in this paper. While Copernicus Publications makes every effort to include appropriate place names, the final responsibility lies with the authors.

Special issue statement. This article is part of the special issue "ICFM9 – River Basin Disaster Resilience and Sustainability by All". It is a result of The 9th International Conference on Flood Management, Tsukuba, Japan, 18–22 February 2023.

Acknowledgements. I would like to express my sincere gratitude to The 9th International Conference on Flood Management (ICFM9) on "River Basin Disaster Resilience and Sustainability by All – Integrated Flood Management in the Post COVID-19 Era" organized by International Centre for Water Hazard and Risk Management (ICHARM) under the Auspices of UNESCO, Public Works Research Institute (PWRI), Secretariate team encourage to submit article on the Catastrophic flood disaster of Melamchi Nepal on 16 June 2021, and allocating time for my presentation in the session 1 on 20 February 2023, to present my views on ICFM9.

I am also grateful to Editor-in chief Toshio Koike, Executive Director, ICHARM, and IAHS, Miho Ohara Guest editor, and anonymous reviewers for their valuable suggestion and creative inputs and funder IAHS for Post-conference publication in the PIAHS.

Financial support. This research has been supported by the International Association of Hydrological Sciences (PIAHS grant).

Review statement. This paper was edited by Miho Ohara and reviewed by two anonymous referees.

References

Collins, B. D. and Jibson, R. W.: Assessment of existing and potential landslide hazards resulting from the April 25, 2015 Gorkha, Nepal earthquake sequence (ver. 1.1, August 2015): U.S. Geological Survey Open-File Report 2015–1142, 50 pp., https://doi.org/10.3133/ofr20151142, 2015.

Ministry of Home Affairs (MoHA): Nepal Disaster Report, The Hazardscape and Vulnerability, Kathmandu Nepal, 2009.

Ministry of Home Affairs (MoHA): Nepal Disaster Report, Kathmandu Nepal, 2015.

Ministry of Home Affairs (MoHA): Nepal Disaster Report, 2017: The Road to Sendai, Kathmandu Government of Nepal, ISBN 978-9937-0-4217-8, 2018.

Nepal Flying Labs: Drone Based Rapid Assessment of Melamchi Flood 2021, https://storymaps.arcgis.com/stories/5df39824b5604c80be8dfc429be93cfa (last access: 14 July 2024), 2021.

Takamatsu, M., Karelia, H. D., Oommen, T., and Dahal, R. K.: Melamchi Flood Disaster in Nepal: Damage and Risk Quantification with Drone Survey, Satellite-Based Land Displacement Analysis, and 2D Flood Modeling, World Bank Group, Washington, D.C., http://documents.worldbank.org/curated/en/099600006212237293/P16057809e269a0cf096ab044bc77400259 (last access: 14 July 2024), 2021.

Proc. IAHS, 386, 333–338, 2024
https://doi.org/10.5194/piahs-386-333-2024

Open Access PIAHS

Regional Characteristics of shape and stagnation of rainfall in Japan using Radar/Raingauge Analyzed Precipitation product

Yuta Ohya[1,2] **and Tomohito J. Yamada**[3]

[1]Graduate school of Engineering, Hokkaido University, Sapporo, 060-8628, Japan
[2]Hokkaido Research Organization, Sapporo, 060-0819, Japan
[3]Faculty of Engineering, Hokkaido University, Sapporo, 060-8628, Japan

Correspondence: Yuta Ohya (y_ohya@eis.hokudai.ac.jp)

Received: 1 August 2023 – Revised: 20 September 2023 – Accepted: 24 February 2024 – Published: 13 August 2024

Abstract. Recently, floods and sediment disasters caused by line-shaped rainbands (LRBs) have increased in frequency following events such as the torrential rainfall around Tokyo of September 2015, which flooded the Kinu River, Japan. During this event, the shape of the rainfall area matched the topology of the river basin, causing water levels to rise rapidly and resulting in a serious disaster. Therefore, the shape and duration of heavy rainfall patterns are critical factors that must be considered for flood prevention. In this study, we examined features of LRBs in river basins across Japan using long-term radar raingauge data, and investigated relationships between rainfall intensity, shape, duration, and quantity.

1 Introduction

Japan is located in the East Asian monsoon region, where rainfall is concentrated in summer/autumn season due to typhoons and the Baiu. Except for meso-α-scale or larger rainfall events associated with these seasonal patterns, most heavy rainfall in Japan is caused by mesoscale convective systems (MCSs), 40 %–60 % of which are long and narrow (Tsuguchi and Kato, 2014). Line-shaped rainbands (LRBs) are composed of multiple linearly organized cumulonimbus clouds that produce stagnant rainfall over the same area for several hours. However, the size of these rainfall areas varies widely, ranging from several hundred to several thousand square kilometers, which is on the scale of river basins in Japan. Many of LRBs have influenced the flow of first-order rivers as well as smaller rivers in Japan, such as torrential flooding of the Kinu River near Tokyo in September 2015 and the Kuma River in Kyushu in July 2020. These events caused serious damage due to rainfall stagnation over river basins.

The physical mechanisms of LRBs have been investigated using numerical models and Doppler radar observations (e.g., Kato, 1998, 2006, 2020; Yoshizaki et al., 2000). Statistical analysis of radar rainfall data was first conducted by Bluestein and Jain (1985), who analyzed an 11-year dataset in the USA and classified 52 linear MCSs into five classes (broken line, back-building, broken area, embedded area, and other). Later, Houze et al. (1990) and Parker and Johnson (2000) reclassified linear MCSs into two or three classes, based on a combination of convective and stratiform clouds. Schumacher and Johnson (2005) analyzed 76 MCSs that occurred during a 3-year period for the three classes proposed by Parker and Johnson (2000), plus two frequently occurring, unclassified MCS types. Based on these classification systems, regional and seasonal trends of other MCSs were analyzed by Stevenson and Schumacher (2014) and Haberlie and Ashley (2019). Feng et al. (2016) used a 35-year observational dataset to show an increasing trend in the number of MCSs in recent years. Furthermore, Fairman et al. (2016) and Ashley et al. (2019) investigated features relating to LRBs in the United States. Prein et al. (2017) applied a climate projection model to show that the frequency and amount of rainfall will increase in a future climate.

Shimura et al. (2000) classified 37 heavy rainfall events in the Japanese Kanto region over a 4-year period into frontal organization and multiple- and single-cell movement classes. Seko (2010) categorized back-building, squall line, and back-and side-building MCS rainfall in the USA using previous case analyses and classification systems. Yamada et al. (2012) examined the spatiotemporal distribution and environmental characteristics of LRBs in northern Japan based on 21 years of observations. Tsuguchi and Kato (2014) showed that among 286 heavy rainfall events throughout Japan during a 15-year observation period, 64.8 % involved LRBs. Unuma and Takemi (2016) investigated the rainfall shape, duration, amount, and environmental characteristics of 4133 events in Japan using an 8-year observation dataset.

In this study, we investigated the relationship between rainfall duration and shape associated with LRBs in Japan using long-term, nationwide observational datasets. LRBs were extracted using an algorithm proposed by Hirockawa et al. (2020) and revised with some thresholds by Ohya and Yamada (2023; OY22 hereafter).

2 Methodology

2.1 Datasets

In this study, we used the Radar/Raingauge Analyzed Precipitation product (RA; Nagata, 2011), which is a composite of C-band radar rainfall data corrected using ground-based rain gauge data managed by the Japan Meteorological Agency. The spatial resolution was 5 km when data delivery began in 1988, but was increased to 2.5 km in 2001, and to 1 km in 2006. These changes are improvements due to the expanded radar and ground rain gauges used. In this study, to analyze long-term trends, we standardized the resolution through bilinear interpolation to 5 km, and data from 1988 to 2019 were used.

2.2 LRBs definition

A flowchart of the LRBs extraction process is shown in Fig. 1. This method was based on that of Hirockawa et al. (2020), using a threshold proposed by OY22. First, closed curves that satisfy 40 mm per 3 h before the rainfall event were extracted, and an area of 500–15 000 km^2 was selected as the target area of intense rainfall as a meso-β-scale criterion. If the area overlap threshold was ≥ 40 % of the rainfall area in the previous 3-h rainfall period, both were considered to be the same event, indicating stagnation. For each 3 h rainfall area, the aspect ratio of the major axis to the minor axis was calculated as an indicator of the shape of the area. In this study, we adopted the OY22 LRBs definition to apply statistical analyses to many regions. Thus, LRBs were defined as having an aspect ratio of ≥ 2.5, a stagnation time of ≥ 4 h, and a maximum grid rainfall of 35 mm h^{-1} in all heavy rainfall events. In this study, heavy rainfall events are considered to be those that satisfy the rainfall and area thresholds of the algorithm; LRBs are assumed to satisfy the aspect ratio and duration time criteria for such events.

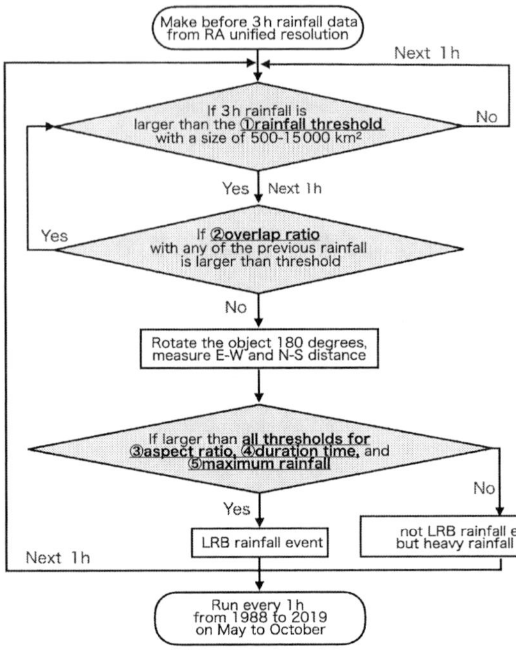

Figure 1. Flowchart of the line-shaped rainband (LRBs) extraction process used in this study.

3 Results

Through LRBs extraction, we obtained 15 119 heavy rainfall events, among which 5240 were LRBs (Fig. 2). The locations where LRBs occurred during May–October from 1988 to 2019 are plotted in Fig. 2a, along with average rainfall and LRBs orientation according to longitude and latitude. LRBs also occurred in northern parts of Japan such as the Tohoku and Hokkaido regions, although they were relatively common in Kyushu and western Japan (Fig. 2a). The frequency distributions of these rainfall events according to longitude and latitude (Fig. 2b, e) indicate that LRBs frequency was higher in the west and south, and lower in the east and north. The average rainfall amount was higher for LRBs in the south than for those in the north, but not higher for LRBs in western Japan than for those in the east (Fig. 2c, f). In the northwest Kyushu region, the southwestern part of Japan, more than half of all rainfall areas appeared in June and July, indicating that LRBs oriented in the east–west and southwest-northwest directions were significantly more frequent in this region. Many of the LRBs extracted in this study appear to have been generated during the Baiu season, although heavy rainfall events extracted in the Chubu, southern Kii, and Kanto regions (see Fig. 4 for the detail locations) mainly occurred from August to September, corresponding to the typhoon season; many of these events were directly caused by typhoons and topography. Around 140° E, particularly near 37–38° N, larger proportions of LRBs were oriented in the

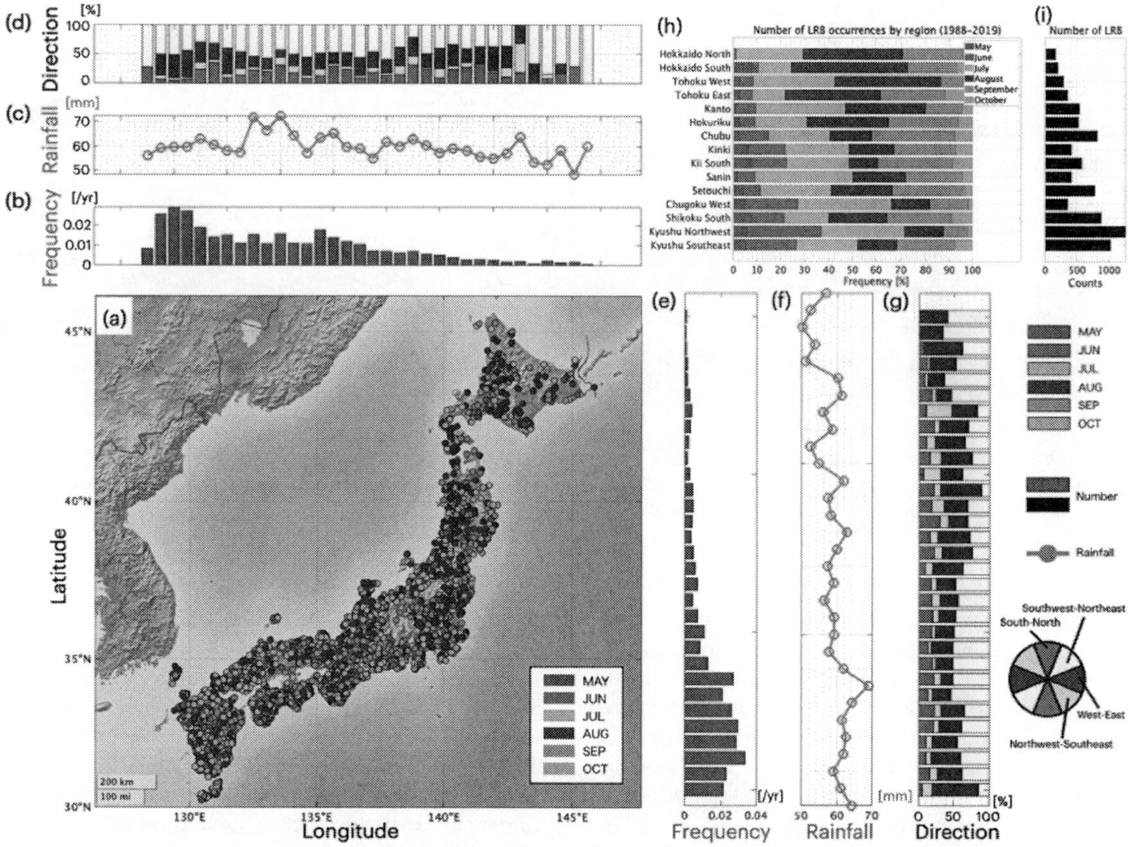

Figure 2. Distribution of LRBs during May–October period from 1988 to 2019 according to **(b)** latitude and **(e)** longitude, normalized by land area. Average rainfall according to **(c)** latitude and **(f)** longitude. Rainband orientation according to **(d)** latitude and **(g)** longitude. **(h)** Frequency of LRBs occurrence by month and total LRBs numbers for each region.

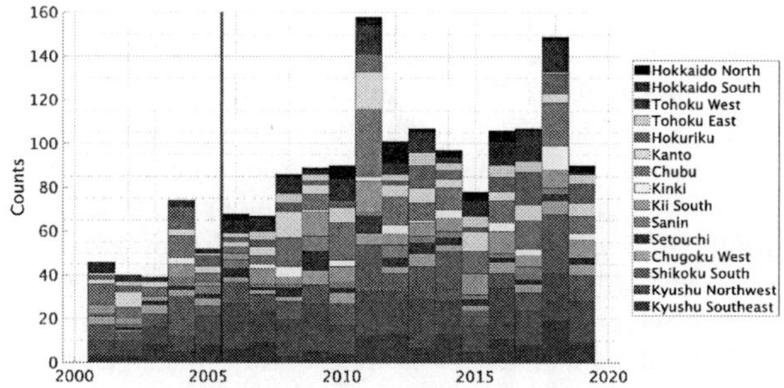

Figure 3. Interannual variation in LRBs occurrence by region. Colors indicate the regions where the center coordinates are located.

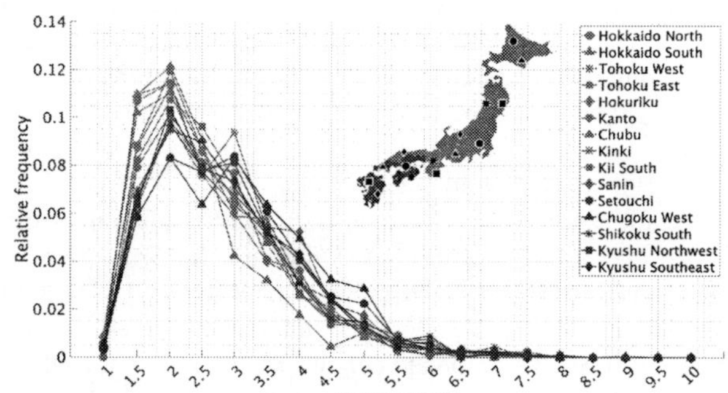

Figure 4. Frequency of LRBs aspect ratios by region. Colors indicate the regions where the center coordinates are located.

northwest-southeast direction (green color, Fig. 2d, g), which is typical of LRBs generated from the Sea of Japan toward the Hokuriku and Tohoku regions.

The numbers of LRBs that occurred in each region during May–October from 1988 to 2019 are plotted in Fig. 3. The annual mean was 101, with a standard deviation of 20. Years with particularly high LRBs occurrence included 1993, 1998, 2011, and 2018, which correspond to years with many heavy rainfall events nationwide due to typhoons and Baiu fronts as large-scale disturbances. Although few events occurred from

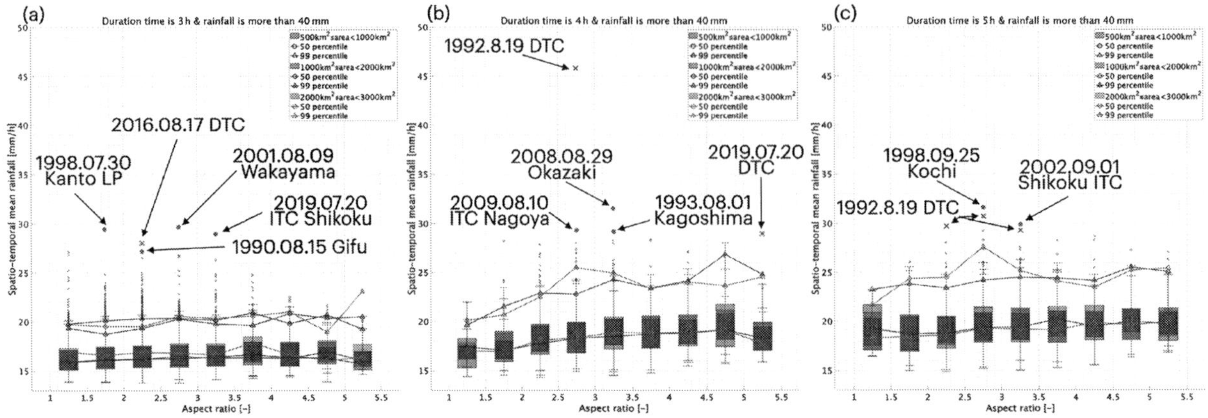

Figure 5. Box and whisker plots of the relationship between rainfall intensity and aspect ratio. Lines represent means, box borders represent 25th and 75th percentiles, whiskers represent 1.5× the interquartile range, and dots and crosses represent outliers. The dates and regions of the top five rainfall events are indicated for rainfall durations of **(a)** 3 h, **(b)** 4 h, and **(c)** 5 h. Precipitation directly caused by tropical cyclones are indicated by dashes.

2001 to 2007, the spatial resolution of RA changed in 2001. Although this study interpolated the data to standardize the resolution, the analysis results may have been affected by the resolution change. Similarly, the spatial resolution changed in 2006, such that simple comparisons based on the same resolution are possible only after 2006. The Mann–Kendall test rejected our results for the Kinki and southern Chugoku regions ($P < 0.05$), indicating an increasing trend.

The frequency distributions of heavy rainfall aspect ratios for each region are shown in Fig. 4. The aspect ratio indicates the ratio of the major axis to the minor axis of the rainfall area. A larger aspect ratio represents a more elongated shape, while a smaller aspect ratio represents a more blocky shape. Empirically, an aspect ratio of 2.5 or higher is often associated with LRBs. In all regions, aspect ratios of 2 to 2.5 showed the highest frequency, and significantly fewer events had larger aspect ratios, indicating a high degree of linearity. Large aspect ratios were more frequent in southern regions (blue, Fig. 4) than in all other regions.

Relationships between aspect ratio and rainfall intensity according to date, location, and duration for 7424 rainfall events that occurred from May to October during 1988–2019 are shown in Fig. 5. For average rainfall events, there was no significant relationship between aspect ratio and rainfall intensity. However, many extreme rainfall events had aspect ratios of 1–3.5. For example, some strong, short-duration heavy rainfall events showed a linear relationship with aspect ratio, such as torrential rains of 101 $\mathrm{mm\,h^{-1}}$ recorded at Kurisugawa site, in Kii South region, in August 2001, and the 92 $\mathrm{mm\,h^{-1}}$ rainfall event recorded in Yokohama site, near Tokyo, in July 1998, which was the largest rainfall event to have occurred since the station was established. More than 60 % of all rainfall area corresponded to localized torrential rains of 1000 $\mathrm{km^2}$ or smaller. Among 4 h rainfall events, both the median and 95th percentile values indicate that rainfall intensity increased with the aspect ratio (Fig. 5b), and extreme rainfalls, including the 8.1 disaster in Kyushu, in 1993, and the Okazaki city torrential rainfall in 2008, in Chubu, showed a linear relationship for aspect ratios of 2–4. Among 5 h rainfall events, no significant relationship with aspect ratio was detected, although the 95th percentile value tended to increase with the aspect ratio at aspect ratios close to 2.5 (Fig. 5c). Similar trends were observed nationally and regionally throughout Japan.

4 Summary

In this study, we examined the nationwide distribution of LRBs using long-term rainfall data for May to October during 1988–2019, as well as relationships between rainfall shape, duration, amount, and intensity. We extracted 15,119 meso-β-scale heavy rainfall events from datasets covering all of Japan during the study period. Among these, approximately 40 % of rainfall events had a duration of ≥ 4 h, and 5240 were LRBs. Both the number of LRBs and rainfall amount were significantly more frequent in the south than in the north, and almost 80 % were in Kyushu and along the coast of the Pacific Ocean (i.e., southern Shikoku, southern Kii, central Japan, and Kanto). The most frequent aspect ratio range was 2.0–2.5, which indicates shape linearity, and this distribution pattern was similar over broad regions of Japan. Significantly more rainbands were oriented in the east-west direction in Kyushu, where rainfall is concentrated during the Baiu season, and in the southwest-northeast direction in other areas. In southern Shikoku and southern Kii, the algorithm implemented in this study extracted heavy rainfall events that were directly caused by typhoons. Among non-stagnant heavy rainfall events, there was no significant relationship between rainfall amount and aspect ratio for many events. Extreme rainfall events were concentrated within an aspect ratio of 2.0–3.5. These results suggest that linear shape may contribute to rainfall intensity, regardless of rainfall duration and area. A future study should investigate the physical properties influencing the precipitation efficiency of LRBs.

Code availability. The software code used in this study is not publicly accessible. The reasons for not making the code available include: The software code contains proprietary algorithms and methods that are part of ongoing research and development projects. Public disclosure at this stage could compromise competitive advantage and intellectual property rights.

Data availability. The primary data utilized in our research are available for purchase from the Japan Meteorological Business Support Center (JMBSC). These data can be accessed through the following link: https://www.jma.go.jp/jma/jma-eng/jma-center/rsmc-hp-pub-eg/techrev/text13-2.pdf (Nagata, 2011). The methodologies employed in analyzing the data are detailed in Ohya and Yamada (2023).

Author contributions. YO: Conceptualization; data curation; formal analysis; investigation; methodology; project administration; software; validation; visualization; writing – original draft; writing-review & editing. TJY: Conceptualization; data curation; funding acquisition; methodology; project administration; resources; software; supervision; validation; writing – original draft; writing-review & editing.

Competing interests. The contact author has declared that neither of the authors has any competing interests.

Disclaimer. Publisher's note: Copernicus Publications remains neutral with regard to jurisdictional claims made in the text, published maps, institutional affiliations, or any other geographical representation in this paper. While Copernicus Publications makes every effort to include appropriate place names, the final responsibility lies with the authors.

Special issue statement. This article is part of the special issue "ICFM9 – River Basin Disaster Resilience and Sustainability by All". It is a result of The 9th International Conference on Flood Management, Tsukuba, Japan, 18–22 February 2023.

Acknowledgements. This study is supported by the Council for Science, Technology and Innovation (CSTI), Cross-ministerial Strategic Innovation Promotion Program (SIP), Development of a Resilient Smart Network System against Natural Disasters, with funding provided by the National Research Institute for Earth Science and Disaster Resilience (NIED), and The Advanced Studies of Climate Change Projection (SENTAN) (grant no. JPMXD0722680734) from the Ministry of Education, Culture, Sports, Science, and Technology (MEXT), Japan.

Financial support. This work was supported by MEXT – Program for the advanced studies of climate change projection (SENTAN) (grant no. JPMXD0722680734) and the Cross-Ministerial Strategic Innovation Promotion Program (SIP) Enhancement of Societal Resiliency against Natural Disasters and KAKENHI (grant nos. 20J2193800, 22H0159402, and 19J0096322).

Review statement. This paper was edited by Daisuke HARADA and reviewed by two anonymous referees.

References

Ashley, W. S., Haberlie, A. M., and Strohm, J.: A climatology of quasi-linear convective systems and their hazards in the United States, Weather Forecast., 34, 1605–1631, 2019.

Bluestein, H. B. and Jain, M. H.: Formation of mesoscale lines of precipitation: Severe squall lines in Oklahoma during the spring, J. Atmos. Sci., 42, 1711–1732, 1985.

Fairman Jr., J. G., Schultz, D. M., Kirshbaum, D. J., Gray, S. L., and Barrett, A. I.: Climatology of banded precipitation over the contiguous United States, Mon. Weather Rev., 144, 4553–4568, https://doi.org/10.1175/MWR-D-16-0015.1, 2016.

Feng, Z., Leung, L. R., Hagos, S., Houze, R. A., Burleyson, C. D., and Balaguru, K.: More frequent intense and long-lived storms dominate the springtime trend in central US rainfall, Nat. Commun., 7, 13429, https://doi.org/10.1038/ncomms13429, 2016.

Haberlie, A. M. and Ashley, W. S.: A radar-based climatology of mesoscale convective systems in the United States, J. Climate, 32, 1591–1606, 2019.

Hirockawa, Y., Kato, T., Tsuguchi, H., and Seino, N.: Identification and classification of heavy rainfall areas and their characteristic features in Japan, J. Meteorol. Soc. Jpn. Ser. II, 98, 835–857, 2020.

Houze, R. A., Smull, B. F., and Dodge, P.: Mesoscale organization of springtime rainstorms in Oklahoma, Mon. Weather Rev., 118, 613–654, 1990.

Kato, T.: Numerical simulation of bandshaped torrential rain observed over the southern Kyushu, Japan on 1 August 1993, J. Meteor. Soc. Jpn., 76, 97–128, 1998.

Kato, T.: Structure of the band-shaped precipitation system inducing the heavy rainfall observed over northern Kyushu, Japan on 29 June 1999, J. Meteor. Soc. Jpn., 84, 129–153, 2006.

Kato, T.: Quasi-stationary band-shaped precipitation systems, named "senjo-kousuitai", causing localized heavy rainfall in Japan, J. Meteor. Soc. Jpn., 98, 485–509, 2020.

Nagata, K.: Quantitative precipitation estimation and quantitative precipitation forecasting by the Japan Meteorological Agency, RSMC Tokyo–Typhoon Center Technical Review, 13, 37–50, 2011 (data available at: https://www.jma.go.jp/jma/jma-eng/jma-center/rsmc-hp-pub-eg/techrev/text13-2.pdf, last access: 13 August 2024).

Ohya, Y. and Yamada, T. J.: Characteristics of line-shaped rainbands regarding duration, shape, and rainfall intensity in northern Japan using the Radar/Raingauge-Analyzed Precipitation Product, IOP Conf. Ser.-Earth Environ. Sci., 1136, 012025, https://doi.org/10.1088/1755-1315/1136/1/012025, 2023.

Parker, M. D. and Johnson, R. H.: Organizational modes of midlatitude mesoscale convective systems, Mon. Weather Rev., 128, 3413–3436, 2000.

Prein, A. F., Liu, C., Ikeda, K., Trier, S. B., Rasmussen, R. M., Holland, G. J., and Clark, M. P.: Increased rainfall volume from future convective storms in the US, Nat. Clim. Change, 7, 880–884, 2017.

Schumacher, R. S. and Johnson, R. H.: Organization and environmental properties of extreme-rain-producing mesoscale convective systems, Mon. Wather Rev., 133, 961–976, 2005.

Seko, H.: Study of the shapes and maintenance mechanisms of meso-β scale line-shaped precipitation systems in the middle-latitudes, J. Meteorol. Res., 62, 1–3, 2010 (in Japanese).

Shimira, K., Hara, H., and Yamada, T.: Study in deneration mechanism and classification of rainfall by the images of radar raingauge in the Kanto plain, P. Hydraul. Eng., 44, 97–102, 2000 (in Japanese).

Stevenson, S. N. and Schumacher, R. S.: A 10-year survey of extreme rainfall events in the central and eastern United States using gridded multisensor precipitation analyses, Mon. Weather Rev., 142, 3147–3162, 2014.

Tsuguchi, H. and Kato, T.: Objective Extraction of Heavy Rainfall Events and Statistical Analysis on their Characteristic Features, Tenki, 61, 455–469, 2014 (in Japanese).

Unuma, T. and Takemi, T.: Characteristics and environmental conditions of quasi-stationary convective clusters during the warm season in Japan, Q. J. Roy. Meteor. Soc., 142, 1232–1249, 2016.

Yamada, T. J., Sasaki, J., and Matsuoka, N.: Climatology of line-shaped rainbands over northern Japan in boreal summer between 1990 and 2010, Atmos. Sci. Lett., 13, 133–138, https://doi.org/10.1002/asl.373, 2012.

Yoshizaki, M., Kato, T., Tanaka, Y., Takayama, H., Shoji, Y., Seko, H., Arao, K., Manabe, K., and members of X-BAIU-98 observation: Analytical and numerical study of the 26 June 1998 orographic rainband observed in western Kyushu, Japan, J. Meteor. Soc. Jpn., 78, 835–856, 2000.

Proc. IAHS, 386, 339–344, 2024
https://doi.org/10.5194/piahs-386-339-2024

An integrated design discharge calculation system for small to mid-sized ungauged catchments in Serbia

Nikola Zlatanović[1] and Mamoru Miyamoto[2]

[1]Faculty of Civil Engineering, University of Belgrade, Serbia
[2]International Centre for Water Hazard and Risk Management (ICHARM), Tsukuba, Japan

Correspondence: Nikola Zlatanović (zzznikola@gmail.com)

Received: 21 July 2023 – Revised: 22 November 2023 – Accepted: 24 February 2024 – Published: 19 August 2024

Abstract. This study deals with the application of freely available datasets on the global and local level coupled with open source GIS software in hydrological modelling, greatly improving cost and time effectiveness of hydrological analyses of ungauged catchments. The automated calculation system was developed as an extremely useful tool for hydrologists for the quick estimation of design discharges. The morphometric module of the calculation system showed excellent results, saving a great deal of time that would otherwise been spent on manual processing of geospatial data. Using the automated calculation system, the methodologies currently in use in hydrological practice in Serbia were evaluated. The distributed velocity method was introduced as a way of calculating hydrograph lag time. The comparison between design discharges calculated by rainfall-runoff analysis and by frequency analysis yielded very good resemblance. Examples of real-world application for over 2500 catchments are shown.

1 Introduction

In Serbia, catchments with area less than $100\,\mathrm{km}^2$ are for the most part ungauged, and a good number of larger catchments remain ungauged. Very often, design discharges are needed for these rivers where no streamflow data is available, for various applications. Examples include river training works for flood protection measures or erosion control, culvert design, small hydropower plants etc.

Current hydrological practice still mainly relies on numerous manual methods of handling data and calculations. This is partly due to the segregation and inadequate form of the necessary data, and partly due to experts lacking the GIS skills needed for the efficient manipulation of spatial data.

The purpose of this study is to: (1) identify datasets that can readily be used for hydrological modelling, to overcome data availability issues, (2) create an integrated design discharge calculation system, to greatly speed up the modelling process, (3) evaluate the existing methodologies used for predicting discharge in ungauged catchments, and (4) introduce modifications to the methodology in order to improve accuracy.

2 Study area

Serbia is located at the crossroads of Central and Southeast Europe, covering the southern part of the Pannonian Plain and the central Balkans. A landlocked country, Serbia borders Hungary to the north; Romania and Bulgaria to the east; Macedonia to the south; and Croatia, Bosnia, and Montenegro to the west; it also borders Albania through the disputed territory of Kosovo.

The study area of this study is the region of Serbia south of the Danube and Sava rivers (Fig. 1). This area is predominantly hilly and mountainous, with a moderately dense network of rainfall and water level gauging stations, making it suitable for rainfall-runoff analysis of small-to-mid sized catchments.

Out of a total of 28 gauged catchments in this region with area less than $500\,\mathrm{km}^2$, 21 have been chosen for the study. The criteria for choosing the catchments were: (1) natural catchment hydrology is dominant, (2) the actual catchment area corresponds to topographic catchment area, (3) sufficient time series of data are available, and (4) there have not been major changes in the catchment during the course of observations that influence catchment runoff.

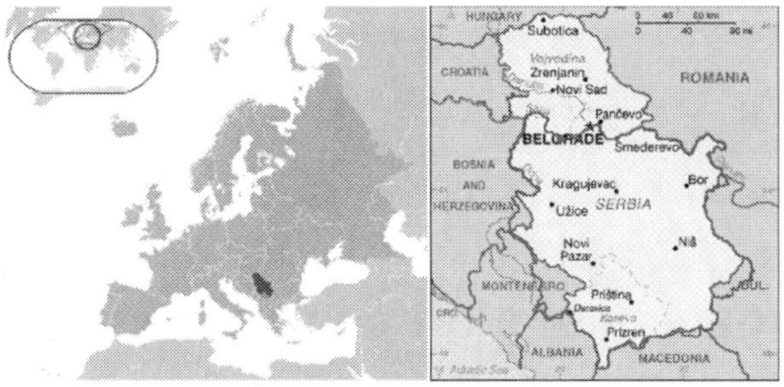

Figure 1. Location of Serbia, with study area outlined in red (basemap source: The World Factbook, 2017).

3 Data

The SRTM DEM (Shuttle Radar Topography Mission Digital Elevation Model) (Rodriguez et al., 2005), with a resolution of 3 arcsec, was downloaded for the entire territory of Serbia, with some margin to account for some transboundary basins. The downloaded original digital elevation model was pre-processed using void-filling interpolation and a combination of sink filling and drainage route deepening techniques in order to obtain a hydrologically correct DEM that is capable of correctly representing riverbeds.

The CORINE (Coordination of Information on the Environment) Land Cover dataset (EEA, 2000), developed by the European Environment Agency (EEA) and made available online, was used in this study. The dataset was downloaded for the entire country and converted to raster form to facilitate calculations.

Soil data was obtained from soil maps. Two different sets of soil maps were used in this study. To represent the currently used methodology, the soil map of scale 1 : 1 000 000 (developed in 1959 by the Yugoslav Society of Soil Science) was used. As an improvement to the methodology, the soil maps of scale 1 : 50 000 (developed between 1958 and 1982 by the Soil Institute of Serbia) were introduced. All maps were manually digitized and classified. The implemented datasets are shown in Fig. 2.

The hydrological and meteorological data used in this study were obtained from the Hydrometeorological Service of Serbia (HMSS). Rainfall data included: mean annual rainfall at 704 rain gauges, maximum daily rainfall at 433 rain gauges, intensity-duration-frequency (IDF) curves at 26 main meteorological stations. Mean annual air temperature data was acquired for 81 stations. Discharge data was acquired in the form of maximum annual instantaneous discharge and mean annual discharge, for the 21 reference catchments, from the beginning of measurement for each discharge station up to 2011. The locations of the used meteorological data are shown in Fig. 3.

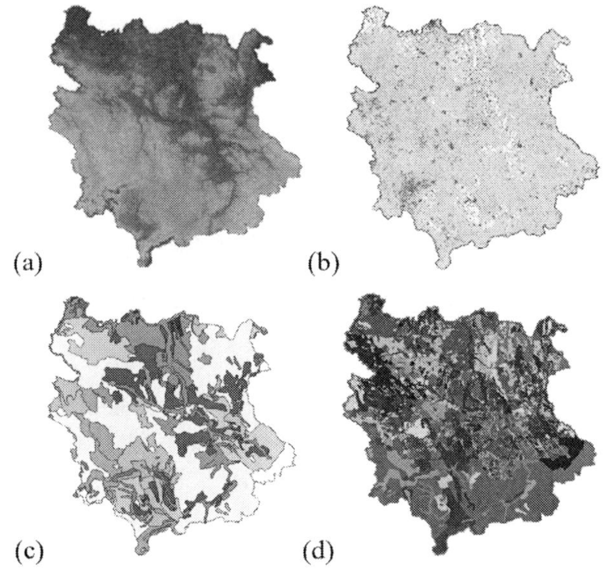

Figure 2. Overview of used datasets: **(a)** Pre-processed DEM, **(b)** CORINE Land Cover, **(c)** digitized soil map of scale 1 : 1 000 000, and **(d)** digitized soil map of scale 1 : 50 000.

4 Theory and methodology

4.1 Automation of Currently Used Methodologies

The Soil Conservation Service (now National Resources Conservation Service) Curve Number method, developed at the United States Department of Agriculture (USDA) during the 1950s, is the method most frequently used for estimating direct runoff from storm rainfall in ungauged catchments in Serbia today. The runoff curve number (CN) is based on the hydrologic soil group, land use, treatment and hydrological condition. Once the characteristic land cover description and hydrologic soil group have been defined, runoff curve numbers may be read from provided tables such as in the National Engineering Handbook (NEH) (NRCS, 2002).

The Curve Number (CN) was estimated using the CORINE Land Cover dataset and the digitized soil map of scale 1 : 1 000 000. The soils were categorized into hydrologic soil groups (A, B, C and D) based on criteria from the Soil Conservation Service and soil characteristics. The classes of the CORINE Land Cover classification system have each been assigned a Curve Number for every hydrologic soil group. The complex Curve Number for the entire catchment was computed using the weighted CN method, following the identification of all hydrologic soil-cover complexes.

The transformation of excess rainfall into direct runoff was simulated using a modified SCS dimensionless unit hydrograph. Jovanović (1989) documented a modification of the SCS synthetic unit hydrograph based on a large number

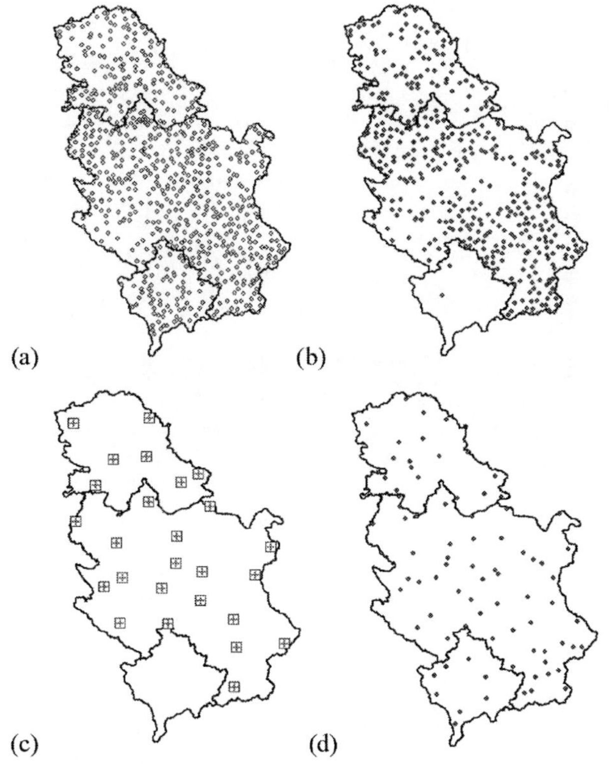

Figure 3. Availability of meteorological data: **(a)** mean annual rainfall, **(b)** maximum daily rainfall, **(c)** IDF curves, and **(d)** mean annual air temperature.

of observed catchments in Yugoslavia, and this method has since grown much in popularity. This method defines the dimensionless unit hydrograph based on the duration of rainfall and certain morphometric characteristics of the catchment (catchment area, longest flow path, flow distance from the catchment centroid to the outlet and the distance-weighted channel slope).The peak discharge for a specific storm event is calculated by approximating the unit hydrograph with an equivalent triangular hydrograph having the same units of time and discharge, thus having the same percent of volume in the rising side of the triangle.

Design rainfall for flood discharge estimation is most often based on the premises that an event with rainfall intensity of a certain probability of occurrence will cause runoff with the same probability of occurrence. These rainfall events were estimated using available IDF (intensity-duration-frequency) curves from the nearest meteorological station for each catchment. When performing calculations, each combination of rainfall duration and intensity (for a certain probability of occurrence) is evaluated, and the combination which yields the highest runoff is chosen.

4.2 Proposal for Improvement of Methodology

In an attempt to improve the results of current methodologies, two modifications were made to the calculations: an update of soil data and the introduction of a distributed velocity method for calculating catchment lag time.

The soil database was updated to include available soil data from soil maps of scale 1 : 50000. All available soil maps were digitized and classified into hydrologic soil groups. As soil maps at this scale do not cover the entire study area, the gaps were substituted by previously used soil data (from the smaller scale 1 : 1000000 maps). This way, all of the study area was covered with the best available soil data.

A more hydraulic and deterministic approach was taken to estimate hydrograph lag time, improving upon current empirical methods relating hydrograph lag to longest flow path length and average slope, length of flow path from the catchment centroid and rainfall duration. The motivation to the suggested improvement was to utilize the computing power of GIS to achieve a more process-based, distributed approach that accounts for spatial variability of flow, while still implementing the unit hydrograph methodology (as a lumped model).

The distributed velocity method for estimating hydrograph lag time developed in this work estimates flow velocities on every flow path within a catchment (in this case in every grid cell), then sums travel times along the hydraulically most distant path. Catchment areas were characterized by driving processes into three flow types: sheet flow, shallow concentrated flow, and open channel flow.

Sheet flow is defined as flow over plane surfaces, which usually occurs in the headwaters of a stream near the ridgeline that defines the catchment boundary. Sheet flow was calculated using a simplified version of Manning's kinematic solution. Shallow concentrated flow was defined as flow that follows sheet flow, but lacks a well-defined channel, and was calculated using an empirical relationship between velocity, land cover and slope. Open channel flow, defined as a stream (or river) with a clearly defined channel, was calculated using Manning's equation. The value for Manning's roughness n was estimated using Jarrett's formula, while the hydraulic radius R was estimated based on the assumption of a trapezoidal channel cross section.

A flow velocity raster is created by calculating the flow velocities in every cell of the catchment based on flow type, slope and estimated roughness. After the flow velocity raster has been calculated, travel times are also computed using the aspect raster. Examples of the flow velocity and time to outlet rasters are shown for the Obnica River in Fig. 4.

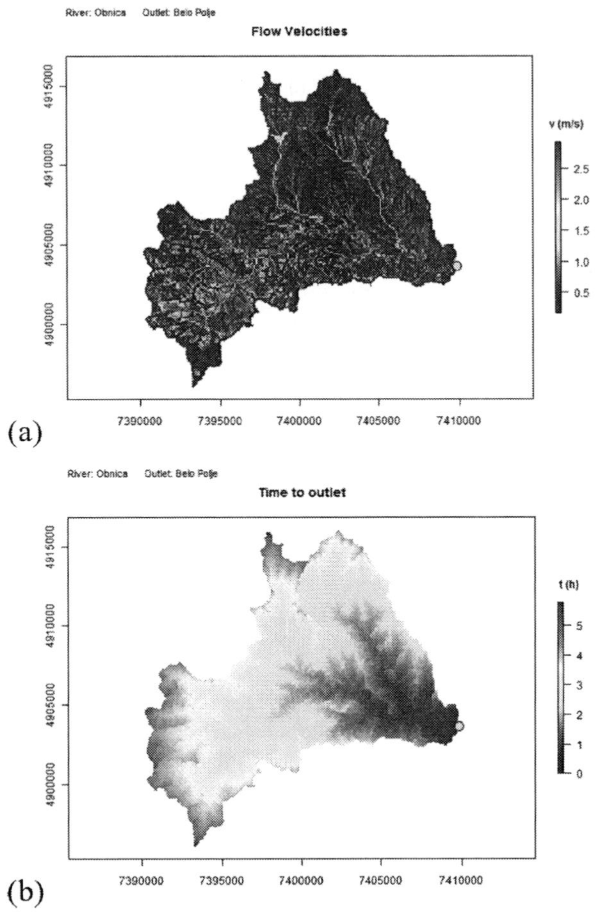

Figure 4. Example of flow velocity **(a)** and time to outlet **(b)** rasters.

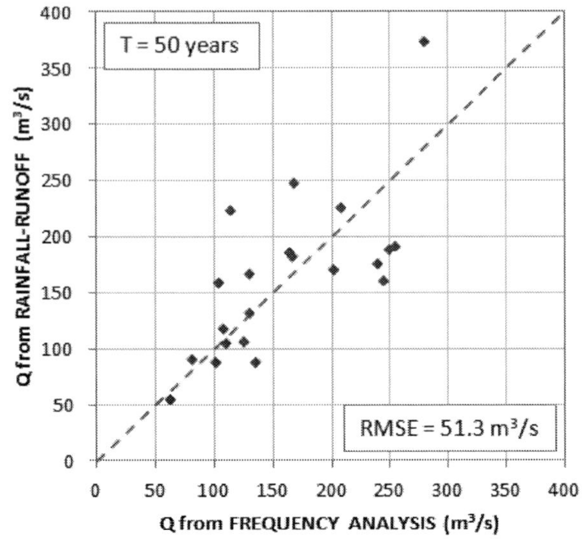

Figure 5. Comparison of calculated peak discharges and statistical design discharges, BEFORE methodology improvements, for 50-year return period.

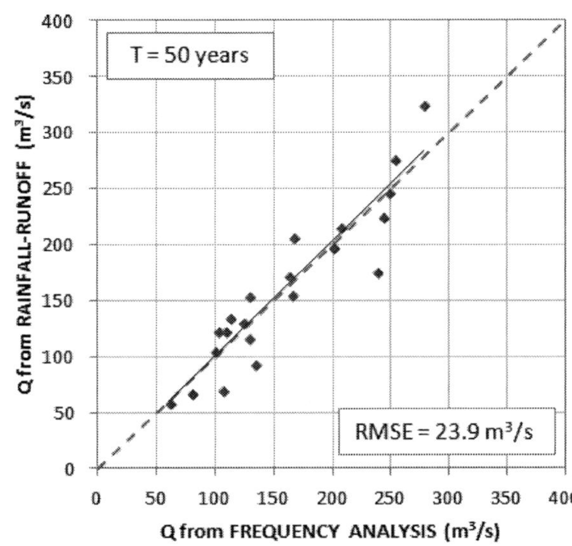

Figure 6. Comparison of calculated peak discharges and statistical design discharges, AFTER methodology improvements, for 50-year return period.

4.3 Validation of Flow Prediction in Reference Catchments

To validate discharge prediction methods applied in ungauged catchments in Serbia, 20, 50, 100 and 1000 year return period discharges predicted in 21 reference (gauged) catchments were compared to results obtained by frequency analysis of annual maximum discharge. To test the impact of improved soil data and the distributed velocity estimation of hydrograph lag time, discharges were predicted using both standard and improved methods. Frequency analysis included Generalized Extreme Value (GEV) and Gumbel (Extreme Value Type I) distributions, where parameters were estimated using L-moments.

5 Results

Comparison of predicted discharges and those derived from frequency analysis (Figs. 5 and 6) indicate that the methodology applied is likely acceptable for deriving design discharges for ungauged catchments in Serbia. Improvements of improved soil data and the distributed velocity estimation of hydrograph lag time greatly reduce residuals (Fig. 7), indicating much better model performance. Greater improvements were associated with lower return periods, while the level of improvement for the 1000-year return period was insignificant. No clear link between rate of improvement and geographic characteristics of the catchments was found.

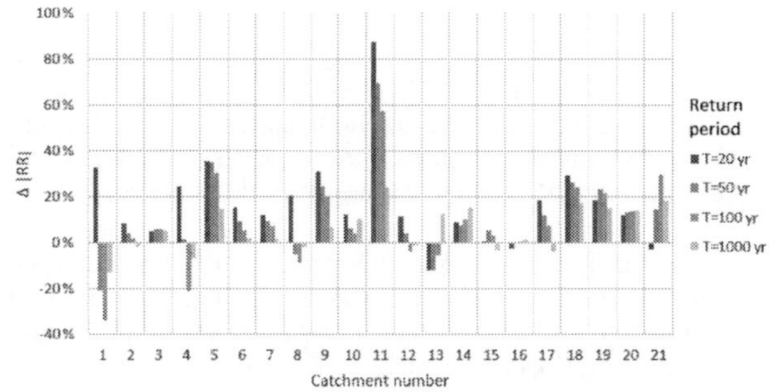

Figure 7. Difference in absolute value of relative residuals for all catchments before and after implementing methodology improvements (greater values indicate more improvement).

6 Example of application

The methodology presented has been successfully applied to calculate design flood flows for over 2500 catchments in Serbia for numerous design projects (roads, railways, pipelines, etc.), with the results shown in Fig. 8 as a plot of specific discharge (defined as the discharge per catchment area) of the 100-year flood and catchment area, comparing calculated results with verified results. The term "verified results" is used for results that correspond to the results of statistical analysis of observed values (where available).

7 Conclusion and recommendation

The automated calculation system presented in this study is one way of creating an extremely useful tool for hydrologists for the quick estimation of design discharges. The morphometric module of the calculation system showed excellent results, saving a great deal of time that would otherwise have been spent on manual processing of geospatial data. This type of automated analysis presented in this study will enable a much quicker hydrologic analysis on multiple catchments, while the methodology improvements introduced in this study will help to provide better estimates of predicted runoff.

Further work could include an even more distributed approach to modelling runoff (as opposed to the lumped unit hydrograph method used in this study) to better take into account the spatial variability of runoff. Unfortunately, more accurate datasets are currently unavailable, most notably soil and rainfall data, which are all limiting factors in achieving a closer resemblance of calculated results to observed values.

Code availability. The methodology presented in this paper has been entirely implemented using open source software, including QGIS (http://qgis.org/, QGIS Development Team, 2022) for data preparation, SAGA GIS (Conrad et al., 2015) for raster analyses and the R programming language (R Core Team, 2021) for running the calculations.

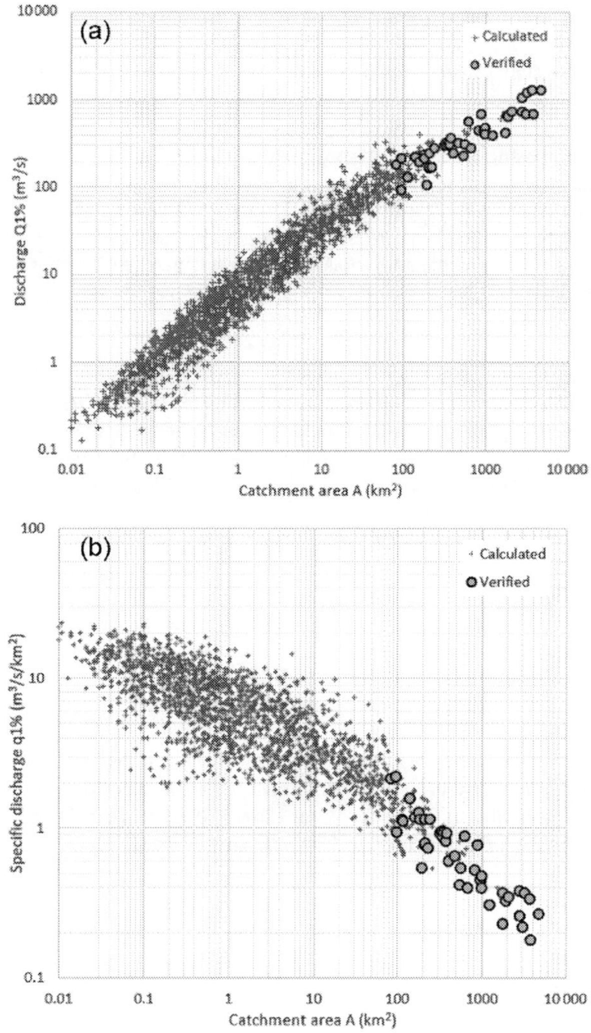

Figure 8. Results of calculation of 100-year floods on over 2500 catchments, presented as discharge values (above) and specific discharge values (below) compared to values verified with statistical analysis.

Data availability. The SRTM Digital Elevation Model (https://doi.org/10.5069/G9445JDF, NASA SRTM, 2013) and CORINE Land Cover dataset (https://doi.org/10.2909/71c95a07-e296-44fc-b22b-415f42acfdf0, EEA, 2020) are open and freely available for download from their respective repositories. Soil maps that were used in this research are available upon request from the Soil Institute of Serbia. Hydrological and Meteorological data are partly available online (Republic Hydrometeorologic Service of Serbia, 2022), while the complete datasets are available upon request due to governmental data policies.

Author contributions. NZ and MM contributed to the design and implementation of the research, to the analysis of the results and to the writing of the manuscript.

Competing interests. At least one of the (co-)authors is a guest member of the editorial board of *Proceedings of IAHS* for the special issue "ICFM9 – River Basin Disaster Resilience and Sustainability by All". The peer-review process was guided by an independent editor, and the authors also have no other competing interests to declare.

Disclaimer. Publisher's note: Copernicus Publications remains neutral with regard to jurisdictional claims made in the text, published maps, institutional affiliations, or any other geographical representation in this paper. While Copernicus Publications makes every effort to include appropriate place names, the final responsibility lies with the authors.

Special issue statement. This article is part of the special issue "ICFM9 – River Basin Disaster Resilience and Sustainability by All". It is a result of The 9th International Conference on Flood Management, Tsukuba, Japan, 18–22 February 2023.

Acknowledgements. The completion of this research paper would not have been possible without the guidance of the staff of the International Centre for Water Hazard (ICHARM), and support from the Japan International Cooperation Agency (JICA).

Review statement. This paper was edited by Kensuke Naito and reviewed by two anonymous referees.

References

Conrad, O., Bechtel, B., Bock, M., Dietrich, H., Fischer, E., Gerlitz, L., Wehberg, J., Wichmann, V., and Böhner, J.: System for Automated Geoscientific Analyses (SAGA) v. 2.1.4, Geosci. Model Dev., 8, 1991–2007, https://doi.org/10.5194/gmd-8-1991-2015, 2015.

European Environment Agency (EEA):Technical Report No. 40, CORINE land cover technical guide – Addendum 2000, prepared by: Bossard, M., Feranec, J., and Otahel, J., European Environment Agency, Copenhagen, Denmark, 2000.

European Environment Agency (EEA): Copernicus Land Monitoring Service information – CORINE Land Cover 2018 (vector), European Environment Agency [data set], https://doi.org/10.2909/71c95a07-e296-44fc-b22b-415f42acfdf0, 2020.

Jovanović, S.: Hydrology, Građevinski priručnik – tehničar, 6, 11–181, IRO Građevinska knjiga, Beograd, 1989 (in Serbian).

NASA Shuttle Radar Topography Mission (SRTM): Shuttle Radar Topography Mission (SRTM) Global, OpenTopography [data set], https://doi.org/10.5069/G9445JDF, 2013.

Natural Resources Conservation Service (NRCS): National Engineering Handbook: Part 630 Hydrology, United States Department of Agriculture (USDA), Washington, D.C., 2002.

QGIS Development Team: QGIS Geographic Information System. Open Source Geospatial Foundation Project, http://qgis.org/ (last access: 21 October 2022), 2022.

R Core Team: R: A language and environment for statistical computing, R Foundation for Statistical Computing, Vienna, Austria, https://www.R-project.org/ (last access: 21 October 2022), 2021.

Republic Hydrometeorologic Service of Serbia: Republic Hydrometeorologic Service of Serbia website, https://www.hidmet.gov.rs/ (last access: 21 October 2022), 2022.

Rodriguez, E., Morris, C.S., Belz, J.E., Chapin, E., Martin, J., Daffer, W., Hensley, S.: An assessment of the SRTM topographic products, Technical Report JPL D-31639, Jet Propulsion Laboratory, Pasadena, California, 2005

The World Factbook: The World Factbook 2017, Washington, D.C., Central Intelligence Agency, https://www.cia.gov/ (last access: 21 October 2022), 2017.

AUTHOR INDEX

Acierto, Ralph Allen 127, 209
Agarwal, Ankit 41
Ahmad, Fauziana 61
Amaguchi, Hideo 133
Antonelli, Tiago 7
Apperloo, Johan 285
Bahramian, Katayoon 237
Bi, Wuxia 141, 307
Biswas, Robin K. 229
Boer, Femke De 173
Brandão, Abderraman R. A. 81
Breedeveld, Joost 285
Bryant, Seth 181
Carrara, Elisabetta 237
Dairaku, Koji 55
Denda, Masatoshi 21, 259
Deng, Xingdong 277
Egashira, Shinji 1, 109, 159, 229
Fava, Maria C. 81
Fava, Maria Clara 41
Filho, Frederico C. M. De Menezes 81
Fortes, Andre Araujo 203
Fujikane, Masakazu 21, 127
Fukuoka, Shoji 47
Ghosh, Mousumi 299
Ghosh, Subimal 299
Goel, Narendra K. 41
Guruge, Thanura Lasantha Babarande 1
Gusyev, Maksym 265
Hanashima, Makoto 189
Harada, Daisuke 109, 159, 229
Hashimoto, Masakazu 203
He, Bingshun 313
Hemakanth, Selvarajah 147
Hemakumara, Sajana Pramudith 147
Herrera, Eugene C. 95, 121
Hoshino, Tsuyoshi 101
Hou, Jiawei 237
Ichikawa, Ken 203
Ichikawa, Yutaka 75
Imamura, Yoshiyuki 133
Ishidaira, Hiroshi 33
Ishii, Yutaro 47
Ishiwatari, Mikio 87
Itagaki, Osamu 153
Ito, Hiroyuki 127
Iwamae, Nobuyuki 165
Jayapadma, Mudalige Miyuru Uthpala 33
Jayasinghe, Roshan Indika 291
Kamoto, Minoru 321
Karmakar, Subhankar 299
Kaushalya, Ranapura Dewage Thilini 147
Kawamoto, Takatoshi 127
Kawamura, Akira 133
Kawamura, Naoaki 197
Kawasaki, Akiyuki 217
Kawasaki, Masaki 13
Kelkar, Saurabh 55
Khadka, Tapendra Bahadur 327
Kiuchi, Nozomu 251
Koike, Toshio 153, 209, 223, 265, 271, 291
Komiyama, Koichi 197
Kottier, Jaap 285
Kowlesser, Akshay 271
Koyabu, Tsuyoshi 259
Kreibich, Heidi 181
Kuang, Minyi 277
Kuhlmann, Leandro G. 7
Kurian, Mathew 41
Lana, Julio C. 7
Leeuw, Annemargreet De 173
Li, Changzhi 313
Li, Qing 313
Lin, Wenqing 141
Lu, Minjiao 197
Macedo, Marina Batalini De 41
Maeda, Yuta 13
Magome, Jun 33
Mangukiya, Nikunj K. 41
Masood, Muhammad 27
Masuya, Shigekazu 69
Mendiondo, Eduardo Mario 41
Merz, Bruno 181
Miyamoto, Mamoru 7, 339
Moiz, Abdul 217
Naito, Kensuke 95, 121
Nakaegawa, Tosiyuki 75
Nakakita, Eiichi 61
Nakamura, Akira 321
Nakano, Taku 251
Nardocci, Adelaide Cássia 41
Nohara, Daisuke 165
Nonaka, Saki 165
Ogura, Masashi 321
Ogura, Takahiro 197
Ohara, Miho 153, 265

Ohya, Yuta 333
Oka, Shigeaki 197
Oliveira, Paulo T. S. 81
Olsson, Jonas 133
Onitsuka, Shun-Ichi 197
Palugaswewa, Bandara 1
Peng, Yuan 277
Phyoe, Pyae Pyae 115
Pickett-Heaps, Christopher 237
Rahman, Md. Majadur 109
Rasmy, Abdul Wahid Mohamed 27, 265
Rasmy, Mohamad 291
Rasmy, Mohamed 127, 223, 271
Razzolini, Maria Tereza 41
Rongen, Guus 69
Rüdiger, Christoph 237
Sano, Hiroaki 189
Sasaki, Daisuke 87
Sasaki, Hidetaka 75
Serrano, Jonathan S. 121
Sharma, Ashutosh 41
Sharples, Wendy 237
Shinya, Takafumi 127
Shrestha, Badri Bhakta 127
Silva, Roberto Fray Da 41
Souma, Kazuyoshi 33
Sugeeshwara, Seenipellage Chaminda 147
Sui, Xintian 277
Suzuki, Akihiro 101
Suzuki, Kazuki 165
Tachikawa, Yasuto 75
Takeshita, Tetsuya 13
Tamakawa, Katsunori 265
Tedla, Mihretab G. 223
Tinumbang, Aulia Febianda Anwar 75
Tromp, Ellen 173, 285
Uchida, Tatsuhiko 115
Udo, Keiko 203
Uemura, Fumihiko 69, 101
Unnithan, Kesav 237
Ushiyama, Tomoki 61, 127, 209
Usuda, Yuichiro 189
Wang, Fan 141
Wang, Jingchen 277
Wang, Weiqi 141
Wickramaarachchi, Thushara Navodani 33
Yamada, Tomohito J. 69, 101, 333
Yamada, Yoriyuki 165
Yamaguchi, Kosei 61
Yamamoto, Takahiro 197
Yamamoto, Taro 101
Yang, Yukui 277
Yorozu, Kazuaki 75
Yoshida, Takatoshi 69
Zhang, Cheng 307
Zhang, Dawei 141
Zhang, Miao 313
Zheng, Yanyan 277
Zhou, Li 223
Zin, Thandar Tun 197
Zlatanovic, Nikola 339

International Association of Hydrological Sciences
Centre for Ecology and Hydrology
Wallingford, Oxfordshire OX10 8BB
United Kingdom

ISBN 979-8-3313-0728-8
9 798331 307288

CURRAN ASSOCIATES INC.
proceedings.com